# Advances in Technology-Rich Science Education

Volume 2

This book series provides a comprehensive and authoritative overview of current research, trends, and innovations that leverage educational technology to improve science teaching and learning. Contributions from a global community of scholars, educators, policymakers, and leaders address formal, non-formal, and informal science education contexts.

It provides a platform for sharing research, practice, development, and policy, highlighting emerging innovations, and examining how technology influences engagement, learning, and workforce readiness. Additionally, it also addresses key challenges and opportunities related to access, equity, inclusion, and digital literacies while fostering critical perspectives and discussion about technology's evolving role in science education.

Focused on the enterprise of science education, the series explores the design and use of technology-enhanced learning environments, applications of artificial intelligence, simulation, and computation, and innovations in online and blended learning, including MOOCs and virtual laboratories. It also considers augmented reality, authentic digital experiences, strategies for supporting diversity and inclusion, research methodologies, workforce development, and the creation and critique of educational policy and leadership. Book proposals for this series may be submitted to the Publishing Editor: Claudia Acuna E-mail: Claudia.Acuna@Springer.com

Laurinda Leite · Luís Dourado · Ana S. Afonso ·
Sofia Morgado · Luísa Carvalho
Editors

# Practical Work in Science Education in the Digital Era

## Challenges and Opportunities

Springer

*Editors*
Laurinda Leite
University of Minho
Braga, Portugal

Luís Dourado
University of Minho
Braga, Portugal

Ana S. Afonso
University of Minho
Braga, Portugal

Sofia Morgado
University of Minho
Braga, Portugal

Luísa Carvalho
Lusophone School of Higher Education, IPLUSO
Lisboa, Portugal

ISSN 3091-339X ISSN 3091-3403 (electronic)
Advances in Technology-Rich Science Education
ISBN 978-3-032-13787-6 ISBN 978-3-032-13788-3 (eBook)
https://doi.org/10.1007/978-3-032-13788-3

This work is funded by national funds through FCT—Science and Technology Foundation, I.P., under the Pluriannual Funding of CIEd—Research Centre on Education of the University of Minho (UID/01661/2025).



This Springer imprint is published by the registered company Springer Nature Switzerland AG
The registered company address is: Gewerbestrasse 11, 6330 Cham, Switzerland

# Foreword

## Growing Up with Practical Work

I would describe my own science education as 'traditional'. Growing up in England in the 1960s and 1970s, I attended a selective, boys-only school which, while state funded, was modelled on one of the UK's most well-known independent schools, Rugby. At the age of 16, everyone took O-level (Ordinary) examinations and then chose three A-levels (Advanced) which we studied for two years before applying to go to university. A-level choices tended to cluster around either science subjects or arts subjects. I chose Mathematics, Physics and Chemistry and ended up taking a degree in Chemistry at the University of Birmingham. Science was synonymous with practical work. The A-level physics examination included an assessment of practical skills—in 1975 it involved a potentiometer—the last time I ever used one. After my degree, I trained as a science teacher at the Centre for Science Education at Chelsea College, part of the University of London. The first lesson that I taught as a student teacher involved practical activity—looking at the allotropes of sulphur (as it was spelled then). More so then than now, perhaps, teaching science has revolved around practical work.

As someone who has worked with pre-service teachers since 1989—coming up for 40 years now—and as someone who visits schools in the UK (and elsewhere in the world), it seems to me that the role of practical work has diminished, particularly in state-funded schools. The reasons for this reduction include the cost of chemicals and apparatus, a lack of trained science technicians to assist teachers and, probably most important, the dominance of external testing that forces teachers to focus on teaching to the test such that practical work becomes a luxury.

While working as a teacher, I studied for a master's degree in Science Education at King's College London (Chelsea College and Queen Elizabeth College merged in 1985 and then amalgamated to form what is now King's College London). I began to read academic papers and books on a range of topics not least of which was the role of practical work. The literature on this topic has continued to grow and this new volume is the latest contribution to the field. In this foreword I will discuss some of the trends and issues that I have seen and been part of as a way of contextualising what you will discover as you read the book's chapters.

I know from my own experience that, at its worst, practical activity in school science can be frustrating and pointless. Practicals are used to 'prove' something that they cannot possible prove—students follow recipes without needing to understand the science they are trying to learn. Students can tell you what they have done but cannot explain why they have done it. Experimental work can reinforce gender stereotypes with boys doing the hands-on activities while girls take notes. I have seen a great deal of bad practice over many years. Derek Hodson (1988, 1990, 1992) a staunch critic, derided poorly planned practical work as "ill-conceived, muddled and of little educational value" (1988, p. 53). However, while I am critical of poor quality practical work, I suspect that some writers on the subject deliberately or otherwise play down the value of practical activity in science education.

The arguments for practical activity in science education are multi-dimensional. Science plays a critical role in our everyday lives and scientists around the world are building knowledge through a blend of theory and experiment. To understand how scientists build knowledge, it is imperative that students experience some aspects of practical work, as long as they are aware of the gulf between the school laboratory and the real-world lab. Too often, school students use antiquated apparatus because that is how we've always done it—mercury thermometers and Bunsen burners offer little to students by way of authenticity. On the other hand, I have seen students using high-level scientific equipment in a Field Studies Council residential centre in the UK and at the Novo Nordisk Foundation's LIFE Centre in Denmark so some opportunities exist to do more authentic practical activity. Studying science without doing practical work would be like studying art without drawing or painting. Done well, practical work adds enormous value to the student experience and provides an opportunity for teachers to develop appropriate pedagogic strategies based on their own experience at school, university and, in some cases, as professional scientists prior to becoming a science teacher.

## Looking Back at What We Knew Almost 20 Years Ago

During the first decade of the 21st century, concern about the quality and amount of practical work in school science grew. Ofsted, who inspect schools in England, reported that scientific enquiry in primary schools was both variable and vulnerable (HMI/Ofsted, 2004a, 2004b). They found that, at secondary (high school) level, the range of investigations that students undertook was narrow. A major reason for this poor state of affairs was teachers' perceptions of assessed coursework. Not for the first time in England, concern about the amount of practical activity in and out of the classroom was shared by many teachers as well as by sections of the science community, industry and business.

As a result of this concern, the Government's STEM High Level Strategy Group asked SCORE (Science Community Representing Education) to develop a strategy to promote high quality practical work. By way of explanation, SCORE was a partnership between the Association for Science Education (ASE), the Biosciences Federation, the Institute of Biology, the Institute of Physics, the Royal Society, the Royal Society of Chemistry and the Science Council. On behalf of SCORE, and in partnership with the ASE, the Royal Society commissioned me to carry out a focused review of research regarding the state of effective practical work in school science (Dillon, 2008). This review was complemented by a teacher survey and a stakeholder consultation.

Compiling the report took up much of 2007, and the final version was published in March 2008—coming up for 18 years ago. I refer to it here because it helps to explain where we have come from in terms of the research into practical work in school science. I began by pointing out that there was confusion in the broader science education community about what was meant by ‘practical work’. This was particularly true in ‘Science in the National Curriculum’ which, at the time, defined what students in England should learn: ‘Practical and enquiry skills’, ‘practical and investigative activities’, ‘independent enquiry’ and ‘experimental work’ (QCA, 2007a, 2007b) were all used without any attempt to define what they meant. In looking for a comprehensive conceptualisation, I was particularly taken by Vince Lunetta’s definition that he used in his review of the literature: ‘learning experiences in which students interact with materials or with secondary sources of data to observe and understand the natural world (for example: aerial photographs to examine lunar and earth geographic features; spectra to examine the nature of stars and atmospheres; sonar images to examine living systems)’ (Lunetta et al., 2007). Some science educators might argue that some of these activities are not what they would call ‘practical work’ but I like the breadth of this definition.

In terms of what teachers thought were the purposes of practical activity in school science, the most common ones were: “to encourage accurate observation and description; to make phenomena more real; to arouse and maintain interest; to promote a logical and reasoning method of thought” (Dillon, 2008, p. 4). The introduction of the National Curriculum in England and Wales had led to four other aims being commonly stated by teachers: “to practice seeing problems and seeking ways to solve them; to develop a critical attitude; to develop an ability to cooperate; for finding facts and arriving at new principles” (p. 4). I suggested that a “discussion about the value, aims and purposes of practical work among stakeholders might be a useful first step in addressing some of the criticisms made by various concerned bodies” (p. 4).

In terms of the impact of practical work, we knew that teachers and students were generally positive. A 2005 NESTA survey (n=510) found that an extraordinary 99% of the sample of UK science teachers believed that enquiry learning had a positive impact (83%—‘very’; 16%—‘a little’) on student performance and attainment (NESTA, 2005, p. 5). Based on their review, Lunetta et al. concluded that “When well planned and effectively implemented, science education laboratory and simulation experiences situate students’ learning in varying levels of inquiry requiring students to be both mentally and physically engaged in ways that are not possible in other science education experiences” (Lunetta et al., 2007, p. 405).

As early as 1992, Dick White and Dick Gunstone (1992) had noted that "students must manipulate ideas as well as materials in the school laboratory" (Lunetta et al., 2007). As well as developing students' understanding of scientific concepts, there was also evidence that practical work can increase "students' sense of ownership of their learning and can increase their motivation (Johnstone and Al-Shuaili, 2001)" (Dillon, 2008, p. 23) and, in terms of the teachers' role—there was evidence that their role "in helping students to compare their findings with those of their peers and with the wider science community is critical (Driver, 1995)" (Dillon, 2008, p. 22). The caveat, of course, was spelled out by Abrahams and Millar (2008) who argued that "teachers need to devote a greater proportion of the lesson time to helping students use ideas associated with the phenomena they have produced, rather than seeing the successful production of the phenomenon as an end in itself".

Many of the issues and challenges that I identified in 2008 will resonate with the chapters in this new volume. However, there is one section of the review where substantial progress might have been made. In "The role that information technology can and might play in supporting teaching and learning in practical work", I examined relevant research. While the internet was invented in 1989 and the world's first website went live in 1991, the digital era was still in its relative infancy.

I left school science teaching in 1989 to take up a role at King's that involved working with trainee teachers and helping to set up a curriculum development project led by the late Rod Watson. He had secured from the Nuclear Electricity Information Group to create a National Environmental Database (NED). Aimed at students aged 12–14, NED had three strands, one for each of the three sciences taught in English schools—biology, chemistry and physics. The biology activity involved monitoring water quality using benthic organisms using a well-established scale. The chemistry activity involved measuring the acidity of rainwater. The physics activity involved using a small piece of plastic that could be used to detect radon gas in the environment. Each activity involved collecting environmental data, entering it into a simple educational tool and then uploading it to a central database managed by King's. Schools could then download data from participating schools from around the country. Needless to say, this process would be much easier these days than it was then. The project was reasonably successful but there were a number of technical difficulties and when funding ceased, NED gradually faded away.

## Into the Digital Era

In the 2008 review, I focused on the use of computer simulations and the use of primary data obtained from the internet or other research sources. At the time, and perhaps more so now, new hardware and software became available virtually every day. Lunetta et al. (2007) pointed out an additional issue:

> To complicate matters, science education studies have not always helped to distinguish between and link important ends (learning outcomes that are sought) and means to those ends (teaching resources and strategies such as specific kinds of investigative activities

> in the laboratory). For example, significant changes in technologies since the 1980s have offered new resources for teaching and learning, but insufficient attention has been directed to examine critically how these new technologies can enhance or confound experiences in the school laboratory. (p. 396)

My recollection of these times was that there was a great sense of optimism that information technology was going to revolutionise science teaching and learning (Barton, 1998; Lunetta, 1998). Exhibitors at science teacher conferences showed off their exciting sensors and simulations (Braund & Reiss, 2006). Peripheral devices could be used by students and teachers to carry out long-term investigations, for example in data-logging experiments (Dori et al., 2004; Friedler et al., 1990; Krajcik et al., 2000; Lunetta, 1998). Software was available that could aid in visualizing data and modelling scientific phenomena (Reiser et al., 2002).

How this technology might make science education more effective was discussed by Robin Millar who noted that "computer-based simulations may also help to reduce the 'noise' of the laboratory bench and focus attention on important aspects of experimental planning and data interpretation (Millar, 1999)" (2004, p. 19). Lunetta et al. (2007, p. 412) summarised recent research findings as follows:

> When teachers and students properly use inquiry empowering technologies to gather and to analyze data, students have more time to observe, reflect, and construct conceptual knowledge that underlies their laboratory experiences. The associated graphics also offer visualization resources that can enhance students' experiences with authentic activities while promoting deeper conceptual understanding (Edelson, 2001). When students have the time and when the activity is valued by the teacher and by high-stakes assessment, students can examine functional relationships and the effects of modifying variables; they can also make and test predictions and explanations. Technologies that offer instantaneous display of data as it is gathered can offer opportunities through which students may be helped to understand systemic functional relationships and more holistic relationships among variables. Using appropriate high technology tools can enable students to conduct, interpret, and report more complete, accurate, and interesting investigations. Such tools can also provide media that support communication, student—student collaboration, the development of a community of inquirers in the laboratory-classroom and beyond, and the development of argumentation skills (Zembal-Saul et al., 2002).

Some researchers were examining how IT might increase the authenticity of scientific inquiry, an issue discussed above. Nelson and Ketelhut (2007) described the developing use of Multi-User Virtual Environments (MUVEs) (see, also, Zacharia, 2007). They concluded that

> (1) with careful design and inclusion of virtual inquiry tools, MUVE-based curricula can successfully support real-world inquiry practices based on authentic interactivity with simulated worlds and tools, (2) Educational MUVEs can support inquiry that is equally compelling for girls and boys, and (3) research on student engagement in MUVE-based curricula is uneven. (p. 265)

They urged researchers to examine the extent to which MUVE-based curriculum could help teachers meet state and national standards with inquiry curricula; and whether scientific inquiry curricula embedded in MUVE environments could help teachers learn how to integrate interactive scientific inquiry into their classroom. (ibid.).

However, those of us who had been involved in science education in the 1980s and 1990s remained somewhat sceptical of the promise held out by the IT advocates—we had heard it all before—and NED had convinced us that what was possible was not what was always wanted by teachers. Lunetta et al. neatly summarised the issues.

> Much evidence now documents that using appropriate technologies in the school laboratory can enhance learning, and important research on learning empowering technologies is the focus of this section. That said, an initial cautionary note is fitting since evidence also documents that inappropriate use of even simple technology tools has interfered with meaningful science learning (Hofstein & Lunetta, 2004; Olson & Clough, 2001). When a device is introduced prematurely, before students have made sense of the underlying science concepts, there is evidence that device or tool may serve as a "black-box" that interferes with students' perceptions of what is happening and hinder their understanding of important scientific ideas. (Lunetta et al., 2007, p. 411)

As well as criticisms of the limitations of new hardware and software, other academics warned of the challenges of using the internet in science education—Marcia Linn argued that: 'The internet provides a rich, confusing, chaotic, informative, persuasive set of scientific information' (2000, p. 785). One possible aid to teachers was suggested by the UK House of Lords Science and Technology Committee who recommended that the 'Government assess the feasibility of a unified and comprehensive central website dedicated to practical work in all the sciences' (2006, p. 33). Such a website, the Committee suggested 'should offer health and safety advice and exemplar practicals that can stimulate students' (ibid, p. 33).

The UK Government's response to this recommendation was very positive: "As one of our 'Next Steps' commitments, the Government has also asked the Secondary National Strategy to promote effective practice in interactive teaching including imaginative use of practical work". (HM Government, 2007, para 6.11). Such pronouncements assume that teachers are willing and able to utilise these new technologies. A blended approach to simulations and laboratory work was found to be effective with Finnish elementary (aged 11–12) students ($n = 66$) by Jaakkola and Nurmi (2007, unnumbered). However, an unpublished study of the impact of a London Science Challenge initiative, found little evidence of sustained, focused use of ICT in science lessons on any significant scale: "While examples of good practice exist, teachers appear to prioritise other approaches to the teaching of science. It is difficult to see the situation changing substantially in the near future despite evidence that appropriate use of ICT in science can lead to higher attainment and more positive attitudes to science" (Dillon, 2008, p. 40).

## This Volume's Contribution to the Field

In this final section, I try to establish how far we have progressed in terms of understanding the affordances of practical work for science education. One might hope to see substantial differences but change in education is often slow and incremental. Indeed, many of the often unsubstantiated assertions made about the role and value of practical work can be found among the chapters: "Practical work has long been a cornerstone of science education, evolving from 19th-century laboratory experiments to modern, inquiry-based approaches. It serves to develop scientific skills, reinforce theoretical knowledge, and foster curiosity by making abstract concepts tangible" (Chapter 2); "Inquiry-based practical work plays a central role in Science Education and is essential for developing students' scientific literacy" (Chapter 7); "Laboratory work is recognized as essential for promoting students' learning in science education" (Chapter 12); "Practical work is crucial in science education" (Chapter 19) and "Pupils are usually eager to "work" in the school laboratory to get the feeling of being scientists" (Chapter 6).

Several chapters explore the limitations of practical work not least Chapter 1 which critically compares and contrasts beliefs about the effectiveness and affective value of practical work in schools with the available research evidence. Chapter 5 looks at how conducting epistemological or historical enquiry in a science classroom differs from practical work in school science?

We know a lot more about the affordances of digital technologies in school science. Chapter 18 for example, provides a systematic review of research on the integration of real and virtual laboratories for inquiry-based learning. By way of caution, this chapter notes that while digital tools, such as virtual labs and simulations, offer promising ways to complement traditional methods and expand access, they cannot fully replicate the sensory, manipulative, and context-rich experiences of real-world experiments. Virtual laboratories, particularly 3D animated simulations, are also discussed in Chapter 13 that looks at benefits such as enhanced conceptual understanding, increased student motivation, and improved accessibility for resource-constrained settings. Chapter 14 examines computer simulations that can act as a pivot, to obtain different configurations of practical work.

Novel approaches including integrating maker education into science education are described in Chapter 4. Chapter 16 looks at augmented reality technologies and their role in school science education. Chapter 17 argues that Immersive Virtual Reality offers unique virtual learning experiences by leveraging 3D visualisation and full-body interactions. The role of educational robotics is discussed in Chapter 25.

Continued interest in the use of modelling in science education is evident in Chapter 8. Models are seen as supporting the epistemic and sensemaking aspects of learning in laboratories, field sites, online environments, and other venues for practical work in science. In Chapter 15, the focus is on computer-based modelling that encompasses both simulations -where students explore complex digital representations of real-world phenomena—and environments in which they build and revise their own computational models. The chapter also looks at the use of emerging AI-driven tools for automated model evaluation. Chapter 3 focuses on the context of practical science and how AI is impacting the nature of practical work on science and what experts recommended about how science education needs to address the emerging changes.

A more radical discussion of the place of practical work in school science takes place in Chapter 9 with a move to reduce practical work to the interaction with equipment and materials, and conceive it as a form of intervention in the world through the mobilization of scientific knowledge. Such a shift helps to address issues that traverse students' lives. This perspective is complemented by Chapter 11 that encourages teachers to focus on using hands-on activities to investigate authentic problems that resonate with students' lived experiences and societal concerns—such as environmental degradation, public health, or social inequality—this approach positions science education as a tool for understanding and transforming the world.

An increasing awareness of the value of learning science outside the classroom is evident in Chapter 10 which identifies that many natural phenomena have to be studied in the place where they occur. The authors argue that studying a phenomenon at the place where it occurs fosters a relationship between the school and the real world, which may increase interest in science, promote contextualized conceptual and procedural learning, encourage positive attitudes towards the environment, and develop real problem-solving abilities. Chapter 12 also addresses this theme in discussing the role of science museums in supporting and extending school laboratory activities.

A focus on social justice is evident in another four chapters. Chapter 19 examines methods related to the inclusion of students from countries and contexts with limited economic resources in science-related practical work. Chapter 20 looks at the critical need for physics laboratories that are both physically accessible and pedagogically inclusive to serve all learners equitably and Chapter 21 discusses current methodological and technological solutions to promote a more inclusive science learning experience for students who are blind or have low vision. Chapter 22 explores the importance of both providing high-quality education for such children from an early age and establishing a network for them, and explores the possibilities of gifted education and talent development from an early age through the use of digital tools. Chapter 23 discusses the potentials and limitations of virtual field trips to promote inclusion in nature-based learning contexts.

New approaches to the assessment of practical work are discussed in Chapter 26 which notes that technology now challenges the primacy of the teacher in the assessment of students' laboratory performance viewed through the lenses of posthumanism and new materialism. Chapter 24 derives some key messages from academic research on practical work in science that have clear implications for practice with a focus on how teachers can plan and evaluate effective practical work.

Ethical issues in research laboratories are addressed in Chapter 27 which describes a novel method to embed education about research integrity. Moving outdoors, Chapter 28 looks at ethical arguments both for and against undertaking school science out-of-doors. Chapter 29 situates the ethics of Extended Reality (XR) for schools within a human rights framework and outlines the risks of these technologies for schools.

Chapters 30 and 31 look at laboratory safety. The former chapter examines the unique laboratory safety risks of Virtual Environments (VLEs) and contrasts them with traditional safety concerns. The chapter proposes the redefinition of laboratory safety in the digital age and offers insights for putting these proposals into practice. The latter chapter also addresses VLEs in terms of data privacy risks and the potential to misuse digital tools and simulations. It focuses on data protection, ethical considerations in virtual experimentation, and strategies for promoting responsible digital citizenship.

Justin Dillon
Institute of Education, UCL
London, UK

## References

Barton, R. (1998). IT in practical work. In J. Wellington (Ed.), Practical work in school science. Which way now? (pp. 237–251). Routledge.

Braund, M., & Reiss, M. (2006). Towards a more authentic science curriculum: the contribution of out-of-school learning. International Journal of Science Education, 28, 1373–1388. https://doi.org/10.1080/09500690500498419

Dillon, J. (2008). A review of the research on practical work in school science. King's College London.

Dori, Y. J., Sasson, I., Kaberman, Z., & Herscovitz, O. (2004). Integrating case-based computerized laboratories into high school chemistry. The Chemical Educator, 9, 1–5.

Driver, R. (1995). Constructivist approaches to science teaching. In L. P. Steffe and J. Gale (Eds.), Constructivism in education (pp. 385–400). Lawrence Erlbaum Associates.

Edelson, D. C. (2001). Learning-for-use: a framework for the design of technology-supported inquiry activities. Journal of Research in Science Teaching, 38(3), 355–385. https://doi.org/10.1002/1098-2736(200103)38:3<355::AID-TEA1010>3.0.CO;2-M

Friedler, Y., Nachmias, R., & Linn, M. C. (1990). Learning scientific reasoning skills in microcomputer-based laboratories. Journal of Research in Science Teaching, 27(2), 173–191.

Her Majesty's Government. (2007). Government response to the House of Lords Science and Technology Committee's report into science teaching in schools. http://www.publications.parliament.uk/pa/ld200607/ldselect/ldsctech/167/167we02.htm

Hodson, D. (1988). Experiments in science and science teaching. Educational Philosophy and Theory, 20(2), 53–66. https://doi.org/10.1111/j.1469-5812.1988.tb00144.x

Hodson, D. (1990). A critical look at practical work in school science. School Science Review, 71(256), 33–40.

Hodson, D. (1993). Re-thinking old ways: Towards a more critical approach to practical work in school science. Studies in Science Education, 22, 85–142. https://doi.org/10.1080/03057269308560022

Hofstein, A., & Lunetta, V. N. (2004). The laboratory in science education: foundations for the twenty-first century. Science Education. 88(1), 28–54. https://doi.org/10.1002/sce.10106

House of Lords Science and Technology Committee (2006). Tenth report of Session 2005–06 science teaching in schools. Retrieved from 05 October, 2025: http://www.publications.parliament.uk/pa/ld200506/ldselect/ldsctech/257/25706.htm#a12

Jaakkola, T., & Nurmi, S. (2007). Fostering elementary school students' understanding of simple electricity by combining simulation and laboratory activities. Journal of Computer Assisted Learning, 24(4), 271–283. https://doi.org/10.1111/j.1365-2729.2007.00259.x

Johnstone, A. H., & Al-Shualili, A. (2001). Learning in the laboratory: Some thoughts from the literature. University Chemical Education, 5, 42–50.

Krajcik, J., Blumenfeld, B., Marx, R., Soloway, E. (2000). Instructional, curricular, and technological supports for inquiry in science classrooms. In J. Minstrell & E. Van Zee (Eds), Inquiring into inquiry: Science learning and teaching (pp. 283–315). American Association for the Advancement of Science Press.

Linn, M. C. (2000). Designing the knowledge integration environment. International Journal of Science Education, 22(8), 781–796. https://doi.org/10.1080/095006900412275
Lunetta, V. N. (1998). The school science laboratory: historical perspectives and centers for contemporary teaching. In P. Fensham (Ed.). Developments and dilemmas in science education (pp. 169–188). Falmer Press.
Lunetta, V. N., Hofstein, A., & Clough, M. P. (2007). Teaching and learning in the school science laboratory. An analysis of research, theory, and practice. In S. K. Abell & N. G. Lederman (Eds), Handbook of Research on Science Education (pp. 393–431). Lawrence Erlbaum Associates.
Millar, R. (1999). Understanding how to deal with experimental uncertainty: A 'missing link' in our model of scientific reasoning? In Conference of the European Science Education Research Association (ESERA), Kiel, Germany (Vol. 31).
Nelson, B. C., & Ketelhut, D. J. (2007). Scientific inquiry in educational multi-user virtual environments. Educational Psychology Review, 19(3), 265–283. https://doi.org/10.1007/s10648-007-9048-1
Office for Standards in Education (Ofsted) (2005a). Science in secondary schools. HMI/Ofsted. Retrieved March 17, 2008 from: http://live.ofsted.gov.uk/publications/annualreport0405/4.2.15.html
Office for Standards in Education (Ofsted) (2005b). Science and mathematics. HMI/Ofsted. Retrieved October 06, 2025 from: http://live.ofsted.gov.uk/publications/annualreport0405/5.14.html
Olson, J. K., & Clough, M. P. (2001). Technology's tendency to undermine serious study: a cautionary note. The Clearing House, 75(1), 8–13. https://doi.org/10.1080/00098650109599225
Qualifications and Curriculum Authority (QCA) (2007a). Science: Programme of study for key stage 3 and attainment targets. www.qca.org.uk/curriculum
Qualifications and Curriculum Authority (QCA) (2007b). Science. Programme of study for key stage 4. www.qca.org.uk/curriculum
Reiser, B. J., Tabak, I., Sandoval, W. A., Smith, B. L., Steinmuller, F., & Leone, A. J. (2001). BGuILE: Strategic and conceptual scaffolds for scientific inquiry in biology classrooms. In S. M. Carver & D. Klahr (Eds.), Cognition and instruction: Twenty-five years of progress. Erlbaum.
White, R. T., & Gunstone, R. F. (1992). Probing understanding. Falmer Press.
Zacharia, Z. C. (2007). Comparing and combining real and virtual experimentation: An effort to enhance students' conceptual understanding of electric circuits. Journal of Computer Assisted Learning, 23(2), 120–132. https://doi.org/10.1111/j.1365-2729.2006.00215.x
Zemball-Saul, C., Munford, D., Crawford, B., Friedrischen, P., & Land, S. (2002). Scaffolding preservice science teachers' evidence-based arguments during an investigation of natural selection. Research in Science Education, 32, 437–463. https://doi.org/10.1023/A:1022411822951

# Preface

Our experiences with practical work date back to our school days, when it was primarily limited to lab work and fieldwork. Concerning lab work, they range from not having the chance of performing or even watching the teacher perform a lab activity to having two times per week, two-hour-long lab classes to having integrated theoretical and practical courses. Fieldwork was seldom performed, partly because the school was seen as the place of learning and partly because a field trip requires time and money that were not available. Later, at the university, we were engaged in lab classes and or fieldwork as part of our undergraduate programs, leading to secondary school science teaching. Our school and university experiences with lab work led us to understand that it is possible to learn and understand science without working in the lab, and also that lab work can be a valuable tool or a waste of time and resources, depending on how theory and practice bind together.

This awareness was a key for making decisions on using or not using the lab during teacher training, afterward, as school teachers, and still later, for some of us, on how to work with prospective teachers on the topic of practical work in science education. Pupils enjoy practical work, which is good, but labs are not playgrounds, and field trips are not entertaining excursions. A critical approach to practical work in school science education was advocated—set the purpose of using it, decide on how to use it to fulfil that purpose, uncover its added value, and use it only if it brings a meaningful added educational value. Conferences organized, projects submitted, and students' supervision aligned with this overall idea, which was strengthened by the research papers, books, and experts we were lucky to come across.

Nowadays, practical work is more varied than it was when we were students, driven by the need to develop the competencies, knowledge, and attitudes required in today's society. This society is increasingly digital, interconnected, and shaped by complex socio-scientific challenges. Schools cannot remain indifferent to these changes if they aim to prepare citizens capable of understanding and acting responsibly in the world. This does not mean abandoning all the physical laboratory and field experiences that remain essential for understanding the core ideas of science. We are assisting a gradual shift in which physical lab work and field work are losing their places on the podium, due to the competition of modern, digitally based practical work and interdisciplinary practical tools, like robotics or 3D printing. We would not like replacement to occur; we would like to think about the different kinds of practical work powering each other.

The most recent books on practical work—more than 10 years old now—started announcing this change, but it was too early to develop it further. Since then, that change has become so big that it requires reconceptualizing the concept of practical work to delocalize it from the physical (wet) lab and the field, to make it encompass all the possibilities of active learning by doing. Acknowledging that this has too broad a scope, we limit the focus of the practical work concept to in-school or in open spaces learning.

To us, active learning through practical work is much more than connecting or moving things around. It is about thinking and inquiring, purposeful planning, doing as planned, critically analysing results, evaluating processes, raising new questions, and enjoying what is being done. It is primarily about minds-on, as minds are the glue that binds theory and practice. Nevertheless, minds-on should align with hands-on and hearts-on, as science is also about doing, and affective attachment is a crucial factor for students to engage in science careers.

Aligned with the previous ideas, this book does not reject the classic conceptualizations of practical work; rather, it places practical work within the challenging framework offered by digitalization and argues that classic and emerging approaches to practical work may power each other. In doing so, it offers a wide and inclusive overview of practical work in science education in the digital era, as we planned, for the benefit of science education in increasingly multicultural and diverse schools.

The comprehensive nature of this book would not be possible without the trust and cooperation of a large set of outstanding authors from all continents, who also acted as referees of other chapters. We are very grateful to all of them. We are also grateful to Professor Justin Dillon for accepting to write the Foreword.

Offering this book in open access, and so enabling all those interested in the topic to access it, was only possible with the funding provided by the Research Centre on Education (CIEd), UID/01661, through national funds of FCT/MCTES-PT. We are grateful to the previous and current CIEd directors for supporting this project.

Finally, we would like to thank Springer for trusting us as editors and the Springer staff for the magnificent support throughout the development of this project.

Laurinda Leite
University of Minho
Braga, Portugal

Luís Dourado
University of Minho
Braga, Portugal

Ana S. Afonso
University of Minho
Braga, Portugal

Sofia Morgado
University of Minho
Braga, Portugal

Luísa Carvalho
Lusophone School of Higher
Education, IPLUSO
Lisboa, Portugal

# Contents

# Editors and Contributors

## About the Editors

**Laurinda Leite** is a full professor of science education and a CIEd's researcher at University of Minho, Portugal. After graduating as a physics and chemistry teacher, she earned a master's (University of London) and a Ph.D. (University of Minho) in science education. She coordinates projects, supervises, and conducts research and has published works in science teaching and learning, namely in the history of science, lab work, and problem-based learning.

**Luís Dourado** is an assistant professor of science education at University of Minho, Portugal, and a CIEd's researcher. He graduated as a biology and geology teacher, and earned a master's and a PhD in science education. His research interests focus on science teaching and learning, namely on fieldwork, problem-based learning, and ecoethics. He is a member of the editorial committee of a few journals, including the Revista Portuguesa de Educação.

**Ana S. Afonso** holds a Ph.D. from the University of Reading, UK, and is an assistant professor in science education and a researcher at CIEd, UMinho, Portugal. Her research interests focus on science learning in and out of school, particularly in science museums. She is a member of the editorial committee of the Revista Portuguesa de Educação and of the editorial board of the International Journal of Science Education.

**Sofia Morgado** graduated as a physics and chemistry teacher, and earned a master's and a Ph.D. in science education (University of Minho). She has been engaged in science education research projects as well as in curriculum development projects, at the University of Minho, Portugal. Author of several papers and book chapters, her research interests concentrate on science teaching and learning, with a focus on problem-based learning and contextualized science learning.

**Luísa Carvalho** graduated in biology and geology, and earned a master's in biology and geology teaching and a PhD in science education. Her research interests focus on sustainability and environmental ethics, areas in which she has been publishing papers. She was involved in a curriculum development program, and currently, is an assistant professor at the Lusophone School of Higher Education, IPLUSO, Portugal, where she teaches science education courses.

## Contributors

**Ian Abrahams** University of Roehampton, Roehampton, UK

**Agustín Adúriz-Bravo** CONICET/Universidad de Buenos Aires, Facultad de Ciencias Exactas y Naturales, CeFIEC-Instituto de Investigaciones Centro de Formación e Investigación en Enseñanza de las Ciencias, Buenos Aires, Argentina

**Ana S. Afonso** CIEd, Institute of Education, University of Minho, Braga, Portugal

**Ivan Au** University of Oxford, Oxford, UK

**Mónica Baptista** UIDEF, Institute of Education, University of Lisbon, Lisbon, Portugal

**Marcilia Elis Barcellos** Federal Center for Technological Education of Rio de Janeiro, Rio de Janeiro, Brazil

**Óscar Bedoya-Cano** Pontifical Bolivarian University, Medellín, Colombia

**Jasodhara Bhattacharya** Independence Science, West Lafayette, IN, USA

**Jonathan Bowers** Michigan State University, East Lansing, MI, USA

**Mustafa Şahin Bülbül** Kafkas University, Kars, Turkey

**Robertson Burgess** University of Newcastle, Newcastle, NSW, Australia

**Cheryl L. Burleigh** Sanford College of Education, National University, San Diego, USA

**Ella Camporeale** University of Newcastle, Newcastle, NSW, Australia

**Luísa Carvalho** Lusophone Higher School of Education, IPLUSO, Lisboa, Portugal

**Ho-Yin Chan** University of Cambridge, Cambridge, UK

**Marcos Corrêa da Silva** Federal Center for Technological Education of Rio de Janeiro, Rio de Janeiro, Brazil

**James de Winter** Faculty of Education, University of Cambridge, Cambridge, UK

**Raimundo Nonato de Medeiros Jr** University of Trás-Os-Montes and Alto Douro, Vila Real, Portugal;
Federal Institute for Education, Science and Technology of Rio Grande Do Norte, Natal, Brazil

**Maria Eugenia Dies Álvarez** University Institute for Research in Environmental Sciences of Aragon, University of Zaragoza, Zaragoza, Spain

**Glauco dos Santos Ferreira da Silva** Federal Center for Technological Education of Rio de Janeiro, Rio de Janeiro, Brazil

**Luís Dourado** CIEd, Institute of Education, University of Minho, Braga, Portugal

**Emil Eidin** University of Wyoming, Laramie, WY, USA

**Sibel Erduran** University of Oxford, Oxford, UK

**João Paulo Fernandes** Federal Center for Technological Education of Rio de Janeiro, Rio de Janeiro, Brazil

**Nikolaos Fotou** School of Education and Communication, University of Lincoln, Lincoln, UK

**Urtza Garay-Ruiz** University of the Basque Country, Bilbao, Spain

**Remeredzayi Gudyanga** Faculty of Education, University of the Free State, Bloemfontein, South Africa

**Peng He** Washington State University, Pullman, WA, USA

**Elisabeth Hildt** Center for the Study of Ethics in Professions, Illinois Institute of Technology, Chicago, IL, USA

**Avi Hofstein** The Weizmann Institute, Rehovot, Israel

**Nor Asniza Ishak** School of Educational Studies, Universiti Sains Malaysia, Gelugor, Penang, Malaysia

**Bruna Karl** Federal University of Rio de Janeiro, Rio de Janeiro, Brazil

**Eric A. Kirk** Utah State University, Logan, USA

**Dionysios Koulougliotis** Department of Environment, Ionian University, Zakynthos, Greece

**Kelly Laas** Center for the Study of Ethics in Professions, Illinois Institute of Technology, Chicago, IL, USA

**Laurinda Leite** CIEd, Institute of Education, University of Minho, Braga, Portugal

**Joaquim Bernardino Lopes** University of Trás-Os-Montes and Alto Douro, Vila Real, Portugal;
CIDTFF, University of Aveiro, Aveiro, Portugal

**Jesús López-Belmonte** University of Granada, Granada, Spain

**Toni Maddock** University of Newcastle, Newcastle, NSW, Australia

**Muaz Mohd Zaini Makhtar** School of Industrial Technology, Universiti Sains Malaysia, Gelugor, Penang, Malaysia

**Silvija Markic** Ludwig Maximilian University of Munich, Munich, Germany

**Isabel Martins** Federal University of Rio de Janeiro, Rio de Janeiro, Brazil

**Jorge Martín-García** University Institute for Research in Environmental Sciences of Aragon, University of Zaragoza, Zaragoza, Spain

**José-Antonio Marín-Marín** University of Granada, Granada, Spain

**Henry Matovu** School of Chemistry, The University of Sydney, Sydney, Australia

**Robin Millar** Department of Education, University of York, York, UK

**Sofia Morgado** University of Minho, Braga, Portugal

**Marco Duarte Naia** University of Trás-Os-Montes and Alto Douro, Vila Real, Portugal;
CEMMPRE, University of Coimbra, Coimbra, Portugal

**Matheus Monteiro Nascimento** Federal University of Rio Grande do Sul, Porto Alegre, Brazil

**Jairo Ortiz-Revilla** University of Burgos, Burgos, Spain

**Katerina Paschalidou** Department of Chemistry, National and Kapodistrian University of Athens, Athens, Greece

**Javier Portillo-Berasaluce** University of the Basque Country, Bilbao, Spain

**Amaia Quintana-Ordorika** University of the Basque Country, Bilbao, Spain

**Umesh Ramnarain** University of Johannesburg, Johannesburg, South Africa

**André Ramos** Unit of Science Teaching, Faculty of Sciences, Porto University, Porto, Portugal

**Sophia Rapti** Department of Education, School of Philosophy and Education, Aristotle University of Thessaloniki, Thessaloniki, Greece

**Pedro Reis** UIDEF - Institute of Education, University of Lisbon, Lisbon, Portugal

**Michael J. Reiss** Institute of Education, University College London, London, UK

**Tiago Ribeiro** Unit of Science Teaching, Faculty of Sciences, Porto University & Interdisciplinary Centre of Marine and Environmental Research, Matosinhos, Portugal

**Rafiza Rosli** School of Educational Studies, Universiti Sains Malaysia, Gelugor, Penang, Malaysia

**Troy D. Sadler** The University of North Carolina, Chapel Hill, NC, USA

**Katerina Salta** Department of Chemistry, National and Kapodistrian University of Athens, Athens, Greece

**Theodosios Sapounidis** Department of Education, School of Philosophy and Education, Aristotle University of Thessaloniki, Thessaloniki, Greece

**Rachael Sharpe** School of Education and Communication, University of Lincoln, Lincoln, UK

**Erica Southgate** University of Newcastle, Newcastle, NSW, Australia

**Manabu Sumida** Faculty of Education, Centre for Gifted Education and Talent Development, Ehime University, Matsuyama, Japan

**Cary A. Supalo** Independence Science, Princeton, NJ, USA

**Clara Vasconcelos** Unit of Science Teaching, Faculty of Sciences, Porto University & Interdisciplinary Centre of Marine and Environmental Research, Matosinhos, Portugal;
Department of Geoscience, Environment and Spatial Plannings, Faculty of Sciences, Porto University, Porto, Portugal

**Eleni Vlachoudi** Department of Education, School of Philosophy and Education, Aristotle University of Thessaloniki, Thessaloniki, Greece

**Mihye Won** Faculty of Education, Monash University, Melbourne, Australia

**Regina H. Yorkgitis** The University of North Carolina, Chapel Hill, NC, USA

# Part I
# Overview of Practical Work in Science Education

# Chapter 1
# A Critical Look at the History of School Practical Work over the Past half a Century

**Ian Abrahams, Nikolaos Fotou, and Avi Hofstein**

## 1.1 Introduction

With the reform in science education in the 1960s, both in the USA and the UK, the focus became to engage school students with investigations, discoveries, inquiry and problem-solving activities. In other words, practical work in the school laboratory became the core of the science learning process and science instruction (Shulman & Tamir, 1973). This belief, held by many within both the science education community and the science community more widely, as to the central importance of practical work, as a means to develop conceptual understanding, motivate students towards science, and to develop both practical skills, critical thinking skills, and problem-solving skills has been challenged relatively infrequently (e.g., Millar & Abrahams, 2009; Hodson, 1993; Hofstein & Lunetta, 2004). This chapter looks at the history of school practical work since 1960 in terms of three themes: (i) its effectiveness in developing conceptual understanding (ii) its affective value and (iii) its role in developing practical skills, critical thinking skills, and problem-solving skills.

---

I. Abrahams (✉)
University of Roehampton, Roehampton, UK
e-mail: ian.abrahams@roehampton.ac.uk

N. Fotou
University of Lincoln, Lincoln, UK
e-mail: NFotou@lincoln.ac.uk

A. Hofstein
The Weizmann Institute, Rehovot, Israel
e-mail: avi.hofstein@weizmann.ac.il



L. Leite et al. (eds.), *Practical Work in Science Education in the Digital Era*, Advances in Technology-Rich Science Education 2, https://doi.org/10.1007/978-3-032-13788-3_1

## 1.2 The Affective Value of School Practical Work

In many countries, such as the UK, with a tradition of practical work in school science its inclusion in the teaching of science is seen as central to the appeal of science. Yet despite its frequent and widespread use in UK schools (Bennett, 2003; Millar, 2004), and the common perception reported amongst teachers that its use *motivates* students (Wellington, 2005), the research shows that, in a strict psychological sense, its affective value is better understood not as motivation but rather as non-enduring situational interest (Abrahams, 2009). Are teachers, we might then ask, using the term 'motivate' in its strict psychological sense or as a 'catch-all' term that interchangeably means, interest, fun, enjoyment, and engagement? Bandura (1986) suggested that teachers are not the only ones to use those terms interchangeably and that the terms 'motivate' and 'interest' have been used, in the science education literature, to mean the same thing even though "there is a major difference between a motive, which is an inner drive to action, and an interest, which is a fascination with something" (p. 243). An example of this can be seen in Lazarowitz and Tamir (1994), who claimed that practical work *motivates* students, citing in support of this Ben-Zvi et al. (1977), Henry (1975), and Selmes et al. (1969), even though those studies focused almost exclusively on the issue of student *interest* rather than motivation. Indeed, of these three citations, only in Henry (1975) is the term 'motivation' actually used, albeit only once, when Henry states, citing no sources or research-based evidence, "[i]n addition, psychological reasons can be proposed which relate to the improved motivation of students by the inclusion of laboratory exercises in the science program" (p. 73).

The fact therefore that situational interest is, unlike motivation or personal interest, unlikely to endure beyond the end of a particular lesson (Hidi & Harackiewicz, 2000) helps to explain why students need to be continuously re-stimulated by the frequent use of practical work. Once this is understood, the reason why many of those students who claim to like practical work also claim to have little, if any, personal interest in science or, as Cerini et al. (2003) reported, lack any motivation to pursue it post compulsion, becomes clearer. For whilst these students do *like* practical work, their reason for doing so appears to be primarily that they see it as being preferable to non-practical teaching techniques in science lessons that they tend to associate with more writing (Hodson, 1990). Indeed, Abrahams (2009) found that even from their first year at secondary school (age 11) the novelty of being in a laboratory environment wore off relatively quickly as students became increasingly disillusioned (Haste, 2004) by the reality of regular school science. This disillusionment being because the reality they experienced in their regular school science contrasted markedly with the practical work, specifically designed to be fun and exciting, that they are frequently introduced to (Ogborn et al., 1996) during visits to secondary schools as primary school students (age 10).

Such findings also provide an explanation as to why, despite the frequent and widespread use of practical work in secondary school science lessons (Bennett, 2003), and a belief in its motivational value (Hodson, 1990), the number of students in the UK choosing to pursue post-compulsory 'A'-levels (age 17–18) in chemistry or physics, arguably the two science subjects that offer the most practical work during Key Stages 3 and 4 (ages 11–16), are generally lower (e.g., UK Government Official Statistics, 2024) than for biology that offers the least practical work.

Indeed, despite the claim that "practical work is absolutely essential in creating enthusiasm" (House of Commons Science and Technology Committee, 2002, Question 514) there is, almost a quarter of a century later, still no research evidence to substantiate it being essential in this, or any other, respect. What the research has in fact found is that a student's decision to pursue science post-compulsion is more likely to be influenced by the personality of their science teachers (Jarvis & Pell, 2005; Reiss 2005), career and/or university aspirations (House of Commons Science and Technology Committee, 2002), perceived relevance (Jenkins & Pell, 2006) rather than whether, and the extent to which, practical work is used in science lessons.

What therefore emerges from a critical review of the literature relating to the affective value of school practical work over the past 65 years is the lack of clarity and consistency in the use of affective terminology both by teachers, science educators, and policy makers. Claiming that students find doing practical work motivating, when what is actually meant, in a strict psychological sense, is that it generates non-enduring situational interest, means that we unintentionally, and unrealistically, inflate the affective value of practical work. Indeed, even though the Twenty First Century Science suite of GCSEs (University of York, 2009) did *not* include any more practical work than alternative suites of GCSE science it generated an average 32% increase in the number of students who opted to pursue science post-compulsion. This illustrates that increased motivation to study science is not necessarily caused by/ correlated with an increase in the amount of practical work used and that it is reasonable to expect student motivation towards science to manifest itself in their choice to study science post-compulsion.

## 1.3 Practical Work for Conceptual Development: A Critical Retrospection of its Effectiveness

Since its introduction in the school science curriculum early in the nineteenth century and despite its initial focus as a means to confirm what had previously been taught, practical work has been regarded as a pedagogical bridge between, what was later defined as, the domain of objects and observables and the domain of concepts and ideas of science (Millar, 2004). Despite its ubiquity, the assumption that practical work reliably enhances conceptual understanding remains contentious.

The pedagogical shift between 1875 and 1920, influenced by seminal work like the German model (Liebig, 1843) emphasising systematic and practice-based instruction, Dewey's (1897) and Armstrong's (1898) experiential and discovery learning respectively, marked a significant change in science education in Europe and North America. This was part of the progressive education movement advocating for experiential learning as a means of fostering understanding which led to the integration of practical work into the school science curricula. Historical analyses, however, suggest that this wide adoption of practical work in these early years was often ritualistic rather than reflective, serving more as a symbol of scientific rigor than a tool for the development of conceptual understanding (Palmer, 2007).

An effort to empirically assess the effectiveness of practical work in science education is first observed in the early 1920s and gained momentum through the 1960s. Cunningham's (1946) comprehensive review of 37 U.S.-based studies and Bradley's (1965) comparison of lecture and individual laboratory methods exemplify this effort. However, this body of research predominantly focused on the domain of objects and observables rather than the domain of ideas. In the relatively few studies (e.g., Scott, 1929) that did attempt to assess effectiveness in the domain of ideas, there was no clear definition of what a science idea or concept is, or how its development of understanding should be measured (Cunningham, 1946). Also, the conceptual understanding was often conflated with factual recall with the focus being on differences between immediate and delayed recall that Bradley's (1965) controlled study showed no effect on. Moreover, the methodological approaches employed in many of these studies have been questioned with common issues including limited evidence of validity or reliability, and no, or inconsistent control of, confounding variables (Hofstein & Lunetta, 1980).

Despite the limited number of empirical studies, and ongoing concerns about the robustness of the evidence base supporting the conceptual understanding benefits, practical work became an integral part of science education in many countries as evidenced in historical and scholarly sources during the 1920s–1960s. The Science Master's Association (1953), for example, strongly advocated for individual or group practical work as part of modernising science education in the UK. The assertion was that "the pupils must… be helped to devise experiments to test their ideas and to form their own conclusions" (Science Master Association, 1953, p. 5). While the process of testing ideas to independently form conclusions is vital for conceptual understanding, it is contingent on the independently formed conclusions ultimately aligning with the scientifically accurate concepts. This very challenge of independent learning was highlighted by Mayer (2004) who reviewed studies published from 1950 to 1980 showing the ineffectiveness of such independent and unguided approaches that instead of conceptual understanding, more often lead to the development of misconceptions.

Another major turning point in the reflection on the teaching of science, and more specifically with regards to the development and use of the practical work, occurred in the early 1960s following the Sputnik launch. This triggered profound curricula reforms in which practical work was seen as providing a means to facilitate a deeper understanding of scientific concepts through hands-on engagement.

Despite this renewed emphasis, educators began to question the effectiveness of practical work realising that its role is not as self-evident as it was thought. Comparative studies from the 1960s to early 1980s (e.g., Coulter, 1966; Ben-Zvi et al., 1977) showed no significant differences in students' achievement between practical work and other alternative teaching methods (for example, demonstrations, discussions, or simulations) and that whilst students were enthusiastic about practical work activities, these experiences did not significantly enhance conceptual understanding (Woolnough, 1976). Bates (1978) summarised this body of research highlighting growing concerns regarding the effectiveness of practical work, including in terms of the development of conceptual understanding, and likewise Hofstein and Lunetta's (1980) review of research studies during this period reinforced these reservations and emphasised the need for further research. In their conclusions, they noted that there was insufficient evidence to make generalisation about the optimal role of practical work in the domain of ideas and regarding the development of conceptual understanding.

By the end of the 1990s, critiques of practical work had become more nuanced with Hodson (1996), in his guest editorial for a journal special issue on practical work, highlighting a paradox of practical work in science education in that it was both "over-used" and "under-used" (p. 756) and the need for a more critical stance towards teaching/learning through practical work. This under-use related mostly to the unrealised potential of practical work to support the development of conceptual understanding (Woolnough, 1991). It was overused in the sense that teachers often believed that its use would achieve all their learning objectives and expecting students to develop an understanding of science concepts through an unguided restructuring of their knowledge conceptual system. This, given the complexity of conceptual development, reflected an overly simplistic use of practical work that ignored its possible pedagogical potential. In a similar vein, Gunstone and Champagne (1990) emphasised the need for time, interaction, and reflection for conceptual understanding to develop, opportunities that are rare in typical school settings across the United States and other countries (e.g., Polman, 1999; Tobin, 1990).

From the close of the twentieth century to the present day, practical work still holds a privileged place in science education in schools. Curricula and policy documents continue to portray practical work as a panacea for a wide range of learning objectives, the development of conceptual understanding included, without any explicit indication that different kinds of practical work might be needed for each objective and the specific pedagogical approaches to achieve these. Therefore, concerns in the preceding decades about the complexity of practical work in the school laboratory for developing conceptual understanding and the limited empirical evidence supporting such claims remain. For example, Hofstein and Lunetta (2004), in their second major review covering research evidence up to 2002, concluded that, despite the long-standing beliefs, strong or consistent evidence that practical work is effective in promoting students' conceptual understanding was lacking. In a similar vein, Abrahams and Millar (2008) found that students struggled to independently connect the domain of ideas (often involving unobservable entities and behaviours) with the domain of objects and observables (things or properties that we can see directly) and despite frequently enjoying practical work, it functions more as a ritual of scientific practice than a tool for conceptual development.

A more recent systematic review by Ferreira and Morais (2020) revealed that the cognitive demands of practical work are frequently underestimated by teachers who emphasise procedures and objectives in the domain of objects and observables over the domain of ideas and, as a result, students often fail to make meaningful connections between the two domains. This disconnection is exacerbated by practicalities of teaching through practical work such as limited time, inadequate resources, large class sizes which reduce practical work to a rote, almost cook-book like exercise—an issue that had already been raised in earlier decades (Hofstein & Lunetta, 2004; Tobin, 1990).

Across decades of reform and reflection, the role of practical work in developing conceptual understanding has remained central, yet contested, in the science education literature. Although we do not dispute that practical work has the potential, if used appropriately, to hold an important role in supporting students' conceptual understanding, its intuitively argued appeal, resulting in a widespread use over the years, has not been supported by robust evidence to demonstrate that, as commonly used, it significantly enhances conceptual development. This enduring gap necessitates further rigorous, concept-sensitive, research that not only evaluates the impact of different forms of, and pedagogical practices (like structured reflection, guided and scaffolded inquiry) within practical work, but also provides clear, actionable, guidance for science teachers on how best to design and implement these activities to maximise effectiveness in the domain of ideas.

## 1.4 A Critical Review of the Effectiveness of Practical Work for Skills Development

Practical work has often been seen as an essential part of school science education with a widely held belief that it offers multifaceted benefits (e.g., Woodley, 2009). Although, as Tampakis (2011) notes, the primary belief up until the mid-twentieth century was that the value of school practical work lay in its role to develop practical skills rather than its use having a central role to play in the development/ enhancement of both critical, and problem-solving, skills. Indeed, it was only following the reform movements of the 1960s–1990s, exemplified by the Nuffield curricula in the UK (e.g., Nuffield Foundation, 1966) and the US (e.g., CHEMS, 1959), that progressive educators expanded the idea of what doing science through practical work meant and proposed, albeit without empirical research to substantiate the claims, that critical thinking and problem solving were important skills that undertaking practical work can develop (Kolovou, 2022). Whilst Jakeways (1986) also noted that practical work had the *potential* to develop such skills there was an acknowledgement that its success in so doing depended heavily on how it was implemented in science teaching and the specific pedagogical approaches used. Similarly, Hofstein and Lunetta (1980) reported that although there was a widely held belief that practical work had the *potential* to provide a unique environment for the development of a range of skills, many research studies did not actually provide clear evidence of its effectiveness in doing so and, as such, such claims remained essentially conjectures. Indeed, empirical research (e.g., Bryce et al., 1983; DES/APU, 1984) has shown that even after being taught science at secondary school, many students still lacked basic observational, and measurement, skills as well as the ability to record accurate measurements and organise data for analysis and interpretation. Although, it has to be noted, that it is highly likely that those same students already had those generic skills to some extent in that it is highly likely that they could already read the time on their watches, put information onto a calendar, read a bus timetable, and discuss the data from weekend sports fixtures.

What also needs to be critically considered is the distinction between what skills can, in an *ideal* situation, be developed using practical work (or any other pedagogical approach) and what skills might actually be developed in a 'real' school laboratory, with students of a range of academic abilities and interests, and teachers who are not always subject specialists. As Hegarty-Hazel (1990) reported, the dominant teaching approach up until the end of the twentieth century was primarily very teacher-directed, offering limited, to no, opportunities for students to design their own experiments, or to develop critical, and problem-solving, skills. As a result, even at the level of practical skills development, students tended to follow highly structured, recipe-style, practical work with clear guidance on which practical skills to apply and how to execute them, leaving little, if any, room for an independent application (Hofstein & Lunetta, 1980).

Indeed, the common practice amongst science teachers in the UK to "focus on the 'hands on' aspects of tasks at the expense of the 'minds on' aspects" (Millar & Abrahams, 2009, p. 63) suggests that whilst practical work *might*, in an ideal situation, be potentially effective as a means of developing a wide range of skills, including critical thinking and problem solving, further research is needed in this area if such beliefs are to be substantiated. The development of such skills is also unlikely to be the case if teachers choose to continue to use highly structured, recipe-style, practical work as a means of ensuring that their students obtain and see the results that they want them to see then, as noted by Jenkins (2011), the potential educational aims of practical work are too easily lost.

Another issue regarding claims about the effectiveness of practical work in the development of practical skills, problem solving and critical thinking skills within the period 1960–1980 is that many studies undertaken in that period did not actually seek to evaluate the effectives of practical work in developing these skills (Bryce & Robertson, 1985). Likewise, in their review of literature of this period, Hofstein and Lunetta (1980) also reported that empirical studies frequently lacked methodological rigor in that they failed to define or measure the skills they aimed to assess, making it difficult to draw firm conclusions about the effectiveness of practical work in developing those skills. Indeed, although the term 'practical skills' or 'practical competencies' are widely referred to in the literature and by science teachers, they are rarely explicitly defined in terms of what these skills actually are (Oliveira & Bonito, 2023) and how they might, most effectively, be validly assessed remains unclear (Abrahams et al., 2013). Furthermore, in terms of high order thinking skills, García-Carmona (2023) notes that traditional science instruction often conflates scientific thinking with critical thinking, without clearly defining or fostering either in practice.

This ambiguity, combined with a lack of student autonomy in designing and independently undertaking practical work, has the potential to further limit the development of such skills. Therefore, whilst it has been suggested by a number of authors that practical work *might* have a role to play in fostering, for example, critical thinking skills (e.g., Kind et al., 2011), metacognitive skills (e.g., Kipnis & Hofstein, 2008; Opstal & Daubenmire, 2017) and argumentative skills (e.g., Hofstein & Kind, 2012) there is also a recognition by those authors that further research is needed to evaluate its effectiveness in the development of such skills i.e., to replace a tentative *might* with a research informed *does*.

## 1.5 Conclusion and Ways Forward

The aim of this chapter has been to show that there still remains, well into the twenty-first century, surprisingly little research evidence to show clearly that practical work, as frequently used in school laboratories, is either effective in terms of developing scientific conceptual understanding, or that it provides an effective means of developing essential life skills such as critical thinking, or problem solving, or that it actually motivates students towards science either in general or as a career. As others have previously suggested (e.g., Hodson, 1996; Schweingruber et al., 2006), there is a need, moving forward, for a more research-based approach to improve our understanding of practical work with an emphasis on robust, empirical, evidence, rather than personal beliefs—however widely those beliefs might be shared and however strongly they might be held. If it can be demonstrated that those who do practical work, compared to those who do not, or who do it for more years, are, for example, statistically significantly more effective at problem solving, or at critical thinking, or at teamwork then there is research evidence to substantiate those beliefs and to justify—given its relative expense—the use of practical work. Looking ahead to the next half a century of science education such an evidence-based approach has the potential to improve the quality of practical work and support science teachers to better understand its potential role and value in the teaching and learning of science.

## References

Abrahams, I. (2009). Does practical work really motivate? A study of the affective value of practical work in secondary school science. *International Journal of Science Education, 31*(17), 2335–2353. https://doi.org/10.1080/09500690802342836

Abrahams, I., & Millar, R. (2008). Does practical work really work? *Studies in Science Education, 44*(1), 1–36. https://doi.org/10.1080/09500690701749305

Abrahams, I., Reiss, M., & Sharpe, R. (2013). The assessment of practical work in school science. *Studies in Science Education, 49*(2), 209–251.

Armstrong, H. E. (1898). *The heuristic method of teaching or the art of making: Children discover things for themselves*. Board of Education Special Reports.

Bandura, A. (1986). *Social foundations of thought and action: A social cognitive theory*. Prentice-Hall.

Bates, G. R. (1978). The role of the laboratory in secondary school science programs. *What Research Says to the Science Teacher, 1*, 55–82.

Ben-Zvi, R., Hofstein, A., Samuel, D., & Kempa, R. F. (1977). Modes of instruction in high school chemistry. *Journal of Research in Science Teaching, 14*, 433–439. https://doi.org/10.1002/tea.3660140507

Bennett, J. (2003). *Teaching and learning science: A guide to recent research and its applications*. Continuum.

Bradley, R. L. (1965). Lecture demonstration versus individual laboratory work in a general education science course. *The Journal of Experimental Education, 34*(1), 33–42.

Bryce, T. G., McCall, J., MacGregor, J., Robertson, I. J., Weston, R. A. J., Bryce, T. G. K., McCall, J., MacGregor, J., Robertson, I. J., & Weston, R. A. J. (1983). *Teacher's guide in TAPS assessment pack*. Heinemann Educational Publishers.

Bryce, T. G., & Robertson, I. J. (1985). What can they do? A review of practical assessment in science. *Studies in Science Education, 12*(1), 1–24. https://doi.org/10.1080/03057268508559921

Cerini, B., Murray, I., & Reiss, M. (2003). *Student review of the science curriculum. Major findings*. Planet Science/Institute of Education University of London/Science Museum. Retrieved September 30, 2025 from http://www.planet-science.com/sciteach/review

CHEM Study. (1959). *CHEM study program: Chemical education material study*. University of California.

Coulter, J. C. (1966). The effectiveness of inductive laboratory, inductive demonstration, and deductive laboratory in biology. *Journal of Research in Science Teaching, 4*(3), 185–186. https://doi.org/10.1002/TEA.3660040314

Cunningham, H. A. (1946). Lecture demonstration versus individual laboratory method in science teaching—a summary. *Science Education, 30*(2), 70–82. https://doi.org/10.1002/sce.3730300204

DES/APU. (1984). *Science in schools: Ages 13 and 15, Research Report No. 3*. DES publications.

Dewey, J. (1897). My pedagogic creed. *The School Journal, 54*(3), 77–80.

Ferreira, S., & Morais, A. M. (2020). Practical work in science education: Study of different contexts of pedagogic practice. *Research in Science Education, 50*, 1547–1574. https://doi.org/10.1007/s11165-018-9743-6

García-Carmona, A. (2023). Scientific thinking and critical thinking in science education: Two distinct but symbiotically related intellectual processes. *Science & Education, 34*, 227–245. https://doi.org/10.1007/s11191-023-00460-5

Gunstone, R. F., & Champagne, A. B. (1990). Promoting conceptual change in the laboratory. In E. Hegarty-Hazel (Ed.), *The student laboratory and the science curriculum* (pp. 159–182). Routledge.

Haste, H. (2004). *Science in my future: A study of values and beliefs in relation to science and technology amongst 11–21 year olds*. Nestle Social Research Programme.

Hegarty-Hazel, E. (1990). *The student laboratory and the science curriculum*. Routledge.

Henry, N. W. (1975). Objectives for laboratory work. In P. L. Gardner (Ed.), *The structure of science education* (pp. 61–75). Longman.

Hidi, S., & Harackiewicz, J. M. (2000). Motivating the academically unmotivated: A critical issue for the 21st century. *Review of Educational Research, 70*(2), 151–179. https://doi.org/10.2307/1170660

Hodson, D. (1990). A critical look at practical work in school science. *School Science Review, 70*(256), 33–40.

Hodson, D. (1991). Practical work in science: Time for a reappraisal. *Studies in Science Education, 19*, 175–184. https://doi.org/10.1080/03057269108559998

Hodson, D. (1993). Re-thinking old ways: Toward a more critical approach to practical work in school science. *Studies in Science Education, 22*, 85–142. https://doi.org/10.1080/03057269308560022

Hodson, D. (1996). Practical work in school science: Exploring some directions for change. *International Journal of Science Education, 18*(7), 755–760. https://doi.org/10.1080/0950069960180702

Hofstein, A., & Kind, P. (2012). Learning in and from science laboratories. In B. Fraser, K. Tobin, & K. McRobbie (Eds.), *International handbook on science education* (pp. 189–207). Springer.

Hofstein, A., & Lunetta, V. N. (1980). *The role of the laboratory in science teaching: Research implications*. National Association for Research in Science Teaching.

Hofstein, A., & Lunetta, V. N. (2004). The laboratory in science education: Foundations for the twenty-first century. *Science Education, 88*(1), 28–54. https://doi.org/10.1002/sce.10106

House of Commons Science and Technology Committee. (2002). *Third report. Science education from 14 to 19*. HMSO. Retrieved September 30, 2025 from http://www.publications.parliament.uk/pa/cm200102/cmselect/cmsctech/508/50802.htm

Jakeways, R. (1986). Assessment of A-level physics (Nuffield) investigations. *Physics Education, 21*, 212–214.

Jarvis, T., & Pell, T. (2005). Factors influencing elementary school children's attitude towards science before, during, and after a visit to the UK national space Centre. *Journal of Research in Science Teaching, 42*(1), 53–83. https://doi.org/10.1002/tea.20045

Jenkins, E. (2011). *The background to practical work in school science. In practical experiments in school science lessons and science field trips: Written evidence submitted to the house of commons science and technology committee*. UK Parliament.

Jenkins, E.W., & Pell, R.G. (2006). *The relevance of science education project (ROSE) in England: A summary of findings*. Centre for Studies in Science and Mathematics Education, University of Leeds, UK.

Kind, P. M., Kind, V., Hofstein, A., & Wilson, J. (2011). Peer argumentation in the school science laboratory-exploring effects of task features. *International Journal of Science Education, 33*(18), 2527–2558. https://doi.org/10.1080/09500693.2010.550952

Kipnis, M., & Hofstein, A. (2008). The inquiry laboratory as a source for development of metacognitive skills. *Int J of Sci and Math Educ, 6*, 601–627. https://doi.org/10.1007/s10763-007-9066-y

Kolovou, M. (2022). Lessons from the past: Reviewing how we teach science, what's changed, and why it matters. *Science & Education, 31*, 855–859. https://doi.org/10.1007/s11191-021-00267-2

Lazarowitz, R., & Tamir, P. (1994). Research on using laboratory instruction in science. In D. L. Gabel. (Ed.) *Handbook of research on science teaching and learning* (pp. 94–130). Macmillan.

Liebig, J. (1843). *Familiar letters on chemistry and its relation to commerce, physiology, and agriculture*. James M. Campbell & Company.

Mayer, R. (2004). Should there be a three-strikes rule against pure discovery learning? The case for guided methods of instruction. *American Psychologist, 59*(1), 14–19. https://doi.org/10.1037/0003-066X.59.1.14

Millar, R. (2004). The role of practical work in the teaching and learning of science. *Paper prepared for the meeting High school science laboratories: Role and vision*. National Academy of Sciences.

Millar, R., & Abrahams, I. (2009). Practical work: Making it more effective. *School Science Review, 91*(334), 59–64.

Nuffield Foundation. (1966). *Nuffield physics ordinary level course*. Nuffield Foundation.

Ogborn, J., Kress, G., Martins, I., & McGillicuddy, K. (1996). *Explaining science in the classroom*. Open University Press.

Oliveira, H., & Bonito, J. (2023). Practical work in science education: A systematic literature review. *Frontiers in Education, 8*, 1–20. https://doi.org/10.3389/feduc.2023.1151641

Opstal, M. P., & Daubenmire, P. L. (2017). *Metacognition in chemistry education: Connecting research and practice.* ACS publications.
Palmer, B. (2007). *Diffusion of knowledge in Britain and America: Some historical comparisons in science education between 1800 and 1920.* ANZCIES 35th Annual Conference, University of Auckland, New Zealand.
Polman, J. L. (1999). *Designing project-based science: Connecting learners through guided inquiry.* Teachers College Press.
Reiss, M. (2005). Importance of affect in science education. In S. Alsop (Ed.), *Beyond cartesian dualism: Encountering affect in the teaching and learning of science* (pp. 17–26). Springer.
Schweingruber, H. A., Hilton, M. L., & Singer, S. R. (2006). *America's lab report: Investigations in high school science.* National Academies Press.
Science Master Association. (1953). Secondary modern science teaching: A Report on the Teaching of Science in Secondary Modern Schools. Murray.
Scott, V. G. (1929). *An experiment in the effect of the individual laboratory versus demonstration method on the pupils' final grade in general science.* (Doctoral dissertation,. University of Washington.
Selemes, C., Ashton, B. G., Meredith, H. M., & Newal, A. (1969). Attitudes to science and scientists. *School Science Review, 51*, 7–22.
Shulman, L. S., & Tamir, P. (1973). Research on teaching in the natural sciences. In R. M. Travers (Ed.), *Second handbook of research on teaching* (pp. 1098–1148). Rand McNally.
Tampakis, K. (2011). The unrecognized mechanism: History of science education in the 19th century. *Almagest, 2*(2), 80–101.
Tobin, K. (1990). Research on science laboratory activities: In pursuit of better questions and answers to improve learning. *School Science and Mathematics, 90*(5), 403–418.
UK Government Official Statistics. (2024). *Provisional entries for GCSE, AS and A level: Summer 2024 exam series.* Provisional entries for GCSE, AS and A level: Summer 2024 exam series—GOV.UK.
University of York. (2009). *Success in encouraging teenagers to study science.* Retrieved September 30, 2025 from www.york.ac.uk/news-and-events/news/2009/encouraging-science.
Wellington, J. (2005). Practical work in the affective domain. In S. Alsop (Ed.), *Beyond cartesian dualism: Encountering affect in the teaching and learning of science* (pp. 99–110). Springer.
Woodley, E. (2009). Practical work in school science-why is it important. *School Science Review, 91*(335), 49–51. https://doi.org/10.3389/feduc.2023.1151641
Woolnough, B. E. (1976). Practical work in sixth-form physics. *Physics Education, 11*(6), 392–397. https://doi.org/10.1088/0031-9120/11/6/305
Woolnough, B. E. (1991). Practical science..

**Ian Abrahams** is professor of science education at the University of Roehampton, UK. Having previously been a teacher of physics and head of department, he undertook a PhD in Education, and then, in 2007, moved into Higher Education. His research interests relate to practical work, misconceptions, Continuing Professional Development (CPD) - in particular the effectiveness of such programmes - and the assessment of practical skills.

**Nikolaos Fotou** is a senior lecturer and the MA of Arts in Education program, lead at the University of Lincoln, UK. He studied physics at the University of Crete and earned his MA and PhD in science education from the University of Leeds. He has taught from primary school to postgraduate programs in various institutions. He advocates for evidence-based education, with research interests including analogical reasoning and interactive pedagogies.

**Avi Hofstein** served more than 55 years in the Department of Science Teaching, The Weizmann Institute of Science, Israel. He was involved in chemistry curriculum development, implementation, and research, mainly in practical work. In 2016 he was awarded the ACS award for his achievement related to understanding of the field of learning in and from chemistry laboratories. In 2017 he was awarded the DCRA by NARST.

# Chapter 2
# Practical Work in Science Education: Foundations, Challenges, and Future Perspectives

**Matheus Monteiro Nascimento and Silvija Markic**

## 2.1 Introduction

Practical work has occupied a central position in science education for over a century, often considered emblematic of science teaching itself. Its historical prestige is rooted in its close association with the practices of professional scientists and the epistemology of empirical inquiry. In countries like the United Kingdom, this tradition has been particularly entrenched, with science education becoming inextricably linked to laboratory-based instruction by the mid-20th century (Hodson, 1996). However, the widespread use of practical work has not always been accompanied by a clear understanding of its pedagogical objectives or its actual effectiveness in fostering scientific literacy.

Historically, practical work was justified on the grounds of mimicking scientific methodology, students were expected to learn science by "doing science." This was reinforced during the post-Sputnik era, notably in the United States, where major curricular reforms such as the Physical Science Study Committee (PSSC), Biological Sciences Curriculum Study (BSCS) and the Earth Science Curriculum Project (ESCP) promoted inquiry-based and laboratory-centered science instruction as a way to build national scientific competence (Oliveira & Bonito, 2023). These programs aimed to move science teaching beyond rote memorization and passive learning, emphasizing authentic engagement with empirical data, hypothesis generation, and argumentation. Nevertheless, while such reforms succeeded in integrating more laboratory and field activities into school science curricula, they also revealed the challenges of operationalizing these ideals within institutional constraints. Hofstein

M. M. Nascimento
Federal University of Rio Grande do Sul, Porto Alegre, Brazil
e-mail: matheus.monteiro@ufrgs.br

S. Markic (✉)
Ludwig Maximilian University of Munich, Munich, Germany
e-mail: s.markic@cup.lmu.de



L. Leite et al. (eds.), *Practical Work in Science Education in the Digital Era*, Advances in Technology-Rich Science Education 2, https://doi.org/10.1007/978-3-032-13788-3_2

and Lunetta (2004) reaffirm that the school laboratory (or hands-on practical work) remains an essential feature in science education because it provides a unique environment where students can observe, manipulate, and investigate scientific phenomena within a social learning context. They emphasize that when laboratory learning is structured effectively, it can foster cooperative interactions, positive attitudes toward science, and cognitive development.

It is important to emphasize that the 'doing science' perspective, which we refer to as practical work (Dillon, 2008), is designated by various terminologies within science education research (Gericke et al., 2023). A significant conceptual overlap is often observed between practical work and terms such as 'laboratory work,' 'hands-on activities,' 'science inquiry,' 'science practices,' and 'experiments.' Furthermore, it is widely acknowledged that practical work is the predominant term in European literature, whereas 'laboratory work' is more commonly used in the United States (Rudolph, 2019).

Despite its longstanding presence, practical work has been critiqued as being simultaneously overused and underused (Hodson, 1996). Overused in the sense that it is often employed indiscriminately, with the assumption that hands-on activity alone guarantees learning; underused because its deeper potential—for fostering epistemological understanding, critical reasoning, and authentic inquiry—is rarely realized. Building on this perspective, Abrahams and Millar (2008) conducted observational studies of 25 typical secondary school science lessons. They demonstrated that practitioners frequently use practical work but emphasize content delivery over genuine scientific inquiry. As a result, many tasks favor "hands-on," but lack "minds-on" engagement. Thus, students rarely generate their own questions, plan investigations, or reflect on outcomes. In response to this contradiction, Hodson (1996) proposed a tripartite framework for reconceptualizing practical work: (1) learning science: developing conceptual and theoretical understanding; (2) learning about science: understanding the nature and methods of science; and (3) doing science: developing skills in scientific inquiry and problem solving. This framework has since become foundational in both research and curriculum design.

The literature reviewed by Oliveira and Bonito (2023) confirms the persistence of these three dimensions while also highlighting the uneven implementation of practical work in educational systems. Their systematic review, encompassing 53 peer-reviewed studies, revealed that the dominant conception of practical work remains "hands-on" manipulation of materials. However, they also observed an increasing recognition of the need to integrate "minds-on" components emphasizing conceptual understanding, hypothesis-driven reasoning, and critical reflection. Despite these advances, the authors note that practical work often lacks alignment with the authentic practices of professional scientists, and many teachers continue to struggle with the epistemological and pedagogical complexities of facilitating open-ended inquiry.

Gericke et al. (2023), in their integrative review of laboratory work in secondary schools, propose a refined typology of practical work based on three overlapping dimensions: (1) learning goals (learning science and learning to do science); (2) science practices (e.g., modeling, analyzing data, constructing explanations); and (3) levels of inquiry (from confirmatory to open inquiry). This typology builds on

Hodson's (1996) earlier framework but introduces a greater specificity in how student activity is structured and scaffolded. Particularly notable is their emphasis on the "level of inquiry" dimension, which concerns the extent of teacher guidance versus student autonomy. They categorize inquiry into three levels: confirmatory (teacher-led), guided (teacher sets the question, students design methods), and open (students define both questions and methods). Their review indicates that guided inquiry tends to be most effective in balancing conceptual understanding and inquiry skills, aligning with previous meta-analyses showing moderate to strong learning gains when some degree of teacher support is retained (Lazonder & Harmsen, 2016).

Despite these theoretical refinements, practical work continues to face persistent implementation challenges. In practice, it is often reduced to confirmatory tasks with predetermined outcomes, undermining its potential to cultivate genuine investigative skills or epistemological awareness. More critically, substantial structural and systemic barriers prevent equitable access to high-quality practical experiences (Fussy et al., 2023). Schools in under-resourced settings frequently lack the infrastructure, materials, or staff training required to implement effective laboratory work. At the same time, students with disabilities—whether physical, sensory, or cognitive—encounter additional obstacles, often stemming from the design of activities, learning environments, and assessment systems that assume a narrow standard of participation and embodiment (e.g., Chasen et al., 2025).

Another critical issue concerns the alignment of practical work with assessment practices and curriculum standards. Oliveira and Bonito (2023) emphasize that one of the persistent limitations of practical work is the misalignment between intended learning goals, classroom implementation, and assessment. Teachers often rely on traditional assessments that privilege content recall over investigative reasoning or procedural fluency. Furthermore, many practical tasks are structured to produce predictable results, reducing student engagement to mechanical execution rather than authentic exploration. As several studies in their review note, such "cookbook" approaches can undermine students' epistemic agency and reinforce a misleading image of science as a fixed body of knowledge rather than a dynamic process of inquiry.

The gap between research-based recommendations and classroom practice is further complicated by institutional constraints and teacher preparedness (Tah, 2025). Teachers face various pressures that often limit the feasibility of implementing open or even guided inquiry approaches (Akuma & Callaghan, 2019). Moreover, teachers themselves may lack sufficient training in the epistemological foundations of science or in the pedagogical strategies required to manage productive uncertainty and student autonomy (Akuma & Callaghan, 2019). Addressing these issues requires systemic investment in teacher education programs that integrate theoretical knowledge with reflective practice, as well as curricular reforms that prioritize depth over breadth in scientific content.

These inequities are further compounded by curricular regimes that prioritize standardized testing and content coverage over experiential learning. In such contexts, practical work risks becoming peripheral—either tokenistic or altogether absent—especially where teachers lack the support or autonomy to implement more open-ended, inquiry-based methods. In response, digital technologies have emerged as promising alternatives or supplements to traditional formats. Virtual labs, simulations, and interactive platforms can expand access to certain kinds of scientific engagement and provide safe, repeatable, and often cost-effective experiences. Yet, as the literature cautions, these digital environments cannot fully replicate the tactile, sensory, and context-rich dimensions of real-world experimentation (Rowe, 2024).

This tension between the promise of digital innovation and the irreplaceable affordances of authentic practical work raises a pressing question for science education: How can we balance the pedagogical benefits of digital tools with the need to ensure equitable, meaningful access to hands-on scientific experiences for all learners? This question is not merely technical, but ethical and political. It demands that we confront longstanding disparities in educational provision, reimagine accessibility beyond minimum compliance, and reconsider the epistemological assumptions embedded in both traditional and digital forms of practical work.

In this chapter, we adopt a critical stance on the status and future of practical work in science education. Drawing on recent systematic reviews and theoretical frameworks, we examine its conceptual evolution, pedagogical functions, and the enduring challenges of implementation. We interrogate the assumptions underpinning both traditional and digital formats and argue for a more nuanced approach—one that acknowledges the affordances and limitations of each while foregrounding the imperative of educational justice. The chapter ultimately asks: *What forms of practical work can foster both epistemically robust and socially inclusive science education in a digitally mediated, structurally unequal world?*

## 2.2 Bridging Gaps or Replicating Divides? Digital Tools and the Future of Inclusive Science Education

The benefits of practical work in science education have long been recognized in the literature. Numerous studies highlight its role in strengthening both procedural and epistemological knowledge by engaging students in tasks that simulate authentic scientific investigation (e.g., Andersson & Enghag, 2017; Šorgo & Špernjak, 2012). Practical work is also valued for its participatory dimension, which promotes deeper cognitive engagement, fosters curiosity, and enhances the retention of scientific concepts (e.g., Di Fuccia et al., 2012; Mkimbili & Ødegaard, 2019; Ye et al., 2021). Practical work also supports the integration of hands-on and minds-on learning by combining the development of manual and procedural competencies with the cultivation of conceptual understanding (e.g., Harrison, 2016; Pols et al., 2021; Wei & Liu, 2018). This dual engagement is particularly relevant in science education, where the relationship between theory and practice is fundamental. Finally, several authors emphasize the contribution of practical work to the development of scientific literacy, particularly in enabling students to engage critically with scientific knowledge and its applications in everyday life (e.g., Abrahams et al., 2014; Kácovský & Snětinová, 2021; Karpin et al., 2014).

Most literature on the limitations of practical work focuses on four broad categories: teacher-related challenges, evaluation-driven distortions, material constraints, and task design. Teacher-related issues often include insufficient professional content knowledge, a lack of training in inquiry-based approaches, and difficulties in managing open-ended or student-centered investigations (Wei & Li, 2018). These are frequently compounded by curricular pressures that prioritize standardized assessments, leading to the instrumentalization of practical work as a means to rehearse known outcomes rather than engage in authentic inquiry (Hodson, 1996; Oliveira & Bonito, 2023). The result is a prevalence of "cookbook" experiments, tasks in which students follow rigid procedures with predetermined results, reducing experimentation to mere compliance (Gericke et al., 2023).

Economic, organizational, and environmental constraints also impose real limits. In many schools, especially those serving marginalized communities, access to well-equipped laboratories, consumable materials, and specialized staff is scarce (Smith et al., 2016). Time constraints, class sizes, and safety concerns further restrict teachers' ability to engage students in meaningful practical work (Di Fuccia et al., 2012). This is particularly troubling when viewed through the lens of inclusive science education. According to UNESCO (2017), inclusion encompasses the right of all learners to a quality education that enables the development of their full potential, regardless of social, economic, or cultural background, special educational needs, disabilities, or gender. In this sense, we adopt a broad understanding of inclusion, wherein all students are entitled to educational opportunities that support their comprehensive development, free from structural, material, or symbolic barriers. Practical work, as traditionally conceived, often fails to meet this standard.

Attempts to address this gap have led some researchers to propose alternatives to traditional laboratory setups. Di Fuccia et al. (2012) for instance, discuss the use of low-cost, safer, and more accessible materials, such as micro-scale chemistry, household substances, or improvised equipment to democratize access to hands-on experiences. These innovations demonstrate that inclusion need not always be digital, sometimes it is a matter of rethinking the material and epistemic assumptions behind the design of practical activities. Nevertheless, the digital era has transformed the landscape of science education. Tools such as virtual laboratories, simulations, augmented reality, and data-logging platforms now offer a variety of opportunities to visualize microscopic processes, simulate dangerous or expensive experiments, and collect or manipulate data in real time. These technologies have the potential to overcome some of the barriers mentioned above, they reduce dependency on physical infrastructure, allow repetition without additional cost, and can be tailored to individual learning needs.

Several studies emphasize the benefits of such digital tools in fostering engagement, enhancing visualization of abstract concepts, and facilitating differentiated instruction (e.g. de Jong et al., 2013; Rutten et al., 2012). Simulations reduce the time required for laboratory setup and execution, allowing students to focus on conceptual learning rather than procedural tasks. Gibbons et al. (2004) found that virtual laboratories saved significant time without compromising learning outcomes. Additionally, virtual environments provide access to experiments that may be impractical or unsafe in real-world settings (Climent-Bellido et al., 2003). For students with physical or sensory disabilities, digital platforms can offer alternative means of interaction with scientific phenomena, potentially increasing participation in ways that traditional laboratory environments cannot.

However, the use of digital tools in science education also presents some risks and limitations. First, there is the issue of access. As recent UNESCO (2021) reports have shown, the global digital divide remains stark. In many low-income or rural schools, reliable internet connections, updated hardware, and adequate digital literacy are far from guaranteed. This digital exclusion replicates the same patterns of educational inequality observed in access to traditional laboratory resources, merely shifting the locus of exclusion from physical to digital infrastructure. Second, even where digital tools are available, their pedagogical use is not always aligned with the goals of inquiry-based learning. Virtual labs, for instance, can easily reproduce the same "cookbook" logic as traditional experiments if they are designed as linear sequences with fixed inputs and outcomes. In such cases, the illusion of interactivity masks the absence of genuine epistemic agency (Gericke et al., 2023). Similarly, simulations that provide perfect data without noise or error can mislead students about the nature of scientific investigation, where uncertainty, iteration, and failure are integral components.

Moreover, the shift to digital platforms often reinforces technocentric assumptions about innovation, sidelining questions of inclusion, accessibility, and critical engagement. As Oliveira and Bonito (2023) note, much of the literature focuses on the cognitive and procedural dimensions of practical work while neglecting its sociocultural and affective dimensions. This is particularly problematic when we consider students with special educational needs, who may require multimodal, embodied, or collaborative forms of engagement not easily replicated in screen-based environments. Another concern lies in teacher readiness. Research consistently shows that the successful integration of digital technologies into science teaching depends less on the availability of tools and more on teachers' pedagogical content knowledge, confidence, and support (Di Fuccia et al., 2012). Without targeted professional development, teachers may either underutilize digital resources or use them in ways that reinforce transmissive models of instruction.

From a philosophical and epistemological standpoint, the replacement of real-world manipulation with digital interaction raises critical questions. Can scientific knowledge, which is in part grounded in embodied engagement with the material world, be authentically constructed through virtual experience alone? While digital tools can complement practical work, they cannot fully replicate the sensory, spatial, and contingent dimensions of laboratory inquiry. For this reason, some scholars advocate for a hybrid model that combines the strengths of digital platforms with efforts to preserve and adapt hands-on experiences in inclusive and context-sensitive ways (de Jong et al., 2013). Such an approach would require systemic changes at multiple levels. Curriculum frameworks must articulate clearer expectations regarding the epistemic functions of both traditional and digital practical work. Assessment systems should value inquiry processes, not just outcomes. Teacher education programs must equip educators with the skills to design and facilitate inclusive practical experiences using both physical and digital resources. And infrastructure investments must prioritize not only technological access, but also the conditions necessary for meaningful pedagogical innovation.

The digital era offers significant potential to address some of the historical limitations of practical work in science education. However, realizing this potential requires a critical stance that resists simplistic narratives of technological salvation. Rather than seeing digital tools as replacements for practical work, they should be understood as part of a broader toolkit, one that must be deployed with attention to issues of inclusion, access, pedagogy, and epistemology. The central question, then, is not whether digital tools can substitute for practical work, but how they can be integrated in ways that expand, not restrict, students' opportunities to engage meaningfully in science.

## 2.3 Towards a more Equitable and Integrated Approach to Practical Work

Addressing the challenges discussed here requires more than technological adaptation, it demands a reconsideration of how practical work is conceptualized, organized, and justified in science education. Although digital tools offer promising avenues to complement traditional formats, their implementation must be situated within broader efforts to ensure that all students, regardless of background, ability, or school infrastructure, can access meaningful and epistemically grounded experiences in science.

While the body of research on practical work in science education is well established, the focus on inclusive science education has increasingly gained traction in recent years. In parallel to the widely adopted Universal Design for Learning (UDL) framework (CAST, 2018), a noteworthy model by Kieferle et al. (2024) emphasizes the relevance of inclusion in science teaching through a combination of pedagogical approaches. These include: (i) context-based learning, (ii) inquiry-based learning, (iii) cooperative learning and (iv) game-based learning. These methods have shown consistent positive effects on students' science learning, particularly in promoting communication, applying knowledge to real-world problems, and engaging with scientific working methods such as experimentation (Kieferle et al., 2024).

Among these, inquiry-based learning stands out as a central pillar for inclusive science education. Research has demonstrated its potential to increase motivation, develop critical thinking, and promote meaningful engagement with scientific content (e.g., Bybee, 2014). When this approach is combined with the UDL framework, it supports accessibility, learner variability, and autonomy, particularly in laboratory settings where barriers to participation often persist. For example, based on UDL principles, practical science activities can be designed to include:

- Pre-lab materials presented in multiple formats (e.g., video, text, diagrams), allowing students to prepare at their own pace and according to their learning preferences.
- Flexible output formats, where students can present lab results through writing, drawings, oral presentations, or video logs.
- Choice in task type, grouping, or difficulty level, which supports autonomy and sustained engagement.

These design elements are empirically supported. In a recent study, Clare Finnegan (2024) found that offering preparatory lab videos significantly increased students' confidence, autonomy, and emotional well-being, particularly among students with diverse needs and learning styles. Moreover, inquiry-based strategies have been shown to foster a deeper understanding of scientific content, especially when connected to authentic, real-world contexts (e.g., Kieferle et al., 2024; Hofstein & Kind, 2012).

Bringing these strands together, we propose a complementary model for inclusive science education "Inclusive Practical Science Education (IPSE)" with a special focus on practical work. The IPSE model integrates UDL principles with inquiry-based and context-driven approaches to promote inclusive learning environments in laboratory settings.

The Inclusive Practical Science Education (IPSE) model is intentionally structured across three interrelated layers. The rationale for distinguishing three layers lies in the need to (i) provide a strong theoretical foundation, (ii) translate this foundation into concrete practices, and (iii) highlight the targeted outcomes of these practices. Each layer is therefore conceptually distinct, but also strongly connected to the others, ensuring that theory, practice, and outcomes remain integrated.

The Inclusive Practical Science Education (IPSE) model is structured across three interrelated layers (Fig. 2.1), each building upon well-established pedagogical foundations to promote inclusion within the context of science laboratory work. At its base, the model draws from the theoretical underpinnings of Universal Design for Learning (CAST, 2018), the Inclusive Science Education framework developed by Kieferle et al. (2024), and the inquiry-based learning paradigm. Together, these approaches provide a robust foundation for designing science instruction that is both accessible and intellectually rigorous.

The second layer focuses on the practical implementation of these principles within laboratory settings. Here, the UDL framework is operationalized through three core dimensions: representation, expression, and engagement. Instructional materials (particularly pre-laboratory resources) are offered in multiple formats, such as videos, textual explanations, and visual diagrams, allowing students to access information according to their individual learning preferences and needs. Equally important is the flexibility in how students demonstrate their understanding. The model encourages diverse forms of expression, enabling learners to present experimental outcomes not only through traditional written lab reports, but also via drawings, oral presentations,

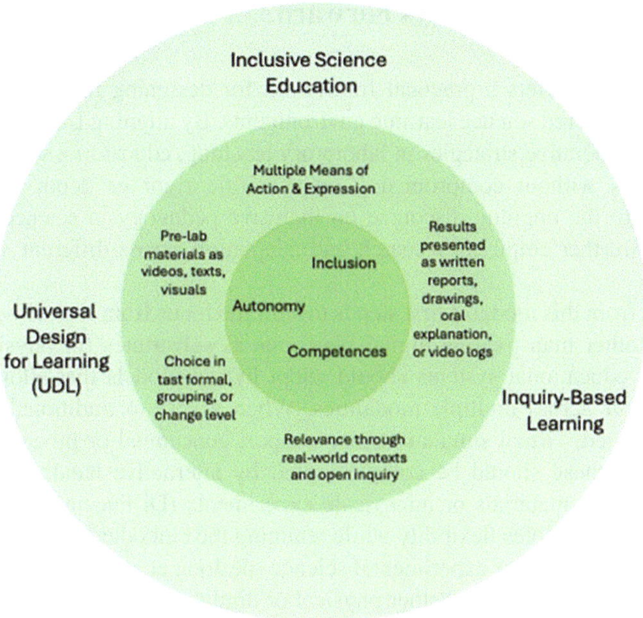


**Fig. 2.1** Inclusive Practical Science Education (IPSE) Model

or multimedia formats such as video logs. Additionally, the model promotes engagement by offering students choices in how they approach tasks, whether through varying levels of difficulty, different groupings, or modes of participation, thereby enhancing their autonomy and motivation.

The third layer highlights the key outcomes that result from applying the theoretical and practical layers. These outcomes—inclusion, autonomy, competence, belonging, and motivation—are not separate from the other layers but emerge directly from the interplay between foundations and practices. Making these outcomes explicit clarifies the contribution of the model and helps educators evaluate its effectiveness.

At its core, the IPSE model aims to foster inclusive and empowering learning environments by supporting five key outcomes: inclusion, autonomy, competence, belonging, and motivation. Through its layered structure, the model ensures that all students can meaningfully participate in and benefit from practical science education. It bridges the gap between theory and practice, offering educators a flexible yet research-informed framework for making laboratory learning both inclusive and effective.

## 2.4 Conclusion and Ways Forward

The IPSE model offers a practical framework for designing inclusive, engaging, and student-centered science learning environments. By aligning UDL with inquiry-based and cooperative strategies in laboratory teaching, educators can meet diverse learner needs without compromising on scientific rigor or depth. This model contributes to the ongoing discourse on inclusive pedagogy in science education and invites further empirical research and adaptation across different educational contexts.

Starting from this model, a first step is to shift the focus from replacing to supplementing. Rather than treating digital platforms as substitutes for physical experimentation, educational systems should adopt hybrid models that allow practical work to occur across multiple modalities. When access to traditional laboratory setups is limited, virtual simulations can support conceptual or procedural understanding, but these should be complemented by alternative hands-on strategies, such as low-cost materials or microscale experiments (Di Fuccia et al., 2012). A hybrid approach enables flexibility while retaining the embodied, manipulative, and situational dimensions of experimental science (de Jong et al., 2013).

Second, practical work, whether physical or digital, should be guided by clearly defined epistemic goals. This implies moving away from task designs that emphasize procedural completion or reproduction of known outcomes. Instead, learning environments should be structured around inquiry-oriented processes such as posing questions, developing explanations, and evaluating evidence. Virtual labs can support such goals when designed with variable inputs, iterative trials, or embedded decision-making opportunities. However, this requires that developers and educators alike resist the simplification of inquiry into sequences of mechanical steps.

Third, inclusion must be addressed not as an add-on, but as a design principle. To serve all students, practical activities, digital or otherwise, should accommodate a diversity of needs and learning profiles. This includes offering multimodal representations, adjustable interfaces, and differentiated pathways through the material. But it also includes creating social and discursive spaces where all learners feel authorized to engage, question, and contribute. Affective and sociocultural dimensions of practical work are often neglected in the literature, yet they are essential for meaningful participation, especially for students who have historically been excluded from science.

Fourth, supporting teachers is critical. Professional development programs must not only introduce new digital tools but also provide sustained opportunities for teachers to reflect on their epistemological commitments, experiment with new pedagogical strategies, and share practices with peers. As Di Fuccia et al. (2012) emphasize, the gap between idealized visions of practical work and actual classroom practices is often a function of institutional and professional constraints. Addressing these requires long-term investment in teacher education, planning time, and collaborative support structures.

Finally, the IPSE model addresses the broader infrastructural and material conditions that shape access to practical work. Digital inclusion cannot be reduced to distributing devices (Ragnedda & Muschert, 2013). Reliable internet, accessible platforms, technical support, and safe learning environments are all prerequisites for equitable participation. The model thus embeds inclusion at multiple levels—pedagogical, social, professional, and infrastructural—and calls for policy efforts to align with curricular reform and inclusive assessment practices.

At its core, the IPSE model affirms that practical work is defined not by format, but by function: its capacity to foster authentic engagement with science as a way of thinking, reasoning, and acting in the world. This perspective positions inclusive practical science education within a broader pedagogical ecology—a dynamic system where quality, inclusion, and epistemic integrity are the guiding values.

## References

Akuma, F. V., & Callaghan, R. (2019). Characterising extrinsic challenges linked to the design and implementation of inquiry-based practical work. *Research in Science Education, 49*(6), 1677–1706. https://doi.org/10.1007/s11165-017-9671-x

Abrahams, I., & Millar, R. (2008). Does practical work really work? A study of the effectiveness of practical work as a teaching and learning method in school science. *International Journal of Science Education, 30*(14), 1945–1969. https://doi.org/10.1080/09500690701749305

Abrahams, I., Reiss, M. J., & Sharpe, R. (2014). The impact of the getting practical: Improving practical work in science continuing professional development programme on teachers' ideas and practice in science practical work. *Research in Science and Technological Education, 32*, 263–280. https://doi.org/10.1080/02635143.2014.931841

Andersson, J., & Enghag, M. (2017). The relation between students' communicative moves during laboratory work in physics and outcomes of their actions. *International Journal of Science Education., 39*, 158–180. https://doi.org/10.1080/09500693.2016.1270478

Bybee, R. W. (2014). NGSS and the next generation of science teachers. *Journal of Science Teacher Education, 25*(2), 211–221. https://doi.org/10.1007/s10972-014-9381-4

CAST. (2018). Universal design for learning guidelines. *Version, 2*, 2. https://udlguidelines.cast.org

Chasen, A., Chapman Tripp, H., & Borrego, M. (2025). Disability and postsecondary fieldwork experiences in the natural sciences: A systematic review. *Journal of Research in Science Teaching, 62*(4), 1006–1039. https://doi.org/10.1002/tea.21989

Climent-Bellido, M. S., Martínez-Jiménez, P., Pontes-Pedrajas, A., & Polo, J. (2003). Learning in chemistry with virtual laboratories. *Journal of Chemical Education, 80*(3), 346. https://doi.org/10.1021/ed080p346

De Jong, T., Linn, M. C., & Zacharia, Z. C. (2013). Physical and virtual laboratories in science and engineering education. *Science, 340*(6130), 305–308. https://doi.org/10.1126/science.1230579

Di Fuccia, D., Witteck, T., Markic, S., & Eilks, I. (2012). Trends in practical work in German science education. *Eurasia Journal of Mathematics, Science and Technology Education, 8*, 59–72. https://doi.org/10.12973/eurasia.2012.817a

Dillon, J. (2008). *A review of the research on practical work in school science* (pp. 1–84). King's College.

Finnegan, C. (2024). Enhancing practical pedagogy: A UDL perspective on the value of videos for practical class preparation among third-level student-teachers in science education. *All Ireland. Journal of Higher Education, 16*(2). https://doi.org/10.62707/aishej.v16i2.851

Fussy, D. S., Iddy, H., Amani, J., & Mkimbili, S. T. (2023). Girls' participation in science education: Structural limitations and sustainable alternatives. *International Journal of Science Education, 45*(14), 1141–1161. https://doi.org/10.1080/09500693.2023.2188571

Gericke, N., Högström, P., & Wallin, J. (2023). A systematic review of research on laboratory work in secondary school. *Studies in Science Education, 59*(2), 245–285. https://doi.org/10.1080/03057267.2022.2090125

Gibbons, N. J., Evans, C., Payne, A., Shah, K., & Griffin, D. K. (2004). Computer simulations improve university instructional laboratories. *Cell Biology Education, 3*(4), 263–269. https://doi.org/10.1187/cbe.04-06-0040

Harrison, M. (2016). Making practical work work: Using discussion to enhance pupils' understanding of physics. *Research in Science and Technological Education, 34*, 290–306. https://doi.org/10.1080/02635143.2016.1173668

Hofstein, A., & Kind, P. M. (2012). Learning in and from science laboratories. In B. Fraser, K. Tobin, & C. McRobbie (Eds.), *Second international handbook of science education. Springer international handbooks of education, 24*. Springer. https://doi.org/10.1007/978-1-4020-9041-7_15

Hofstein, A., & Lunetta, V. N. (2004). The laboratory in science education: Foundations for the twenty-first century. *Science Education, 88*(1), 28–54. https://doi.org/10.1002/sce.10106

Hodson, D. (1996). Practical work in school science: Exploring some directions for change. *International Journal of Science Education, 18*(7), 755–760. https://doi.org/10.1080/0950069960180702

Kácovský, P., & Snětinová, M. (2021). Physics demonstrations: Who are the students appreciating them? *International Journal of Science Education, 43*(4), 529–551. https://doi.org/10.1080/09500693.2020.1871526

Karpin, T., Juuti, K., & Lavonen, J. (2014). Learning to apply models of materials while explaining their properties. *Research in Science and Technological Education, 32*, 340–351. https://doi.org/10.1080/02635143.2014.944494

Kieferle, S., Devetak, I., Essex, J., Hayes, S., Stojanovska, M., Mamlok-Naaman, R., & Markic, S. (2024). A rising tide lifts all boats? The model of differentiation as a tool for diversity in science toward social inclusion. *Journal of Chemical Education, 101*(3), 789–797. https://doi.org/10.1021/acs.jchemed.3c00550

Lazonder, A. W., & Harmsen, R. (2016). Meta-analysis of inquiry-based learning: Effects of guidance. *Review of Educational Research, 86*(3), 681–718. https://doi.org/10.3102/0034654315627366

Mkimbili, S. T., & Ødegaard, M. (2019). Student motivation in science subjects in Tanzania, including students' voices. *Research in Science Education, 49*, 1835–1859. https://doi.org/10.1007/s11165-017-9677-4

Oliveira, H., & Bonito, J. (2023). Practical work in science education: A systematic literature review. *Frontiers in Education, 8*, 1151641. https://doi.org/10.3389/feduc.2023.1151641

Pols, C. F. J., Dekkers, P. J. J. M., & de Vries, M. J. (2021). What do they know? Investigating students' ability to analyse experimental data in secondary physics education. *International Journal of Science Education., 43*, 274–297. https://doi.org/10.1080/09500693.2020.1865588

Ragnedda, M., & Muschert, G. W. (2013). *The digital divide*. Routledge.

Rowe, L. (2024). Comparing learning outcomes and student and instructor perceptions of a simultaneous online versus in-person biochemistry laboratory course. *Journal of Chemical Education, 101*(3), 882–891. https://doi.org/10.1021/acs.jchemed.3c00571

Rudolph, J. L. (2019). *How we teach science: What's changed, and why it matters*. Harvard University Press.

Rutten, N., van Joolingen, W. R., & van der Veen, J. T. (2012). The learning effects of computer simulations in science education. *Computers & Education, 58*(1), 136–153. https://doi.org/10.1016/j.compedu.2011.07.017

Smith, P. S., Trygstad, P. J., & Banilower, E. R. (2016). Widening the gap: Unequal distribution of resources for K–12 science instruction. *Education Policy Analysis Archives, 24*(8). https://doi.org/10.14507/epaa.v24.2207

Sorgo, A., & Špernjak, A. (2012). Practical work in biology, chemistry and physics at lower secondary and general upper secondary schools in Slovenia. *Eurasia Journal of Mathematics, Science and Technology Education, 8*, 11–19. https://doi.org/10.12973/eurasia.2012.813a

Tah, J. K. (2025). Teachers understanding of inclusive education and challenges in implementation in the English education system in Cameroon. *Policy Futures in Education.* https://doi.org/10.1177/14782103251320812

UNESCO. (2017). A guide for ensuring inclusion and equity in education. *United Nations Educational, Scientific and Cultural Organization.* https://doi.org/10.54675/MHHZ2237

UNESCO. (2021). *Reimagining our futures together: A new social contract for education.* UNESCO. https://unesdoc.unesco.org/ark:/48223/pf0000379707

Wei, B., & Liu, H. (2018). An experienced chemistry teacher's practical knowledge of teaching with practical work: The PCK perspective. *Chemisty Education: Research and Practice, 19*, 452–462. https://doi.org/10.1039/c7rp00254h

Ye, J., Mi, S., & Bi, H. (2021). Constructing core teaching competency indicators for secondary school science teachers in China. *Journal of Baltic Science Education, 20*, 389–406. https://doi.org/10.33225/jbse/21.20.389

**Matheus Monteiro Nascimento** is a professor at the Department of Physics of the Federal University of Rio Grande do Sul (UFRGS), Brazil, working in undergraduate and graduate physics teacher training courses. He earned his PhD in physics education at UFRGS. His research interests focus on equity, diversity, and inclusion in science education.

**Silvija Markic** is professor of chemistry education at LMU Munich, Germany. She studied chemistry and mathematics for grammar school, earned her PhD in Bremen, and was a professor at Ludwigsburg University of Education. Her research focuses on teachers' beliefs, PCK, linguistic diversity in science education, cooperative learning, and innovative teaching methods.

# Chapter 3
# Practical Science, Artificial Intelligence and Science Education

**Sibel Erduran, Ho-Yin Chan, and Ivan Au**

## 3.1 Introduction

Scientific research uses a diversity of practices and methods in the pursuit of generating new knowledge (Erduran & Dagher, 2014; Irzik & Nola, 2014). The scientific practices include experimentation and observation while scientific methods can involve hypothesis testing and parameter measurements. Through such practices and methods, scientists construct theories that explain natural and physical phenomena. Understanding such scientific approaches can be considered a key component of scientific literacy (Lemke, 2004; Reiss et al., 2012), a skill highly valued in science education as evidenced by decades of curriculum reforms dedicated to this theme. Traditionally, however, textbooks often feature an over-simplified versions of science that presents scientific methods, for instance, in a linear stepwise fashion (Wei et al., 2024). This depiction of "the scientific method" as hypothesis testing involving experiments in many textbooks has dominated the image of science for decades (McComas, 2008), despite the fact that scientific method is not a single process nor is it exclusively based on experimentation (Brandon, 1994).

Artificial intelligence (AI) is increasingly having a significant impact on how science is done (Erduran, 2023). Such impact is being felt with regards to the way in which scientific practices and methods are conducted. Consider, for example AlphaFold, an AI tool named the "Method of the Year" (2022) and subsequently led to the award of a Nobel Prize. The tool has enabled scientists to predict protein-folding

S. Erduran (✉) · I. Au
University of Oxford, Oxford, UK
e-mail: sibel.erduran@education.ox.ac.uk

I. Au
e-mail: ivan.au@education.ox.ac.uk

H.-Y. Chan
University of Cambridge, Cambridge, UK
e-mail: ho.chan@st-annes.ox.ac.uk



L. Leite et al. (eds.), *Practical Work in Science Education in the Digital Era*, Advances in Technology-Rich Science Education 2, https://doi.org/10.1007/978-3-032-13788-3_3

with great accuracy, already considered to have initiated a revolution in biology (Ewen, 2022). The use of AI is raising key questions about scientific practices, particularly in relation to *how* science is carried out. Scientists themselves are beginning to reflect on how AI is influencing science (Chan & Erduran, 2025). Erduran and Levrini (2024) traced examples of scientific practices impacted by AI. They used the definition of "scientific practices" from key science curriculum standards documents such as NRC (2012) and NGSS (NGSS Lead States, 2013). For example, the authors indicated that the scientific practice of "analysing and interpreting data" was an instance where the molecular dynamics of the virus responsible for COVID-19 was studied. Researchers Casalino et al. (2020) uncovered new biological functions of the spike protein that show different behaviours in the open and closed conformations. This discovery led to a change in the view of glycans in biological systems and inspired new ways to analyse them.

Articulating how AI is impacting scientific practices and methods is relevant for science education given the international attention placed on the teaching and learning of practical science for some time (Abraham & Saglam, 2010). However, there is often lack of clarity about what is meant by "practical science". For example, the terms used include *experimental work, practical and enquiry skills*, *practical and investigative activities* (QCA, 2007), as well as *practical experimentation,* and *practical scientific methods* (Department for Education, 2014). The ambiguity in the language around practical science is compounded with an overemphasis on the experimental method even though scientists also use other methods such as non-manipulative observations and manipulative descriptions (Ioannidou & Erduran, 2021).

In this chapter we focus on specific definitions of scientific practices and scientific methods and collectively refer to them as "practical science". In other words, we seek to clarify how "practical science" can be differentiated and thus clarified. Such clarity will help guide our discussion on how current developments in AI is influencing which aspects of practical science. We begin our discussion with an overview of scientific practices and methods and subsequently turn to a broad coverage of how AI is being applied in science. Some results from Project AI-Vision conducted in the United Kingdom are shared to illustrate how expert scientists and educators consider as pressing issues for the integration of AI in school science.

## 3.2 Practical Science in Science Education

Practical science in science education has a long history with various definitions offered in the literature (e.g., Abraham & Saglam, 2010; Abrahams & Reiss, 2012). "Scientific practices" have been a central theme of recent educational reforms in science education in the USA (e.g., NRC, 2012) and the interest in this theme has also extended to the rest of the world (e.g., Costa & Broietti, 2021). The focus on scientific practices has been a major concern to science educators given the "practice turn" more broadly in science studies (Berland et al., 2016) which highlighted the importance of students' engagement in how scientists *do* science, including not only the manipulative hands-on aspects of practical work but also the mind-on elements. Erduran and Dagher (2014) argue that "scientific practice" is a term that engages skills and knowledge meaningfully to characterise scientific investigations. In this respect, it is more comprehensive than the term "science process skills". In other words, scientific practices do not only involve practicing skills but also refer to making sense of the relationship between skills and underlying scientific content knowledge and epistemology.

Erduran and Dagher (2014) proposed a heuristic to summarise how scientific practices may be approximated for the purposes of science education (Fig. 3.1). The Benzene Ring Heuristic (BRH) includes epistemic, cognitive and social components of scientific practices. This heuristic has been used in supporting science teachers in teacher education programmes (e.g., Saribas & Ozer, 2021). Epistemic components refer to scientific activities, data, real world, model, explanation and prediction. These components also refer to the features of the scientific practices that are involved in knowledge construction. Cognitive and social components respectively refer to representations and reasoning and social dissemination and certification of scientific claims. BRH suggests that all features of scientific practices are related to each other, and these relationships do not have a linear order but rather they are co-constituted and inter-related. Overall, BRH illustrates nonlinear relationships between the epistemic, cognitive and social aspects of scientific practices, and it aims to capture holistic understanding of science.

Another issue in relation to practical science in science education is that often this term is often conflated with scientific methods. For example, DfE (2014) conflates the two (ie. practical science and scientific method) by referring to "practical scientific methods". Many science textbooks have influenced both teachers' and students' views of science through how they present "the scientific method" (Emden, 2021; McComas, 2014; Reiff-Cox, 2020). Historically, there has been a depiction of a single scientific method, which does not reflect the diversity of approaches used by scientists (Ioannidou & Erduran, 2021; Woodcock, 2014). Traditionally, scientific method is considered to be a linear process that starts with a hypothesis to be tested (Brandon,

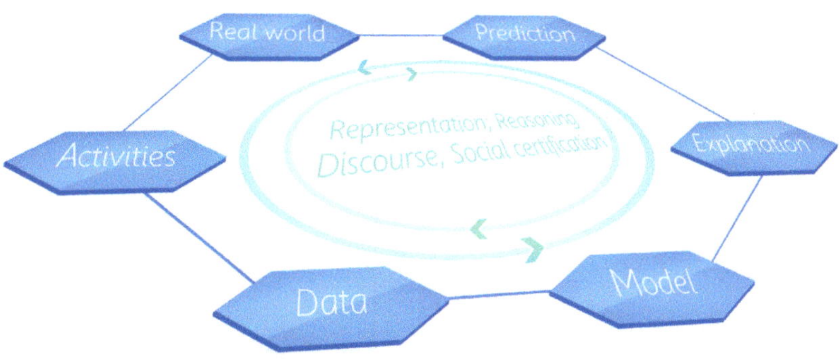


**Fig. 3.1** Benzene Ring Heuristic of Scientific Practices (Erduran & Dagher, 2014)

1994). There are dependent and independent variables to consider, and the hypothesis is verified or rejected using an experiment. However, not all scientific methods are about experimentation. For instance, astronomers cannot manipulate the planets, black holes or other bodies being investigated. Rather, astronomers perform observational studies and may employ hypothesis testing without any experimentation involving manipulation of variables.

Apart from a misconstrued account of the scientific method as being experimental hypothesis testing, many traditional accounts of practical science omit the elements of uncertainty, incompleteness, or modifications that form part of the scientific process (Erduran et al., 2020). They are excluded from the process and students come to believe that the "scientific method" guarantees discovery, unequivocal and reliable conclusions (Wivagg & Allchin, 2002). Science in schools offer little understanding of how scientific knowledge is generated and the extent of uncertainty that underlies the knowledge generation processes (Wei et al., 2021).

Brandon's Matrix (Brandon, 1994, p.64) presents an alternative approach to the "scientific method" through a matrix (Fig. 3.2), recognising that there is no single scientific method, but instead scientists utilise a range of methodological approaches and that these different approaches are essential. This framework has been utilised by science educators internationally for analysis of textbooks (Chen et al., 2024) and examination questions (Cullinane et al., 2019) as well as in teachers' professional development (Cullinane et al., 2023). It has underpinned a funded research and development project on practical science in the United Kingdom (Project Calibrate, 2020). BM describes the nature of a practical investigation, which he identifies as an experiment and/or an observation, and by whether it involves the manipulation of variables, testing of hypotheses or simply measuring parameters (Brandon, 1994). Additionally, BM illustrates how not all investigations in science rely on hypothesis testing and not all observational work is non-manipulative (Brandon, 1994; Emden, 2021).

**Fig. 3.2** Brandon's Matrix of Scientific Methods

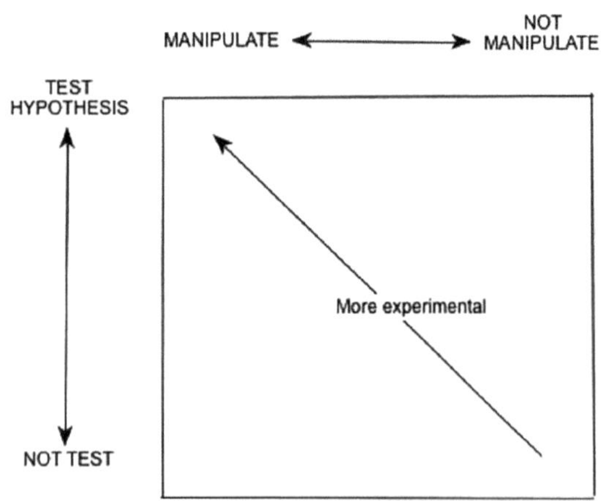

BM provides a framework to aid understanding of methods used in scientific investigations, and how the findings from a range of different methods can be used to explain different scientific concepts and phenomena. In applications to science education, Project Calibrate (2020) utilised BM to promote a *"hands-on, minds-on"* approach to scientific methods and to help students see the purpose of different approaches to practical work. Although BM was initially a theoretical framework, classroom applications of it has been produced for the purposes of science teaching and learning (Erduran & Wooding, 2021). Furthermore, the Project Calibrate team (Cullinane et al., 2023; Ioannidou & Erduran, 2021) reported empirical studies that examined science teachers' understanding of scientific methods, particularly the diversity of scientific methods. BM was used to generate tasks that were used to investigate teachers' understanding of scientific methods. Fifty-six science teachers (25% male, 75% female) from different regions in the UK responded to an online survey in Ioannidou and Erduran's (2021) study. The results showed that majority of the teachers showed satisfactory understanding of basic components of BM; however, more than half of the sample held a naïve understanding of scientific methods.

Focusing on example frameworks such as BRH (Erduran & Dagher, 2014) and BM (Brandon, 1994) can potentially help differentiate and clarify the key definitions underpinning practical science. In drawing out implications for science education, it is important to have clearly articulated definitions of the aspects of practical science which are being discussed. Furthermore, having robust frameworks may be helpful in recognising what scientists themselves might be alluding to in how practical work that they do is being influenced by AI in the contemporary era. In the next section, we review some recent trends in the impact of AI on practical science before focusing on the case of a research project that has drawn from expert views. In both cases, BRH (Erduran & Dagher, 2014) and BM (Brandon, 1996) are used to clarify how practical science in the age of AI works and what the implications are for revisiting practical science in school science.

## 3.3 The Influence of Artificial Intelligence on Practical Science

Artificial Intelligence (AI) has evolved from a futuristic concept into an essential tool across many domains, and practical science is no exception. For example, the prestigious scientific organisation the Royal Society in the United Kingdom has released a report that showcases how AI is changing the nature and method of scientific research (The Royal Society, 2024). From biology to physics, chemistry to environmental studies, AI is reshaping how scientists collect data, make predictions, analyze complex systems, and accelerate innovation (OECD, 2023; World Economic Forum, 2024). Its impact is not just theoretical; AI is now embedded into the everyday practices and workflows of laboratories, field research, and industrial applications (Bianchini et al., 2022). In practical science, the ability to collect and analyze large volumes of data efficiently is critical. AI technologies excel at handling large, complex datasets, often referred to as "big data" (Savage, 2024). This capability is transforming how scientists approach experimental design, data processing, and pattern recognition. For instance, in genomics, AI algorithms can scan millions of DNA sequences to identify mutations related to specific diseases. Tools like AlphaFold mentioned earlier (Method of the year, 2022) use deep learning to predict the 3D structure of proteins with remarkable accuracy—a task that was once time-consuming and error-prone. The breakthrough in protein modeling has significantly sped up drug discovery and our understanding of biological functions (Ren et al., 2023).

Similarly, in environmental science, AI-powered remote sensing systems analyze satellite imagery to monitor deforestation, climate change, and biodiversity (Maqbool et al., 2025). These systems can detect subtle changes in land use or temperature patterns far more quickly and accurately than traditional methods. AI-driven automation is revolutionizing laboratory workflows. Robotic systems guided by AI can perform repetitive tasks such as pipetting, mixing chemicals, or growing bacterial cultures. This not only increases efficiency but also reduces human error, making scientific research more reliable. Scientists can use voice-controlled AI assistants to document experiments in real-time, streamlining note taking and data logging. More advanced applications involve autonomous labs—fully automated research environments where AI plans, executes, and evaluates experiments without human intervention. This is particularly useful in chemistry, where AI can test thousands of compound combinations to analyse materials, for instance pollutants (e.g., Jin et al., 2024).

When viewed from the framework of the Benzene Ring Heuristic, the recent developments in AI in science highlights the increasing trend in automation of many data analysis techniques. The conventional emphasis on data collection and analysis seems to be left to AI tools whilst scientists are focusing more design and modelling (Sreenivasan & Suresh, 2024). In this respect, the communicative and epistemic dimensions of BRH come to the foreground for the purposes of science teaching and learning when many hands-on activities can be picked up by machines. Indeed, a major benefit of AI in practical science is that AI can simulate and model complex phenomena much faster than traditional computational methods (Wang, 2024). In fields like physics, climate science, and materials engineering, AI models help scientists test hypotheses and make predictions based on existing data. For example, in materials science, AI algorithms can predict the properties of new compounds before they are synthesized. In climate modelling, AI enhances our ability to predict weather patterns, sea-level rise, and natural disasters (Royal Society, 2024). By integrating data from multiple sources—satellites, sensors, historical records—AI models provide more accurate and localized forecasts, supporting scientific research. When considered from the view of Brandon's Matrix discussed earlier (Brandon, 1994), such examples point to the ways through which hypothesis testing is now being guided by AI tools and how modelling is being integrated into hypothesis testing.

One of the most transformative applications of AI in practical science is in healthcare and biomedical research (Mizna et al., 2025). AI enables personalized medicine by analyzing an individual's genetic makeup, lifestyle data, and health records to tailor treatments (Johnson et al., 2021). Machine learning algorithms are used to detect diseases at early stages, often before symptoms appear. For instance, AI models trained on medical images can identify early signs of cancer, retinal diseases, or neurological disorders with accuracy comparable to expert radiologists (Pinto-Coelho, 2023). AI also plays a vital role in epidemiology. During the COVID-19 pandemic, AI was used to model the spread of the virus, assess risks, and guide public health interventions (Mann et al., 2022). Such tools continue to aid in tracking disease outbreaks and evaluating the effectiveness of treatments and vaccines (Taherdoostt & Ghofrani, 2024).

AI can also enhance the integrity and reproducibility of scientific research. By automating data analysis and documentation, AI reduces human bias and error (Uzzi et al., 2024). Moreover, AI systems can flag inconsistencies, detect fraudulent data, and suggest corrections, improving the overall quality of scientific publications (Khalifa & Albadawy, 2024). These aspects of scientific practices highlight the need for robust peer review systems as articulated through the Benzene Ring Heuristic framework (Erduran & Dagher, 2014). However, the use of AI also introduces new ethical challenges. Black-box algorithms may produce accurate results without offering explanations, raising concerns about transparency and interpretability (Zednick, 2022). This is especially critical in fields like medicine and environmental policy, where decisions must be justified with clear rationale. To address such concerns, scientists are increasingly focused on explainable AI which aims to make AI systems more transparent and their decisions more understandable to humans (Agarwal et al., 2025). This not only helps build trust but also aligns AI usage with scientific principles of openness and verifiability.

Finally, AI is blurring the lines between disciplines, leading to the emergence of hybrid scientific fields like computational biology, AI chemistry, and astroinformatics (Berman et al., 2025). These interdisciplinary domains combine expertise in computer science, statistics, and domain-specific knowledge to solve complex scientific problems. For instance, in astronomy, AI helps analyze data from telescopes to identify exoplanets, black holes, and other celestial phenomena. Traditional methods would take years to process this volume of data, but AI systems can do it in days. Such developments highlight the need to revisit the silos in which many school subjects are still taught around the world. In other words, subjects such as geography, science and mathematics are taught separately in isolation, rather than in an integrated fashion that illustrates how professional scientists themselves engage in multidisciplinary research.

The preceding discussion highlights some recent examples of innovation in scientific research in the age of AI. Some of the examples have been related back to the conceptual frameworks on scientific practices and scientific methods discussed earlier in the context of the Benzene Ring Heuristic (Erduran & Dagher, 2014) and Brandon's Matrix (Brandon, 1994). Although the literature reviewed on the impact of AI on scientific research as well as education (e.g., Yang et al., 2024) is now increasing in volume, science education is yet to focus on this aspect of AI (Erduran, 2023; Erduran & Levrini, 2024) and hence, the chapter's discussion so far makes a contribution to this area of research in science education. A further aspect of this line of work is how expert scientists and educators themselves might view what then should be included in the science curriculum about the impact of AI on practical science. Project AI-Vision is one example project that aimed to explore scientists' and educators' views and it will be discussed in the next section.

## 3.4 Expert Views about AI's Impact in Science: Insights from Project AI-Vision

Given the recent trends on the impact of AI on scientific research, there is interest in exploring how scientists themselves consider the issue of AI in scientific research (Chan & Erduran, 2025). In a survey conducted by *Nature*, 1600 scientists were asked how useful they thought AI tools would become for their fields in the next decade. More than half expected the tools to be 'very important' or 'essential' (Van Noorden & Perkel, 2023). However, scientists also expressed strong concerns about the ethical dimensions of AI use in research. In Project AI-Vision carried out at University of Oxford, we conducted both literature reviews and empirical studies to examine scientists' and educators' views about the impact of AI on practical science (Erduran et al., 2025). The aim was to better understand how to distil information to inform the teaching and learning of practical science in schools. By examining published scientific research concurrently with empirical investigation of scientists' and educators' views, we also aimed to establish some coherence for what can be selected and prioritised for inclusion in science education.

When we investigated published articles for how scientists are referring to practical science, we observed some resonance with the way that we have discussed scientific practices and methods earlier in this chapter. For example, scientists stated that they engage in disciplined inquiry by utilising a variety of observational, investigative, and analytical methods to generate reliable evidence."[T]here are systematic experiments done with artificial intelligence in order to test out how artificial intelligence can help us to keep public places safe" (Byrne, 2024, np).

As the quote illustrates, some remarks were also cognizant of highlighting the public utility of scientific research. Furthermore, they specified diverse cognitive, epistemic, and investigative approaches used in scientific inquiry, a theme that coheres with the Benzene Ring Heuristic (Erduran & Dagher, 2014). Self-supervised learning, which allows models to be trained on vast amounts of unlabelled data, and geometric deep learning create designs, such as small-molecule drugs and proteins, by analysing diverse data modalities, including images and sequences (Wang et al., 2023).

Observational, investigative, and analytical methods and their role in producing reliable evidence have been highlighted along with the multidisciplinary nature of contemporary scientific research. Cancer treatment is no longer the domain just of oncologists. It now also involves specialists in microbiology, artificial intelligence, diet and nutrition, genomics, bioinformatics and computing (Tiwari et al., 2025).

Furthermore, given AI automates routine scientific methods including hypothesis generation, there is a direct link to Brandon's Matrix (Brandon, 1994). The narratives of scientists seem to suggest that human roles shift toward oversight and interpretation of AI outputs. The references to different scientific practices and methods in scientists' narratives reinforce the need to have clearly articulated frameworks for practical science in school science. They also provide specific examples that complement our discussion in Sect. 3 of this chapter.

As part of Project AI-Vision, we approached expert scientists and educators to understand their views on the impact of AI in science and what suggestions they have for school science. We recruited 12 experts (8 male, 4 female) whose research focus on some use of AI. The experts had expertise in STEM, education and philosophy. They were invited to participate in this study because they have extensively researched AI and/or the application of it in science. For instance, scientists regularly and actively use AI in their research for purposes ranging from articulating and facilitating ideas, generations of codes (e.g., for use in R and Python), algorithm or model selection, to the development of AI-driven approaches to infectious disease surveillance and public health interventions. Educators, on the other hand, also use AI for brainstorming, enhancing training in English for Academic Purposes, and some coding for qualitative data analysis. Philosophers work on the philosophical aspects of AI technologies and their role in knowledge production.

Convenience sampling was employed by recruiting participants from British (8) and Chinese (3) universities. The purpose of this aspect of the study was to gain some understanding of any variations between the viewpoints of the different experts as a preliminary step, not necessarily to be able to generalise to these expert communities. Although the sample is small, the outcomes of the investigation sheds light on a new emerging area of research for science education and future studies can adapt a similar methodology to conduct a larger scale study for more representative outcomes.

The results reported here are part of a larger study that investigated the different experts' baseline views in the first instance and subsequently engaging all participants with a Delphi study to reach a consensus on a range of issues. Delphi methodology is defined as "systematic solicitation and collection of judgements on a particular topic through a set of carefully designed sequential questionnaires interspersed with summarized information and feedback of opinions derived from earlier responses" (Delbecq et al., 1975, p. 10). These characteristics mean the Delphi method is particularly suitable for decision-making or social forecasting, thus has been widely chosen for social science research (Landeta, 2006). As a first step in the Delphi study, the participants were asked a set of questions such as *"What is the most important way AI is changing how science is done?"* and *"Which AI tools can best support science education in representing how science works?"*. Subsequently, their responses were then collated and clustered thematically and were represented to the whole sample to see to what extent they agreed with them. Here we focus on a set of outcomes to showcase both the participants' responses and their relative agreement with them. In a future study, we will report on the full analysis of the Delphi study.

When asked what they thought about the most important ways in which AI is changing how science is conducted, majority of participants selected the option that specified transformation of data-driven science (Fig. 3.3).

Consistent with the literature review, the participants agreed to a further extent that AI transforming scientific processes through automation and speed (Fig. 3.4).

An aspect that had not explicitly come up in our literature search was about visualisation. This could include the 3-D modelling of protein structures as illustrated by the famous AlphaFold example referred to earlier. The experts considered the role of scientific visualisation being enhanced through AI in investigations (Fig. 3.5).

Despite acknowledging the value of AI in scientific research, some scientists were cautious and critical. One scientist (*S1*) was critical of the quality of the output generated by AI and warned that: "[…] their propensity for producing gibberish

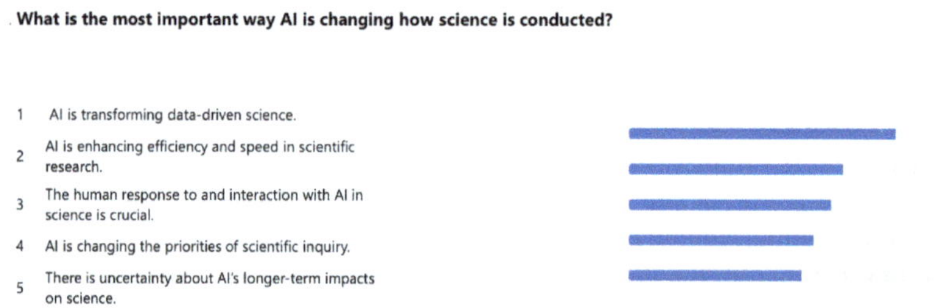


**Fig. 3.3** Expert views on how AI is changing how science is conducted

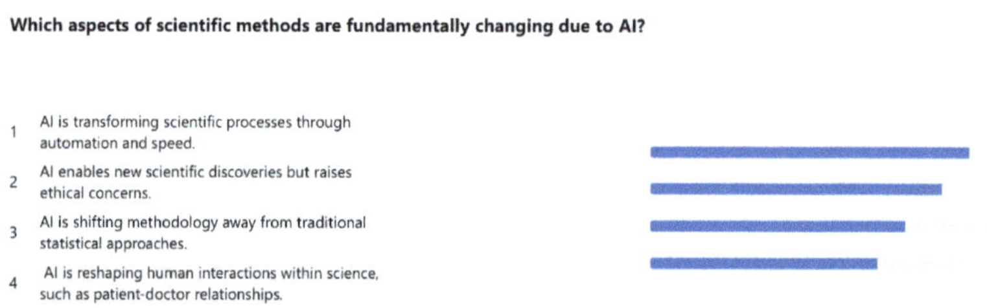


**Fig. 3.4** Expert views on how scientific methods are changing due to AI

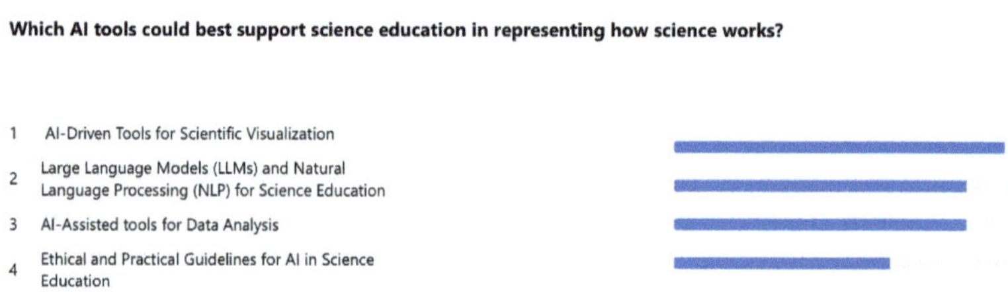


**Fig. 3.5** Expert views on how AI tools can support science education in representing how science works

backed up with plausible-looking chains of reasoning. Their output cannot be trusted, and the danger is that students do not gain the skill to sort rubbish from fact".

This comment speaks to the shared concern of how an overreliance on AI in education, in relation to the integration of AI application in the curriculum, could compromise students' problem-solving skills and their ability to differentiate facts from plausible accounts of the reality generated by AI. Such comments are consistent with concerns reported in the literature. For example, Zhai & Wibowo (2023) examined the effects of over-reliance on AI dialogue systems on students' cognitive abilities. Through a systematic review, they demonstrated that over-reliance stemming from ethical issues of AI impacts cognitive abilities, as individuals increasingly favor fast and optimal solutions over slow ones constrained by practicality. The authors argued that such tendency explained why users prefer efficient cognitive shortcuts, or heuristics, even amidst the ethical issues presented by AI technologies. In a similar vein, educator participants in the Project AI-Vision research (E2 and E4, respectively) not only agreed with the need for cognitive engagement but also the need to ensure educational equity in access to AI use. "Who is able to benefit from well-design AI use? Who is likely to be left behind? New AI technologies seem likely to be monetised in some way, possibly through subscriptions for the latest versions (e.g., Open AI). This has the potential to create a tiered system where wealthier institutions, and high socioeconomic contexts, get to experience greater benefits from AI". "It may be relevant at University, but in my view, students need to know how to do research without AI first in order appreciate the strengths and limitations and develop the skills to evaluate and interpret the outputs of AI".

## 3.5 Conclusion and Ways Forward

As our discussion has illustrated, the influence of AI on practical science is profound and far-reaching. By enhancing data analysis, automating experiments, and enabling new forms of discovery, AI is not just accelerating scientific progress—it is fundamentally transforming how science is done. The report "AI for Science" published by *Nature* argues that a paradigm shift is already taking place in research (Nature Research Intelligence, 2025). The report explored how artificial intelligence is transforming scientific discovery and it covers significant advances across disciplines—such as mathematics, life sciences and physical sciences—while highlighting emerging paradigms and strategies shaping the future of science through intelligent innovation. The narratives of the experts in Project AI-Vision reported in the chapter resonated with the transformative nature of AI in scientific research. While challenges remain, particularly around ethics and transparency, the benefits of AI-driven science are already evident in fields as diverse as medicine, climate science, and materials engineering. Similar concerns were also raised in our sample that included reference to original scientific research published in peer-reviewed science journals.

The developments on the impact of AI in scientific research point to the need in science education to have clarity about the conceptualisation of practical science for the purposes of school science. In this chapter, we have used two example conceptual frameworks on scientific practices and scientific methods to provide some nuance on what is meant by "practical science" and traced them in some literature and in expert scientists' as well as educators' views. The exploration was indicative, and it was not intended to be representative. Future research can focus on systematic literature reviews and larger samples of participants in empirical studies for further nuances about how science is being transformed by AI and what such transformation implies for the teaching and learning of science. Such future studies will benefit from having robust frameworks on practical science such as those exemplified in this chapter so that they can capture how science education needs to consider how it covers practical science in schools. The fast-developing transformations in science in the age of AI can begin to provide insight into how science curricula ought to be reformed to address the changing demands on future citizens and scientists.

**Funding** The research was supported by John Fell Fund (JFF) at the University of Oxford (0015479) and the GATES project (Grenoble ATtractiveness and ExcellenceS), funded by National Research Agency (ANR) under the France 2030 program (ANR-22-EXES-0001) both awarded to Sibel Erduran.

## References

Abrahams, I., & Reiss, M. (2012.) Practical work: Its effectiveness in primary and secondary schools in England. *Journal of Research in Science Teaching, 49*(8) pp. 1035–1055.

Abraham, I., & Saglam, M. (2010). A study of teachers' views on practical work in secondary schools in England and Wales. *International Journal of Science Education, 32*(6), 753–768. https://doi.org/10.1080/09500690902777410

Agarwal, M. K., Bhardwaj, R., Tiwary, G., Aljohani, A. A., Kawatra, R., & Das, A. (2025). Developing explainable artificial intelligence models for space science applications. *Space: Science & Technology*, 5. https://doi.org/10.34133/space.0255

Berland, L. K., Schwarz, C. V., Krist, C., Kenyon, L., Lo, A. S., & Reiser, B. J. (2016). Epistemologies in practice: Making scientific practices meaningful for students. *Journal of Research in Science Teaching, 53*(7), 1082–1112. https://doi.org/10.1002/tea.21257

Berman, G., Williams, K., & Cohen, E. (2025). The benefits of being between (many) fields: Mapping the high-dimensional space of AI research. *Big Data & Society, 12*(1). https://doi.org/10.1177/20539517241306355

Bianchini, S., Müller, M., & Pelletier, P. (2022). Artificial intelligence in science: An emerging general method of invention. *Research Policy, 51*, 104604. https://doi.org/10.1016/j.respol.2022.104604

Brandon, R. (1994). Theory and experiment in evolutionary biology. *Synthese, 99*, 59–73. https://doi.org/10.1007/BF01064530

Byrne, D. (2024). How artificial intelligence can help to keep us safe. *Nature*. https://doi.org/10.1038/d41586-024-02902-8

Casalino, L., Gaieb, Z., Goldsmith, J., Hjorth, C. K., Dommer, A., Harbison, A., Fogarty, C., Barros, E., Taylor, B. C., McLellan, J., Fadda, E., & Amaro, R. (2020). Beyond shielding: The roles of glycans in the SARS-CoV-2 spike protein. *ACS Central Science, 6*, 1722–1734. https://doi.org/10.1021/acscentsci.0c01056

Chan, H. Y., & Erduran, S. (2025). How scientists' narratives on AI signal a new era for Science education abstract. *Research in Science Education*. https://doi.org/10.1007/s11165-025-10307-4

Chen, B., Liu, H., Xu, Y., Yang, Q., Wu, H., & Li, Z. (2024). An analysis of the changes in the representation of scientific methods in laboratory work in Chinese senior high school chemistry textbooks. *Journal of Chemical Education, 101*(5), 1941–1950. https://doi.org/10.1021/acs.jchemed.3c01331

Costa, S. L., & Broietti, F. (2021). Scientific practices in science education: A systematic review of research characteristics and trends. *Ensino & Pesquisa*. https://doi.org/10.33871/23594381.2021.19.3.203-219

Cullinane, A., Hillier, J., Childs, A., & Erduran, S. (2023). Teachers' perceptions of Brandon's matrix as a framework for the teaching and assessment of scientific methods in school science. *Research in Science Education, 53*, 193–212. https://doi.org/10.1007/s11165-022-10044-y

Cullinane, A., Erduran, S., & Wooding, S. J. (2019). Investigating the diversity of scientific methods in high-stakes chemistry examinations in England. *International Journal of Science Education, 41*(16), 2201–2217. https://doi.org/10.1080/09500693.2019.1666216

Delbecq, A. L., van de Ven, A. H., & Gustafson, D. H. (1975). *Group techniques for program planning: A guide to nominal group and Delphi processes*. Scott.

Department for Education. (2014). National curriculum in England: Science programmes of study. *Key stage 4*.

Emden, M. (2021). Reintroducing the scientific method to introduce scientific inquiry in schools? *Science & Education, 30*, 1037–1073. https://doi.org/10.1007/s11191-021-00235-w

Erduran, S., Au, I., & Chan, H. (2025, August 25–29). *Expert views on the use of artificial intelligence in science and implications for science education*. European Science Education Research Association Conference.

Erduran, S. (2023). AI is transforming how science is done. Science education must reflect this change. *Science, 382*(6677). https://doi.org/10.1126/science.adm9788

Erduran, S., El Masri, Y., Cullinane, A., & Ng, D. (2020). Assessment of practical science in high stakes examinations: A thematic analysis focusing on high performing countries. *International Journal of Science Education, 42*(9), 1544–1567. https://doi.org/10.1080/09500693.2020.1769876

Erduran, S., & Dagher, Z. (2014). *Reconceptualizing the nature of science for science education: Scientific knowledge, practices and other family categories*. Springer.

Erduran, S., & Levrini, O. (2024). The impact of artificial intelligence on scientific practices: An emergent area of research for science education. *International Journal of Science Education, 46*(18), 982–1989. https://doi.org/10.1080/09500693.2024.2306604

Erduran, S., & Wooding, S. (2021). A project calibrate approach to summative assessment of practical science. *School Science Review, 102*(381), 71–77.

Ewen, C. (2022). What's next for AlphaFold and the AI protein-folding revolution. *Nature, 604*, 234–238. https://doi.org/10.1038/d41586-022-00997-5

Irzik, G., & Nola, R. (2014). New directions for nature of science research. In M. Matthews (Ed.), *International handbook of research in history, philosophy and science teaching* (pp. 999–1021). Springer.

Jin, H., Kong, F., Li, X., & Shen, J. (2024). Artificial intelligence in microplastic detection and pollution control. *Environmental Research, 262*(1), 119812. https://doi.org/10.1016/j.envres.2024.119812

Johnson, K. B., Wei, W. Q., Weeraratne, D., Frisse, M. E., Misulis, K., Rhee, K., Zhao, J., & Snowdon, J. L. (2021). Precision medicine, AI, and the future of personalized health care. *Clinical and Translational Science, 14*(1), 86–93. https://doi.org/10.1111/cts.12884

Khalifa, M., & Albadawy, M. (2024). Using artificial intelligence in academic writing and research: An essential productivity tool. *Computer Methods and Programs in Biomedicine Update, 5*. https://doi.org/10.1016/j.cmpbup.2024.100145

Landeta, J. (2006). Current validity of the Delphi method in social sciences. *Technological Forecasting and Social Change, 73*(5), 467–482. https://doi.org/10.1016/j.techfore.2005.09.002

Lemke, J. L. (2004). The literacies of science. In E. W. Saul (Ed.), *Crossing borders in literacy and science instruction* (pp. 33–47). International Reading Association.

Ioannidou, O., & Erduran, S. (2021). Beyond hypothesis testing. *Science & Education, 30*, 345–364. https://doi.org/10.1007/s11191-020-00185-9

Maqbool, A. I., Zaib-un-Nisa, Kiayani, S., & Safdar, A. (2025). Artificial intelligence in environmental science and engineering: A review of applications, challenges, and sustainable futures. *Spectrum of Engineering Sciences, 3*(6), 861–876.

Mann, D. M., Chen, J., Chunara, R., Testa, P. A., & Nov, O. (2022). Artificial intelligence applications used in the clinical response to COVID-19: A scoping review. *Journal of Medical Internet Research, 24*(6), e26549. https://doi.org/10.2196/26549

McComas, W. F. (2008). Seeking historical examples to illustrate key aspects of the nature of science. *Science & Education, 17*, 249–263. https://doi.org/10.1007/s11191-007-9081-y

McComas, W. F. (2014). Scientific method (scientific methodology). In W. F. McComas (Ed.), *The Language of Science Education* (pp. 93–93). Sense Publishers.

Method of the year 2021: Protein structure prediction. (2022). *Nature Methods, 19*(1), 1. doi:https://doi.org/10.1038/s41592-021-01380-4.

Mizna, S., Arora, S., Saluja, P., Das, G., & Alanesi, W. A. (2025). An analytic research and review of the literature on practice of artificial intelligence in healthcare. *European Journal of Medical Research, 30*, 382. https://doi.org/10.1186/s40001-025-02603-6

National Research Council (NRC) (2012). A Framework for K-12 Science Education: Practices, Crosscutting Concepts, and Core Ideas. The National Academies Press.

Nature Research Intelligence. (2025). AI for Science 2025. https://www.nature.com/articles/d42473-025-00161-3

NGSS Lead States. (2013). *Next generation science standards: For states, by states*. The National Academies Press.

OECD (2023). *Artificial Intelligence in Science: Challenges, Opportunities and the Future of Research*, OECD Publishing. https://doi.org/10.1787/a8d820bd-en.

Pinto-Coelho, L. (2023). How artificial intelligence is shaping medical imaging technology: A survey of innovations and applications. *Bioengineering, 10*(12), 1435. https://doi.org/10.3390/bioengineering10121435

Project Calibrate. (2020). Key Research Findings. University of Oxford. https://doi.org/10.5287/bodleian:g7vkbr5g9

Qualifications and Curriculum Authority. (2007). *Science: Programme of study for key stage 3 and attainment targets*. www.qca.org.uk/curriculum

Reiff-Cox, R. (2020). Exchanging the myth of a step-by-step scientific method for a more authentic description of enquiry in practice. In W. F. McComas (Ed.), *Nature of science in science instructions: Ration-ales and strategies* (pp. 127–140). Springer.

Reiss, M., Abrahams, I., & Sharpe, R. (2012). *Improving the assessment of practical work in school science*. Gatsby Foundation.

Ren, F., Ding, X., Zheng, M., Korzinkin, M., Cai, X., Zhu, W., Mantsyzov, A., Aliper, A., Aladinskiy, V., Cao, Z., Kong, S., Long, X., Liu, B. H. M., Liu, Y., Naumov, V., Shneyderman, A., Ozerov, I. V., Wang, J., Pun, F. W., Polykovskiy, D. A., Sun, C., Levitt, M., Aspuru-Guzik, A., & Zhavoronkov, A. (2023). AlphaFold accelerates artificial intelligence-powered drug discovery: Efficient discovery of a novel CDK20 small molecule inhibitor. *Chemical Science, 14*, 1443–1452. https://doi.org/10.1039/D2SC05709C

Royal Society. (2014). Science in the age of AI: How artificial intelligence is changing the nature and method of scientific research. London. ISBN: 978-1-78252-712-1

The Royal Society. (2024). *Science in the age of AI: How artificial intelligence is changing the nature and method of scientific research*. The Royal Society.

Savage, N. (2024). AI decodes the calls of the wild. *Nature*. https://doi.org/10.1038/d41586-024-04050-5

Saribas, D., & Ozer, F. (2021). Action research in a teacher education program: Probing into pre-service elementary teachers' understandings of scientific practices and teaching scientific practices. *Journal of Education for Teaching, 48*(2), 197–213. https://doi.org/10.1080/02607476.2021.1985937

Sreenivasan, A., & Suresh, M. (2024). Design thinking and artificial intelligence: A systematic literature review exploring synergies. *International Journal of Innovation Studies, 8*(3), 297–312. https://doi.org/10.1016/j.ijis.2024.05.001

Taherdoost, H., & Alireza Ghofrani, A. (2024). AI's role in revolutionizing personalized medicine by reshaping pharmacogenomics and drug therapy. *Intelligent Pharmacy, 2*(5), 643–650. https://doi.org/10.1016/j.ipha.2024.08.005

Tiwari, A., Mishra, S., & Kuo, T. R. (2025). Current AI technologies in cancer diagnostics and treatment. *Molecular Cancer, 24*, 159. https://doi.org/10.1186/s12943-025-02369-9

Uzzi, B., et al. (2024). Using artificial intelligence to predict replicability of scientific studies. *Proceedings of the National Academy of Sciences, 121*(14), e2302155121. https://doi.org/10.1073/pnas.2302155121

Van Noorden, R., & Perkel, J. M. (2023). AI and science: What 1,600 researchers think. *Nature, 621*, 672–675. https://doi.org/10.1038/d41586-023-02980-0

Wang, D. (2024). How I'm using AI tools to help universities maximize research impacts. *Nature, 630*, 794–794. https://doi.org/10.1038/d41586-024-02081-6

Wang, H., Fu, T., Du, Y., Gao, W., Huang, K., Liu, Z., Chandak, P., Liu, S., Van Katwyk, P., Deac, A., Anandkumar, A., Bergen, K., Gomes, C. P., Ho, S., Kohli, P., Lasenby, J., Leskovec, J., Liu, T. Y., Manrai, A., & Zitnik, M. (2023). Scientific discovery in the age of artificial intelligence. *Nature, 620*, 47–60. https://doi.org/10.1038/s41586-023-06221-2

Wivagg, D., & Allchin, D. (2002). The dogma of 'the' scientific method. *The American Biology Teacher, 69*(9), 645–646. https://doi.org/10.1662/0002-7685(2002)064[0645:TDOTSM]2.0.CO;2

Wei, B., Jiang, Z., & Gai, L. (2021). Examining the nature of practical work in school science textbooks: Coverage of the diversity of scientific methods. *Science & Education, 31*, 943–960. https://doi.org/10.1007/s11191-021-00294-z

World Economic Forum (2024). *AI for scientific discovery*. In *Top 10 Emerging Technologies of 2024*. https://www.weforum.org/publications/top-10-emerging-technologies-2024/in-full/1-ai-for-scientific-discovery/

Yang, Y., Sun, W., Sun, D., & Salas-Pilco, S. Z. (2024). Navigating the AI-enhanced STEM education landscape: A decade of insights, trends, and opportunities. *Research in Science & Technological Education*, 1–25. https://doi.org/10.1080/02635143.2024.2370764

Woodcock, B. A. (2014). The scientific method as myth and ideal. *Science & Education, 23*, 2069–2093.

Zednik, C. (2022). Scientific exploration and explainable artificial intelligence. *Minds and Machines, 32*(2), 219–239. https://doi.org/10.1007/s11023-021-09583-6

Zhai, C., Wibowo, S., & Li, L. D. (2024). The effects of over-reliance on AI dialogue systems on students' cognitive abilities: A systematic review. *Smart Learning Environments, 11*, 28. https://doi.org/10.1186/s40561-024-00316-7

Zhai, C., & Santoso Wibowo, S. (2023). A systematic review on artificial intelligence dialogue systems for enhancing English as foreign language students' interactional competence in the university, *Computers and Education: Artificial Intelligence*. https://doi.org/10.1016/j.caeai.2023.100134

**Sibel Erduran** is a professor of science education and fellow of St Cross College at University of Oxford, UK. She is a fellow of the American Educational Research Association and editor of International Journal of Science Education. Previously, she served as the president of the European Science Education Research Association and the editor-in-chief of Science & Education. Her research interests focus on the infusion of epistemic practices of science in science education.

**Ho-Yin Chan** is a research associate at the University of Cambridge, UK, who investigates how transportation science is not merely a technical pursuit but a socially embedded form of life, shaped by language, politics, and cultural practices. He is particularly interested in how the meaning and function of science in transport emerge from the ordinary, yet often overlooked, grammar of everyday practices and institutional life.

**Ivan Au** is a doctoral researcher in the Department of Education, University of Oxford, UK. He has recent concurrent roles in two projects on EAL students' academic attainment in the UK and the role of artificial intelligence in science education. Previously, he has taken up lecturerships in a number of universities in Hong Kong. His research interests include artificial intelligence, digital pedagogy, and the psycholinguistics of word meaning.

# Chapter 4
# Maker Education and Makerspaces: Challenges and Opportunities for STEAM Education

**Amaia Quintana-Ordorika, Urtza Garay-Ruiz, and Javier Portillo-Berasaluce**

## 4.1 Introduction

The roots of maker education reveal deep connections with the core principles of teaching and learning, emphasizing learning by doing and interdisciplinary integration, particularly through the Science, Technology, Engineering, Arts, and Mathematics (STEAM) approach as well as a culture of collaboration and shared knowledge acquisition. Unlike conventional teaching approaches, maker education is based on active methodologies and collaborative projects. This enables students to engage more deeply with their learning experiences. However, ongoing support is needed to effectively implement it due to the demands of contemporary education (Godhe et al., 2019). One of its well-known strengths is that it holds educational potential, and that it has the capacity to support future-oriented and meaningful learning (Gilbert, 2017). This educational approach is based on students applying knowledge to solve open-ended, real-world problems rather than closed, simulated ones, as in traditional education (Doppelt et al., 2008). The maker movement follows the idea that people learn best by doing, and on the concept of learning through practice. This involves working with tools and materials to create, approaching problem solving with a playful mindset; and applying scientific principles to design, build, and invent (Martinez & Stager, 2013). This chapter begins by exploring the theoretical background of maker education, focusing specifically on makerspaces. It then discusses

A. Quintana-Ordorika (✉) · U. Garay-Ruiz · J. Portillo-Berasaluce
University of the Basque Country, Bilbao, Spain
e-mail: amaia.quintana@ehu.eus

U. Garay-Ruiz
e-mail: urtza.garay@ehu.eus

J. Portillo-Berasaluce
e-mail: javier.portillo@ehu.eus



L. Leite et al. (eds.), *Practical Work in Science Education in the Digital Era*, Advances in Technology-Rich Science Education 2, https://doi.org/10.1007/978-3-032-13788-3_4

the potential opportunities and challenges of integrating maker-based approaches into education. Finally, it provides an overview of some general pedagogical strategies about how this emerging technology could be incorporated into education.

## 4.2 Theoretical Foundations of Maker Education

In pedagogical terms, the origins of maker education can be traced back to the ideas of John Dewey and Seymour Papert. Dewey argued that active participation was essential to meaningful learning experiences (Martinez & Stager, 2013). Papert emphasised the importance of construction over passive knowledge transmission. Furthermore, he stressed the importance of enabling learners to design and create artefacts that align with their interests, thereby fostering deeper knowledge acquisition (Papert, 1986). Dougherty (2013) considered the transformative potential of the maker movement in education, viewing it as both a challenge and an opportunity. Halverson and Sheridan (2014) observed that this emerging pedagogy blurs the boundaries between formal and informal learning, prompting a broader view of where and how learning takes place.

Maker education is an approach for fostering students' motivation to learn, strengthening their belief in their own capabilities such as self-efficacy, and supporting their understanding of concepts across multiple disciplines (Jia et al., 2021). Moreover, it fosters critical thinking, initiative, collaboration, experimentation and other skills, positioning learners as active problem solvers and owners of their learning processes (Blikstein, 2013; Chu et al., 2017; Halverson & Sheridan, 2014; Martinez & Stager, 2013).

## 4.3 Makerspaces

Makerspaces are collaborative environments where people with shared interests come together to work on projects by sharing tools, resources, and knowledge (Fleming, 2015, Roslund & Rogers, 2014). They focus on specific areas such as electronics, robotics, carpentry, sewing, laser cutting, or programming; alternatively, they may combine several disciplines, thus promoting the development of multiple skills (Roslund & Rodgers, 2014). Although both maker education and makerspaces have been associated with STEM education, their core principles extend beyond merely teaching subject content to encompass the development of essential global skills and competencies (Hughes & Kumpulainen, 2021).

Makerspaces operate under key principles such as collaboration, the constant exchange of ideas, and practical creativity. Their most significant features include stimulating learning through experience and discovery, creating opportunities for learning and promoting active cooperation between participants, as well as fostering a culture focused on creation rather than consumption (Hatch, 2014).

These spaces have become increasingly popular in schools, with some being incorporated into classrooms and others being located in libraries or in designated areas (Rouse & Rouse, 2022). What began as exclusive spaces for technological innovation are now transforming the way communities learn, create, and solve problems. In relation to school makerspaces, maker tasks are usually carried out in maker laboratories or makerspaces. Therefore, it is crucial to equip teachers with the knowledge to optimally utilise these environments and facilitate interaction and learning through proper design; teachers should also be able to implement maker activities effectively within the curriculum (Rouse et al., 2025).

## 4.4 Current Research Trends in Maker Education and Makerspaces

Research has highlighted a significant rise in interest in teacher training programmes focused on maker education. However, it should be noted that the research is still fragmented. Programmes vary widely in scope and design: many focus solely on theory, while others focus on acquiring practical skills or combine theory and practice. Ideally, these programmes would combine theoretical knowledge with practical training in the maker methodological approach and teach educators how to incorporate maker education into the curriculum (Quintana-Ordorika et al., 2024).

Within studies that focus on training future teachers to design and plan according to maker principles, the impact of providing them with first-hand experience in makerspaces has been analysed. In one study, direct participation in this learning environment prepared prospective teachers to teach in similar contexts, and the results indicated that including preparation in informal STEM teaching could be beneficial in an elementary teacher education programme (Douglass & Verma, 2022). In another study that trained future teachers to design and plan according to maker principles, it was observed that by engaging directly in collaborative environments between students and teachers, there was an opportunity to confront misconceptions about the planning process and become truly prepared to teach. Interaction in dynamic environments allows both students and teachers to share in the process of learning and creating, which fosters the development of specific, practical skills. This hands-on approach enriches their training and also lays the foundation for a lasting impact on their future educational work (Shively et al., 2021). Another example of teacher training in maker environments is a learning programme based in a university makerspace, designed to familiarise future teachers with tools and technologies applicable in their future classrooms. Through this immersive experience in a community of practice, participants not only gained confidence in the use of maker technologies, but also developed a mindset towards playful, collaborative and iterative learning. However, the results indicated that teacher trainees need more opportunities to observe and experience how primary school students engage in maker processes in the classroom (Heredia & Fisher, 2022).

## 4.5 Opportunities of Maker Education and Makerspaces

The field of education is constantly transforming and by 2030, it is expected that knowledge of concepts alone will have limited value and that proper development of competencies and skills will be necessary. In light of recent technological advances, educational practices could be reformed by adopting innovative teaching and learning approaches that promote STEAM education and the development of twenty-first-century skills (Gratani & Giannandrea, 2022). Consequently, education could focus on fostering students' capacity for creative, independent, critical and collaborative thinking, while considering their social context and self-awareness. This approach could align with the principles of maker education, which emphasises learning through experimentation, problem-solving, and interdisciplinarity and were makerspaces have the potential to foster the development of a variety of twenty-first century skills (Iwata et al., 2020). Teachers who have adopted this emerging methodology have embraced a teaching philosophy that engages students in twenty-first century learning (Freiman, 2020).

Research has shown that one of the main goals of maker education is to prepare students for the challenges of the twenty-first century. The so-called '4C' skills; critical thinking, creativity, communication and collaboration, are widely recognised as fundamental learning outcomes within STEM education (Weng et al., 2022). In academia, critical thinking is defined as the ability to reflect clearly and rationally on problems, with the aim of forming well-founded judgements. This process involves adopting a novel perspective and establishing connections between different areas of knowledge (Hu & Guo, 2021).

The concept of creativity is based on two fundamental elements: novelty and appropriateness (Atkinson, 2000). People with creative thinking skills are able to use their minds freely to produce new ideas, possibilities and inventions. These skills involve using innovative approaches to complete tasks, identify solutions or design inventions (Ozkan & Umdu Topsakal, 2021). Research has shown that adopting a design thinking approach in maker education can enhance students' creative potential, particularly in terms of their creative tendencies, creative thinking and design performance (Chen et al., 2025). It can also foster innovative pieces of work (Tan et al., 2024). Focusing on studies within the STEAM context it has also been analyzed that maker tasks can enhance collective creativity and that there is a linear relationship between collective creativity and the success of maker projects (Timotheou & Ioannou, 2021).

Collaboration has also been a matter of study in this context finding out that this emerging pedagogy can foster collaboration (Doppelt et al., 2008). Collaboration involves individuals using their combined talents, experiences and intelligence to achieve a shared goal (Lin et al., 2020). In an increasingly interconnected world, the importance of collaboration is expected to grow (Griffin & Care, 2015).

Finally, communication is defined as the process by which information, including thoughts, questions, ideas and solutions, is exchanged between individuals (Owens & Hite, 2022). These skills can be developed through problem-solving activities, an area in which maker education has been shown to be effective (Sheffield et al., 2017). A study analysing communication found that the design and layout of the makerspace physical learning environment was an important factor in facilitating the exchange of information among participants (Braga & Guttmann, 2019).

In addition to the ideas previously outlined, maker education could offer a number of opportunities to promote STEAM due to the interdisciplinary nature of makerspaces. These spaces encourage hands-on creation across subject areas that are often treated separately. They also offer flexibility in learning formats, and they support learning by engaging individuals with real-world, meaningful problems, allowing them to gain new knowledge and derive meaning from the experience (Vuorikari et al., 2019).

Focusing on the improvement in students' skills, it must be highlighted that the link between maker activities and STEM has been researched by Bevan (2017) concluding that STEM-rich making activities promoted the learning of contents among students. Another more recent case study showed that maker activities can enhance STEM knowledge and skills (Tabarés & Boni, 2023). Furthermore, active involvement in maker tasks also improves STEM comprehension (Li & Todd, 2019). In the current society, this opportunity is vital for aspects related to education, innovative approaches and initiatives, and industry (Taylor, 2016). Notably, maker education has been observed to not only strengthen knowledge in various areas, but also to promote a more active attitude towards problem solving and facilitate greater participation in the learning process (Chu et al., 2017).

## 4.6 Challenges of Maker Education and Makerspaces

The integration of makerspaces into STEAM education has received increasing attention due to the opportunities it offers to formal education. However, the practical implementation of makerspaces and maker education within educational institutions, particularly STEAM contexts, presents a number of significant challenges.

Some studies have highlighted how maker education has been translated into formal educational settings as the step-by-step construction or fabrication of objects rather than an iterative design process (Resnick & Rosenbaum, 2013). Teacher educators should avoid reducing maker education to a simple set of technical tools and activities used in the classroom to support students' content development. This limited view overlooks key fundamental aspects of the maker approach, such as iterative design, collaboration, and experimentation as part of the learning process (Cohen, 2017). Focusing solely on technological resources may result in overlooking the transformative potential of these environments, which not only seek to acquire knowledge, but also to develop key skills such as problem solving, creativity, critical thinking and collaboration among other skills.

Another significant obstacle is limited access to makerspace resources, preventing educators from acquiring the necessary materials and equipment to complete maker projects (Stevenson et al., 2019). This issue is exacerbated in environments with limited technological infrastructure or in rural areas, where dedicated, equipped spaces are scarce. However, a promising strategy for overcoming these limitations could be to train and support maker-mentors, local young people or community members who can co-facilitate maker events for younger participants. These mentors extend the reach and sustainability of maker initiatives in rural settings and benefit from the experience themselves (Nixon et al., 2021). A viable solution could be considering that these learning environments do not always include just technological and digital devices, as some designs and prototypes can be created using traditional materials and conventional tools (Vuorikari et al., 2019). Moreover, a travelling or mobile makerspace could also be used to democratise access to resources, as it would be a version of a traditional makerspace. Mobile makerspaces are designed to provide innovative learning experiences for diverse communities. Its development involves a comprehensive process encompassing design, construction, implementation and evaluation to ensure it is accessible, functional and tailored to the needs of users. The purpose of a travelling makerspace is to make education more accessible, and to promote digital and manufacturing skills in different contexts. This approach impacts not only participants, but also strengthens the capacity of teachers and facilitators to integrate technological tools and innovative methodologies into their educational practices. However, the continuous evaluation of the travelling makerspace's impact is essential for its improvement and sustainability over time (Compeau, 2018).

Another critical issue is the need for adequate training for educators. Some teachers may not have received the necessary training to be effective in these learning environments, or they may simply not know how to integrate this innovative approach into the curriculum. There are also challenges in training future teachers, which highlights the need to prepare the next generation of educators adequately for maker teaching. Some research demonstrates the challenges involved in integrating making into teaching, which can limit the effectiveness of this approach in the classroom (Cheng et al., 2020). One of the main challenges is a lack of technical expertise, raising significant concerns about encountering technical issues during activities (Stevenson et al., 2019). This lack of technical competence can make teachers feel insecure when using advanced tools and devices, which could affect the quality of the learning experience. Another obstacle could be that some curricular constraints may prevent teachers from designing maker activities that align with specific learning standards and curricular objectives. Teachers must balance the creativity and experimentation inherent in the maker approach with established academic requirements, which requires considerable effort to integrate both aspects coherently and effectively (Jones, 2021). This process requires a thorough understanding of maker education and detailed pedagogical planning to transform these activities into structured, measurable learning opportunities.

One promising response to the challenge of teachers being unprepared for maker education would be to implement targeted professional development initiatives. Some studies have addressed this need by designing and implementing virtual professional development activities focused on making for primary and secondary school teachers (Chen & Cao, 2022). Previous research shows that these activities significantly improved teachers' knowledge of maker education and contributed to transforming their attitudes and beliefs. Professional development practices in a virtual format are beneficial as they enable teachers to gain hands-on experience of making and develop a deeper understanding of the capabilities of maker education and the learning processes involved. Furthermore, making activities that are specific to the school setting have been found to enhance the development of pedagogical skills associated with this approach, as well as improving attitudes and beliefs about the use of maker technology in the classroom (Chen & Cao, 2022).

To implement making remotely, it might be beneficial to provide teachers with access to a variety of tools and ensure an immersive experience that simulates the authenticity of a makerspace. This allows teachers to experience the benefits and challenges of this pedagogical approach first-hand. Building networks of educators and collaborating with specialised institutions could be key strategies for sustaining the model (Lock et al., 2020).

## 4.7 Conclusion and Ways Forward

This chapter has focused on the opportunities afforded by maker education and makerspaces, particularly with regard to STEAM education. However, it has also identified gaps that must be addressed to advance further, as well as potential solutions.

Looking ahead, research should continue to investigate the impact of maker-oriented teacher training on learning processes and outcomes. It should also examine how makerspaces can be considered as future learning environments for educational innovation. Overall, maker education shows possibilities for STEAM education, but it is important to recognise that ongoing training and research are vital considerations together with collective continuous efforts.

**Funding** This study was funded by the European Union-European Innovation Council STEAM-Brace project-Grant Agreement nr.101132652. Views and opinions expressed are however those of the author(s) only and do not necessarily reflect those of the European Union or the European Innovation Council. Neither the European Union nor the granting authority can be held responsible for them.

## References

Atkinson, S. (2000). Does the need for high levels of performance curtail the development of creativity in design and technology project work? *International Journal of Technology and Design Education, 10*(3), 255–281. https://doi.org/10.1023/A:1008904330356

Bevan, B. (2017). The promise and the promises of making in science education. *Studies in Science Education, 53*(1), 75–103. https://doi.org/10.1080/03057267.2016.1275380

Blikstein, P. (2013). Digital fabrication and making in education: The democratization of invention. In J. Walter-Herrmann & C. Büching (Eds.), *Fab Lab* (pp. 203–222). Cultural and Media Studies. doi:https://doi.org/10.1515/transcript.9783839423820.203

Braga, M., & Guttmann, G. (2019). The knowledge networks in a makerspace: The topologies of collaboration. *International Journal of Science and Mathematics Education, 17*(S1), 13–30. https://doi.org/10.1007/s10763-019-09954-7

Chen, P., Wang, R., & Ma, Y. (2025). Promoting junior high school students creativity in maker education: The design thinking approach. *Thinking Skills and Creativity, 56*, 101764. https://doi.org/10.1016/j.tsc.2025.101764

Chen, Y., & Cao, L. (2022). Promoting maker centred instruction through virtual professional development activities for K-12 teachers in low income rural areas. *British Journal of Educational Technology, 53*(4), 1025–1048. https://doi.org/10.1111/bjet.13183

Cheng, L., Antonenko, P. D., Ritzhaupt, A. D., Dawson, K., Miller, D., MacFadden, B. J., Grant, C., Sheppard, T. D., & Ziegler, M. (2020). Exploring the influence of teachers' beliefs and 3D printing integrated STEM instruction on students STEM motivation. *Computers & Education, 158*, 103983. https://doi.org/10.1016/j.compedu.2020.103983

Chu, S. L., Schlegel, R., Quek, F., Christy, A., & Chen, K. (2017). «*I make, therefore I am*»: The effects of curriculum-aligned making on childrens self-identity. *Proceedings of the 2017 CHI Conference on Human Factors in Computing Systems*, 109–120. doi:https://doi.org/10.1145/3025453.3025458

Cohen, J. (2017). Maker principles and technologies in teacher education: A national survey. *Journal of Technology and Teacher Education, 25*(1), 5–30. Society for Information Technology & Teacher Education. Retrieved May 14, 2025 from https://www.learntechlib.org/primary/p/172304/

Compeau, S. (2018). Designing a mobile makerspace: A strategy for increasing diversity by offering engineering outreach workshops to underrepresented youth. *Proceedings of the Canadian Engineering Education Association (CEEA).* https://doi.org/10.24908/pceea.v0i0.13106

Doppelt, Y., Mehalik, M. M., Schunn, C. D., & Krysinski, D. (2008). Engagement and achievements in design-based learning. *Journal of Technology Education, 19*(2), 21–38.

Dougherty, D. (2013). The maker mindset. In M. Honey & D. E. Kanter (Eds.), *Design, make, play: Growing the next generation of STEM innovators* (1st ed.). Routledge.

Douglass, H., & Verma, G. (2022). Examining STEM teaching at the intersection of informal and formal spaces: Exploring science pre-service elementary teacher preparation. *Journal of Science Teacher Education, 33*(3), 247–261. https://doi.org/10.1080/1046560X.2021.1911456

Fleming, L. (2015). *Worlds of making: Best practices for establishing a makerspace at your school.* Corwin.

Freiman, V. (2020). Issues of teaching in a new technology-rich environment: Investigating the case of New Brunswick (Canada) school makerspaces. In Y. Ben-David Kolikant, D. Martinovic, & M. Milner-Bolotin (Eds.), *STEM teachers and teaching in the digital era: Professional expectations and advancement in the 21^st^ century schools* (pp. 273–292). Springer. https://doi.org/10.1007/978-3-030-29396-3_15

Gilbert, J. (2017). Makerspace in education-disruptive technology or just another passing fad? *Teachers Work, 14*(2), 80–89. https://doi.org/10.24135/TEACHERSWORK.V14I2.232

Godhe, A.-L., Lilja, P., & Selwyn, N. (2019). Making sense of making: Critical issues in the integration of maker education into schools. *Technology, Pedagogy and Education, 28*(3), 317–328. https://doi.org/10.1080/1475939X.2019.1610040

Gratani, F., & Giannandrea, L. (2022). Towards 2030. Enhancing 21st century skills through educational robotics. *Frontiers in Education, 7*, 955285. https://doi.org/10.3389/feduc.2022.955285

Griffin, P., & Care, E. (Eds.). (2015). *Assessment and teaching of 21st century skills: Methods and approach.* Springer. https://doi.org/10.1007/978-94-017-9395-7

Halverson, E., & Sheridan, K. (2014). The maker movement in education. *Harvard Educational Review, 84*(4), 495–504. https://doi.org/10.17763/haer.84.4.34j1g68140382063

Hatch, M. (2014). *The maker manifesto*. McGraw-Hill.

Heredia, S. C., & Fisher, M. (2022). Makers-in-residence: An apprenticeship model for supporting pre-service elementary teachers to adopt making tools and technologies. *Tech Trends, 66*(5), 760–770. https://doi.org/10.1007/s11528-022-00751-8

Hu, W., & Guo, X. (2021). Toward the development of key competencies: A conceptual framework for the STEM curriculum design and a case study. *Frontiers in Education, 6*, 684265. https://doi.org/10.3389/feduc.2021.684265

Hughes, J. M., & Kumpulainen, K. (2021). Editorial: Maker education: Opportunities and challenges. *Frontiers in Education, 6*, 798094. https://doi.org/10.3389/feduc.2021.798094

Iwata, M., Pitkänen, K., Laru, J., & Mäkitalo, K. (2020). Exploring potentials and challenges to develop twenty-first century skills and computational thinking in K-12 maker education. *Frontiers in Education, 5*, 87. https://doi.org/10.3389/feduc.2020.00087

Jia, Y., Zhou, B., & Zheng, X. (2021). A curriculum integrating STEAM and maker education promotes pupils' learning motivation, self-efficacy, and interdisciplinary knowledge acquisition. *Frontiers in Psychology, 12*, 725525. https://doi.org/10.3389/fpsyg.2021.725525

Jones, W. M. (2021). Teachers' perceptions of a maker-centered professional development experience: A multiple case study. *International Journal of Technology and Design Education, 31*(4), 697–721. https://doi.org/10.1007/s10798-020-09581-2

Li, X., & Todd, R. J. (2019). Makerspace opportunities and desired outcomes: Voices from young people. *The Library Quarterly, 89*(4), 316–332. https://doi.org/10.1086/704964

Lin, K.-Y., Yu, K.-C., Hsiao, H.-S., Chang, Y.-S., & Chien, Y.-H. (2020). Effects of web-based versus classroom-based STEM learning environments on the development of collaborative problem-solving skills in junior high school students. *International Journal of Technology and Design Education, 30*(1), 21–34. https://doi.org/10.1007/s10798-018-9488-6

Lock, J., Redmond, P., Orwin, L., Powell, A., Becker, S., Hollohan, P., & Johnson, C. (2020). Bridging distance: Practical and pedagogical implications of virtual makerspaces. *Journal of Computer Assisted Learning, 36*(6), 957–968. https://doi.org/10.1111/jcal.12452

Martinez, S. L., & Stager, G. (2013). *Invent to learn: Making tinkering and engineering in the classroom*. Constructing Modern Knowledge Press.

Nixon, J., Stoiber, A., Halverson, E., & Dando, M. (2021). Making makers: Tracing stem identity in rural communities. *J-PEER, 11*(1), 157-9288.1296. https://doi.org/10.7771/2

Owens, A. D., & Hite, R. L. (2022). Enhancing student communication competencies in STEM using virtual global collaboration project-based learning. *Research in Science & Technological Education, 40*(1), 76–102. https://doi.org/10.1080/02635143.2020.1778663

Ozkan, G., & Umdu Topsakal, U. (2021). Exploring the effectiveness of STEAM design processes on middle school students' creativity. *International Journal of Technology and Design Education, 31*(1), 95–116. https://doi.org/10.1007/s10798-019-09547-z

Papert, S. (1986). *Constructionism: A new opportunity for elementary science education. A Proposal to the National Science Foundation*. Massachusetts Institute of Technology, Media Laboratory, Epistemology and Learning Group.

Quintana-Ordorika, A., Garay-Ruiz, U., & Portillo-Berasaluce, J. (2024). A systematic review of the literature on maker education and teacher training. *Education Sciences, 14*(12), 1310. https://doi.org/10.3390/educsci14121310

Resnick, M., & Rosenbaum, E. (2013). Designing for tinkerability. In M. Honey (Ed.), *Design, make, play* (pp. 163–181). Routledge.

Roslund, S., & Rodgers, E. P. (2014). *Makerspaces*. Cherry Lake Publishing.

Rouse, R., & Rouse, A. G. (2022). Taking the maker movement to school: A systematic review of preK-12 school-based makerspace research. Educational Research Review, 35, Article 100413. doi:https://doi.org/10.1016/j.edurev.2021.100413.

Rouse, R., Torralba, J., & Espinoza Mora, E. A. (2025). Characterizing the foundations of school makerspace research and preparing for the future. *Tech Trends*. https://doi.org/10.1007/s11528-025-01103-y

Sheffield, R., Koul, R., Blackley, S., & Maynard, N. (2017). Makerspace in STEM for girls: A physical space to develop twenty-first-century skills. *Educational Media International, 54*(2), 148–164. https://doi.org/10.1080/09523987.2017.1362812

Shively, K., Hitchens, C., & Hitchens, N. (2021). Teaching severe weather: Examining teacher candidates early field experience in a makerspace environment. *Journal of Education, 201*(3), 198–209. https://doi.org/10.1177/0022057420908061

Stevenson, M., Bower, M., Falloon, G., Forbes, A., & Hatzigianni, M. (2019). By design: Professional learning ecologies to develop primary school teachers' makerspaces pedagogical capabilities. *British Journal of Educational Technology, 50*(3), 1260–1274. https://doi.org/10.1111/bjet.12743

Tabarés, R., & Boni, A. (2023). Maker culture and its potential for STEM education. *International Journal of Technology and Design Education, 33*(1), 241–260. https://doi.org/10.1007/s10798-021-09725-y

Tan, J., Kinshuk, & Wu, L. (2024). The effectiveness of design thinking on K-12 school students' creativity in a maker curriculum. *Educational Technology Research and Development, 72*(2), 1091–1110. https://doi.org/10.1007/s11423-023-10332-y

Taylor, B. (2016). Evaluating the benefit of the maker movement in K-12 STEM education. *Electronic International Journal of Education, Arts, and Science, 2*, 1–22.

Timotheou, S., & Ioannou, A. (2021). Collective creativity in STEAM making activities. *The Journal of Educational Research, 114*(2), 130–138. https://doi.org/10.1080/00220671.2021.1873721

Vuorikari, R., Ferrari, A., & Punie, Y. (2019). *Makerspaces for education and training: Exploring future implications for Europe*, Publications Office of the European Union. https://doi.org/10.2760/946996

Weng, X., Cui, Z., Ng, O.-L., Jong, M. S. Y., & Chiu, T. K. F. (2022). Characterizing students' 4C skills development during problem-based digital making. *Journal of Science Education and Technology, 31*(3), 372–385. https://doi.org/10.1007/s10956-022-09961-4

**Amaia Quintana-Ordorika** is a PhD, lecturer and researcher of the University of the Basque Country (UPV/EHU), at the Faculty of Education, Spain. She is a lecturer of the subject of Information and Communication Technologies in Primary and Early Childhood Education degrees. She is a member of the WebLearner Educational Technology research group, her research focuses on maker education and makerspaces, Collaborative Online International Learning (COIL) projects, innovative education and emerging technologies.

**Urtza Garay-Ruiz** PhD from the University of the Basque Country (UPV/EHU), is currently Vice-Rector for Undergraduate Studies and Digital Transformation, at the same university in Spain. She is principal investigator of the WebLearner research group funded by the Basque Government (Spain) and second prize in the STEAM Euskadi rewards in 2024. Her lines of research are linked to educational technology, massive open online courses, and maker education.

**Javier Portillo-Berasaluce** doctor in telecommunications engineering, is professor at the University of the Basque Country (UPV/EHU), the Faculty of Education in Spain, and a member of the WebLearner research group. He has taught courses on programming, microprocessors, digital control, system engineering, digital competencies for educators, and asynchronous tools for collaboration in education. His research interests focus on PLE and PLN, digital competences for educators, and generative AI for education.

# Chapter 5
# Practical Work to Teach the Nature of Science to Science Teachers

**Agustín Adúriz-Bravo**

## 5.1 Introduction

The aim of this chapter is to discuss the role of practical activities designed to teach a relatively recent component of the science curriculum prescribed both for students in their schooling and teachers in their professionalisation: the *nature of science* (NOS) (Jenkins, 2013). The concept of NOS is used to refer to a conglomerate of metascientific knowledge on the epistemic and methodological aspects, *and* the social and cultural constraints, characterising science as a human endeavour (García-Carmona, 2025). NOS content in the curriculum is intended to acquaint students with valid *(meta)theoretical models* (usually from the philosophy and history of science) on the ideas, practices and values underpinning scientific knowledge construction (Adúriz-Bravo, 2014; Matthews, 2000).

There is consensus nowadays within didactics of science (science education as a discipline) that NOS is *indispensable* to foster scientific literacy *and* to support meaningful science teaching (Boven et al., 2025). Historically, NOS appeared to help science teachers deliver content *together with* the scientific processes—as a way to stimulate students' thinking *on* science and enable critical assessment (Yacoubian, 2020). Teaching NOS entails inviting students to explicitly and critically reflect on science *as product and process* (Namakula & Akerson, 2024)—a sociocultural activity that deploys practices to give meaning to the world that produces theories as complex artefacts.

To face this task, science teachers require *autonomous* understanding of NOS, together with knowledge on strategies devised to teach it (*pedagogical content knowledge of NOS*: Schwartz & Lederman, 2002). Didactics deems competence in

A. Adúriz-Bravo (✉)
CONICET/Universidad de Buenos Aires, Facultad de Ciencias Exactas y Naturales, CeFIEC-Instituto de Investigaciones Centro de Formación e Investigación en Enseñanza de las Ciencias, Buenos Aires, Argentina
e-mail: aadurizbravo@cefiec.fcen.uba.ar



L. Leite et al. (eds.), *Practical Work in Science Education in the Digital Era*, Advances in Technology-Rich Science Education 2, https://doi.org/10.1007/978-3-032-13788-3_5

NOS crucial for science teachers; it considers that this component in teacher education *coordinates* and *modulates* the different variables in their teaching, therefore becoming the most determinant factor towards students' comprehension of scientific contents and of science as thinking (Abd-El-Khalick & Lederman, 2000).

It is also contended that teachers' views on NOS determine the nature of the practical work they propose in school science (Ma et al., 2025). During NOS-enriched teacher education, it is shown how different conceptions of the classroom interventions (experimental, observational, theoretical, simulation-based…) on the material world relate to *epistemological models of science as knowledge to transform reality* (Adúriz-Bravo, 2020a). The underlying thesis is that teachers' views on what science is direct their choices of which laboratories, field trips, observations, simulations, etc. to include in their science lessons (Demirel et al., 2023).

NOS reflection on science contents and practices is performed with the aid of conceptual tools from various social sciences and humanities—ranging from classical metasciences (philosophy, history and sociology of science) to other kinds of metatheoretical knowledge (science studies, feminist epistemologies, decolonial studies…) (Erduran, 2014). Accordingly, science teachers need a comprehensive preparation conveying the *multidisciplinary* character of this content they are required to teach.

The fact that NOS combines the use of markedly diverse theoretical references also has *direct bearings on the kind of practical activities* that teachers and teacher educators can put into action so that this curriculum component—characterised by its metatheoretical function (White et al., 2011)—is learnt. Findings in this chapter are framed by my contention that the historically constituted praxis of those disciplines can *inform* practical NOS teaching.

Since the 1990s, didactics of science has issued a number of proposals on effective ways to teach NOS—welding nature of scientific knowledge and practices (Lederman & Lederman, 2019). These existing proposals are adjusted to target audiences: I am particularly interested in prospective and in-service science teachers for compulsory education. Most available approaches assume the need of performing *practical work on NOS* (cf. Helms, 1998), but they have various understandings of what NOS practice supposes. I adhere to a view of practical NOS teaching whose aim is to promote science teachers' meta-reflections allowing their active intervention on sensitive issues connected to *science in society* (Piliouras et al., 2018).

With this in mind, a number of *ways forward* can be formulated for research in NOS teaching: 1. which are effective ways to achieve NOS learning, and what is the role of practice?; 2. how does conducting historical—epistemological enquiry contribute to science teachers' NOS education, and what are the differences between enquiry—based NOS and traditional school investigations?; 3. what tools do teachers need in order to assess and eventually *participate* in science, and how can this kind of assessment be *practiced*? An objective for this chapter is to elucidate some facets of these matters; for this, I will propose and exemplify theoretical categories to characterise practical NOS work, understood as the *exercise of metadiscourse*.

I will focus in the kind of activities that empower teachers for critical action—NOS practice during learning based on enquiry, modelling and argumentation

(Adúriz-Bravo, 2020a). I will also briefly identify possible relations between school NOS practices and the usual science practices (labs, field trips, problem guides, investigations, etc.).

The next section explores a range of practical activities designed to enhance science teachers' understanding of NOS, thereby enabling them to implement formative "school scientific activity" (Izquierdo-Aymerich & Adúriz-Bravo, 2003) in their classrooms. It has been proposed that, through investigations, historical-case analyses, paper-and-pencil tasks, role-playing of debates, mass-media evaluation and didactical planning—among other activities—, science teachers can internalise key NOS dimensions such as: nature of scientific theories; observation and experimentation; evidence and inference; methodologies; and values and ethics (Erduran & Dagher, 2014).

For a practical approach to NOS to be valuable, it should focus on bridging the gap between metatheoretical knowledge, *applicable* understandings of such knowledge, and effective pedagogies to teach it. The NOS to be learnt by science teachers should ultimately contribute to improved science education *for all* through introducing more consistent and nuanced conceptions of science.

Proposals for "practical NOS" should be tailored to their audiences, but the overall aim is to improve teachers' "grasp" of NOS teaching, equipping them with didactical knowledge needed to introduce this content in their classes. Successful practical NOS activities that are based on *intellectually active*, collective, experiential and reflective learning (Adúriz-Bravo, 2020b) can make more tangible the abstract aspects of this curriculum component, encouraging teachers to internalise and subsequently impart a more updated theoretical perspective on science.

Despite the formative importance recognised for NOS, many science teachers are reported to possess incomplete, outdated or erroneous views on science, often perceiving it as a fixed set of discovered *facts* rather than an evolving, argumentative, creative human endeavour (Amador-Rodríguez et al., 2023). These positivistic and scientistic representations in teachers can hinder their competence to teach science as enquiry (Abd-El-Khalick & Lederman, 2000) and to foster critical thinking in their students; hence the urgent need to research around ways of improving science teachers' NOS preparation.

Transposing NOS constructs (as presented in the literature accessible to practitioners) into *realistic* teaching practices is still a challenge in science teacher education. Implementing such a transposition requires making teachers critically analyse NOS aspects through practical activities fostering more *regulated* understandings. This chapter proposes categories to group practical activities potentially promoting science teachers' engagement with the core elements of NOS and building in them professional knowledge to introduce those elements into their teaching.

## 5.2 Practical Teaching of NOS: Typology of Activities

Effective practical instruction in NOS should be grounded on a post-constructivist paradigm that emphasises active learning—individual and cooperative intervention on the world accompanied with reflection before, during and after. Active practical NOS education requires that science teachers vividly *experience* the set of scientific processes understood as enquiry. This NOS teaching oriented towards *enquiry on the notion of science* is then similar to school investigations insofar both activities are triggered by relevant questions, problems or projects, and both require careful *planning* and *regulation* of actions, and rigorous assessment of the concepts proposed as answers to the trigger.

Practical NOS activities should resort to school scientific investigations, introduce historical contexts, promote collective discussion, critically manage circulating information, and employ key metascientific concepts. These types of activities, among others reviewed in this section, invite teachers to question their previous (often reductive) notions about what science is, helping to align their epistemological understandings with developments of 1980s and 1990s (cf. Eflin et al., 1999). The aim is to nurture in science teachers a more sophisticated view of science as an evidence-based, transformative, socially-influenced activity (Aslan & Taşar, 2013).

The literature reports a spectrum of practical activities explicitly addressing the need to professionalise teachers in NOS[1]. Those activities are designed to be implemented within science teacher education courses and professional development programmes. On the basis of an informal systematisation—performed during my 30-year-long career-of over 200 published teaching materials, I propose here a typology of practical NOS activities. Each type is technically described; then, *epitomic exemplars* devised by authors from didactics of science are provided.

### *5.2.1 Conceptual Exercises*

NOS "exercises" are tasks with well-defined prompts given to science teachers orally or in writing. Teachers solve them partially as typical paper-and-pencil tasks, and partially through rich oral exchange with colleagues in teacher education classes. More frequently than with traditional exercises in school science, "correction" at the end is done in a plenary moderated by the teacher educator; only occasionally through comments and markings on each teacher's written production.

This type of reflective activities requires the *refined* use of key metascientific concepts, and provide situations for science teachers to articulate their understandings of NOS and to employ those concepts to model one—or a few—particular features of science. Individual work and group discussions on topics like worldviews, nature of scientific explanation, scientists' ethics, or the tension between creativity and rigour help deepen teachers' conceptualisation of the multifaceted nature of science-in-the-making, and of the structure and validity of its theories.

*Most extended instructional objective for NOS exercises:* To deepen science teachers' understanding of particular aspects of NOS through reading, reflection and dialogue on exemplars of metatheoretical analysis.

*Possible didactical strategies*: Presentation of "good questions" inviting teachers to reflect and discuss around a particular aspect of science. Engineered prompts that are first individually responded and then shared with a small group of teachers and the whole classroom to contrast perspectives. Focussed readings explicating key NOS concepts to be used to understand the issue at stake. Carefully administered interventions of the teacher educator to model the use of those concepts to address the question posed.

*Theoretical elements underneath*: The approach is concept-based—it requires contrasting diverse conceptions (common sense, previous acquisitions, social representations, ad-hoc constructions, "sanctioned" knowledge from metasciences). Science teachers' intellectual and discursive activity to respond to the prompt-including search for pertinent information from different sources—promotes self-regulation: teachers are asked to compare their own views on the problem under examination with those of their colleagues and with "normative" answers provided by the specialised literature. The teacher educator *scaffolds* these processes through exemplifying the use of metascientific constructs.

*Epitomic examples:* A sophistication of the draw-a-scientist-task is used by Acosta-Paz and colleagues (2024) to discuss the overrepresentation of experiments in the social image of science. The quest to manufacture the first precision pendulum clocks (as reconstructed in Matthews, 2000) has been used to examine Karl Popper's "falsificationist" approach to the scientific method. Specific excerpts of Charles Darwin's classic *On the origin of species* (curated in Adúriz-Bravo & González Galli, 2002) are suggested as valuable original scientific texts to show science teachers the mechanics of *abductive reasoning* at the basis of scientific explanation.

### 5.2.2 *Puzzles, Enigmas and Mysteries*

In these activities, a wide range of "puzzling" materials—from "black boxes" and other intriguing contraptions to detective novels—are used as *analogues* for the process of scientific problem-solving to inspect the methodologies of science. This aspect of NOS has the highest number of "misconceptions" reported in the literature (Cetin, 2012); it results essential to eradicate from school science the infamous *scientific method*—an impoverished representation of scientists' practices.

In NOS puzzle-solving, science teachers are invited to resolve *well-contextualised*, non-scientific situations going out of the ordinary with no straightforward answers. In the way to providing satisfactory solutions, teachers are requested to reflect on the methodological *paths* that they take and on the nature of the solutions constructed. This kind of work serves as a solid metaphor of what scientists actually do when they conduct research.

*Most extended instructional objective for NOS puzzles:* To acquaint science teachers with theoretical models on some of the cognitive and linguistic procedures of science.

*Possible didactical strategies:* Presentation of puzzling situations to be solved, not directly related to scientific contents, and often with more than one possible answer. Instructions that are first individually followed and then tackled in small groups of fellow teachers (and eventually the whole classroom) to contrast approaches and solutions. Teacher educator's interventions to organise students-teachers' productions and making links with scientific processes that can be mapped to what was done.

*Theoretical elements underneath:* The approach is model-based and directed to NOS problem-solving—science teachers activate a "map of the world" seen as potential candidate to give an answer to the proposed puzzle. Teachers' intellectual, discursive—and eventually material—activity in order to solve the problem is structured as a piece of enquiry outside the realm of science, thus lowering cognitive load and enhancing interest and motivation. Intensive work done on the analogues provides foundations for a meaningful transfer to the analysis of scientific practices (Adúriz-Bravo, 2024).

*Epitomic examples:* Standard black boxes, and more sophisticated versions thereof, have been used by Lederman and Abd-El-Khalick (1998) to make reasoning visible and help science teachers distinguish between observation and inference. Roald Dahl's short story "The landlady" is introduced in Adúriz-Bravo (2014) to present the semanticist view of scientific models. Tutankhamun's dagger made out of "metal from the sky" (Yuhas, 2016) is often presented as an enigma to learn about scientific hypotheses.

### 5.2.3 Historical Case Studies

With the study of selected cases from the history of science, conveyed through narratives known as "science stories", teachers have the opportunity to examine *science as a whole* (Allchin, 2011), in a more integrated manner than with exercises and puzzles. History-based NOS work takes into account social and cultural aspects of science much more significantly than the previous types of practical activities.

Analysing pivotal scientific discoveries and inventions that occurred in the long and rich past of the scientific disciplines *through the lens of key concepts and models from the metasciences* is probably the most inspected teaching path into the nature of science. Historical case studies prove to be powerful to illuminate NOS aspects such as inference from results, creativity and scientific errors, but also, and more importantly, social significance of science, and political, economic, cultural and even religious influences on the scientific complex.

*Most extended instructional objective of NOS historical cases:* To understand how scientific knowledge develops in specific places and times, heavily influenced by sociocultural factors.

*Possible didactical strategies:* Reading of original historical materials, narratives and case studies in different formats (written, oral, multimedia; cinema can be especially interesting). Identification of key NOS elements in the cases. Discussion around different metascientific reconstructions of how scientific ideas are generated and accepted, and then evolve and eventually perish.

*Theoretical elements underneath:* Case studies set the production of science in socio-political contexts; the purpose is to understand science as an evolving human activity open to critique and subject to change, embedded in culture, and influenced by societal factors. The main didactical model under historical case studies relies on the use of narratives in science education (Revel Chion & Adúriz-Bravo, 2022). A reported benefit of this pedagogy is the introduction of the "human factor" in school science, which enhances motivation and understanding.

*Epitomic examples:* Allchin (2011) uses Christian Eijkman's studies on beriberi to examine, among other issues, the "[g]rowth of knowledge through small cumulative additions versus through major conceptual reinterpretations" (p. 533). Guerra and Abalada (2024) historicise the work of Eddington's team in Brazil during the 1919 solar eclipse to make local contributors visible. Klassen (2009) created a short story on the physicist Louis Slotin's death in a nuclear accident during the Manhattan project—"The dragon's revenge"—to explore with science teachers the axiological intricacies of the scientific enterprise.

### 5.2.4 Enquiry-Based Activities

It is contended that one of the most effective ways to teach science *coupled with NOS* is through school scientific enquiries that mirror actual science practices (Lederman et al., 2013). In this approach, teachers are asked to conduct stylised experiments, observations or investigations during which they can reflect on the strategies and assumptions involved.

Carefully selected and engineered scientific "experiences" help show teachers that scientific knowledge does not *empirically* and *inductively* emerge from collected data, but is rather derived from theoretical reconstruction of evidence. The epistemological notion of model can be introduced to understand how scientists explain the phenomena that they encounter when interrogating the world.

*Most extended instructional objective of NOS enquiry:* To make science teachers experience core processes of scientific research showing that scientific knowledge arises from systematic interrogation of nature with the use of models.

*Possible didactical strategies:* Conduction of simplified (often experimental) pieces of school scientific investigation where planning is central and procedures are made explicit. Engagement of teachers in the design, hypotheses formulation, data collection, analysis of results, and establishment of conclusions. Discussions after the experience on the nature of scientific methods and on the tentative status of scientific knowledge. Parallelism with famous episodes of science.

*Theoretical elements underneath:* There is abundant literature on enquiry as a pedagogy that, by making students immerse into practices that are similar to those in scientists' science, achieves higher degrees of motivation, engagement and learning gains than traditional teaching. In enquiry-based science education, it is recommended that "hands-on" activities be accompanied by explicit reflection that shows the laboriously conquered tools of scientific thinking.

*Epitomic examples:* A chemical lab experience titled "Mystery powders" (Bianchini & Colburn, 2000) is used to teach the NOS tenet that "no one right way necessarily exists to solve a problem in science" (p. 187). The problem of how the ancient Egyptian pyramids were constructed has been used as "blueprint" of scientific inquiry (Elbih et al., 2022; Rodríguez-Arteche et al., 2022). Schizas et al. (2024) propose to simulate the Avery-MacLeod-McCarty experiment on genetics to learn about the relations between experimentation and hypothesising.

### 5.2.5 Debates, Role-Playing and Simulations

These kinds of practical NOS activities, based on the competence of scientific argumentation, can function as realistic emulation or enactment of the processes of *knowledge negotiation* in science. Key epistemic tasks performed within the scientific community (consensus building, confrontation of opposing models, peer review, diffusion of new ideas, ethical vigilance) are introduced through this approach where teachers participate in toy models of science assuming various roles.

In this type of NOS practical work, communication and persuasion, and sometimes controversy, are the constituent elements; teachers "live" these elements by engaging in the representation of plausible scenarios depicting specific communities of practitioners. Teachers' participation in such scenarios requires intensive previous work to understand the views and motivations of the actors involved.

*Most extended instructional objective of NOS debates:* To make science teachers experience some of the rhetorical, social and ethical stances of the scientific activity.

*Possible didactical strategies:* Construction of scenarios of cognitive and social relevance, including the design of the (real or fictional) characters that will be involved. Teachers' preparation for the session in which they will interact with each other. Metatheoretical reflection before, during and after the execution of the task. Connections with actual cases in contemporary science. Strong moderation of the sessions by the teacher educator.

*Theoretical elements underneath:* The use of debates and controversies in science education has been extensively documented, pondered and justified (Dunlop & Veneu, 2019). A tried-out pedagogy involves the discussion of socially acute issues in science and technology as a way to develop citizenship. This approach gives teachers the opportunity to screen publicly available information, develop and defend their own positions, and evaluate counter-arguments and alternative views.

*Epitomic examples:* Simonneaux (2011) reports the use of unsettled issues in human health to discuss the feasibility and acceptability of human gene therapy.

Puig and Ageitos (2022) resort to debates around vaccination to support decision-making in citizen science. Howes and Cruz (2009) propose to their college students role-playing TV interviews to researchers in order to discuss the social image of scientist.

### 5.2.6 *Other Types of Practical Activities*

*Critical analysis of mass media and social networks:* Science teachers can be asked to review newspaper or blog articles, popularisation programmes, documentaries, or posts in social networks on specific scientific issues. Critical evaluation of scientific information available to the public helps teachers identify well-established consensuses, on-going controversies, misinformation, and even fake news. The role of media in shaping public understanding of science rapidly becomes the NOS issue under scrutiny.

*Socioscientific issues:* Defined as "dilemmas that arise due to the complex relationship that exists between science and society" (Viehmann, 2024, n/p.), they have been hailed as a powerful entry-point to learning NOS. In my typology, socioscientific issues such as climate change, nuclear energy, euthanasia, racism, etc. can be located at the centre of exercises, problems, cases, investigations or debates.

*Productions around NOS topics:* Some authors report proposals where science teachers are required to elaborate extended writings (essays, investigation reports, bibliographical reviews…). There are also activities where teachers produce other genres with heavily multisemiotic character: infographics, videoclips, blogs, scale models, artistic performances, etc.

*"NOS field trips":* Here I include visits to science museums, centres and facilities, and interviews with real scientists.

*NOS lesson planning:* A last type of activity, which I deem particularly relevant for teachers' NOS education, is the *design of didactical units* (teaching-learning sequences). Student-teachers (or eventually active teachers in qualification programmes) are requested-usually as part of their summative evaluation—to plan a number of science lessons for their (future) students that explicitly include activities on the nature of science. The teacher educator assesses the pertinence and quality of such activities, especially taking into account the theoretical *rationale* that teachers provide for their inclusion.

## 5.3 Conclusion and Ways Forward

Science teachers' understandings of NOS impact their ability to foster science enquiry skills, critical thinking, epistemic and informational literacy, and even desirable values, among students. Teachers who perceive science as an edifice of discovered laws struggle to encourage enquiry—or context-based learning in their classrooms. Conversely, teachers who understand some of the epistemic and social dimensions of science can create learning environments that are more influential for students. This makes NOS education one of the most important contributions to teachers' professional knowledge, and hence the urgency of conducting research on this issue.

I am convinced that practical activities are crucial to build science teachers' comprehension of the complex facets of NOS. Experts consider the types of practical NOS teaching revised in this chapter effective strategies for the *internalisation* of NOS. Embedding these activities into teachers' professional development seems to be the way to foster durable changes in science education directed towards the promotion of a scientific literacy rooted in an understanding of the role of science in contemporary society.

The ways forward from this chapter, then, require from us teacher educators systematically integrating practical NOS into teacher training programmes. The deeper understanding of NOS they can provide improves teachers' proficiency *and confidence* to teach more complex scientific issues together with their "nature".

Another line that remains open for the near future, in tight connection with the rest of this volume, is how practical work in NOS interacts with practical work in school science (systematic observations, laboratories, science visits, etc.). The latter is shaped by the conceptions of NOS held by curricula, textbooks and teachers, and at the same time shapes students' learning of science and NOS. It is then necessary to ascertain how teachers can design practical proposals in which the NOS component is *monitored* in detail.

## Note

1. See activities in: Lederman & Abd-El-Khalick 1998; Estany & Casacuberta 2000; Irwin 2000; Matthews 2000; Adúriz-Bravo 2014; Eastwood et al. 2012; Erduran & Dagher 2014; McComas 2017; Wooding et al. 2020; Kranjc Horvat et al. 2022; Science Learning Hub 2023; Berkeley University of California 2025; The Open University 2025; NCSE n.d.

## References


Abd-El-Khalick, F., & Lederman, N. G. (2000). The influence of history of science courses on students views of nature of science. *Journal of Research in Science Teaching, 37*(10), 1057–1095. https://doi.org/10.1002/1098-2736(200012)37:10<1057::AID-TEA3>3.0.CO;2-C

Acosta Paz, J.D., Pujalte, A.P., & Adúriz-Bravo, A. (2024). Students views of science and of scientists outside the laboratory. In A.A. Viveiro, M.C. de Senzi Zancul, & P. Neves da Rocha (Eds.), *Teaching science in primary schools* (pp. 175–199). Editora FE-Unicamp. https://editora.fe.unicamp.br/index.php/fe/catalog/view/66/177/889

Adúriz-Bravo, A. (2014). Teaching the nature of science with scientific narratives. *Interchange, 45*(3–4), 167–184. https://doi.org/10.1007/s10780-015-9229-7

Adúriz-Bravo, A. (2020a). Contributions to the nature of science: Scientific investigation as inquiry, modeling, and argumentation. In C. N. El-Hani, M. Pietrocola, E. F. Mortimer, & M. R. Otero

(Eds.), *Science education research in Latin America* (pp. 394–425). https://doi.org/10.1163/978 9004409088_017
Adúriz-Bravo, A. (2020b). Innovative strategies for science teaching. In L. Leite, E. Oldham, A. S. Afonso, F. Viseu, L. Dourado, & M. H. Martinho (Eds.), *Science and mathematics education for 21st century citizens* (pp. 107–131). Nova Science Publishers.
Adúriz-Bravo, A. (2024). Scientific, didactical and analogical models in science teaching. In M. R. Quintanilla Gatica & A. Adúriz-Bravo (Eds.), *Science teaching and a new teacher culture* (pp. 41–58). Springer. https://doi.org/10.1007/978-3-031-50004-6_3
Adúriz-Bravo, A., & González Galli, L. (2022). Darwin's ideas as epitomes of abductive reasoning in the teaching of school scientific explanation and argumentation. In L. Magnani (Ed.), *Handbook of abductive cognition* (pp. 1–37). Springer. https://doi.org/10.1007/978-3-030-68436-5_50-1
Allchin, D. (2011). Evaluating knowledge of the nature of (whole) science. *Science Education, 95*(3), 518–542. https://doi.org/10.1002/sce.20432
Amador-Rodríguez, R., Ospina-Quintero, N., & Adúriz-Bravo, A. (2023). Teachers views on the nature of science: An epistemological analysis. *Australian Journal of Teacher Education, 48*(4), 78–95. https://doi.org/10.14221/1835-517X.6023
Aslan, O., & Taşar, M.F. (2013). How do science teachers view and teach the nature of science?: A classroom investigation. *Eğitim ve Bilim, 38*(167), 65-80. https://doi.org/10.15390/ES.2013.1103
Bejarano, N. R. R., Adúriz-Bravo, A., & Bonfim, C. S. (2019). Natureza da ciência (NOS): Para além do consenso [Nature of Science (NOS): Beyond Consensus]. *Ciência e Educação, 25*(4), 967–982. https://doi.org/10.1590/1516-731320190040008
Berkeley University of California. (2025). *Understanding science: Teaching guides*. https://undsci.berkeley.edu/for-educators/teaching-guides/
Bianchini, J. A., & Colburn, A. (2000). Teaching the nature of science through inquiry to prospective elementary teachers: A tale of two researchers. *Journal of Research in Science Teaching, 37*(2), 177–209. https://doi.org/10.1002/(SICI)1098-2736(200002)37:2<177::AID-TEA6>3.0.CO;2-Y
Boven, L., De Schrijver, J., Pinxten, R., & Van Petegem, P. (2025). Explicating explicit reflection in nature of science education: A systematic literature review. *Frontiers in Education*, 10, Article 1577021. doi:https://doi.org/10.3389/feduc.2025.1577021
Cetin, P. S. (2012). Pre-service science teachers' conceptions of scientific method. *Journal of Teacher Education and Educators, 1*(2), 257–274. https://dergipark.org.tr/en/download/article-file/646207
Demirel, Z. M., Sungur, S., & Çakıroğlu, J. (2023). Science teachers views on the nature of science and its integration into instruction. *Science & Education, 32*(5), 1401–1433. https://doi.org/10.1007/s11191-022-00409-0
Dunlop, L., & Veneu, F. (2019). Controversies in science. *Science & Education, 28*(6–7), 689–710. https://doi.org/10.1007/s11191-019-00048-y
Eastwood, J. L., Sadler, T. D., Zeidler, D. L., Lewis, A., Amiri, L., & Applebaum, S. (2012). Contextualizing nature of science instruction in socioscientific issues. *International Journal of Science Education, 34*(15), 2289–2315. https://doi.org/10.1080/09500693.2012.667582
Eflin, J. T., Glennan, S., & Reisch, G. (1999). The nature of science: A perspective from the philosophy of science. *Journal of Research in Science Teaching, 36*(1), 107–116. https://doi.org/10.1002/(SICI)1098-2736(199901)36:1<107::AID-TEA7>3.0.CO;2-3
Erduran, S. (2014). Beyond nature of science: The case for reconceptualising 'science' for science education. *Science Education International, 25*(1), 93–111. https://www.icaseonline.net/sei/mar ch2014/p11.pdf
Erduran, S., & Dagher, Z.F. (2014). *Reconceptualizing the nature of science for science education.* Springer. https://doi.org/10.1007/978-94-017-9057-4
Estany, A., & Casacuberta, D. (2000). *Manual de prácticas de filosofía de la ciencia [Practical Guide to the Philosophy of Science].* Crítica.

García-Carmona, A. (2025). The non-epistemic dimension, at last a key component in mainstream theoretical approaches to teaching the nature of science. *Science & Education, 34*(3), 1149–1165. https://doi.org/10.1007/s11191-024-00495-2

Guerra, A., & Abalada, P. (2024). Invisible problem-solvers in the record of the 1919 total eclipse in Sobral by the British team. *Centaurus, 66*(1–2), 165–188. https://doi.org/10.1484/J.CNT.5.143683

Helms, J. V. (1998). Learning about the dimensions of science through authentic tasks. In J. Wellington (Ed.), *Practical work in school science* (pp. 126–152). Routledge. https://doi.org/10.4324/9780203062487

Horvat, A. K., Boselli, M., Chatzidaki, P., Dahlkemper, N., Duggan, R., Durey, G., Herff, N., Molaro, D., Scheerer, G. W., Schmeling, S., Thill, P. G., Wiener, J., Woithe, J., & Zoechling, S. (2022). The mystery box challenge: Explore the nature of science. *Science in School: The European Journal for Science Teachers*, 59 Article. https://scienceinschool.org/wp-content/uploads/2022/08/Issue59_MysteryBox.pdf

Howes, E. V., & Cruz, B. C. (2009). Role-playing in science education: An effective strategy for developing multiple perspectives. *Journal of Elementary Science Education, 21*(3), 33–46. https://doi.org/10.1007/BF03174721

Irwin, A. R. (2000). Historical case studies: Teaching the nature of science in context. *Science Education, 84*(1), 5–26. https://doi.org/10.1002/(SICI)1098-237X(200001)84:1<5::AID-SCE2>3.0.CO;2-0

Izquierdo-Aymerich, M., & Adúriz-Bravo, A. (2003). Epistemological foundations of school science. *Science & Education, 12*(1), 27–43. https://doi.org/10.1023/A:1022698205904

Jenkins, E. W. (2013). The 'nature of science' in the school curriculum: The great survivor. *Journal of Curriculum Studies, 45*(2), 132–151. https://doi.org/10.1080/00220272.2012.741264

Klassen, S. (2009). The construction and analysis of a science story: A proposed methodology. *Science & Education, 18*(3–4), 401–423. https://doi.org/10.1007/s11191-008-9141-y

Lederman, N. G., & Abd-El-Khalick, F. (1998). Avoiding de-natured science: Activities that promote understandings of the nature of science. In W. F. McComas (Ed.), *The nature of science in science education* (pp. 83–126). Kluwer. https://doi.org/10.1007/0-306-47215-5_5

Lederman, N.G., & Lederman, J.S. (2019). Teaching and learning nature of scientific knowledge: Is it déjà vu all over again? *Disciplinary and Interdisciplinary Science Education Research, 1*, Article 6. https://doi.org/10.1186/s43031-019-0002-0

Lederman, N. G., Lederman, J. S., & Antink, A. (2013). Nature of science and scientific inquiry as contexts for the learning of science and achievement of scientific literacy. *International Journal of Education in Mathematics, Science and Technology, 1*(3), 138–147. https://doi.org/10.18404/IJEMST.19784

Ma, M., Lu, Z., Dai, P., Chen, H., Yang, D., Wang, H., & Geng, M. (2025). A framework for assessing practical activities in science education. *Journal of Contemporary Educational Research, 9*(4), 147–154. https://doi.org/10.26689/jcer.v9i4.9887

Matthews, M.R. (2000). *Time for science education.* Kluwer. doi:https://doi.org/10.1007/978-94-011-3994-6

McComas, W. F. (2017). Understanding how science works: The nature of science as the foundation for science teaching and learning. *School Science Review, 98*(365), 71–76. https://www.researchgate.net/publication/318912990_Understanding_how_science_work_The_nature_of_science_as_they_foundation_for_science_teaching_and_learning

Namakula, E. K., & Akerson, V. L. (2024). Explicit incorporation of the nature of science (NOS) in an undergraduate preservice teacher science content course: Action research. *International Journal of Research in Education and Science, 10*(2), 241–278. https://doi.org/10.46328/ijres.3354

NCSE-National Center for Science Education (n.d.). *Nature of science lesson sets.* https://ncse.ngo/nature-science-lesson-sets

Piliouras, P., Plakitsi, K., Seroglou, F., & Papantoniou, G. (2018). Teaching explicitly and reflecting on elements of nature of science: A discourse-focused professional development program with

four fifth-grade teachers. *Research in Science Education, 48*(6), 1221–1246. https://doi.org/10.1007/s11165-016-9600-4
Puig, B., & Ageitos, N. (2022). Critical thinking to decide what to believe and what to do regarding vaccination in schools. In B. Puig & M. P. Jiménez-Aleixandre (Eds.), *Critical thinking in biology and environmental education* (pp. 113–132). Springer. https://doi.org/10.1007/978-3-030-92006-7_7
Rodríguez-Arteche, I., Ros, G., Rodríguez-Laguna, M.T., & Martínez-Aznar, M.M. (2022). Tensiones docentes en el proyecto STEAM Maquinando en el Antiguo Egipto [Teacher tensions in the STEAM project 'Engineering in Ancient Egypt']. In A. Benarroch (Ed.), *30 Encuentros Internacionales de Didáctica de las Ciencias Experimentales* [30 International Meetings on Experimental Science Teaching] (pp. 671-677). Universidad de Granada. http://apice-dce.com/wp-content/uploads/2022/10/Actas-30-EDCE.pdf
Schizas, D., Psillos, D., & Asimopoulos, S. (2024). Advancing scientific literacy: Integrating nature of science and nature of scientific inquiry in teaching Griffith's and Avery-Mac Leod-McCarty experiments. *Interdisciplinary Journal of Environmental and Science Education*, *20*(4), Article e2421. https://doi.org/10.29333/ijese/15522
Schwartz, R., & Lederman, N. G. (2002). It's the nature of the beast: The influence of knowledge and intentions on learning and teaching nature of science. *Journal of Research in Science Teaching, 39*(3), 205–236. https://doi.org/10.1002/tea.10021
Science Learning Hub. (2023). *Tenets of the nature of science.* https://www.sciencelearn.org.nz/resources/413-tenets-of-the-nature-of-science
Simonneaux, L. (2011). Students' reasoning on socioscientific issues and socially acute questions. In D. J. Boerwinkel & A. J. Waarlo (Eds.), *Genomics education for decision making* (pp. 73–81). Utrecht University. https://www.researchgate.net/publication/239854026_Genomics_education_for_decision_making_proceedings_of_the_second_invitational_workshop_on_genomics_education_2-3_December_2010_Utrecht_The_Netherlands
The Open University (2025). *How can students learn about the nature of science?* https://www.open.edu/openlearn/education-development/teaching-secondary-science/content-section-3
Viehmann, C., Fernández Cárdenas, J. M., & Reynaga Peña, C. G. (2024). The use of socioscientific issues in science lessons: A scoping review. *Sustainability, 16*(14), 5827. https://doi.org/10.3390/su16145827
White, B. Y., Collins, A., & Frederiksen, J. R. (2011). The nature of scientific meta-knowledge. In M. S. Khine & I. Saleh (Eds.), *Models and modeling: Cognitive tools for scientific enquiry* (pp. 41–76). Springer. https://doi.org/10.1007/978-94-007-0449-7_3
Wooding, S., Cullinane, A., & Erduran, S. (2020). Supporting the teaching of scientific methods in practical science. *University of Oxford.*. https://www.education.ox.ac.uk/wp-content/uploads/2024/06/Teaching_scientific_methods.pdf
Yacoubian, H. A. (2020). Teaching nature of science through a critical thinking approach. In W. F. McComas (Ed.), *Nature of science in science instruction* (pp. 199–212). Springer. https://doi.org/10.1007/978-3-030-57239-6_10
Yuhas, A. (2016, June 2). Dagger in Tutankhamun's tomb was made with iron from a meteorite. *The Guardian.* https://www.theguardian.com/world/2016/jun/01/dagger-king-tut-tomb-iron-meteorite-egypt-mummy

**Agustín Adúriz-Bravo** holds a masters in Physics Education (University of Buenos Aires) and a PhD in Didactics of Science (Autonomous University of Barcelona). He is currently associate professor and superior researcher of CONICET at the CeFIEC Institute, University of Buenos Aires, Argentina. His area of research concerns the contributions of philosophy of science to science education, focusing on teacher education. He has published over 300 papers and chapters.

# Part II
# Physical Practical Work in Science Education

# Chapter 6
# Physical Laboratory-Based Practical Activities in School: A Literature Review centred on Students

**Sofia Morgado, Luís Dourado, and Laurinda Leite**

## 6.1 Introduction

For a long time, the physical laboratory was seen as the key tool for science and science education. Scientists relied on the research laboratory to solve research problems or gather evidence to support their theoretical propositions; science teachers relied on (or felt guilty for not being able to use) the school laboratory to teach science to their students. Although preferences may vary with science disciplines and school levels (Sharpe & Abrahams, 2020), some students, especially the younger ones (Abrahams, 2009), are eager to enter the laboratory to experience the thrill of being scientists. Many of them view science as a practical subject with visual or sonic spectacular 'experiments'. However, when the novelty wears off, students' enjoyment in the lab seems to decrease (Abrahams, 2009).

Science is taught not only because it is a practical subject but also because it is a theoretical subject of value for students (Millar, 1998; Osborne & Allchin, 2024). In this sense, science education is not just about watching or performing spectacular 'experiments'. It encompasses learning science, learning about science, and learning how to do science (Hodson, 1988). Each of these dimensions leads to the development of different competencies and has different requirements and dependencies on the laboratory, with some (namely the third) being more dependent on it than others.

A challenge of using the lab for teaching is that there is unseen science behind the spectacular observations. Hands-on or observational approaches may be appealing

S. Morgado (✉) · L. Dourado · L. Leite
CIEd-University of Minho, Braga, Portugal
e-mail: sofiamorgado@ie.uminho.pt

L. Dourado
e-mail: ldourado@ie.uminho.pt

L. Leite
e-mail: lleite@ie.uminho.pt



L. Leite et al. (eds.), *Practical Work in Science Education in the Digital Era*, Advances in Technology-Rich Science Education 2, https://doi.org/10.1007/978-3-032-13788-3_6

to students, but they are insufficient for them to learn science concepts and theories (Abrahams & Millar, 2008; Kluge, 2014). Appropriate guidance and debate of ideas are required (Howe & Smith, 1998; Kirschner et al., 2006; Solomon, 1999; Osborne & Allchin, 2024). Missing the link between observation and theories (students' and scientists' theories) impairs students from making sense of the data they collect through observation (Solomon, 1999; Abrahams & Millar, 2008). Those data may depend on students' previous experiences (Millar et al., 2002) and constitute an obstacle to learning, which may reduce their enjoyment of learning in the laboratory (Solomon, 1980), especially when this is no longer a novelty for them (Abrahams, 2009).

Some schools may not have lab facilities to allow students to perform 'experiments' in science. Although this may be frustrating for some students, it may not be necessary to reach some learning outcomes. Also, some 'experiments' may not require a lab; they can be performed in a classroom (given that safety conditions are ensured) and or rely on everyday materials (Gericke, 2023; Leite & Dourado, 2013). In addition, not all activities performed in a lab or using lab equipment and materials are, in fact, experiments (Wellington, 1998b), as some may not require hypothesis testing (Howe & Smith, 1998) and control and manipulation of variables (Wellington, 1998b). There is nothing wrong with it, but experimental and non-experimental activities promote the development of different competencies (Gericke et al., 2023). Differentiating those concepts helps to increase the match between the structure of the activity and the objectives to be attained.

The usual lab activities have progressively integrated digital technologies (e.g., sensors, graphic calculators with LaunchPad Boards, microscopes with digital cameras, etc.). Integrating technologies in the lab activities should not mean replacing physical labs. Rather, it means powering physical lab activities in terms of both learning potential and students' motivation (Dyrberg et al., 2017). Considering the educational potential and the evolution of physical school laboratories, it is worth exploring the extent to which laboratory-based science teaching actually makes a difference in students' education and enjoyment. This chapter presents a critical overview of how research on school practical laboratory-based activities, published in Scopus and or Web of Science-indexed journals and focusing on students, evolved and is influenced by school level. It also discusses future directions for lab-based science education and research in the digital era.

## 6.2 From Practical Work to Laboratory-Based Practical Activities

For the last 45 years, the words practical work and laboratory work have been used alternately in texts, including books (Solomon, 1980; Woolnough & Allsop, 1985; Woolnough, 1991; Wellington, 1998a; Leach & Paulsen, 1999; Psillos & Niedderer, 2002; Abrahams, 2011), related to teaching science in the school lab. All these books address the same object: laboratory activities as a tool for school science teaching and learning. One of the designations they use (practical work) has a broader scope than the other (laboratory work) and does not indicate the object it focuses on. Hodson (1988) argued that not all practical activities are lab activities. If one assumes that practical activities are a "planned intervention in the material world" (Abrahams, 2011, p.1), one can think of some activities that can take place in the physical science lab and others that do not occur within it (Hodson, 1988); some can be performed outside the classroom or the school, namely in open spaces, science centres, or industrial contexts (Leite & Dourado, 2013) and others can take place in a variety of labs that have flourished with educational technological advancements, including maker labs (Kyle et al. 2018), fabrication labs (Fonda & Canessa, 2016), robotics labs (Chiazzese et al., 2018), virtual reality labs (Ikhsan et al., 2020), and virtual labs (Dyrberg et al., 2017). However, most practical activities performed in those laboratories do not integrate physical laboratory tools. Some laboratories (e.g., virtual labs) may be more useful for enhancing physical lab activities in the pre-lab phase (Dyrberg et al., 2017) than for replacing them.

Although the authors of the reference books and papers mentioned above tried to clarify the meanings of the words they used and distinguish the concepts they encompass (e.g., Abrahams, 2011; Hodson, 1988; Wellington; 1998b), recent studies continue to use lab work and practical work as synonymous, even though, in some countries, preferences can be observed. Practical work is common in the UK and German contexts, and laboratory work is common in the USA (Hofstein, 2017). In other countries (e.g., Australia, Portugal, Spain, Sweden), the two words coexist, with some authors preferring one and others the other.

We might agree with Gericke et al. (2023) that laboratory work or practical laboratory work would be more appropriate than the other designations to name "practical science activities involving students manipulating and/or observing real objects, including empirical data collection within authentic science-teaching situations" (p. 246), which are the object of this chapter. Nevertheless, those designations may suggest that they address a single entity, while (as will be discussed in the next section) lab activities encompass different types of activities with a diversity of aims and structures. Therefore, we would like to argue that 'laboratory-based practical activities' or simply 'laboratory activities' would be more appropriate to address what students do when they manipulate and or observe real laboratory objects and materials (alone or integrated with digital technologies) or everyday materials, in a lab or a classroom.

In school contexts, experimental work and investigation are often used interchangeably with laboratory work and practical work. Hodson (1988) argued that some laboratory work is not experiments; doing experiments is engaging in closely controlled events (Hodson, 1988), and not all lab activities (e.g., measuring the pH of a solution) require control and manipulation of variables. This requirement underlies some school lab activities, but the use of a step-by-step, ready-to-follow worksheet often hinders it. The use of this kind of worksheet prevents students from engaging in the cognitively demanding and educationally valuable processes of control and manipulation of variables (Osborne, 2015).

Woolnough and Allsop (1985) considered that science practical work comprises investigations, which they conceptualize as problem-solving activities. They argue that carrying out an investigation requires, among others, the mastery of lab skills associated with exercises (e.g., operating a multimeter), which those authors consider another type of lab activity. An investigation is an open activity that starts with a question to be answered (Crawford, 2014); answering it requires designing and implementing a problem-solving strategy, and assessing the solution(s) obtained as well as the strategy used to reach it (if a solution exists). Some problems (e.g., what affects the speed of a chemical reaction) require control and manipulation of variables to be solved; others do not (e.g., what are the development stages that the bean plant goes throu gh from the time the bean is sown until it produces pods with new beans?), and therefore may be considered non-experimental investigations. In recent research, these concepts are not completely clear. Kluge (2014) uses the term experiments to refer to research projects; Kranz et al. (2023) mix the concepts of experiment and investigation, even though they do it in the context of inquiry-based learning.

Following the previous discussion, it can be argued that scientific experiments are investigations, but not all investigations are experiments (Leite & Dourado, 2013), in the sense that not all of them require hypothesis testing through control and manipulation of variables. Problem-solving and investigations can occur in the lab, but not all problems are solved in the lab (or with lab materials), and not all investigations take place in a science lab. Similarly, all science problem-solving activities are investigations, but some of them do not take place in the lab. Then, it is worth differentiating lab-based experiments, lab-based investigations, and lab-based problem-solving activities. Conceptual accuracy is relevant for highlighting what different activities demand from students and allow them to learn, which in turn depends on the degree of openness of the activity (Leite, 2001), and on the existence (or not) and the features of a lab worksheet.

## 6.3 Matching Laboratory Activities to Learning Objectives

Hodson (1988) argues that lab activities are necessary to learn how to do science, are not enough for learning science, and are inappropriate for learning about the nature of science. As no theory development path can be replicated in school, having students analyse the historical development of ideas would be more beneficial for learning about science than having them work in school labs (Hodson, 1994). However, the lab may allow students to develop some practical skills (Schwichow et al., 2016) required to understand how science is produced.

As far as learning science is concerned, students need to learn updated science content. To achieve this goal, they do not need to rediscover knowledge discovered by past scientists, and it would not be viable, because, as Millar (1998) points out, what they do in school does not compare to what scientists have done. In addition, lab activities show what happens but do not show why it happens (Woolnough & Allsop, 1985). Making sense of what happens and explaining why it happens requires creativity, imagination, and insight (Ogborn et al., 1996). This is because there is no direct path between experiments and theory. Rather, there is a complex interplay between them (Leach, 1999), with theory both informing experiments and emerging from them, and being influenced by the observer's previous ideas and expectations (Millar et al., 2002).

Learning how to do science encompasses mastering basic skills (e.g., preparing apparatus, operating equipment and measurement devices, and making a microscopic preparation) and becoming familiar with scientific methodology. Doing school lab activities is appropriate for developing practical skills (Schwichow et al., 2016), but it is not the same as doing science research (Hofstein & Lunetta, 2004); it only enables students to get a flavour of what doing science research is. Thus, this goal may be better reached through internships in research laboratories (Roth et al., 2009) or by engaging students in open-ended investigations (Guillon & Seré, 2002) or research projects (Lewis, 2002).

To analyse the educational potential of laboratory activities, it is useful to think about them as having three steps: before putting into practice the lab procedure (preparation phase), implementation of the lab procedure (dealing with apparatus), and after the implementation of the lab procedure (data analysis, conclusions, and critical reflection and assessment of what was done and achieved). As Osborne (2015) pointed out, "no practical work should be undertaken without a pre- and post-activity" (p. 23). Considering who performs the lab procedure, lab activities can be classified as: lab demonstrations, if the teacher implements the lab procedure and the students observe it actively (with minds-on); or students' lab activities (hands-on, and hopefully, mind-on), if students, individually or in small groups, perform it themselves (Millar et al., 2002) with or without guidance (e.g., a worksheet). Minds-on engaging demonstrations may be enough for students to learn science; hands-on student activities are required for them to develop technical (manipulative) skills; appropriate teacher mediation or guidance is needed for students to develop other competencies relevant to learn how to do science.

A lab activity may contribute to one or more learning outcomes. However, it is worth identifying the main intended learning outcome to design, perform, and assess it accordingly. Then, it is possible to classify lab activities according to the nature of the main learning outcome. Synthesizing and adding to the proposals of some authors, Leite and Dourado (2013) consider exercises, activities for getting a feeling of the phenomena, illustrative activities, activities to uncover what happens if/when…, Preview-Observe-Explain-Reflects (POER) with or without procedure given, and investigations. It is worth emphasizing that school science experiments are a category that arises from another classification criterion. They may intersect some of these categories, namely those related to learning conceptual knowledge and scientific methodology, depending on how open or guided the experiments are.

## 6.4 Methodology

To attain the research objective, a systematic literature review was conducted. The review followed the PRISMA (Preferred Reporting Items for Systematic reviews and Meta-Analyses) methodology (Page et al., 2021). PRISMA is useful to guide the reporting of why a systematic review was done, what methods were used, and what results were obtained. According to PRISMA, a process of systematic review should be organized into four phases: identification (searching databases to find records); screening (reviewing titles and abstracts for relevance); eligibility (assessing the full-text articles against inclusion/exclusion criteria); and inclusion (analysing the final set of studies included in the review) to extract data.

For the identification phase, a search of original research articles published in Scopus and or the Web of Science (WoS) databases was conducted. These databases are comprehensive (Pranckutė, 2021) in terms of the journals they include, and complementary (Kumpulainen & Seppänen, 2022) as some journals are indexed by one and not by the other. However, some journals are indexed by both databases (Pranckutė, 2021), and this overlap needs to be addressed in the next stage. Other databases are more comprehensive in indexing non-journal materials (Wilder & Walters, 2021), which are not considered in the type of literature review to be conducted. As the literature review concentrates on empirical research articles only, the two selected databases seem fully appropriate. In addition, they allow the use of convenient filters, which facilitate searching.

Regarding the search criteria, two sets of descriptors (one including words used to address lab activities, and the other including the science disciplines) were combined with the Boolean operator AND. From this combination, papers on non-physical labs, higher education, and non-science research areas were excluded using the Boolean operator NOT. Some additional filters were used, too. They were selected to provide a set of original empirical papers on school students' learning from and opinions on and reactions to physical laboratory activities, published in peer-reviewed journals, and written in English, French, Spanish, and Portuguese (which are the languages mastered by the authors). Research focusing on preschool children was disregarded to limit the revision to school (primary and secondary school). Exclusion of preschool and higher education is due to the large amount of research in the area and limitations of space in the chapter.

The following combinations of keywords, in all fields, were used:

- ("practical work" OR "practical activit*" OR "experimental work" OR "experimental activit*" OR "lab* work" OR "lab* activit*") AND
- ("science education" OR "science learning" OR "science teaching" OR "chemi* education" OR "Physics education" OR "Biolog* education" OR "Geolog* education" OR "chemi* learning" OR "Physics learning" OR "Biolog* learning" OR "Geolog* learning" OR "chemi* teaching" OR "Physics teaching" OR "Biolog* teaching" OR "Geolog* teaching") NOT
- (virtual OR "higher education" OR health OR online OR comput* OR engineering OR medic*).

No time limit was established, even though, before 2000, few papers were available online. For that period, papers with full text freely unavailable online were included if the hard versions of the journals were available from the university library only. The searches were done in the month of May, 2025.

The results of the search were 1224 titles for Scopus and 500 titles for the WoS. This led to a total of 1724 papers. As some journals are indexed in both databases, the first search follow-up action consisted of checking the papers' references to eliminate repetitions (n = 202). A total of 1522 papers remained for screening. Screening was based on a reading of the title and the abstract. When this was not enough to decide on the eligibility of the article, other parts of the paper (e.g., objectives and methodology) were read and submitted to the inclusion and exclusion criteria. Papers focusing on scientific research (n = 179), different disciplines (n = 30), other science issues (n = 461), higher education (n = 140), school textbooks (n = 23), pedagogical proposals (n = 106), or teachers (n = 105), as well as position papers/literature reviews (n = 107), and paper with full text unavailable (n = 228) were excluded. Ultimately, 143 journal articles were considered consistent with the inclusion criteria and remained for review. The two oldest articles reviewed were published in 1928, in the Journal of Chemical Education, and 1981, in Research in Science Education. The most recent articles were published in 2025 (two articles, one in Physical Review Physics Education Research and the other in Physics Education).

Eight questions (presented in the next section) guided the authors in identifying the relevant data to extract from the selected papers. An Excel table was created to record the data and facilitate later analysis. To start, the authors performed a joint analysis of four randomly selected articles to ensure that everyone was following the same rules. Afterwards, one of the authors analysed the first half of the papers, and another author analysed the second half. The three authors discussed some difficult cases to ensure the reliability of the analysis. Decisions were taken by consensus.

Data in the Excel file were used to draw tables containing data relevant to answering the research questions. The articles are grouped into three time periods to enable the uncovering of possible trends in the research. They are also separated by school level to allow some comparison of research carried out in primary and secondary schools. In doing so, this review contributes to the knowledge base, as other recent reviews (e.g., Gericke et al., 2023) do not simultaneously inform about possible changes over time and compare primary and secondary education.

## 6.5 Findings

This section is organized around the eight research questions that guided data extraction.

*Question 1: When and where were the research studies done?*

Considering three time periods, the percentages of papers are: 13, 3% before 2000; 39,2% between 2000 and 2012 (limits included), and 47,5% after 2012 (up to the month of May, 2025). Before 2000, only a small part of papers were available online, which means that the number of papers reviewed is small and below the number of papers published in that period. Despite that, there is a tendency for research to increase over the two more recent periods. Almost half of the papers report research carried out in Europe (46, 1%), mainly the UK and Sweden, followed by Asia (20,3%), especially China and Singapore, America (18,2%), primarily the USA, Oceania (9,1%), mainly Australia, and Africa (6,3%), spread by several countries. Research in Europe, America, and Africa shows a tendency to increase, whereas in Oceania it shows a tendency to decrease. Research in Asia did not change during the two most recent periods, and no paper focusing on Antarctica was reviewed. These results indicate that some continents and countries account for more research than others.

*Question 2: What school levels and samples were focused on in the research?*

The majority of the studies reviewed (80,4%) concentrated on secondary school (7th–12th grade), 60,1% of which dealt with upper secondary school (grades 10 to 12); 13,2% of the studies reviewed concentrated on primary school (grades 1 to 6), and 6,3% on both levels (in Asia, Europe, and Oceania). Considering the two most recent periods, there is a tendency for research to increase over time in either primary or secondary schools, with a duplication in the former case (5 to 10 studies) and a moderate increase (44 to 56 studies) in the latter case. Studies dealing with both grade levels decreased from 7 to 2. So far, primary school has attracted little attention from researchers, probably because it is not seen as a rich research environment since most primary schools do not have laboratories, or little science is taught at this level.

Concerning samples, 25,9% of the studies include small samples, lower than or equal to 30 students, being 9,1% lower than 10 students; 32,2% use samples between 31 and 120 students (limits included); 9,8% use samples higher than 550; seven studies do not make explicit the dimension of the sample and the others are in the interval 121–550 students. The largest sample (n = 5194) was used to study secondary school students' views on several teaching approaches, including practical work. The smallest sample (n = 2) was used in a qualitative study, focusing on how students interact when working in a lab. The primary school studies use samples between 11–30 and 261–350 students, and half of them took place within a single school. As for secondary school, 42,6% of the studies took place in a single school, but about one third involved two or more schools, and others took place at the national or international levels; some studies do not provide clear information on where data were collected; large sample studies seem to show a decreasing tendency. These results indicate that most studies were obtained in a specific context with reduced samples, which limits the external validity of the studies.

*Question 3: What are the aims of the research studies reviewed, and to what extent did their aims change over time?*

About half of the studies (49,6%) focus on the affective dimension (i.e., attitudes, motivation, interest), 35,7% on conceptual learning, and 23,8% on procedural learning (e.g., using measuring instruments, operating equipment, etc.); only 9,8%

focus on epistemological issues, namely on the features of scientific inquiry. Some studies deal with one of these issues, but others deal with two or more issues. There seems to be a slight tendency for studies focusing on the affective or procedural dimensions to be increasing over time, in secondary school; no clear tendency is observed in primary school. The affective dimension is important because it interferes with learning. As there is quite a lot of consensual research on the topic, it may be worth concentrating efforts on other targets, namely in primary school.

*Question 4: What science disciplines and themes were focused on in the studies reviewed?*

General sciences (33,6%), chemistry (25,9%), physics (16,1%), and biology (11,9%) are the disciplines that concentrated more studies. Geology and STEM were the object of one and two studies, respectively. A few studies (4,9%) dealt with several disciplines. Primary school studies tend to concentrate on general science. Considering the number of papers in the different periods, an increasing tendency is noted for physics only. Therefore, research efforts should focus on the less researched disciplines (like Geology/Earth Sciences) and the recent STEM area. Research should also be undertaken to get comparative views of the different school levels and disciplines.

Concerning science themes dealt with by studies focusing on conceptual issues (i.e., science learning), there is a variety of themes, some of them common to primary and secondary schools, like plants, animals, kinematics, optics, electricity, thermodynamics, materials, chemical reactions, acids and bases, and rocks. Some studies concentrating on the affective dimension focus on the relationship between lab activities and science, in general (e.g., interest in science), or one of the themes referred to above (e.g., interest, motivation, or attitudes). Argumentation and social interaction were studied within the context of themes like chemical reactions and acids and bases. The science themes addressed are basic themes that may be common to science curricula worldwide. They are complemented with affective themes that are relevant but traditional.

*Question 5: What research paradigms were adopted in the research studies reviewed, and to what extent did their aims change over time?*

Concerning the research paradigm, 28, 0% of the studies reviewed are qualitative and 68,5% are quantitative studies, the remaining being mixed methods studies. About half of the articles (52, 4%) report on the use of statistical tests, and 14% used descriptive statistics. Comparing the percentage of qualitative and quantitative studies in the different periods, there is a tendency for quantitative research to increase. The majority (104, that is 72,7%) of the articles reviewed report research dealing with lab-based teaching interventions, in one or more groups, approached qualitatively or quantitatively; a part of the 104 studies (28,8%) used inferential statistics to compare groups and assess the effect of the intervention. Qualitative research draws mainly on content analysis of students' productions (e.g., reports) and interactions during group work. A small part of the studies dealing with conceptual learning used statistical analysis to conclude about the impact of the interventions. This fact weakens the research conclusions and the external validity of the studies.

*Question 6: What lab-based approaches were addressed in the studies, and to what extent did they change over time?*

Of the 104 studies that include teaching interventions, nine papers (8,6%) state that the lab procedure was used to illustrate/confirm concepts or ideas previously taught to students and other 5,8% report on the use of exercises (e.g., to learn how to operate the Bunsen burner or light microscope); many papers (41,3%) report on the use of investigations, 13,5% used guided worksheets leading to the intended result, and 13,5% used POER-based worksheets, to promote students' conceptual change. The remaining do not provide enough information for the reader to ascertain the type of activity used. There is a good level of papers focusing on investigations, but research may show a picture that does not correspond to what is going on in the normal classroom.

Only 13 (out of 104) papers mentioned the use of digital tools (e.g., Arduino, cellular phone, and sensors), and 19 mentioned the use of daily life materials (e.g., toothpaste, bottles, and credit cards), alone or combined with lab materials. Even though it is highly probable that in the other studies, only traditional lab material was used, 25% of the articles involving teaching interventions do not provide clear information on that. In most studies (55,8%), students implemented the procedure in small groups, and in 15,4% they did it individually; about one fourth of the papers (23,1%) do not make it clear who implements de procedure; in the remaining cases (5,8%), it is the teacher or the teacher with some students that do it, as a demonstration for the class. Before 2000, no reference to individual students' implementation of the lab procedure was found; the same applies to implementation by the teacher, after 2012. Assuming that activities continue being performed, this indicates a big role change that leads students to the centre of the teaching and learning process, which seems to be a promising result.

*Question 7: What and how were data collected in the studies reviewed?*

The majority of the papers reviewed (59,4%) used a single technique to collect data; the remaining used combinations of two, three, or even four techniques: questionnaire/knowledge test, interview, observation, and analysis of documents, namely lab reports. Opinion, motivation, or interest questionnaires and knowledge tests were the types of data collection instruments most used, both in the whole set of reviewed articles (63,6%) and in the subset of papers (n = 104) reporting an intervention study (57,7%); 41,3% of the articles reporting an intervention used knowledge tests. Some studies used: interviews to better understand students' opinions about the relationship between lab and science and attitudes towards lab-based classes; observation to study, for instance, students' interaction within small work and with teacher, students´ argumentation, and how students perform a lab technique (e.g., dissecting a calf eye); and document analysis, namely lab reports, to study students' argumentative skills, or mastery of scientific methodology. Various techniques were used alone or in combination with two or more techniques; unfortunately, interviews were not used in the intervention studies reviewed. They would enable researchers to go deeper into the issues under study. Also, several intervention studies missed pre–/post-test and or experimental/control group comparison, which are severe weaknesses of the analysis.

*Question 8: What findings were obtained with lab-based intervention studies, and to what extent do they depend on time and school level?*

Most studies dealing with science learning (51 out of 58) report positive effects of the lab-based interventions. However, only 13 of them included control and experimental groups, with the experimental groups achieving more conceptual learning, showing more retention, or changing their alternative conceptions in a more extensive way than the control groups, irrespective of school level and period of time. Four (out of 58) studies report negative effects attributed to the resistance to changing the usual lab practice. Most experimental interventions are characterized by student-centred lab use, based on the implementation of activities with different levels of openness (but with the predominance of investigations) that require students to make decisions under the teacher's supervision and guidance; control group interventions include closed worksheet-based activities that guide students to the intended answer. Most studies omit details on the pedagogical approach adopted, which impairs an understanding of the lab procedure and the way it is put into practice, as well as why, for example, there was no difference between groups (two studies) or no evolution (one study).

Almost all studies dealing with procedural knowledge indicate that students learn from investigations (40 out of 43), even though only four of them use an experimental design with a control group. However, they indicate that students show some difficulties in planning lab procedures, manipulating materials (especially in primary school), analysing data, and even following a lab worksheet; no gender differences were reported. The remaining three studies report a negative effect of the interventions on students' procedural learning, which was attributed to the novelty of the investigation-based interventions for students and their lack of procedural knowledge related to planning procedures, using equipment, and analysing data.

Concerning the epistemological dimension, no study reports regression, but the two (out of six) studies that used a control and experimental group design report no evolution. Participants in the single-group studies held traditional views of the nature of science, independent of gender; in constructivist lab environments, they developed an understanding of the central role of ideas in the learning process, and were able to test and revise their ideas. However, details on measurement and the combination of techniques used to collect data are seldom provided, which prevents the reader from making a fair judgment of the robustness of the research reported.

The majority of the studies dealing with the affective dimension (39 out of 43) report positive effects of the interventions on students' opinions, interest, enjoyment, etc. However, only 13 (out of 39) used a research design that enables them to show that students in the experimental groups became more favourable to practical activities, increased their motivation and interest, and improved their attitudes towards science. Two single-group studies report a negative effect of the interventions, caused by students' dislike of performing investigations due to a lack of guidance. Even though studies show that students tend to enjoy and prefer lab-based approaches over lecture-based ones, they also suggest that older students struggle to see the relevance of laboratory activities to daily life. This means that lab activities may need to be contextualized to make more sense for students.

## 6.6 Conclusion and Ways Forward

This literature review suggests that researchers' interest in lab-based practical activities as an educational tool for students has been increasing, both in primary and secondary schools. Despite small differences between school levels and periods, researchers report positive effects on learning science, learning about science, and learning how to do science, as well as on the affective dimension. Some studies focusing on the affective dimension used very large samples and relied on data collection procedures consistent with the research problems, which enables their authors to reach a generalizable conclusion; others drew on small samples for a deep understanding and reached their aim; others omitted information relevant to the reader to judge about their internal consistency. A considerable amount of the intervention studies missed pre–/post-test comparison and/or used quite small convenience samples, made of one or a few classes, which reduces their scientific significance even when statistical significance was achieved. Some of those pieces of research also lack detailed information on the teaching approach(s), namely, on the features of the lab activity, who implemented the lab procedure, and whether it was implemented consistently with the intended learning goals (in a few cases, authors assume it was not), or the data collection techniques and instruments and/or conditions they were applied.

Before 2000, few papers were accessible online, a fact that conditioned the analysis of the evolution of the features of the research. The reduced number of studies focusing on primary school also imposes some conditioning on the comparison between school levels. The screening of the articles may have led to some undue exclusion. However, this review suggests that lab-based activities may be a valuable educational tool, especially those with a higher degree of openness, as they have more power to activate students' motivation and to promote integrated learning of the diverse dimensions of science education, given that the teacher provides appropriate guidance (Kirschner et al., 2006) to them, according to school level and students' use to lab activities. Digital technologies, mentioned in a reduced number of studies, should be more present in the school lab. If integrated with traditional lab material, digital technologies may promote students' motivation and educational potential of the lab-based activities, namely for students with disabilities. In this case, success may depend on how technology selection and use match the demands of students' disabilities (Mikropoulos & Iatraki, 2023). In any case, care should be taken to avoid digital technologies transforming lab activities into a sort of black box that overrides natural phenomena. A good balance between traditional lab equipment, digital technology, and everyday materials may strengthen the link between science and daily life, with digital technologies powering it (Dyrberg et al., 2017; Kluge, 2014) or releasing students from mechanical tasks. Additionally, digital technologies can promote sustainability by reducing the amount of laboratory materials used. However, this should not be done at the expense of developing important manipulative skills, which require hands-on activities. Their integration may promote a more balanced science education while making it inclusive not only for low achievers but also for gifted students, who are often neglected. However, the integration of digital technologies in the school's physical lab needs to be studied in relation to relevant overarching pedagogical approaches and considering the existence of different types of lab activities. Mixed-methods research should be acknowledged to study the impact of that integration, as quantitative approaches provide a big picture, and qualitative approaches allow a deep understanding of how the integration of those technologies can promote students' learning, especially for those with some sort of disability.

It may not be worth doing 'one more study' about students' attitudes, interests, or motivation towards lab activities, but it may be worth understanding why their interest in lab activities tends to decrease with schooling (Sharpe & Abrahams, 2020). The (de)contextualization of lab activities, the relationship between lab activities and assessment practices, including national exams, and higher education admission requirements, may need to be carefully analysed. This may help to reorient physical lab-based practical classes not just to please students but to create a learning environment that uses technologically enriched lab activities when they are valuable for specific and explicit purposes, in such a way that students may perceive their relevance and fully engage in them, with minds, hands, and hearts-on, for the benefit of their science education.

**Acknowledgments** Thanks are due to CIEd, Research Centre on Education of the Institute of Education of the University of Minho (UID/01661/2025), which funded this work through national funds of FCT/MCTES-PT, namely the Pluriannual Program for the Funding of R&D Units.

## Supplementary Material

The list of articles reviewed and tables with data that support the findings section are available from https://doi.org/10.34622/datarepositorium/Z6IXMH.

## References

Abrahams, I. (2009). Does practical work really motivate? A study of the affective value of practical work in secondary school science. *International Journal of Science Education, 31*(17), 2335–2353. https://doi.org/10.1080/09500690802342836

Abrahams, I. (2011). *Practical work in secondary science.* Continuum International Publishing Group.

Abrahams, I., & Millar, R. (2008). Does practical work really work? A study of the effectiveness of practical work as a teaching and learning method in school science. *International Journal of Science Education, 30*(14), 1945–1969. https://doi.org/10.1080/09500690701749305

Chiazzese, G., Arrigo, M., Chifari, A., Lonati, V., & Tosto, C. (2018). Exploring the effect of a robotics laboratory on computational thinking skills in primary school children using the Bebras Tasks. *TEEM'18: Proceedings of the Sixth International Conference on Technological Ecosystems for Enhancing Multiculturality* (pp. 25–30). doi:https://doi.org/10.1145/3284179.3284186

Crawford, B. (2014). From inquiry to scientific practices in the science classroom. In N. G. Lederman & S. K. Abell (Eds.), *Handbook of research on science education* (Vol. 2, pp. 515–541). Routledge.

Dyrberg, N., Treusch, A., & Wiegand, C. (2017). Virtual laboratories in science education: Students motivation and experiences in two tertiary biology courses. *Journal of Biological Education, 51*(4), 358–374. https://doi.org/10.1080/00219266.2016.1257498

Fonda, C., & Canessa, E. (2016). Making ideas at scientific fabrication laboratories. *Physics Education, 15*(6). https://doi.org/10.1088/0031-9120/51/6/065016

Gericke, N., Högström, P., & Wallin, J. (2023). A systematic review of research on laboratory work in secondary school. *Studies in Science Education, 59*(2), 245–285. https://doi.org/10.1080/03057267.2022.2090125

Guillon, A., & Seré, M.-G. (2002). The role of epistemological information in open-ended investigative labwork. In D. Psillos & H. Niedderer (Eds.), *Teaching and learning in the science laboratory* (pp. 121–138). Kluwer Academic Publishers.

Hodson, D. (1988). Experiments in science and science teaching. *Educational Philosophy and Theory, 20*(2), 53–66.

Hodson, D. (1994). Hacia un enfoque más crítico del trabajo de laboratório [Towards a more critical approach to laboratory work]. *Enseñanza de las Ciências, 12*(3), 299–313.

Hofstein, A., & Lunetta, V. (2004). The laboratory in science education: Foundations for the twenty-first century. *Science Education, 88*(1), 28–54. https://doi.org/10.1002/sce.10106

Hofstein, A. (2017). The role of laboratory in science teaching and learning. In K. S. Taber & B. Akpan (Eds.), *Science education* (pp. 357–368). Sense Publishers.

Howe, C., & Smith, P. (1998). Experimentation and conceptual understanding in school science. In J. Wellington (Ed.), *Practical work in school science: Which way now?* (pp. 220–236). Routledge.

Ikhsan, J., Sugiyarto, K., & Astuti, T. (2020). Fostering student's critical thinking through a virtual reality laboratory. *iJIM, 14*(8), 183–184. https://doi.org/10.3991/ijim.v14i08.13069

Kirschner, P., Sweller, J., & Clark, R. (2006). Why minimal guidance during instruction does not work: An analysis of the failure of constructivist, discovery, problem-based, experiential, and inquiry-based teaching. *Educational Psychologist, 41*(2), 75–86. https://doi.org/10.1207/s15326985ep4102_1

Kluge, A. (2014). Combining laboratory experiments with digital tools to do scientific inquiry. *International Journal of Science Education, 36*(13), 2157–2179. https://doi.org/10.1080/09500693.2014.916456

Kranz, J., Baur, A., & Möller, A. (2023). Learners challenges in understanding and performing experiments: A systematic review of the literature. *Studies in Science Education, 59*(2), 321–367. https://doi.org/10.1080/03057267.2022.2138151

Kumpulainen, M., & Seppänen, M. (2022). Combining web of science and Scopus datasets in citation-based literature study. *Scientometrics, 127*, 5613–5631. https://doi.org/10.1007/s11192-022-04475-7

Kyle, A., Carpezza, M., & Covich, C. (2018). Hk maker lab: An engineering design summer program for high school students. *Journal of STEM Outreach, 1*(1). https://doi.org/10.15695/jstem/v1i1.5

Leach, J., & Paulsen, A. (1999). *Practical work in science education.* Roskilde University Press.

Leach, J. (1999). Students understanding of the co-ordination of theory and evidence in science. *International Journal of Science Education, 21*(8), 789–806. https://doi.org/10.1080/095006999290291

Leite, L., & Dourado, L. (2013). Laboratory activities, science education and problem-solving skills. *Procedia-Social and Behavioral Sciences, 106*, 1677–1686. https://doi.org/10.1016/j.sbspro.2013.12.190

Leite, L. (2001). Contributos para uma utilização mais fundamentada do trabalho laboratorial no ensino das ciências. In H. Caetano & M. Santos (Org.). *Cadernos Didácticos de Ciências.* Ministério da Educação-Departamento do Ensino Secundário, 79–97.

Lewis, J. (2002). The effectiveness of mini-projects as a preparation for open-ended investigations. In D. Psillos & H. Niedderer (Eds.), *Teaching and learning in the science laboratory* (pp. 139–150). Kluwer Academic Publishers.

Mikropoulos, T., & Iatraki, G. (2023). Digital technology supports science education for students with disabilities: A systematic review. *Education and Information Technologies, 28*, 3911–3935. https://doi.org/10.1007/s10639-022-11317-9

Millar, R. (1998). Rhetoric and reality: What practical work in science education is really for. In J. Wellington (Ed.), *Practical work in school science: Which way now?* (pp. 16–31). Routledge.

Millar, R., Tiberghien, A., & Le Maréchal, J. (2002). Varieties of labwork: A way of profiling labwork tasks. In D. Psillos & H. Niedderer (Eds.), *Teaching and learning in the science laboratory* (pp. 9–20). Kluwer Academic Publishers.

Ogborn, J., Kress, G., Martins, I., & McGillicuddy, K. (1996). *Explaining science in the classroom.* Open University Press.

Osborne, J. (2015). Practical work in science: Misunderstood and badly used? *School Science Review, 96*(357), 16–24.

Osborne, J., & Allchin, D. (2024). Science literacy in the twenty-first century: Informed trust and the competent outsider. *International Journal of Science Education, 1–22.* https://doi.org/10.1080/09500693.2024.2331980

Page, M., Moher, D., Bossuyt, P., Boutron, I., Hoffmann, T., Mulrow, C., Shamseer, L., Tetzlaff, J., Akl, E., Brennan, S., Chou, R., Glanville, J., Grimshaw, J., Hróbjartsson, A., Lalu, M., Li, T., Loder, E., Mayo-Wilson, E., McDonald, S.,.... McKenzie, J. (2021). PRISMA 2020 explanation and elaboration: Updated guidance and exemplars for reporting systematic reviews. *BMJ, 372*(160), 1–36. https://doi.org/10.1136/bmj.n160

Pranckutė, R. (2021). Web of ˙science (WoS) and Scopus: The titans of bibliographic information in today's academic world. *Publications, 9*(12), 1–59. https://doi.org/10.3390/publications9010012

Psillos, D., & Niedderer, H. (2002). Issues and questions regarding the effectiveness of labwork. In D. Psillos & H. Niedderer (Eds.), *Teaching and learning in the science laboratory* (pp. 21–30). Kluwer Academic Publishers.

Roth, W.-M., van Eijck, M., Hsu, P.-L., Marshall, A., & Mazumder, A. (2009). What high school students learn during internships in biology laboratories? *The American Biology Teacher, 71*(8), 492–496. https://doi.org/10.1662/005.071.0808

Schwichow, M., Zimmerman, C., Croker, S., & Hartig, H. (2016). What students learn from hands-on activities. *Journal of Research in Science Teaching, 53*(7), 980–1002. https://doi.org/10.1002/tea.21320

Sharpe, R., & Abrahams, I. (2020). Secondary school students attitudes to practical work in biology, chemistry and physics in England. *Research in Science & Technological Education, 38*(1), 84–104. https://doi.org/10.1080/02635143.2019.1597696

Solomon, J. (1980). *Teaching children in the laboratory*. Croom Helm.
Solomon, J. (1999). Envisionment in practical work. Helping pupils to imagine concepts while carrying out an experiment. In J. Leach & A. Paulsen (Eds.), *Practical work in science education* (pp. 60–74). Roskilde University Press.
Wei, B., Jiang, Z., & Gai, L. (2022). Examining the nature of practical work in school science textbooks: Coverage of the diversity of scientific methods. *Science & Education, 31*, 943–960. https://doi.org/10.1007/s11191-021-00294-z
Wellington, J. (1998a). *Practical work in school science: Which way now?* Routledge.
Wellington, J. (1998b). Practical work in science: Time for a reappraisal. In J. Wellington (Ed.), *Practical work in school science: Which way now?* (pp. 3–15). Routledge.
Wilder, E., & Walters, W. (2021). Using conventional bibliographic databases for social science research: Web of science and Scopus are not the only options. *Scholarly Assessment Reports, 3*(1), 1–17. https://doi.org/10.29024/sar.36
Woolnough, B., & Allsop, T. (1985). *Practical work in science*. Cambridge University Press.
Woolnough, B. (1991). *Practical Science*. Open University Press.

**Sofia Morgado** graduated as a physics and chemistry teacher, and earned a masters and a PhD in science education (University of Minho). She has been engaged in science education research projects as well as in curriculum development projects, at the University of Minho, Portugal. Author of several papers and book chapters, her research interests concentrate on science teaching and learning, with a focus on problem-based learning and contextualized science learning.

**Luís Dourado** is an assistant professor of science education at University of Minho, Portugal, and a CIEds researcher. He graduated as a biology and geology teacher, and earned a masters and a PhD in science education. His research interests focus on science teaching and learning, namely on fieldwork, problem-based learning, and ecoethics. He is a member of the editorial committee of a few journals, including the Revista Portuguesa de Educação.

**Laurinda Leite** is a full professor of science education and a CIEds researcher at University of Minho, Portugal. After graduating as a physics and chemistry teacher, she earned a masters (University of London) and a PhD (University of Minho) in science education. She coordinates projects, supervises, and conducts research and has published works in science teaching and learning, namely in the history of science, lab work, and problem-based learning.

# Chapter 7
# Inquiry-Based Practical Work in Science Education

**Mónica Baptista and Jairo Ortiz-Revilla**

## 7.1 Introduction

Recognizing the importance of science in our lives, several authors emphasize the need to develop a scientific attitude in students, that is, the skill to critically analyze the knowledge produced, distinguish between scientific knowledge and common-sense knowledge, and evaluate different claims by examining their limitations and strengths, among other aspects (Rudolph, 2024). This implies not only developing knowledge of science (concepts, facts, and theories that make up a fundamental body of scientific knowledge), but also knowledge about science (understanding how science works, how knowledge is constructed, and the nature of scientific knowledge itself), which enables students to adopt a critical stance toward what is presented to them and to make informed decisions (McComas et al., 2020). To achieve these goals, it is important to create teaching and learning situations in which students take an active role in constructing their own knowledge, while also being challenged in their usual ways of reasoning, arguing, and gathering and using information to think and produce knowledge (Arega et al., 2025).

Practical work has evolved throughout the history of Science Education (Akuma & Callaghan, 2018). Its origins trace back to the 19th century, when science subjects began their systematic integration into school curricula across various countries (Çalik & Karataş, 2023; DeBoer, 2006). For science to become an integral part of the school curriculum, it had to be presented as a subject distinct from others. Indeed, it began with observations that led to general principles unlike subjects such as Grammar and Mathematics.

---

M. Baptista (✉)
UIDEF, Institute of Education, University of Lisbon, Lisbon, Portugal
e-mail: mbaptista@ie.ulisboa.pt

J. Ortiz-Revilla
University of Burgos, Burgos, Spain
e-mail: jortizr@ubu.es



L. Leite et al. (eds.), *Practical Work in Science Education in the Digital Era*, Advances in Technology-Rich Science Education 2, https://doi.org/10.1007/978-3-032-13788-3_7

Initially, practical work served to confirm concepts previously introduced by the teacher in a deductive manner. This traditional perspective treated practical work as an extension of theoretical lessons, which were mostly focused on content delivery, and where practical activities played a secondary role (Toplis & Allen, 2012). More recently, the processes and skills involved in science have gained importance and have been incorporated into science curricula (Watson et al., 2004), as seen in the U.S. curricula of the 1990s (National Research Council [NRC], 1996), where there is a strong emphasis on inquiry-based practical work. This approach has since spread to the curricula of other countries such as Canada, Spain, Portugal, Taiwan, among others (García-Carmona, 2020; Leite, 2001).

Unlike traditional practical work - which often involves following fixed procedures to confirm known results - inquiry-based practical work encourages students to formulate questions, design experiments, collect and analyze data, and draw evidence-based conclusions. As Abrahams and Reiss (2012) emphasize practical work only becomes truly effective when it engages the mind as well as the hands.

The concept of inquiry-based practical work has evolved over time and has taken on a variety of definitions. Today, despite the different perspectives among authors, there is a consensus that it involves the use of scientific knowledge and science processes, enabling students to develop both procedural and conceptual understanding (Ramaila, 2024). It can help students learn how to do science and learn about science, while also promoting questioning, planning, evidence collection, evidence-based explanations, and communication. Therefore, it is a highly valuable educational experience in today's world, as it prepares students with tools to face various situations in their daily lives (Akuma & Callaghan, 2018).

This chapter comprises three parts. The first part includes a discussion about how inquiry-based practical work is defined today and its role in Science Education. The second part focus on its implementation within the school context, namely on the key challenges teachers encounter (including students' assessment) and the main benefits for students' learning. The third part of the chapter provides a conclusion and ways forward.

## 7.2 Defining Inquiry-Based Practical Work in Science Education

Inquiry-based practical work has received increasing attention in the Science Education community over the last four decades, being defined in different ways by various authors (Crawford, 2014). Through an extensive analysis of approximately 1500 books and scientific journals, Aulls and Shore (2008) identified that, beginning in the 1960s, studies most commonly associated the terms "discovery" and "problem solving" with inquiry-based practical work. The term "discovery" was used to denote activities involving searching, inventing, locating, finding, uncovering mysteries, and materializing concepts. In contrast to problem solving, discovery appears more intuitive and mysterious, seemingly dependent on everyday experience. Studies that connected inquiry-based practical work to the concept of problem emphasized the presence of problematic questions that students must resolve through various scientific processes.

In the educational literature, inquiry-based practical work assumes various forms depending on the degree of openness and guidance involved. Schwab (1960) identified three distinct levels of inquiry: structured inquiry, in which both the research question and the method are provided by the teacher; guided inquiry, where the teacher provides the question but not the method; and open inquiry, in which students are responsible for formulating both the question and the method. Along similar lines, Wellington (2000) characterizes different types of inquiry-based practical activities that vary in terms of openness and teacher guidance. These tasks can differ significantly: some have clearly defined answers, while others do not; some may be completed within minutes, whereas others require extended periods of time; and while some involve abstract scenarios, others are embedded in real-world contexts. The author further emphasizes that inquiry-based work exists on a continuum ranging from teacher-driven to student-driven approach. At one end of the spectrum, teachers formulate the guiding questions; at the other, students take the lead in designing their own investigative questions.

Several authors emphasize that when students engage in inquiry-based practical work, they identify problems using logical and critical thinking while considering alternative explanations. This approach actively involves students in defining their own path toward finding one or more solutions. It has significant potential to promote students' understanding of phenomena and the development of skills (Carlson et al., 2003). Indeed, inquiry-based practical work involves the exploration of materials and the natural world. This process is student-led, driven by curiosity, interest, and perseverance to understand and solve problems. Students learn through questioning and making predictions, formulating hypotheses, conducting experiences and developing models (Ash & Klein, 2000).

The perspective of Ash and Klein (2000) aligns with that of the National Research Council (NRC, 2012), which defines inquiry-based practical work as scientific practices encompassing multifaceted tasks, including: making observations; posing questions; conducting research using books and other information sources; planning investigations; doing experiences; reviewing existing knowledge about the experience; using tools to analyze and interpret data; exploring, predicting, and responding to questions; and communicating results. Additional characteristics of inquiry-based practical work include student engagement with scientific questions, prioritizing evidence to answer these questions, using evidence to develop explanations and connect them to scientific knowledge, and communicating and justifying those explanations. These characteristics place students at the center of their learning experience, value scientific activity through the development of scientific explanations, and support argumentation and communication skills.

Lederman (2006) argues that inquiry-based practical work encompasses traditional scientific processes while requiring the integration of these processes with scientific knowledge, reasoning, and critical thinking. He defined inquiry-based practical work in two dimensions: do inquiry and know about inquiry. Do inquiry refers to students' skills to carry out scientific investigation, which includes formulating investigable questions, designing and conducting experiments or studies, collecting and analyzing data, drawing conclusions based on evidence, and communicating results. Do inquiry involves actively engaging in the scientific processes, as a scientist would. Know about inquiry refers to the theoretical and conceptual understanding of what scientific investigation is, including understanding that scientific inquiry does not follow a fixed set of steps, recognizing that the approach depends on the specific question being asked, and knowing that there are different types of investigation (descriptive, correlational, experimental). It also involves understanding that real science is more varied and flexible than the simplified image often presented in textbooks and the media.

Despite these potential benefits for student learning, for inquiry-based practical work to effectively contribute to knowledge acquisition and stimulate student engagement, several conditions must be met: discovering what students already know about the natural world and using those ideas as starting points for investigations; employing students' existing ideas as baselines for the growth of their knowledge; inviting students to share their ideas, interests, questions, and suggestions throughout the investigative process; helping students establish connections between their ideas and external world phenomena; and assisting them in reflecting on their learning (Carlson et al., 2003).

Maconi et al. (2008) discuss inquiry-based practical work in relation to processes, content, strategies, and contexts. Regarding processes, inquiry-based practical work allows students to be guided by their own curiosity, interest, and ability to understand observations or solve problems. Students begin by questioning what intrigues and surprises them, then relate content to strategies for addressing the questions posed. Through this process, they develop transferable procedural skills applicable to other contexts, with the learning of processes being directed and controlled by the students themselves. As for content, it serves as a vehicle for developing those processes. Inquiry-based practical work facilitates content learning and its application by relating it to the phenomenon under study. Regarding strategies, students may find their own ways to sequence processes. They rarely follow a linear progression - often revisiting earlier stages, reformulating questions, and transitioning from predictions to hypotheses - thereby expressing their thinking processes. Finally, in terms of context, inquiry-based practical work promotes information seeking and encourages connections with other members of the school community. Therefore, it is essential for teachers to create stimulating environments with accessible resources both within and beyond the school setting.

## 7.3 Implementation of Inquiry-Based Practical Work Within the School Context

The effectiveness of implementing inquiry-based practical work in Science Education has been the focus of studies over the past decades, demonstrating that it is contingent upon a variety of contextual factors, such as appropriate guidance, collaborative learning, and authentic contexts (Lazonder & Harmsen, 2016). The effectiveness of appropriate guidance relies on teachers' skills to recognize the moments when students require their intervention and support in order to make progress in the tasks, while at the same time preserving the inquiry process and students' autonomy, as well as preventing frustration when students are unable to advance (Van de Pol et al., 2013). It is therefore important to establish a balance-particularly when students are not yet familiar with inquiry-based practical work-between teacher guidance and direct instruction, as necessary (Lazonder & Harmsen, 2016). In this regard, Martin-Hansen (2002) has already pointed out that students' previous experience with practical activities influences their ability to participate effectively in more open and autonomous inquiry processes, indicating that as students gain more experience and confidence, they can progress toward more open forms of inquiry, where they have greater control over the research process. Collaborative learning is another factor that contributes to the effectiveness of inquiry-based practical work (Johnson & Johnson, 2009). When tasks involve working collaboratively, students are required to discuss, interact, negotiate meanings, and share knowledge construction, benefiting from multiple perspectives to solve complex problems (Dillenbourg, 1999). Similarly, authentic contexts are also associated with the effectiveness of inquiry-based practical work. When tasks establish explicit connections with real-world problems and students' lived experiences, inquiry-based approaches encourage students to explore phenomena that resonate with their interests and experiences, making learning more relevant by providing an answer to the question why it is necessary to learn science (Belova et al., 2015; Gilbert, 2006).

The implementation of inquiry-based practical work within school contexts has been consistently associated with a range of challenges for teachers, including insufficient preparation for inquiry-based instruction, difficulties in balancing guidance with students' autonomy, and persistent uncertainties regarding assessment practices. In addition, structural constraints such as limited instructional time, rigid curricular frameworks, and restricted access to appropriate materials pose further barriers to its effective implementation.

Akuma and Gaigher (2021) focused on contextual challenges related to the implementation of inquiry-based practical work. Through a systematic review, they identified obstacles teachers face across four instructional phases: initiation, planning, implementation, and summative evaluation. For instance, they noted that teachers often struggle with designing open-ended tasks, facilitating meaningful reflection, and assessing students' inquiry processes. Despite these difficulties, the review confirms that inquiry-based practical work strengthens scientific literacy and increases student engagement when contextualized within real-world issues.

In another contribution, Akuma and Callaghan (2019a) examined the intrinsic teaching challenges of inquiry-based practical work, categorizing them according to similar instructional phases. They highlight persistent issues such as teacher resistance to inquiry-based methods, lack of pedagogical knowledge to structure open inquiry, and difficulty assessing student outcomes. However, their review also supports the notion that overcoming these challenges can foster deeper conceptual understanding and promote critical thinking skills among students.

As indicated, insufficient teacher preparation is one of the most frequently cited barriers to the successful implementation of inquiry-based practical work. Lunetta et al. (2007) emphasize that many educators feel inadequately trained to lead open-ended inquiry activities, impacting their confidence and classroom effectiveness. Moreover, balancing student autonomy with teacher guidance remains a delicate pedagogical challenge, as Colburn (2000) explains, requiring teachers to skillfully scaffold learning without taking control. Traditional assessment methods often fall short in capturing the multifaceted competencies students develop during inquiry, making evaluation a persistent difficulty (Hofstein & Lunetta, 2004). Time constraints and rigid curricular demands limit opportunities for in-depth experiences (Millar, 2004). Additionally, limited access to appropriate materials and insufficient institutional support are commonly reported obstacles that reduce teachers' motivation (Wieman et al., 2008). Structural and cultural barriers embedded within school systems further complicate adoption. Even though teacher education programs have made advances, systemic issues such as curriculum pressure and lack of time continue to challenge implementation (Blanchard et al., 2010). Poorly equipped laboratories also undermine the quality of practical work (Akuma & Callaghan, 2019b).

At the same time, the benefits of inquiry-based practical work are consistently emphasized. It encourages deeper engagement with scientific practices, cultivates transferable skills such as collaboration and argumentation, and enhances motivation through meaningful and authentic tasks. When implemented effectively, inquiry-based practical work not only improves conceptual understanding but also supports the development of students as scientifically literate individuals (Crawford, 2007).

Oliveira and Bonito (2023) conducted a comprehensive literature review of 53 international studies to analyze how practical work is conceptualized and applied in Science Education. They proposed eight conceptual dimensions (hands-on skills, mobilization of skills, minds-on skills, inquiry-based learning, learning through everyday experiences, scientific communication, accessible alternative to learning, studies not addressing the concept of practical work), five advantages (development of research-based learning skills, emphasis on active student participation in the learning process, development of relevant knowledge about practical skills and conceptual understanding, development of scientific literacy, and preparation of students for practical assessments), and five limitations of practical work (teacher concerns and professional content knowledge issues, distortion of purpose triggered by evaluation processes, economic, organizational and environmental constraints, descriptive learning tasks in science, and motivational effects. The authors emphasize the importance of integrating both hands-on and minds-on activities, arguing that meaningful learning occurs when students engage in experimental design and reflective thinking. They conclude that when students are active participants in inquiry processes - posing questions, testing hypotheses, and analyzing data - learning becomes more relevant.

Inquiry-based practical work is widely acknowledged for its capacity to engage students actively in scientific practices, which fosters deeper and longer-lasting learning (Bell et al., 2005). This approach enhances essential social skills such as communication and collaboration, crucial for students' development in science education (Crawford, 2007). Furthermore, it increases motivation by encouraging students to explore authentic, relevant questions that connect science with real-world contexts, which Minner et al. (2010) highlight as key to maintaining engagement. Inquiry-based practical work also supports the development of scientific reasoning and complex problem-solving skills, enabling more sophisticated cognitive growth (Klahr & Nigam, 2004). Scientific literacy, a fundamental objective of education, is significantly promoted through inquiry-based practical work which emphasis on constructing knowledge rather than memorizing facts (Donovan & Bransford, 2005). Additionally, inquiry-based practical work encourages self-efficacy and metacognitive skills, as students reflect on their learning process and develop autonomy (Zacharia & Olympiou, 2011). Problem-solving and critical thinking, vital for success in advanced science learning, are also nurtured through this method (Hmelo-Silver et al., 2007). Recent evidence further supports that inquiry-based practical work promotes effective application of scientific methods and critical thinking in real-world contexts (Windschitl et al., 2018). Lastly, the approach improves collaboration and scientific communication, supporting a more comprehensive and lasting understanding of complex scientific concepts (Chen et al., 2025).

## 7.4 Conclusion and Ways Forward

Overcoming the challenges faced by teachers in the implementation of inquiry-based practical work and maximizing its benefits for students requires ongoing investment and adaptation to the specificities of diverse teaching and learning contexts (Hofstein & Lunetta, 2004). A successful implementation of inquiry-based practical work depends on thorough teacher preparation, enabling them to act as facilitators of their students' learning while also being capable of adapting and developing innovative practices aligned with their specific educational settings (Crawford, 2007). This raises key questions: How can professional development programs support teachers in developing the necessary skills to implement inquiry-based practical work, considering the specificities of different educational contexts? Which stakeholders can support the implementation of inquiry-based practical work, and how can its sustainable implementation be ensured? Addressing these questions outlines potential ways forward.

- A practice-centered professional development program-in which teachers have the opportunity to deeply plan their lessons by, for instance, analyzing inquiry-based practical work tasks and discussing research findings related to this approach; to collect data on the effects of inquiry-based practical work on their students' learning; to analyze and reflect on the collected data; and to improve their practice based on reflection-contribute both to their knowledge development and to helping them address the challenges associated with implementing inquiry-based practical work (Capps & Crawford, 2013). Through this process, teachers are expected to become increasingly autonomous in designing and implementing inquiry-based practical work, and more responsive to the needs of their students and the characteristics of their educational environments (Chen et al., 2024).
- The successful implementation of inquiry-based practical work also depends on collaboration among key stakeholders-teachers, researchers, and policymakers. However, some authors have pointed out that this collaboration often faces significant obstacles (Henderson & Dancy, 2008; Torres-Hernández & Balcazar-Vargas, 2024). So, it is necessary to recognize that their perspectives may diverge: while teachers may prioritize immediate applicability in the classroom, researchers tend to focus on methodological rigor and innovation, and policymakers on large-scale feasibility and outcomes in terms of education policy. Although we agree that these issues can be an obstacle, we see it as an important way forward that these differences can also be perceived as a catalyst for debate and mutual enrichment, provided that constructive dialogue aimed at improving Science Education is promoted.
- Educational policies do not always reflect or support the needs identified by educators and researchers, leading to a misalignment between curriculum guidelines and the necessary conditions for effective implementation in schools (Chen et al., 2024). Reforms are frequently developed without the involvement of professional communities and researchers despite the importance of their engagement for successful and lasting educational change. Without such involvement, traditional practices are likely to persist. Therefore, it is important to have institutional mechanisms that promote collaboration among the various stakeholders, ensuring that educational reforms are co-constructed and adequately supported to enable sustainable implementation of inquiry-based practical work.

**Acknowledgments** This work was supported by National Funds through FCT-Portuguese Foundation for Science and Technology, I.P., under the scope of UIDEF-*Unidade de Investigação e Desenvolvimento em Educação e Formação*, UIDB/04107/2020, https://doi.org/10.54499/UIDB/04107/2020.

## References


Abrahams, I., & Reiss, M. J. (2012). Practical work: Its effectiveness in primary and secondary schools in England. *Journal of Research in Science Teaching, 49*(8), 1035–1055. https://doi.org/10.1002/tea.21036

Akuma, F. V., & Callaghan, R. (2018). Teaching science through inquiry-based instruction - a systematic review. *Journal of Research in Science Teaching, 55*(5), 665–690. https://doi.org/10.3389/feduc.2023.1170487

Akuma, F. V., & Callaghan, R. (2019a). Characterising extrinsic challenges linked to the design and implementation of inquiry-based practical work. *Research in Science Education, 49*, 1677–1706. https://doi.org/10.1007/s11165-017-9671-x

Akuma, F. V., & Callaghan, R. (2019b). A systematic review characterizing and clarifying intrinsic teaching challenges linked to inquiry-based practical work. *Journal of Research in Science Teaching, 56*(5), 619–648. https://doi.org/10.1002/tea.21516

Akuma, F. V., & Gaigher, E. (2021). A systematic review describing contextual teaching challenges associated with inquiry-based practical work in natural sciences education. *Eurasia Journal of Mathematics, Science and Technology Education, 17*(12), em2044. https://doi.org/10.29333/ejmste/11352

Arega, T., Habtamu, A., & Birhanu, M. (2025). Constructivist instructional approaches: A systematic review of evaluation-based evidence for effectiveness. *Review of Education, 13*(1). https://doi.org/10.1002/rev3.70040

Ash, D., & Klein, K. (2000). Inquiry in the informal learning environment. In J. Minstrell & E. H. van Zee (Eds.), *Inquiring into inquiry learning and teaching in science* (pp. 216–240). American Association for the Advancement of Science.

Atkin, J. M., & Karplus, R. (1967). Discovery or invention? *The Science Teacher, 34*(5), 45–51.

Aulls, M. W., & Shore, B. M. (2008). *Inquiry in education: Overcoming barriers to successful implementation* (Vol. 1).

Bell, R. L., Smetana, L., & Binns, I. (2005). Simplifying inquiry instruction. *The Science Teacher, 72*(7), 30–33.

Blanchard, M. R., Southerland, S. A., Osborne, J. W., Sampson, V. D., Annetta, L. A., & Granger, E. M. (2010). Is inquiry possible in light of accountability? A quantitative comparison of the relative effectiveness of guided inquiry and verification laboratory instruction. *Science Education, 94*(4), 577–616. https://doi.org/10.1002/sce.20390

Borko, H. (2004). Professional development and teacher learning: Mapping the terrain. *Educational Researcher, 33*(8), 3–15. https://doi.org/10.3102/0013189X03300800

Capps, D. K., & Crawford, B. A. (2013). Inquiry-based professional development: What does it take to support teachers in learning about inquiry and science? *International Journal of Science Education, 35*(12), 1947–1978. https://doi.org/10.1080/09500693.2012.760209

Carlson, J., Humphrey, G., & Reinhardt, K. (2003). Weaving inquiry into the fabric of the school. *Science and Children, 40*(7), 38–43.

Chen, F., Zhang, Y., & Chen, G. (2025). Exploring the effects of technology-supported collaborative inquiry and students ICT competency on scientific literacy and subject knowledge in rural science classrooms. *Education and Information Technologies*. https://doi.org/10.1007/s10639-025-13512-w

Chen, L., Wang, M., & Zhang, Y. (2024). Teacher beliefs and practices in inquiry-based science education: A longitudinal study. *Journal of Science Teacher Education, 35*(2), 123–145. https://doi.org/10.29329/ijpe.2020.228.7

Colburn, A. (2000). An inquiry primer. *Science Scope, 23*(6), 42–44. https://www.jstor.org/stable/43180086

Crawford, B. A. (2007). Learning to teach science as inquiry in the rough and tumble of practice. *Journal of Research in Science Teaching, 44*(4), 613–642. https://doi.org/10.1002/tea.20157

Crawford, B. A. (2014). From inquiry to scientific practices in the science classroom. In N. G. Lederman & S. K. Abell (Eds.), *Handbook of research on science education* (Vol. 2, pp. 515–541). Routledge. https://doi.org/10.4324/9780203097267-36

Deboer, G. E. (2006). Historical perspectives on inquiry teaching in schools. In L. B. Flick & N. G. Lederman (Eds.), *Scientific inquiry and nature of science* (pp. 17–35). Springer. https://doi.org/10.1007/978-1-4020-5814-1_2

Dillenbourg, P. (1999). What do you mean by collaborative learning? In P. Dillenbourg (Ed.), *Collaborative-learning: Cognitive and computational approaches* (pp. 1–19). Elsevier.

Donovan, M. S., & Bransford, J. D. (Eds.). (2005). *How students learn Science in the classroom.* National Academies Press.

García-Carmona, A. (2020). From inquiry-based science education to the approach based on scientific practices. *Science & Education, 29*, 443–463. https://doi.org/10.1007/s11191-020-00108-8

Gilbert, J. K. (2006). On the nature of context in chemical education. *International Journal of Science Education, 28*(9), 957–976. https://doi.org/10.1080/09500690600702470

Henderson, C., & Dancy, M. H. (2008). Physics faculty and educational researchers: divergent expectations as barriers to the diffusion of innovations. *American Journal of Physics, 76*(1), 79–91. https://doi.org/10.1119/1.2800352

Hmelo-Silver, C. E., Duncan, R. G., & Chinn, C. A. (2007). Scaffolding and achievement in problem-based and inquiry learning: a response to Kirschner, Sweller, and Clark (2006). *Educational Psychologist, 42*(2), 99–107. https://doi.org/10.1080/00461520701263368

Hofstein, A., & Lunetta, V. N. (2004). The laboratory in science education: foundations for the twenty-first century. *Science Education, 88*(1), 28–54. https://doi.org/10.1002/sce.10106

Klahr, D., & Nigam, M. (2004). The equivalence of learning paths in early science instruction: effects of direct instruction and discovery learning. *Psychological Science, 15*(10), 661–667. https://doi.org/10.1111/j.0956-7976.2004.00737.x

Lazonder, A. W., & Harmsen, R. (2016). Meta-analysis of inquiry-based learning: Effects of guidance. *Review of Educational Research, 86*(3), 681–718. https://doi.org/10.3102/0034654315627366

Lederman, N. G. (2006). Syntax of nature of science within inquiry and science instruction. In L. B. Flick & N. G. Lederman (Eds.), *Scientific inquiry and nature of science: implications for teaching, learning, and teacher education* (pp. 301–317). Springer. https://doi.org/10.1007/978-1-4020-5814-1_14

Leite, L. (2001). O ensino das ciências e a perspetiva da alfabetização científica. *Revista Portuguesa de Educação, 14*(2), 87–106.

Lunetta, V. N., Hofstein, A., & Clough, M. P. (2007). Learning and teaching in the school science laboratory: an analysis of research, theory, and practice. In S. Abell & N. Lederman (Eds.), *Handbook of Research on Science Education* (pp. 393–441). Routledge.

Maconi, F., Aulls, M. W., & Shore, B. M. (2008). *Inquiry in education: Inquiry-based instruction in the classroom* (Vol. 2).

Martin-Hansen, L. (2002). Defining inquiry: Exploring the many types of inquiry in the science classroom. *The Science Teacher, 69*(2), 34–37. https://www.jstor.org/stable/24154746

McComas, W. F., Clough, M. P., & Nouri, N. (2020). Nature of science and classroom practice: A review of the literature with implications for effective NOS instruction. In W. F. McComas (Ed.), *Nature of science in science instruction* (pp. 67–111). Springer. https://doi.org/10.1007/978-3-030-57239-6_4

Millar, R. (2004). The role of practical work in the teaching and learning of science. *National Academy of Sciences.*

Minner, D. D., Levy, A. J., & Century, J. (2010). Inquiry-based science instruction - what is it and does it matter? Results from a research synthesis years 1984 to 2002. *Journal of Research in Science Teaching, 47*(4), 474–496. https://doi.org/10.1002/tea.20347

National Research Council. (1996). *National science education standards.* National Academies Press.

National Research Council. (2000). *Inquiry and the national science education standards: A guide for teaching and learning.* National Academies Press.
National Research Council. (2012). A framework for K-12 science education: Practices, crosscutting concepts, and core ideas. *The National Academies Press.* https://doi.org/10.17226/13165
Oliveira, H., & Bonito, J. (2023). Practical work in science education: A systematic literature review. *Frontiers in Education, 8,* 1151641. https://doi.org/10.3389/feduc.2023.1151641
Ramaila, S. (2024). Systematic review of inquiry-based learning: assessing impact and best practices in education. *F1000Research, 13,* Article 1045. https://doi.org/10.12688/f1000research.155367.1
Rudolph, J. L. (2024). Scientific literacy: Its real origin story and functional role in American education. *Journal of Research in Science Teaching, 61*(4), 789–812. https://doi.org/10.1002/tea.21890
Schwab, J. J. (1960). Inquiry, the science teacher, and the educator. *The School Review, 68*(2), 176–195. https://doi.org/10.1086/442536
Toplis, R., & Allen, M. (2012). I do and I understand? Practical work and laboratory use in United Kingdom schools. *Eurasia Journal of Mathematics, Science & Technology Education, 8*(1), 3–9. https://doi.org/10.12973/eurasia.2012.812a
Torres-Hernández, N., & Balcazar-Vargas, M. P. (2024). Barriers and facilitators in implementing inquiry-based practical work: A systematic review. *Studies in Science Education, 60*(1), 89–125. https://doi.org/10.1186/s13012-018-0789-7
Van de Pol, J., Volman, M., Oort, F., & Beishuizen, J. (2013). Scaffolding student learning: A micro-analysis of teacher-student interaction. *Learning and Instruction, 25,* 41–54. https://doi.org/10.1016/j.lcsi.2012.12.001
Watson, R., Swain, J., & McRobbie, C. (2004). Students discussions in practical scientific inquiries. *International Journal of Science Education, 26*(1), 25–45. https://doi.org/10.1080/0950069032000072764
Wellington, J. (2000). *Teaching and learning secondary science: Contemporary issues and practical approaches.* Routledge.
Wieman, C. E., Adams, W. K., & Perkins, K. K. (2008). PhET: Simulations that enhance learning. *Science, 322*(5902), 682–683. https://doi.org/10.1126/science.1161948
Windschitl, M., Thompson, J., & Braaten, M. (2018). *Ambitious science teaching.* Harvard Education Press.
Zacharia, Z. C., & Olympiou, G. (2011). Physical versus virtual manipulative experimentation in physics learning. *Learning and Instruction, 21*(3), 317–331. https://doi.org/10.1016/j.learninstruc.2010.03.001

**Mónica Baptista** is full professor at the Institute of Education and researcher at UIDEF, University of Lisbon (IE-ULisboa), Portugal. She holds a PhD in science education, from the IE-ULisboa. She is deputy director of the IE-ULisboa and has been teaching courses on STEM education and science education. Her research interests focus on inquiry-based science education, STEM education, scientific explanations, and lesson study.

**Jairo Ortiz-Revilla** is an assistant professor of didactics of experimental sciences in the Department of Specific Didactics, University of Burgos, Spain. He holds a PhD in Education. His main lines of research are competency development, science teaching and learning in compulsory education through active methodologies and integrated approaches, teacher training, and the nature of sciences.

# Chapter 8
# Engaging in the Practical Work of Science with Multiple Models

**Regina H. Yorkgitis, Troy D. Sadler, and Eric A. Kirk**

## 8.1 Introduction

In a graduate-level science teaching methods course, we invited aspiring pre-service teachers to observe what happens when a seemingly "empty" water bottle is covered with a deflated latex balloon. The water bottle was placed in a container of warm water (and the balloon expanded) and then in ice water (and the balloon contracted again). Next, using a whiteboard and markers, the pre-service teachers created a model of the demonstration to explain the mechanisms that make this science phenomenon occur. The pre-service teachers worked in small groups to create their models. Then, they revised and adjusted their initial drawings based on feedback from the course instructor and peers. This brief example exemplifies how the practice of scientific modeling can be integrated into science learning environments. In this case, the learners (i.e., the pre-service teachers) created diagrammatic models that focused on the mechanism of a phenomenon that they observed through a simple demonstration.

Through this chapter, we aim to provide an overview of modeling and describe its value-especially when different kinds are used together-as a tool for engaging students in practical work. First, we define models and modeling, describing how modeling can be used as a tool for sensemaking. Next, we provide an overview of the various functions and goals of different modeling practices, considering how practicing scientists utilize models. The latter part of the chapter outlines

R. H. Yorkgitis (✉) · T. D. Sadler
The University of North Carolina, Chapel Hill, USA
e-mail: reginayo@unc.edu

T. D. Sadler
e-mail: tsadler@unc.edu

E. A. Kirk
Utah State University, Logan, USA
e-mail: eric.kirk@usu.edu



L. Leite et al. (eds.), *Practical Work in Science Education in the Digital Era*, Advances in Technology-Rich Science Education 2, https://doi.org/10.1007/978-3-032-13788-3_8

four different model types-mechanistic, causal loop, computational, and participatory. Each of these models has distinct advantages for use in secondary classrooms; however, we will close by describing how using different models together-an approach we employed in MMP - can offer students rich opportunities for sensemaking. Throughout the chapter, we present examples from our work on the Multiple Models Project, where U.S. secondary students (typically ages 11–18) engaged in a variety of modeling practices.

## 8.2 Models and Modeling

Models are abridged representations of science phenomena and systems that can help generate understandings, explanations, and predictions. Modeling, the act of creating, using, and gradually revising models, is a central practice for scientists (Ke et al., 2021; Lehrer & Schauble, 2006; Nersessian, 2008; Schwarz et al., 2009). Students can benefit from modeling because it can aid in the development of conceptual understanding, cultivate metacognition, promote collaboration, and support reasoning about dynamic systems (Louca & Zacharia, 2012). Through modeling, students actively refine their understanding of concepts by creating, critiquing, and revising models. A range of methods and media are used to create models, including drawings and computer simulations (Nersessian, 2008). Modeling does not always require advanced technology or expertise to create as reflected in the description of pre-service teachers creating models. Situating modeling centrally within classroom activities and discourse can promote conceptual understandings of important science ideas (Passmore et al., 2014) and provide students with cognitive tools of their own design to use for sensemaking about how and why systems work (Zangori & Forbes, 2015). The process of creating, reviewing, and amending models empowers learners to participate in authentic science inquiry. In this way, modeling is a key dimension of practical work that supports epistemic and sensemaking goals.

Modeling can be a useful exercise when the goal is to support deep thinking and sensemaking among students. For example, in our recent work in a curricular unit focused on the spread of infectious disease (Elsner et al., 2023), the goal was for middle school students (ages 10–13) to figure out how viral diseases can be transmitted between individuals. In an early lesson within the unit, teachers asked student groups to create a model, using whiteboards and markers, to explain how the flu is transmitted. One student group demonstrated their understanding of the flu through arrows and labels, showing a sick individual coughing and expelling germs and a healthy person seemingly inhaling the germs through their nose and becoming sick (see Fig. 8.1). A second group also showed that the flu was transmitted to another person by coughing, as indicated by the text at the bottom of the model pictured (see Fig. 8.2). The student products look quite different from one another, but each represents the ways in which students were making sense of disease transmission.

Teachers can use models such as these as a basis for recognizing misconceptions or gaps in students' interpretations of disease transmission, and students can revise their

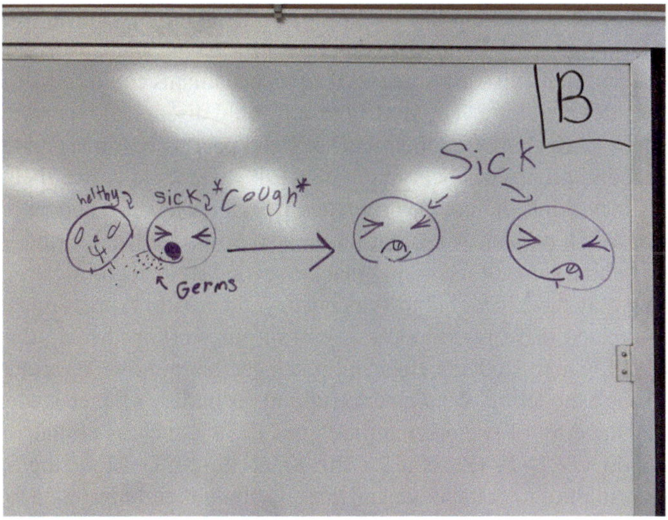


**Fig. 8.1** A student-created mechanistic model of the flu

**Fig. 8.2** A student-created mechanistic model of the flu with stick figures

models based on feedback. For the first group, a teacher might confirm with students that their interpretation of the model was correct by asking students to explain their model. A teacher might also ask students if any mechanisms could cause someone to get the flu besides coughing. The student group could discuss and then create a new version of their model based on their teacher's and peer's questions and feedback.

For the second group, teachers might also ask students to revise their model to show where they think the cough originates from in a healthy person and how it progresses in a sick person. A teacher might also probe students further to explain the meaning behind some of the shapes that the second student group created on their model, which may have meaning to students beyond artistic expression.

Students created these two models as part of our work in the Multiple Models Project (MMP), where students engaged in science learning with various types of models. Through modeling, we aimed to support secondary science students' (ages 11–18) understanding of socio-scientific issues (SSIs), such as climate change, e-cigarettes, and COVID-19 (Ke et al., 2020; Ke et al., 2021). Modeling SSIs offers students opportunities to collaborate and directly apply their knowledge to decision-making processes in their own lives (Ke et al., 2021; Lesnefsky et al., 2023; Sadler et al., 2019; Sadler et al., 2017). As we discuss the epistemic functions of modeling and, later, the different epistemic aims of different models, we will provide several other examples from MMP and consider how multiple models can be used in tandem to achieve this range of goals.

### *8.2.1 Epistemic Functions of Modeling: Representation, Explanation and Prediction*

The ways in which individuals interact with models and what they do with models are referred to as modeling practices (Berland et al., 2016; Gouvea & Passmore, 2017; Schwarz et al., 2009). Intentionally considering modeling practices that students engage with in science classrooms matters because they can be used to support student sensemaking, defined by the U.S. National Science Teaching Association (2025) as the process of "actively trying to figure out how the world works."

Modeling practices can be framed within three broad categories: (1) designing and revising, (2) evaluating, and (3) using models. This framing incorporates the natural intersections between modeling practices and two sets of epistemic dimensions of modeling practices outlined by Gouvea and Passmore (2017): epistemic aims and epistemic criteria.

Epistemic aims describe the knowledge-related goals users have when engaging in modeling such as representing, explaining, or forming predictions about a phenomenon (Berland et al., 2016). These aims shape what models an individual might select, as well as how they might use them. Individual models may be used in service of multiple aims; however, it is often the case that particular models accomplish some aims and not others. For example, whereas the models described in the previous section are well suited for helping learners represent and explain how a disease is transmitted, they are less effective for supporting predictions when compared to computer-based models which allow users to adjust conditions and run dynamic simulations (Passmore et al., 2014; Pluta et al., 2011; Schwarz et al., 2009).

Epistemic criteria are the principles used to assess and evaluate models. Epistemic criteria can be applied to all models, regardless of the specific forms that models take. The criteria used to evaluate models include alignment with evidence, meeting the intended epistemic aims, and addressing the intended audience (Gouvea & Passmore, 2017; Pluta et al., 2011).

Reflecting on what constitutes a well-designed model can help people understand why and how a phenomenon happens (Passmore et al., 2014; Pluta et al., 2011; Schwarz et al., 2009). Considering these epistemic dimensions through modeling practices of designing, revising, and evaluating sets students up for a more nuanced understanding of science concepts and systems. In our opening example to this chapter, we described pre-service science teachers using modeling practices to unpack a demonstration of thermal expansion and contraction. Pre-service teachers watched a demonstration, created the model, received peer feedback, and then revised the model.

After reflection, they iterated on their initial model.

The iterative and flexible nature of modeling practices, along with the epistemic dimensions that fuel them, makes them ideal for sensemaking. Through modeling practices, learners-such as the pre-service teachers creating whiteboard drawings-aim to explain air expansion and evaluate their models based on how well they align with epistemic criteria. They then use these models to apply their understanding to different scenarios. This process can support students in sensemaking about science phenomena and, ultimately, help them comprehend how to effectively engage in scientific work.

### 8.2.2 *Models as Fundamental to the Work of Scientists*

An important rationale for featuring models within science education contexts pertains to the positioning of models and modeling in the work of professional scientists (Gilbert & Osborne, 1980; National Research Council, 2012; Passmore et al., 2014). Modeling is a fundamental dimension of scientific work. In fact, some scholars have suggested that creating and using models is the core work of science (Schwarz et al., 2009). Given the significance and prominence of modeling for the work and progress of science, modeling in science learning opportunities is a natural priority for science education. A similar argument can be made for practical work in science education. Practical work such as laboratory experiments, fieldwork, and computational investigations is a fundamental dimension of the work of scientists; therefore, it should also be represented in science education. We take these arguments one step further: not only is it the case that modeling and practical work are central to science, but modeling and practical work are reciprocally dependent on each other for advancing science (Manz & Georgen, 2023).

Scientists use models to inform and advance their practical work, and the data and findings that emerge from their practical work inform model construction and use. Therefore, science education should not only create opportunities for practical work and modeling, but science learners should also have opportunities to integrate modeling into their practical work. This section assumes a line of argument that justifies science education recommendations based on the practices of professional scientists. We find this to be a compelling argument and one that serves as the basis for important policy documents such as the Next Generation Science Standards (NGSS Lead States, 2013). However, this is certainly not the only justification for practical work, as demonstrated by other contributions in this volume, or modeling as suggested by the opening sections of this chapter.

### *8.2.3 Modeling as a Part of Practical Work*

To demonstrate the link between modeling and practical work in a science education context, we highlight an example from recent work in middle school science classrooms (students ages 10–13). This example stems from a curricular unit in which students consider the energy demands of their communities, and the carbon dioxide emissions associated with fossil fuel use (Sadler et al., 2025). Throughout the unit, students engage in iterative modeling of the carbon cycle. Early in the unit, teachers guide students through the creation of carbon cycle models based on the students' prior knowledge. Most of the students we have worked with on this task know that $CO_2$ can be found in the atmosphere and that the burning of fossil fuels to power things like cars and factories adds more $CO_2$ to the atmosphere. Many students also know that atmospheric $CO_2$ has something to do with warming. Teachers then use these incomplete models to invite students to further explore the movement of carbon through practical work.

This practical work begins by introducing students to a new piece of laboratory equipment, a $CO_2$ sensor, that they will use in a couple of activities. The first of these activities helps to familiarize students with the sensor and encourages them to brainstorm places and processes that might show changes in $CO_2$ levels over time. Small groups of students go around the classroom and/or school to monitor $CO_2$ levels. Students use the sensor in a variety of places-some of the places that yield interesting results include next to someone who is talking, in a closed classroom with many students, next to the school's computer server, and by a car's exhaust system.

After collecting these data, students share their findings as a class and then consider whether these data can inform any changes to their models. One addition that is typically made in this iteration of modeling is the inclusion of humans. Later in the unit, students perform a laboratory activity in which they place spinach leaves that have been soaked in water, along with a $CO_2$ sensor, in a sealed, transparent container. They place this setup in a few different environments, including one trial in bright sunlight

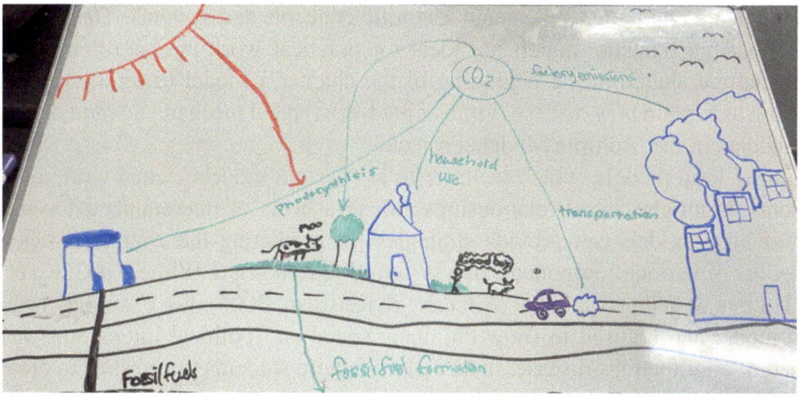


**Fig. 8.3** A student-created model of the carbon cycle

(or grow lights) and another trial in darkness. The experiment directs student attention to outcomes associated with photosynthesis-when the spinach leaves are exposed to light, they photosynthesize and consume $CO_2$. This result becomes another useful piece of information that can inform revisions to the students' models. By the end of the unit, students will have completed three iterations of their carbon cycle models, with the final two versions shaped by data they collect in practical work with the $CO_2$ sensors. Figure 8.3 presents a student model after their practical work.

## 8.3 Different Kinds of Models: Mechanistic, Causal Loop, Computational, Participatory

Scientists and students use different types of models, including mechanistic, causal loop, computational, and participatory models. Recognizing the differences between these models is important because different models support distinct modeling practices since they align with different epistemic aims (Berland et al., 2016). Through our work in the Multiple Models Project, we have used a variety of different models. First, we outline the epistemic aims and modeling practices associated with the four different models and consider how different kinds of models can best be used for practical work. Next, we will consider how multiple models can be used in coordination with one another.

Mechanistic models are used across science domains to explain the "how and why" (i.e., the mechanism) behind phenomena (Forbes et al., 2015; Gray & Rogan-Klyve, 2018; Ke et al., 2021). Mechanistic models may be particularly accessible to students because they require little technical expertise to create initially and can be easily revised later (Ke et al., 2021; Stroupe, 2014). Mechanistic models can support the modeling practice of explaining previously unseen mechanisms (Berland et al., 2016). This modeling practice is an important task for scientists wishing to communicate and better understand scientific concepts and systems. Therefore, we argue that mechanistic models are ideal for practical work in classroom settings. The example shared at the beginning of the chapter-a model using a stick-figure drawing to explain how disease is transmitted-is a typical mechanistic model created by students in the Multiple Models Project.

Causal loop models, which we refer to as "system models" when working with students, highlight causal relationships among aspects of interconnected systems. Causal loop models can provide strategies for predicting the characteristics and behaviors of various components of a system (Frederiksen & White, 2002; Ke et al., 2021; Levy & Wilensky, 2008; Wilensky & Reisman, 2006). For example, a causal loop model can be used to show the many social and political interactions in the spread of an infectious disease. In an activity where students were asked to create a causal loop model about a mystery infectious disease, they used arrows to indicate the causal relationship between the symptoms of the mystery disease, considering that it would cause people to be hospitalized, which would then cause hospitals to become overwhelmed (see Fig. 8.4).

Scientists and students can also use computational models, which simulate complex systems through dynamic controls that computer technology provides. Computational models provide opportunities to illustrate aspects of a science phenomenon that models created through other media may not be able to achieve.

**Fig. 8.4** A student-created causal loop model of a mystery disease

Models created or represented with computer-based technology might be valuable in exemplifying the limits of models (Ke et al., 2021; Wilensky & Reisman, 2006). The limits imposed by the nature of computational models allow for rule-based modeling (Levy & Wilensky, 2008; Wilensky & Reisman, 2006). For instance, a computer model of disease spread in a population may have purposefully set controls that ensure there will be consequences for not taking preventative measures, such as the population wearing a mask or properly washing hands.

In our work, we often prioritize models that students can make and create by hand rather than computational models. Computational models may require specific technical skills and can, therefore, present practical challenges for students in revising and editing models. However, the dynamic affordances of computational models make it possible for students to test different scenarios, generate data, and engage in hypothesis testing. Under the right circumstances, it is also possible to engage students in coding so that they can revise or even create their own computational models. This requires some degree of technical expertise on the part of the students and their teacher and typically takes more time in the classroom. Despite these limitations, computational modeling can be a powerful learning opportunity on its own or an effective complement to other forms of modeling, like mechanistic and causal loop models, for which there are lower barriers for entry.

Participatory models are a final form of models that we use in our work. In this form of modeling, students become part of the representation by becoming agents within the system that is being modeled. This form of participatory modeling helps to make abstract mechanisms more concrete for students by allowing them to become part of the model. Participatory modeling approaches include role-playing activities and agent-based simulations (Levy & Wilensky, 2008). Participatory modeling involves students embodying agents in a system to enact the kinds of rules that drive agent-based simulations. This form of modeling makes phenomena more tangible and visible to students, fostering a stronger understanding of a concept or system behaviors. For example, Levy and Wilensky (2008) observed students engaging in a participatory modeling activity, acting as deer-some as sick, some as sick but unaware, and some as healthy. Through this participatory model, students shared thoughts about their predictions for the spread of disease among wildlife. In the Multiple Models Project, we engaged students in a participatory model in which they represented individuals within a community and exchanged a liquid solution when they came in contact with one another through a series of controlled rounds of interaction. Some of the students unknowingly began the exercise with an "infected" solution and, as the simulation unfolded, they infected other students through the liquid exchange. At the end of the simulation, we added an acid/base indicator to each student's liquid to determine who ended up with the infection. In subsequent trials of the simulation, we introduced rules that mimicked prophylactic strategies such as mask wearing and vaccines, that the students were able to compare rates of disease spread under these various conditions. This made it possible for the students to explore the topic, i.e., disease transmission, through participation in a model of disease spread.

## 8.4 Considering Phenomena and Issues through Multiple Models

Scientists rarely rely on a single model because each one reveals only part of a complex system. By nature, all models come with constraints and provide a limited representation of a particular phenomenon they simplify and represent (Ke et al., 2021). Cartwright (1999) describes the exploration of reality through modelling as a process that uses a collection of overlapping patches of models, where patches have different shapes and sizes, leaving some spaces uncovered. The metaphor suggests that a model alone cannot create a complete picture of a science phenomenon. Since a single model can only explain limited aspects of a mechanism, system, or phenomenon, models used by scientists may be most effective when used as a "patchwork" or with other models. Using multiple models is, therefore, an effective strategy for scientists and learners of science.

Complex science phenomena often have multiple dimensions (e.g., scientific, economic, social, political), and leveraging the affordances of different models together can be used to explore those dimensions (Ainsworth, 2008; Frederiksen & White, 2002; Ke et al., 2021). For example, although a mechanistic model can support a conceptual understanding of disease spread from person to person, it does not address other aspects of infectious disease spread and prevention, such as political and social systems that may exacerbate or lessen outbreaks. Where a mechanistic model may successfully show the mechanism of disease spread, another model, such as a causal loop model, would be necessary to show how disease spread can impact social systems. In our research in the Multiple Models Project, we have documented these patterns. When presented only with a mechanistic modeling experience, students' ideas about how a virus spread were improved, but the exercise did not necessarily help them to consider broader dimensions of disease spread (Elsner et al., 2024). In contrast, when the learning experience coupled mechanistic and causal loop models, students demonstrated a wider range of ideas including ethical questions and issues associated with social justice (Lesnefsky et al., 2025).

This approach of using more than one model to explain science phenomena is considered the multiple models approach (Ke et al., 2021; Sadler et al., 2019). We argue that using multiple models in tandem is both an authentic reflection of the practical work of scientists as well as a meaningful classroom teaching strategy. Linking multiple models instead of a single model allows learners to engage in tasks that more closely mirror the experiences of scientists and facilitate a more comprehensive grasp of science phenomena (Xu et al., 2025).

Using multiple models is not only an authentic science practice; it is likewise an effective science learning strategy. Engaging learners with various models is necessary to demonstrate to students the diverse approaches that scientists can take to understand a phenomenon (Oh & Oh, 2011). Through multiple models, students can better understand the real-world complexity that scientists negotiate by exploring how different concepts interrelate, a key component of scientific literacy (Frederiksen & White, 2002; Ke et al., 2021; Oh & Oh, 2011).

Using multiple models in concert gives students a more holistic understanding of multifaceted science phenomena (Frederiksen & White, 2002; Ke et al., 2021; Oh & Oh, 2011). Leveraging multiple models can provide a foundation for students to participate in scientific argumentation (Ke et al., 2021), support students in transferring knowledge to new circumstances (Frederiksen & White, 2002), nurture students' ability to communicate scientific ideas effectively (Lehrer & Schauble, 2012), and can promote informed decision-making especially when a scientific issue intersects with social issues (Ke et al., 2021).

Students may initially struggle to view single models as means to generate new knowledge, perceiving them only as representations of existing knowledge. The process of creating and revising multiple models allows students to view models as usable tools that are subject to change, rather than just static products (Schwarz et al., 2009; Manz & Georgen, 2023). When using multiple models, students may notice the contradictions between models, which can challenge their preconceived and perhaps erroneous ideas about a phenomenon. By considering the drawbacks of different model types and confronting misconceptions revealed through multiple models, students can refine their understanding and develop stronger explanations about the underlying phenomena (Lehrer & Schauble, 2012; Manz & Georgen, 2023).

Multiple models may be best scaffolded to begin sharing a single model with students and then provide students with time to perceive the limitations of the model. Eventually, students become unsatisfied with one model's limitations-at this point, it makes logical sense to present an alternative model (Ke et al., 2021; Schwarz et al., 2009). As students add different model types to their toolkit, they can leverage ideas across models and create a more complete picture of a science issue.

## 8.5 Conclusion and Ways Forward

Models vary in form and function, and the intention with which a scientist uses a model can impact the epistemic dimension employed. Creating and revising multiple models together enables scientists to gather a more comprehensive understanding of complex science systems and concepts in their practice. Science students also can engage in and benefit from the work of science by building knowledge from dynamic and varied models.

Through our work in the Multiple Models Project, students utilized various model types, including mechanistic, causal loop, computational, and participatory models, together to consider science concepts, such as infectious disease spread and carbon dioxide emissions. Multiple models support the development of student reasoning by presenting an opportunity to center students as active contributors in constructing their understanding of science concepts and systems, thereby engaging in the practical work of science.

**Acknowledgements** This material is based upon work supported by the National Science Foundation under Grants DRL-2101083 and DRL-2201192. Any opinions, findings, and conclusions or recommendations expressed in this material are those of the author(s) and do not necessarily reflect the views of the National Science Foundation.

## References

Ainsworth, S. (2008). The educational value of multiple representations when learning complex scientific concepts. In J. Gilbert, M. Reiner, & M. Nakhleh (Eds.), *Visualization: Theory and practice in science education* (Vol. 3, pp. 191–208). Springer. https://doi.org/10.1007/978-1-4020-5267-5_9

Berland, L. K., Schwarz, C. V., Krist, C., Kenyon, L., Lo, A. S., & Reiser, B. J. (2016). Epistemologies in practice: Making scientific practices meaningful for students. *Journal of Research in Science Teaching, 53*(7), 1082–1112. https://doi.org/10.1002/tea.21257

Cartwright, N. (1999). *The dappled world: A study of the boundaries of science.* Cambridge University Press.

Elsner, J. N., Kirk, E. A., Xu, Z., Zangori, L. A., Ke, L., & Sadler, T. D. (2024, March, 17–20). *Promoting model-informed reasoning through engagement with multiple models.* Annual meeting of the National Association for Research in Science Teaching (NARST), Denver, CO.

Elsner, J., Sadler, T., Kirk, E., Rawson, R., Friedrichsen, P., & Ke, L. (2023). Using multiple models to teach about COVID-19. *The Science Teacher, 90*(3), 40–45. https://doi.org/10.1080/00368555.2023.12294205

Forbes, C. T., Zangori, L., & Schwarz, C. V. (2015). Empirical validation of integrated learning performances for hydrologic phenomena: 3rd-grade students model-driven explanation-construction. *Journal of Research in Science Teaching, 52*(7), 895–921. https://doi.org/10.1002/tea.21226

Frederiksen, J. R., & White, B. Y. (2002). Conceptualizing and constructing linked models: Creating coherence in complex knowledge systems. In P. Brna, M. Baker, K. Stenning, & A. Tiberghien (Eds.), *The role of communication in learning to model* (pp. 69–96). Lawrence Erlbaum Associates.

Gilbert, J., & Osborne, R. (1980). The use of models in science and science teaching. *European Journal of Science Education, 2*(1), 3–13. https://doi.org/10.1080/0140528800020103

Gouvea, J., & Passmore, C. (2017). Models of versus models for: Toward an agent-based conception of modeling in the science classroom. *Science & Education, 26*(1–2), 49–63. https://doi.org/10.1007/s11191-017-9884-4

Gray, R., & Rogan-Klyve, A. (2018). Talking modelling: Examining secondary science teachers modelling-related talk during a model-based inquiry unit. *International Journal of Science Education, 40*(11), 1345–1366. https://doi.org/10.1080/09500693.2018.1479547

Ke, L., Sadler, T. D., Zangori, L., & Friedrichsen, P. (2020). Students perceptions of engagement in socio-scientific issue-based learning and their appropriation of epistemic tools for systems thinking. *International Journal of Science Education, 42*(8), 1339–1361. https://doi.org/10.1080/09500693.2020.1759843

Ke, L., Sadler, T. D., Zangori, L., & Friedrichsen, P. J. (2021). Developing and using multiple models to promote scientific literacy in the context of socio-scientific issues. *Science & Education, 30*(6), 589–607. https://doi.org/10.1007/s11191-021-00206-1

Lehrer, R., & Schauble, L. (2006). Scientific thinking and science literacy. In K. Renninger, I. Sigel, W. Damon, & R. Lerner (Eds.), *Handbook of child psychology: Child psychology in practice* (6th ed., pp. 153–196). John Wiley & Sons, Inc.

Lehrer, R., & Schauble, L. (2012). Seeding evolutionary thinking by engaging children in modeling its foundations. *Science Education, 96*(4), 701–724. https://doi.org/10.1002/sce.20475

Lesnefsky, R. R., Elsner, J. N., Kirk, E. A., Ke, L., & Sadler, T. D. (2025). Exploring resources and reasoning practices in socioscientific system modeling for justice-centered science education. *CBE-Life Sciences Education, 24*(4), 10. https://doi.org/10.1187/cbe.24-03-0045
Lesnefsky, R. R., Ke L. Sadler, T. D., & Kirk, E.A. (2023). How middle school students engage in modeling viral transmission. In *Proceedings of the International Conference of the Learning Sciences (ICLS) 2023*. International Society of the Learning Sciences. https://doi.org/10.22318/icls2023.140516
Levy, S. T., & Wilensky, U. (2008). Inventing a mid level to make ends meet: Reasoning between the levels of complexity. *Cognition and Instruction, 26*(1), 1–47. https://doi.org/10.1080/07370000701798479
Louca, L. T., & Zacharia, Z. C. (2012). Modeling-based learning in science education: Cognitive, metacognitive, social, material and epistemological contributions. *Educational Review, 64*(4), 471–492. https://doi.org/10.1080/00131911.2011.628748
Manz, E., & Georgen, C. (2023). Interlocking models in classroom science. *Science Education, 107*(6), 1399–1434. https://doi.org/10.1002/sce.21806
National Research Council. (2012). *A framework for K-12 science education: practices, crosscutting concepts, and core ideas.* National Academies Press.
National Science Teaching Association. (2025). *Sensemaking.* https://www.nsta.org/sensemaking
Nersessian, N. J. (2008). Creating scientific concepts. *MIT Press.* https://doi.org/10.7551/mitpress/7967.001.0001
NGSS Lead States. (2013). Next generation science standards: For states, by states. *The National Academies Press.* https://www.nextgenscience.org/
Oh, P. S., & Oh, S. J. (2011). What teachers of science need to know about models: An overview. *International Journal of Science Education, 33*(8), 1109–1130. https://doi.org/10.1080/09500693.2010.502191
Passmore, C., Gouvea, J. S., & Giere, R. (2014). Models in science and in learning science: Focusing scientific practice on sense-making. In M. Matthews (Ed.), *International handbook of research in history, philosophy, and science teaching.* Springer. https://doi.org/10.1007/978-94-007-7654-8_36
Pluta, W. J., Chinn, C. A., & Duncan, R. G. (2011). Learners epistemic criteria for good scientific models. *Journal of Research in Science Teaching, 48*(5), 486–511. https://doi.org/10.1002/tea.20415
Sadler, T. D., Foulk, J. A., & Friedrichsen, P. J. (2017). Evolution of a model for socio-scientific issue teaching and learning. *International Journal of Education in Mathematics, Science and Technology, 5*(2), 75–87. https://doi.org/10.18404/ijemst.55999
Sadler, T. D., Friedrichsen, P., & Zangori, L. (2019). A framework for teaching for socio-scientific issue and model-based learning (SIMBL). *Educação e Fronteiras/Education and Borders, 9*(25), 8–26.
Sadler, T. D., Xu, Z., & Fortus, D. (2025). Restructuring the science curriculum around grand challenges. *International Journal of Science Education*, 1–20. https://doi.org/10.1080/09500693.2025.2490777
Schwarz, C. V., Reiser, B. J., Davis, E. A., Kenyon, L., Achér, A., Fortus, D., Shwartz, Y., Hug, B., & Krajcik, J. (2009). Developing a learning progression for scientific modeling: Making scientific modeling accessible and meaningful for learners. *Journal of Research in Science Teaching, 46*(6), 632–654. https://doi.org/10.1002/tea.20311
Stroupe, D. (2014). Examining classroom science practice communities: How teachers and students negotiate epistemic agency and learn science-as-practice.*Science Education, 98*(3), 487–516. https://doi.org/10.1002/sce.21112
Wilensky, U., & Reisman, K. (2006). Thinking like a wolf, a sheep, or a firefly: Learning biology through constructing and testing computational theories-an embodied modeling approach. *Cognition and Instruction, 24*(2), 171–209. https://doi.org/10.1207/s1532690xci2402_1

Xu, Z., Elsner, J. N., Kirk, E. A., Sadler, T. D., & Zangori, L. (2025, March, 23–26). *Using multiple models to learn population-level viral transmission.* Annual meeting of the National Association for Research in Science Teaching (NARST), National Harbor, MD.
Zangori, L., & Forbes, C. T. (2015). Exploring third-grade student model-based explanations about plant relationships within an ecosystem. *International Journal of Science Education, 37*(18), 2942–2964. https://doi.org/10.1080/09500693.2015.1118772

**Regina H. Yorkgitis** Yorkgitis is a PhD student in education at the University of North Carolina, at Chapel Hill, USA, where she studies the role of literacy in secondary science classrooms. She holds an MEd from Vanderbilt University.

**Troy D. Sadler** is a professor of science education and serves as the Thomas James Distinguished Professor in Experiential Learning at the University of North Carolina at Chapel Hill School of Education, USA. His research focuses on how learners negotiate complex socio-scientific issues and how these issues may be used as contexts for science learning.

**Eric A. Kirk** is a postdoc in the Instructional Technology and Learning Sciences department at the Utah State University Emma Eccles Jones College of Education and Human Services. Kirk's research centers around how learners leverage science practices to understand socioscientific issues and how these learning experiences can support learners in becoming informed members of pluralistic society.

# Chapter 9
# The Rain Gauge as a Limit-Act: Practical Work from the Freirean Perspective

**Marcos Corrêa da Silva, Glauco dos Santos Ferreira da Silva, João Paulo Fernandes, Isabel Martins, Bruna Karl, and Marcilia Elis Barcellos**

## 9.1 Introduction

This chapter presents the theoretical bases as well as the empirical results of a pedagogical experience carried out in the context of a course called "Scientific Development, Globalization, and Environmental Impacts," which was integrated into the Youth and Adult Education[1] (EJA herein) curriculum from a state run secondary school in the city of Petrópolis, in the mountainous region of Rio de Janeiro, Brazil. This activity was connected to two funded projects,[2] which involved science education researchers, graduate students, teacher educators, in-service and pre-service physics and mathematics schoolteachers, and secondary school students. Its main goal was to develop horizontal methodologies and didactic materials for the teaching of socio-scientific issues following a Science Technology Society (STS) approach. The funding included a budget to buy equipment for schools, scholarships and

M. C. da Silva (✉) · G. dos Santos Ferreira da Silva · J. P. Fernandes · M. E. Barcellos
Federal Center for Technological Education of Rio de Janeiro, Rio de Janeiro, Brazil
e-mail: marcos.silva@cefet-rj.br

G. dos Santos Ferreira da Silva
e-mail: glauco.silva@cefet-rj.br

J. P. Fernandes
e-mail: joao.fernandes@cefet-rj.br

M. E. Barcellos
e-mail: marcilia.barcellos@cefet-rj.br

I. Martins · B. Karl
Federal University of Rio de Janeiro, Rio de Janeiro, Brazil
e-mail: isabelgrmartins@gmail.com

B. Karl
e-mail: brunakarl@outlook.com



L. Leite et al. (eds.), *Practical Work in Science Education in the Digital Era*, Advances in Technology-Rich Science Education 2, https://doi.org/10.1007/978-3-032-13788-3_9

stipends that allowed for the participation of schoolteachers and students to carry out the project's activities.

This text describes how the project articulated principles of Freirean education, co-teaching and problem-solving methodologies, to develop participants' critical understandings and actions to overcome an extreme situation related to torrential floods which commonly affect the region. Following a Freirean perspective, the course syllabus was not pre-defined or imposed by a group of experts, but rather constructed in horizontal dialogue amongst participants, based on the problematization of local realities. The construction of a rain gauge was understood as a limiting act (Freire, 2005) giving new meaning to the concept of practical work, in dialogue with other approaches to conceptualizing practical work in science education.

## 9.2 Framing Elements: Freirean Perspective, Co-Teaching and Practical Work

The educational perspective that guides our activities is based on three pillars. The first is the Freirean pedagogy, more specifically, the concepts of limit-situation (*situação limite*), limit-act (*ato limite*) and untested feasibility (*inédito viável*). Situation limits are a way to describe oppressive realities that prevent people from realizing their potential. While they may have strong social and historical roots, they can also be thought of as a frontier for human action.

> For Vieira Pinto, the "limit-situations" are not "the impassable boundaries where possibilities end, but the real boundaries where all possibilities begin"; they are not "the frontier which separates being from nothingness, but the frontier which separates being from nothingness but the frontier which separates being from being more." (Freire, 2005, p. 99).

Thus, they may give rise to limit-acts, which are seen as an active transformative response aimed at overcoming impositions and constraints. Reflecting upon limit-situations is, therefore, more than passive understanding of reality but adopting an active role in changing it. The analysis of limit-situations involves the identification and investigation of generative themes, which would lead to an awakening of critical consciousness.

These philosophical bases are at the basis of Freirean pedagogy and its adult literacy methods for Young and Adult Education. Freire's well known literacy experience in Angicos, Rio Grande do Norte, Brazil, stands out as a landmark for Brazilian education (Freire, 2024). It involves the acquisition of reading and writing skills from a critical analysis of certain aspects of the community's culture (Mayo, 2008). From his perspective, education must consider the relationship between students and their cultural environment, orienting them toward transformation and emancipation. In this sense, "authentic education is not carried on by "A" for "B" or by "A" about "B," but rather by "A" *with* "B," mediated by the world—a world which impresses and challenges both parties, giving rise to views or opinions about it" (Freire, 2005, p. 93).

According to Freire (2005) relevant aspects of the relationship between subjects and the world may be concealed in limit-situations.

> When the themes are concealed by the limit-situations and thus are not clearly perceived, the corresponding tasks - people's responses in the form of historical action - can be neither authentically nor critically fulfilled. In this situation, humans are unable to transcend the limit-situations to discover that beyond these situations - and in contradiction to them - lies an *untested feasibility* (p. 102).

In practice, the methodology used for promoting an enhanced perception of the limit-situations involves the process of encoding and decoding images in circles of thematic reduction. These are produced through dialogue between educators and students and between students and educators. Freire (2005) states that:

> In contrast with the antidialogical and non-communicative "deposits" of the banking method of education, the program content of the problem-posing method - dialogical par excellence - is constituted and organized by the students' view of the world, where their own generative themes are found. The content thus constantly expands and renews itself. The task of the dialogical teacher in an interdisciplinary team working on the thematic universe revealed by their investigation is to "re-present" that universe to the people from whom she or he first received it - and "re-present" it not as a lecture, but as a problem. (p. 109)

Such views on the importance of dialogue and critical collective reflections are also found on co-teaching models, the second pillar. Co-teaching entails the possibility of less hierarchical relationships between subjects, institutions and knowledge systems which are usually asymmetrically positioned in society like, for instance, teachers and learners; schools and universities; popular and scientific knowledge. (Tobin, 2006) enjoy learning is an increase in the power-to-act in the real world, characterized by an increase in the actions available to the individual. On the one hand, co-teaching redefines teaching by transforming the work of the teacher, usually recognized from an individual perspective, into a collective one. On the other hand, collaboration between co-teachers is a process of negotiation and tension that reflects the asymmetrical power relations that exist between universities and schools. Traditionally, universities are recognized as centers of knowledge production, while schools are represented as fields of application of this knowledge (Dumrauf & Cordero, 2017; Zeichner, 2010). Thus, co-teaching models involve disrupting such established hierarchies through the proposition and use of horizontal methodologies.

In practice, it requires creating both conceptual and contextual environments where co-teachers can develop the habit of attentive listening and a spirit of openness to negotiating differences. Such attitudes pave the way to new relationships based on respect, trust, horizontal dialogue, affection, and the circulation of social roles may emerge (Corrêa, 2021; Silva et al., 2023). In this way, co-teaching models also encourage participation and protagonism of the actors involved in the development of collectively planned activities. Most often co-teachers come from multidisciplinary backgrounds, have different degrees of professional experience, diverse implications and roles in the teaching and learning processes, and belong to different age groups. Such diversity, instead of representing an obstacle, becomes a source of alterity and allows new social positionings and new meanings to emerge. Consequently, co-teaching, as a horizontal methodology, facilitates dialogue between subjects and between them and the world, as it allows educators with different backgrounds to participate in the same space and time. When their knowledge is brought into dialogue, it paves the way, together with students, for a critical understanding of the existence of extreme situations (Tobin, 2006).

This openness to dialogue is a prerequisite for those who practice co-teaching, as is the awareness that this horizontal dialogue is not pre-established but rather needs to be constructed. This construction begins with the recognition of the power differential that historically exists in the relationships between educators and students, which, in our case, is deepened by the context of extension in the university-school relationship Scantlebury et al., 2008). Therefore, educators in co-teaching launch themselves into the world, alongside students, with the conviction that they need to open spaces for others, who historically remain silent. In the pursuit of authentic dialogue, there is always the risk that, faced with the silence of others, dialogue will become anti-dialogue, where educators begin to dictate rules and content based on prior perceptions of the students' reality, without first initiating a process of thematic investigation (Silva et al., 2018).

The process of attentive and horizontal listening was grounded on a sincere interest in understanding, through the eyes of the individuals living in the city of Petrópolis, what elements manifest as obstacles to "being more". Thus, it is possible to say that the activity begins with inquiry. However, the bases for such inquiry are not pre-established: they reflect the openness to dialogue that underpins co-teaching. In the context of the co-teaching environment, diversity and plurality were valued in so far as they stimulate conditions for multiple readings of the world and, in this way, constitute a requirement for dialogue to take place. It is in and through dialogue that critical reflection about reality can emerge and lead to the perception of limit-situations, in which structural conditions may hinder subjects' potentials and power to act. As said before, limit-situations can lead to limit-acts, that is a form of intervention in the world resulting from the transition from actual consciousness to possible consciousness. In the context of science teaching and learning science, acting upon the world involves inquiry and interaction with the physical world.

This recognition leads to the third pillar of our investigation, that is, the nature and the potential of practical work, reconfigured as an element of an essential step in Paulo Freire's pedagogical practices: "the reading of the world". Practical work has been considered central to science education, given its relationship with the process of constructing scientific knowledge. By identifying practical work with experimental activity, we can consider its importance in the constitution of modern Western science. However, the role of practical work in science, as well as in Science Education, has long been problematized. For instance, rationalist perspectives have challenged naive empiricist ideas that scientific models are direct—and universal—reflections of reality. More recently, curriculum approaches to practical work in science education have sought to overcome the hands-on versus minds-on dichotomy. Debates about the nature of science and of the processes of knowledge production also discuss experiments as a constitutive element of the scientist's work (Allchin, 2013; Hodson, 1994; Millar, 2004; Osborne, 2015; Wellington, 1998).

From an epistemological perspective, experimentation is a stage of research that represents the synthesis of the theory–practice relationship. When planning experimental activity, the researcher uses equipment designed to reveal dimensions of the phenomenon, the magnitudes capable of describing its behavior being predicted in the theoretical plan and measurable, that is, quantified, using some instrument. Thus, designing an experiment requires the planning and operation of a set of devices that, taken together, constitute the synthesis of theory and practice. Experimentation, and consequently, practical work, constitutes a form of intervention in the world guided by theory. As historian and philosopher of science Alexandre Koyré put it, "experimentation is a teleological process whose end is determined by theory" (Koyré, 1991, p. 272).

Oliveira and Bonito (2023) have discussed practical work in terms of the kinds of skills they involve and the conditions under which they are mobilized. They include inquiry-based learning, learning through everyday experiences, promoting scientific communication, and developing both accessible low-cost alternatives to science teaching and learning.

> Overall, the systematic literature review reveals that the concept of practical work is prominently assumed as a process that goes beyond allowing the development and improvement of practical skills associated with handling of laboratory material. It is above all, a method that allows the students to understand the nature of science by involving them in activities that mimic the action of scientific research processes. This understanding results from the construction and mobilization of scientific knowledge, through critical thinking capable of raising hypotheses and formulating ways to test them, simultaneously enabling the theoretical and conceptual reflection of the phenomena in question, in a minds-on approach. (Oliveira & Bonito, 2023, p. 14).

Usually, practical work in science education is seen as a mediating instrument between science as it is produced, and science as intended to be taught. While this perspective has inspired science materials which explored the role of practical work to educate a future generation of scientists in the context of the Cold War and the Space Race (Oliveira & Bonito, 2023), it neglects an analysis of its potential to educate future citizens. Moreover, when we consider contemporary challenges faced by science education, especially in young democracies of the Global South, educating for citizenship acquires different contours. It requires more than the grasp of concepts, laws, theories, and the acquisition of experimental skills; it goes beyond the understanding of internal aspects of science, requiring the connection of these elements of science with social and cultural aspects.

Science education, from the perspective of citizenship education, highlights the importance of scientific knowledge as a tool for interpreting the world. Learning science allows students to relate to the world differently, as they can appropriate a language that has become hegemonic and whose mastery empowers forms of intervention in the world, still dominated by a few (Santos, 2007). Learning science, with the goal of empowering individuals to act in the world, gives science education a sociopolitical dimension, little explored in discussions about the importance of practical work in citizenship education (Kauano & Marandino, 2021).

Considering practical work as a form of intervention in the world, we can consider not only the meaning of this intervention for the individuals who produce it but also the impacts of this intervention in real life. This is particularly relevant for individuals who find themselves in situations of social vulnerability and environmental risk. In this scenario, attentive listening is necessary, highlighting the sociopolitical dimension of science education and the possibility of constituting practical work as a limiting-act.

Then, for practical work to gain this dimension, which imbues it with an emancipatory character, it must be a form of acting on the world capable of overcoming obstacles to the full development of individuals, when they perceive, through a process of thematic investigation, what constitutes a limit-situation. The idea of acting connects with the idea of acting, acting on the world, transforming it and transforming oneself (Freire, 2005). Practical work only assumes the dimension of a limiting act when it emerges as a form of intervention in the world proposed by individuals who engage in authentic dialogue with each other, necessarily horizontal, by not allowing the voices of the excluded to be silenced, nor by allowing the voices of those in historically privileged positions to echo alone.

## 9.3 The Experience of Creating the Discipline "Scientific Development, Globalization and Environmental Impacts"

The perspectives we have described so far were instrumental to guide the group's activities in the creation and implementation of the discipline. It ran from March to June in 2023 and was conducted throughout 15 weekly sessions, each one of which lasted for two hours. The first session consisted of a meeting between the co-teachers and the 30 students in the class. The group of educators consisted of a history teacher and a geography teacher from the elementary school, a postgraduate student in science education, three physics undergraduates, and a physics professor from the university. The students were mostly male working adults, between 20 and 50 years old.

At this point, it was emphasized that the development of the course would be collective, with the active participation of all involved—educators and students alike—proposing an authentic dialogue in which educators do not position themselves as authorities with exclusive control over the knowledge to be transferred to students, because it is "it is to the reality which mediates men, and to the perception of that reality held by educators and people, that we must go to find the program content of education" (Freire, 2005, p. 96).

The dialogue begins with the search for programmatic content through a process of thematic investigation which was carried out in a Conversation Circle, guided by the following questions:

- "What is your motivation and what do you hope to achieve by attending Young and Adult Education courses?"
- "What is the role of the school in the community?"
- "Do you feel as if you belong to the community? What are the main problems and difficulties faced within it?"
- "What is your relationship with this territory?"

Acting in co-teaching, educators were hand in hand students and referenced in the world. The group included history, physics, and geography in-service and pre-service teachers and EJA students who jointly defined the theme that would serve as a guiding thread for the process of reading the world to be experienced in the course.

The discussion took place in the school auditorium and was attended by most students, who demonstrated engagement, promoted debates among themselves, and respected speech turns of their peers and educators. The diversity in experience and interests was evident in testimonies and in the problematization of actual—lived and living—conditions. Aspects such as reasons why studies had to be interrupted at a certain point in life or the role of schooling in attaining goals in life were present in the debate and were instrumental for the following steps of the activities.

Following this initial debate, the co-teachers proposed a reality-coding activity for the students, which involved approaching concrete situations they experience in their daily lives by selecting images which were seen as representative of the problems involved in them. Images were accompanied by a brief explanatory text (Fig. 9.1). This reality coding activity exercise was used to start a discussion about students' views on the main challenges facing the municipality of Petrópolis. Images were shared with the whole class and, following a decoding exercise, several themes emerged: irregular occupations in risk areas, the occurrence of natural disasters, such as large landslides, issues related to public transportation, urban mobility, and social vulnerability.

During the decoding process, there were questions about the origin, allocation, and management of public resources connected to the prevention, mitigation and solution of the problems identified. As a response to this demand, co teachers planned and developed a session focusing on analyzing the Transparency Portal (https://www.petropolis.rj.gov.br/transparencia/#gsc.tab=0), a governmental digital tool that allows public and free access to detailed information on the use of public budgets at the municipal, state, and federal levels.

Throughout the activity, the public nature of resources administered by the State was emphasized, highlighting that such funds belong to the community, which legitimizes the right—and duty—of civil society to monitor, oversee, and question the management of these resources, especially regarding the actions of public and political agents. Co-teachers also sought to emphasize the social role that each individual can and should play within a collective context. This perspective included, for example, the importance of voting, oversight, and active participation.

Following this activity, the thematic analysis process continued with answering a reflective questionnaire. The questions were formulated to encourage students to reflect on their social and political role, fostering the development of a critical stance

**Fig. 9.1** Examples of images brought by the students to the activity

toward collective issues, and encouraging civic practices of research, monitoring, and accountability for public action.

Among the proposed questions, three were related to social problems identified by the students. They were selected for a more in-depth qualitative analysis of the responses. Environmental issues emerged spontaneously and with high frequency in the participants' responses. Based on the systematization and analysis of the collected discursive content, a new reality coding was proposed in the form of a word cloud representing the most recurring themes was created (Fig. 9.2).

In the word cloud showed in Fig. 9.2, "flood" (*enchente)* was the most frequently mentioned word in students' responses. Together with "landslides" (*deslizamento*), "river drenching" (*dragagem dos rios*), "works" (*obras*); "rubbish collection" (*coleta de lixo*) and "inspection" (*fiscalização*), it conjures public perception of as a limiting situation, that is of a city "squeezed" between rivers and mountains, which grew in a disorderly manner and faces environmental problems of socio-historical nature and that are linked to inefficient public resource management policies (Fernandes et al., 2024).

**Fig. 9.2** Word cloud created from students' responses to the questionnaire

Motivated by the emergence of students' responses to socio-environmental disasters, co-teachers invited a representative from the Specialized Group on Natural Disasters (GEDEN) to discuss aspects related to natural disasters. During this visit, the class was introduced to a low-cost measuring instrument for monitoring rainfall levels in the city: a homemade rain gauge. Students were informed that this simple tool could have saved the lives of many people during the tragedy that occurred in the city had it been available to be used in households.

At this point, many students questioned why this information had not been widely disseminated in Petrópolis, especially in areas which were more likely to be affected by the floods and their consequences. This question led to critical reflection about the possibilities of building and using the tool considering its potential to effect significant changes to the existential situation of the population of Petrópolis who live in high-risk areas.

## 9.4 The Rain Gauge as a Limit-Act

After the activity with the GEDEN representative, co-teachers suggested that students to go into the science lab to build the rain gauge as a practical project.

The instructions for constructing the rain gauge were taken from the Petrópolis City Hall website and the sticker used is distributed free of charge by the city's Civil Defense, which provided them for us to carry out construction at the school in the following meetings (Fig. 9.3).

On the day of the activity, students were asked to bring empty, smooth plastic bottles. Before beginning the construction of the instrument, some conceptual dimensions were introduced and triggered relevant discussions. It was possible to discuss,

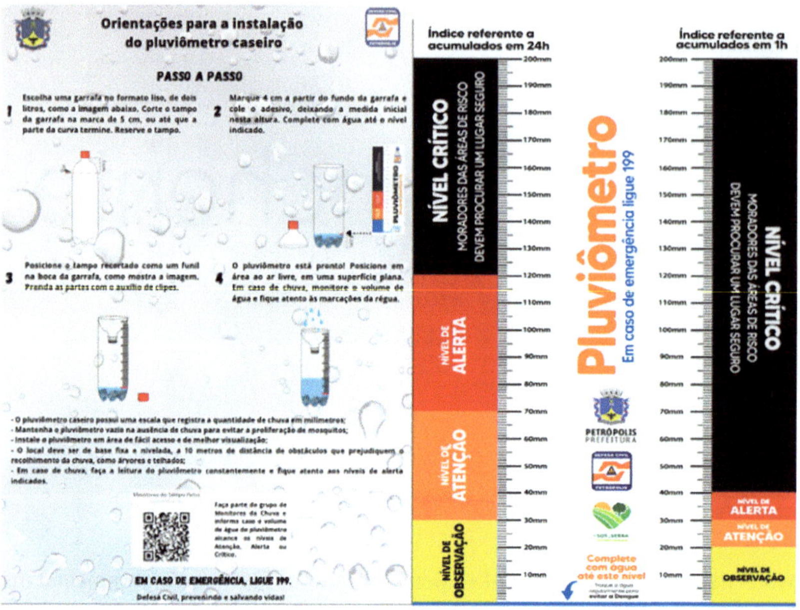


**Fig. 9.3** Instructions and scale stickers for building a homemade rain gauge

for instance, the impact of water on the soil and its relationship to landslides. Also, the group explored visible signs of ground movement that can trigger landslides, such as trees with curved trunks, among others.

The homemade rain gauge was made with a plastic bottle and a sticker that measures the amount of rainfall per hour, as shown in Fig. 9.4. After constructing the tool, co-teachers went on to explore procedural dimensions of the activity, by providing guidance on how to install the equipment and perform rainfall measurements.

The rain gauge emerged to address the extreme situation, not in its entirety, but in part, as it enabled the local population to monitor the risk of rainfall and, as a result, became an important tool for protecting life. Thus, construction of the rain gauge was introduced as a practical project, enabling the team of educators to discuss conceptual and procedural dimensions that connect science learning to the development of practical skills and action in the world, using an instrument that translates nature's behavior into measurement.

In this sense, the rain gauge takes on the dimension of a limit-act. With the perception of the limit-situation, which involves themes and points out tasks, the students were able to expand their real awareness of the world, bringing the rain gauge as a possible solution, which previously went unnoticed. Thus, according to Freire (2005), "in the last analysis, the *themes* both contain and are contained in *limit-situations*; the tasks they imply require *limit-acts*" (p. 102).

**Fig. 9.4** Homemade rain gauge built with low-cost materials

## 9.5 Conclusion and Ways Forward

In this chapter, we presented a pedagogical experience where practical work was re-signified in the context of a co-teaching activity founded on both Freirean methodologies and co-teaching principles. The rain gauge was an activity that mobilized hands-on and minds-on skills (Oliveira & Bonito, 2023) that changed the way students engaged with the knowledge being produced. This was possible, in the context of the limit-situation identified, by reconfiguring the role of practical activity as an important element not only in the process of conscientization but also in the active transformation of reality. Thus, apart from promoting the development of students' procedural knowledge skills, the practical activity allowed for STS principles to be realized and contextualized in the process of Freirean thematic investigation of the limit-situation.

The rain gauge's construction was also a context to widen both co-teachers' and students' access to information and institutional agencies and resources (e.g. interactions with civil servants who are responsible for public policy, access to resources provided by the Petrópolis City Hall and Civil Defense, such as evacuation and rain gauge construction manuals, graduated information labels to be attached to bottles used to collect rainwater and indicate the risk level and the necessity of leaving the affected areas as shown in Fig. 9.3). In this sense, there is a political dimension implied in the relationships between subjects, knowledge and institutions, which constitute practical activity. Thus, the rain gauge serves as an instrument, a material synthesis of various dimensions: scientific, historical, political, and economic. These are the elements which allow us to characterize the construction and use of the rain gauge as a limit-act, that is, as a response to the challenge posed by a limit-situation, as a tool for transforming students' reality. Furthermore, it expands pedagogical possibilities for practical work beyond those already found in the literature, by revealing its sociopolitical dimension.

From this perspective, problem-solving education, focused on the critical analysis of reality, begins with authentic dialogue, in which knowledge is not imposed by one group on another, but rather constructed in communion between people and between them and the world. The experience described reinforcing the potential of Freirean pedagogy, carried out in a context of co-teaching methodology, to help us understand that conceptual, procedural and attitudinal skills can be fostered in contexts which do not neglect or ignore relevant aspects of sociopolitical reality.

**Acknowledgements** FAPERJ (*Fundação Carlos Chagas de Amparo à Pesquisa do Estado do Rio de Janeiro*) and CNPq (*Conselho Nacional de Desenvolvimento Científico e Tecnológico*) are gratefully acknowledged for financial support.

## Notes

1. Youth and Adult Education (EJA) is one of the modalities of Brazilian basic education, aiming to provide formal education for students who, for various reasons, did not have access to or did not continue their studies at the appropriate age (UNESCO, 2023).
2. The first project, called "Futuring with Science", was funded by FAPERJ - Rio de Janeiro State Research Agency. The second project, "Curriculum Challenges for Teacher Training: Insurgent Demands and Contemporary Institutional Experiences," has been funded by CNPq - Brazilian Science and Technology Research Council.

## References

Allchin, D. (2013). *Teaching the nature of science: Perspectives & resources.* SHiPS Education Press.

Corrêa, M. (2021). *Codocência e estágio curricular supervisionado: Uma abordagem discursiva da articulação entre universidade e escola na formação de professores [Coteaching and internship: A discursive approach to the articulation between university and school in teacher training]* [Unpublished doctoral dissertation]. Federal University of Rio de Janeiro.

Dumrauf, A., & Cordero, S. (Coords.). (2017). *Tramas entre escuela y universidad: Formación docente, innovación e investigación colaborativa [Articulation between school and university: Teacher training, innovation and collaborative research].* Edulp.

Fernandes, J. P., et al. (2024). Mobilização do movimento Ciência-Tecnologia-Sociedade e das questões sociocientíficas sob lentes freireanas na relação estabelecida entre universidade-escola [Mobilization of the Science-Technology-Society movement and Socio-Scientific Issues under a Freirean lens in the relationship established between university-school]. *Indagatio Didactica, 16*(3). https://doi.org/10.34624/id.v16i3.38523

Freire, P. (2005). *Pedagogy of the oppressed*. Continuum International Publishing Group.
Freire, P. (2024). Discurso do professor Paulo Freire em Angicos, ao encerramento do curso de alfabetização de adultos [Speech by Professor Paulo Freire in Angicos, at the closing of the adult literacy cours]. *Germinal: Marxismo e Educação em Debate, 16*(2), 485–491. https://doi.org/10.9771/gmed.v16i2.64079
Hodson, D. (1994). Hacia un enfoque más crítico del trabajo de laboratorio [Towards a more critical approach to laboratory work]. *Enseñanza De Las Ciencias, 12*(3), 299–313.
Kauano, R. F., & Marandino, M. (2021). Paulo Freire in science education: Trends and articulations between scientific literacy and the STSE movement. *Revista Brasileira de Pesquisa em Ed. em Ciências, 22*, 1–27. https://doi.org/10.28976/1984-2686rbpec2022u549575
Koyré, A. (1991). *Estudos de história do pensamento científico [Studies on the history of scientific thought]* (2.ª ed.). Forense Universitária.
Mayo, P. (2008). Paulo Freire and adult education. In A. A. Abdi & D. Dip (Eds.), *Global perspectives on adult education* (pp. 95–107). Palgrave Macmillan. https://doi.org/10.1057/9780230617971_6
Millar, R. (2004).The role of practical work in the teaching and learning of science. Paper prepared for the Committee: High School Science Laboratories: Role and Vision,National Academy of Sciences, Washington, DC.
Oliveira, H., & Bonito, J. (2023). Practical work in science education: A systematic literature review. *Frontiers in Education, 8*. https://doi.org/10.3389/feduc.2023.1151641
Osborne, J. (2015). Practical work in science: Misunderstood and badly used? *School Science Review, 96*(357), 16–24.
Santos, B. S. (2007). Para além do pensamento abissal: Das linhas globais a uma ecologia de saberes [Beyond abyssal thinking: From global lines to an ecology of knowledge]. *Novos Estudos CEBRAP, 79*, 71–94. https://doi.org/10.1590/S0101-33002007000300004
Scantlebury, K., Gallo-Fox, E. J., & Wassell, B. (2008). Coteaching as a model for preservice secondary science teacher education. *Teaching and Teacher Education, 24*(4), 967–981. https://doi.org/10.1016/j.tate.2007.10.008
Silva, G., Barcellos, M., & Souza, E. (2018). Mind the gap between science, teaching and education: Challenges for enacting bold and transformative visions in science education. In L. Bryan & K. Tobin, *Critical issues and bold visions for science education* (pp. 13–28). Brill Sense. https://doi.org/10.1163/9789004389663_002
Silva, G. S. F., et al. (2023). Graph analysis of an expanded co-teaching activity in the context of physics teacher education. In K. Plakitsi & S. Barma (Eds.), *Sociocultural approaches to STEM education* (Vol. 21, pp. 199–217). Springer. https://doi.org/10.1007/978-3-031-44377-0_10
Tobin, K. (2006). Learning to teach through coteaching and cogenerative dialogue. *Teaching Education, 17*(2), 133–142. https://doi.org/10.1080/10476210600680358
UNESCO. (2023). *UNESCO strategy for youth and adult literacy and its action plan (2020–2025)*. UNESCO. https://unesdoc.unesco.org/ark:/48223/pf0000387424
Wellington, J. (1998). *Practical work in school science: Which way now?* Routledge.
Zeichner, K. (2010). Rethinking the connections between campus courses and field experiences in college- and university-based teacher education. *Journal of Teacher Education, 61*(1–2), 89–99. https://doi.org/10.1177/0022487109347671

# Chapter 10
# Field-Based Practical Activities for School Science Learning: A Literature Review centred on Students

**Luís Dourado, Laurinda Leite, and Luísa Carvalho**

## 10.1 Introduction

Science is about the natural world. It takes natural phenomena as objects of study and gives meaning to them. Thus, the goal of science is the construction of meaningful models that afford the best understanding of natural phenomena, even acknowledging the recognized limitations of those models and knowledge (Osborne & Pimentel, 2023). Science education involves teaching and learning models and meanings that are of enduring value for students. Some school science topics have a direct relationship with nature and natural phenomena that are observable with the naked eye or with the help of some easy-to-use tools. In these cases, it is possible to learn directly from them and about them. To do so, a spatial shift to outside the classroom is required, although it is not enough. It is necessary to consider the educational possibilities that being outdoors and in nature can open up to students, including a more responsive way of encountering and engaging with the world (Skea & Fulford, 2021) and the possibility of learning science in a contextualized manner (Ayotte-Beaudet et al., 2023; Giamellaro et al., 2022).

Some phenomena can be physically or virtually reproduced in school for study purposes; others are too large to be reproduced indoors or too slow (compared to human life) to observe as they evolve from their beginnings. Reproducing a phenomenon decontextualizes it, withdrawing the framework for its complete and

L. Dourado (✉) · L. Leite
CIEd - University of Minho, Minho, Portugal
e-mail: ldourado@ie.uminho.pt

L. Leite
e-mail: lleite@ie.uminho.pt

L. Carvalho
Lusophone Higher School of Education, IPLUSO, Lisboa, Portugal
e-mail: luisa.carvalho@ipluso.pt



L. Leite et al. (eds.), *Practical Work in Science Education in the Digital Era*, Advances in Technology-Rich Science Education 2, https://doi.org/10.1007/978-3-032-13788-3_10

accurate understanding and reliable interpretation. This is why science should be taught and learned not only in school, but also in a variety of out-of-school places, which may be near or far from the school. These include the natural environment where natural phenomena occur (Hamidy et al., 2024), as well as man-made or man-intervened places (Lima et al, 2010) that use scientific knowledge for leisure (Moll, 2010), goods production, or other purposes.

To study out-of-school phenomena, students need to 'go there', where phenomena occur. Several words have been used to name those 'educational displacements', namely: study visit (e.g., Falk & Dierkering, 2000) which emphasizes the relationship with (in)formal learning environments; excursion (e.g., Compiani & Carneiro, 1993) that highlights its affective/playful role; place-based learning (e.g., Yemini et al., 2025) that puts the chosen place at the centre; field trips (e.g., Orion, 1993), which focus on the way to the phenomena rather than the phenomena itself; fieldwork (e.g., Dummer et al., 2008) that stresses the actions in the field and the practical nature of this educational resource. At first glance, fieldwork may seem a single or homogeneous entity. However, fieldwork encompasses a diversity of practical field activities with different purposes (Skea & Fulford, 2021), structures, and aims (Dourado & Leite, 2013). To perform activities 'out there', students need to make a short or long trip (Del Carmen, 1999), depending on where they can do the planned field-based practical work, including one or more field activities.

Field activities encompass a set of procedures focused mainly on observation and inquiry (Dummer et al., 2008), fostering a dialogue between school knowledge and the real environment, which leads to cognitive (or conceptual), affective, and procedural learning (DeWitt & Storksdieck, 2008; Guilherme et al., 2016). They constitute a potentially rich pedagogical tool in several disciplines, including the sciences, allowing contact with the real world and converting the surrounding environment into a teaching and learning resource (Henriques et al., 2012).

Like other educational resources, the way field activities are structured and implemented at a given moment should be consistent with science education research findings and curriculum guidelines. Using field activities means an additional workload for teachers, the allocation of school resources, and the occupation of students' time (Lock, 2010). Even if inexpensive or cost-free field trips are considered (Behrendt & Franklin, 2014), performing field activities should be more than letting students have a good time. Garner and Eilks (2015) stress that linking outer- and inner-school learning is the biggest challenge for teachers. Hence, it is worth considering how field activities and their use have been evolving, and to what extent school students are learning from them.

This chapter aims to present a critical overview of the research on field-based practical activities for school science learning, focusing on students and considering three time periods. This review differs from others that consider a single period (DeWitt & Storksdieck, 2008; Mann et al., 2022), a single discipline (Lock, 2010), a specific group of students (Chasen et al., 2025), a specific kind of field trip (Behrendt & Franklin, 2014) or a region (Remmen & Iversen, 2023), and, in doing so, it adds to the present knowledge base. Future directions for research studies on field-based science education in the digital era are also discussed.

## 10.2 The Educational Value of Field Activities

The destination of a field trip can be within walking distance of the school (Wells et al., 2015) or may be far from it (Dummer et al., 2008; Hamidy et al., 2024; Lock, 2010). The former is more advantageous (Rowe & Humphries, 2004) as it does not cost money, or require too much time, causes reduced interference in the school timetable and operation, and is easier to get permission from students' parents or legal representatives; the latter is more demanding in terms of logistical issues (Reiss & Braund, 2004), requiring permission from students' parents or legal representatives, funding for transportation (by bus, train, or boat) and possible reorganization of the school timetable. In any case, safety (Reiss & Braund, 2004) and ethical considerations (Telford et al., 2024) must be taken into account.

From an educational viewpoint, the distance to the destination is not important; what is important is what students do in the field and the appropriateness of the destination (Braund & Reiss, 2004). Especially when the distance is long, students should carry out as many activities as possible, these being relevant to one or more school subjects (Ortega-Torres & Moncholí-Pons, 2021). In the latter case, fieldwork may range from multidisciplinary to cross-disciplinary (Dourado & Leite, 2013), according to the degree of integration of knowledge from diverse subjects in the activities to be performed. From a motivational point of view, students may prefer destinations far from school, as the trip serves as a get-together opportunity for them. There is nothing wrong with choosing field trips or places that students enjoy, as enjoyment can favour learning (Dettweiler et al., 2015). However, concerns with students' enjoyment should not outweigh learning possibilities.

Field activities can promote several kinds of learning in an integrated manner (e.g., Braund & Reiss, 2004; King & Glackin, 2010; Mann et al., 2022; Remmen & Iversen, 2023), which means that a single activity can lead to the development of more than one of the following learning areas:

- Conceptual learning, if they foster interactions between school science concepts and students' everyday (natural and social) environments;
- Procedural learning, if they require students to use physical or digital tools and techniques to study the phenomena;
- Epistemological learning, if the activities promote students' awareness of the breadth, diversity, and complexity of the environment and the phenomena that characterize it, as well as the strategies to uncover those features and understand how they work together;
- Attitudinal learning, if they allow students to recognize and value nature and commit to its rational and sustainable use;
- Ethical behaviours toward the natural environment, if activities promote a culture of critical observation of perceived reality in favour of better living conditions and environmental preservation;
- Student autonomy development, if they stimulate students' ability to acquire knowledge about the phenomena, and to relate school content to daily life, without depending on the teacher;

- Peers' collaboration, if they require small group work or teamwork, which fosters the development of interpersonal and communication skills;
- Digital skills, as nowadays several digital technologies can be used to collect, register, and analyse data, and may also facilitate inclusive fieldwork;
- Social intervention skills, if they raise students' awareness of the challenges facing nature and prepare students to consciously intervene as active citizens against the environmental threats they may encounter.

Despite the list above, the field should not be seen as 'the best' place for teaching. As some authors argue (DeWitt & Storksdieck, 2008, p. 181), the field offers "opportunities for exploration, discovery, first-hand and original experiences", but it does not show the theory underlying nature. Hence, learning from field activities should start and continue in the classroom and or in the school laboratory.

## 10.3 Learning from Practical Field Activities

Field trips are necessary to undertake fieldwork, which may involve one or more field activities. These require careful planning and preparation due to the costs involved and the diversity of phenomena that characterize the landscape where the fieldwork will be carried out. Planning also needs to take students' diversity into account for the sake of inclusive fieldwork (Chasen et al., 2025).

Results that can be achieved with a field trip depend on the type(s) of activities it includes and how they are implemented. Several authors (e.g., Orion, 1993; Pedrinaci et al., 1994), concluded that the outcomes of field trips can be influenced by several factors, including: objectives of the field trip and activities; logistical preparation of the field trip; previous preparation of observation activities, as well as their quality; novelty of the location for the students; students' prior knowledge and interest on the visit; structure of the field activities to be performed; students' engagement in the field activities; social context of the field trip. Therefore, when preparing fieldwork (Orion & Hofstein, 1994), as any other kind of practical work (Osborne, 2015), there are three key stages to consider, taking as reference the activities to be done: preparation of the field trip, including the planning of the activities to be performed, which is crucial for successful learning; implementation of activities in the field, as planned; and post-visit tasks and assessment, to consolidate previous learning and get feedback on the fieldwork carried out.

The activities carried out in the field can aim at diverse learning outcomes, but attaining those outcomes may depend on factors like the degree of openness of the activities and the way they are implemented and integrated into the teaching sequence (DeWitt & Storksdieck, 2008; Orion & Hofstein, 1994). They may complement previous teaching or enable contact with nature, to become familiar with some aspect of it, or take nature as a learning object, acting as a starting point for learning to continue at school (Dourado & Leite, 2013), either in the lab or in the classroom.

Based on those factors, some authors developed taxonomies of field activities. For example, Compiani and Carneiro (1993) classified field activities according to their learning aims; Pedrinaci et al. (1994) based their classification on the teaching approach used to implement field activities. Dourado and Leite (2013) integrated those taxonomies and classified field activities based on their underlying learning objectives, the tasks to be carried out by the teacher and the students, and the guidance provided to students. They obtained seven types of field activities:

- Motivating, to raise students' interest in an issue to be studied (e.g., observe a river where dead fish were found to study pollution);
- Training, to promote learning of techniques and the mastery of practical skills (e.g., using a compass to measure a slope and the direction of a geological plan; operating a drone to make pictures of the landscape);
- Illustrative, to consolidate previously acquired knowledge (e.g., observe rocks in the school surroundings and relate their features to those previously studied in the classroom);
- Guided observation, to 'discover' or ascertain something, based on a worksheet (e.g., observe the process of industry ammoniac production; identify its phases following a worksheet; identify animals, based on a worksheet with animal features);
- Inductive, to observe and interpret a part of the environment to discover conceptual knowledge (e.g., compare plants that grow on sun-exposed and sun non-exposed areas to conclude on their dependency on sun exposure);
- Problem-posing, to develop an investigative spirit (e.g., observe a place where the iron parts are corroded and identify the issues that deserve to be studied or worked out, to learn more or solve the problem later on);
- Problem-solving, to investigate the process of solving an identified problem (e.g., find out why the water from a given fountain is murky).

The "fundamental purpose of practical work is to help students make links between the world of ideas and the real world of objects and events" (Osborne, 2015, p. 19). Therefore, undertaking fieldwork is not necessarily better than not doing it; its educational value depends on whether field activities are organized in such a way that they facilitate the link and thus make a meaningful difference in terms of learning outcomes attained. However, it should be noted that learning ideas (i.e., conceptual knowledge) through fieldwork requires procedural knowledge (Osborne, 2015), which also needs to be explicitly taught and theoretically supported (Gott & Roberts, 2008). Therefore, one type of field activity is not better than another; goodness depends on how well the type of activity selected and implemented fits the learning aims. As Osborne (2015) would argue, the power of any activity is wasted if its purpose is unclear to students and if it does not lead to an improvement in teaching practice.

The more student-centred and open the field activities are, the more they demand from students and the more integrative learning power they have. Students are diverse, and fieldwork needs to be inclusive (Chasen et al., 2025). The integration of digital technologies in fieldwork needs to be considered, not to replace fieldwork, as they effectively did during the COVID-19 pandemic (Stagg et al., 2022), but rather to enhance its value for all students.

A considerable amount of research on school science field activities is available so far. The success of these activities cannot be taken for granted and may be influenced by internal (e.g., their structure) and external factors (e.g., the way teachers implement them). By presenting a review of research on school science field-based practical activities, focusing on students, this chapter provides an overview of how research evolved and how fieldwork contributes to students' learning in school science. Opposite to other reviews, this chapter has a broad scope, as it encompasses the four sciences (i.e., biology, chemistry, geology/earth sciences, and physics), and the primary and secondary school levels, and looks at evolution over three time periods, as, for instance, nowadays teachers can make use of devices that could not be imagined thirty years ago.

## 10.4 Methodology

The methodological approach selected to answer the research question is a literature review informed by the PRISMA (Preferred Reporting Items for Systematic reviews and Meta-Analyses) methodology (Page et al., 2021). This methodology encompasses a four-step sequence: identification (searching databases to find records); screening (reviewing titles and abstracts for relevance); eligibility (assessing the full-text articles against inclusion/exclusion criteria); and inclusion (analysing the final set of studies included in the review). Two databases were searched due to their relevance in the field of Social Sciences and Education: Scopus and Web of Science databases. These databases are comprehensive (Pranckutė, 2021) in terms of the journals they include, and complementary (Kumpulainen & Seppänen, 2022). Some journals are indexed by one and not by the other, but others are indexed in both. Articles published in Scopus and or Web of Science-indexed journals were reviewed, considering three time periods (before the year 2000, between 2000 and 2012, limits included, and after 2012) and two school levels (primary and secondary education), to uncover possible trends in research carried out in the area.

Concerning the identification phase, two related strings were searched: one for the Scopus and another for the Web of Science database. The strings used the same keywords and Boolean operators, adjusted to the technical requirements of each database. The keywords comprise the diverse terms used in the literature to address fieldwork. Some filters, like type of text (i.e., article), availability of full text (i.e., only full text available), and language (i.e., English, French, Spanish, and Portuguese, that is, the languages the authors can read), were also introduced. No time frame was introduced, as we would like to have a large timespan to uncover possible trend changes over time. Research papers on higher education were not considered because higher education differs from the other education levels and should be analysed separately, which would lead to an overload. Articles that are unavailable online were only considered if printed versions of the journal issues could be found in the university library.

The following combinations of keywords, in all fields, were used:

- ("field work" OR fieldwork OR "field trip*" OR "field activit*" OR outdoor) AND
- ("science education" OR "science teaching" OR "science learning" OR "Biolog* education" OR "Geolog* education" OR "Physics education" OR "Chemi* education" OR "Geosciences education" OR "Biolog* teaching" OR "Geolog* teaching" OR "Physics teaching" OR "Chemi* teaching" OR "Geosciences teaching" OR "Biolog* learning" OR "Geolog* learning" OR "Physics learning" OR "Chemi* learning" OR "Geosciences learning") NOT
- (virtual OR online OR comput* OR digital OR health OR "higher education" OR museum).

The online searches were conducted in May 2025. They provided two sets of empirical papers on preschool and school students' learning and opinions on and reactions to field activities, published and indexed, before the search was conducted, in peer-reviewed journals, with full text available and written in English, French, Spanish, and Portuguese.

The search in the Scopus database originated 1531 articles, while the search in the WoS database originated 387 articles. Some journals are indexed in both databases, which means that some articles appeared in both searches. Then, the results of the two searches were compared to eliminate repetitions. By comparing the references, it was noticed that 64 articles were common to both searches. Eliminating duplication resulted in a total of 1854 articles to submit to a screening process, based on a reading of the title and abstract. In cases where the two elements were not informative enough to decide on inclusion or exclusion, other parts of the article were analysed (eligibility phase). Then, 1784 articles were excluded because they focus on scientific research (579), non-science disciplines (728), higher education (72), or pedagogical proposals (60), or are revision papers (69), or focus on museums, zoos, and or botanical gardens (6), or the full text is not available (270). The screening resulted in 70 articles, three of which focus on pre-school children. As three articles are too few to be compared with the primary and secondary school ones, they were excluded. Hence, only the 67 articles addressing primary and secondary school students were considered for the review. The two oldest articles reviewed were published in 1994 in the Journal of Research in Science Teaching and in 1996 in the International Journal of Science Education. The most recent ones were published in 2025 (two) in the International Journal of Educational Research and the Multidisciplinary Science Journal.

Having identified the set of articles to be reviewed, a process of relevant data extraction was performed. This process was guided by the eight specific research questions presented in the findings section. Before starting data extraction, an Excel file was created to record the data extracted. Then, the authors analysed four randomly selected papers separately and discussed the results of the analysis to ensure that everyone was conceptualizing the categories in the same way and using the same rules.

Then, one of the authors extracted data from all papers, and another one extracted data from a sample of 10 papers. The results of the double analysis were similar, which supports the reliability of the analysis. Some difficult cases (e.g., study design, teaching approach) encountered in the other studies were discussed by all authors and solved by consensus. The consensual results recorded in the Excel file were then organized into tables that contain data relevant to answering each research question.

## 10.5 Findings

The objective of the study was split into eight specific questions that guided data extraction from the selected articles and were used to structure this section.

*Question 1: When and where were the research studies done?*

Three articles reviewed were written before the year 2000, 24 between 2000 and 2012, and 40 after 2012. Even considering that, before 2000, many papers were unavailable, the number of articles published in Scopus and WoS journals has been increasing over time. The continents where more research was undertaken are Europe (41,8%), mainly in Germany and Portugal, and America (37,3%), mainly in USA. The remaining studies were carried out in Asia (13,4%), mainly in Israel and South Korea, and Africa (7,5%), in three countries (Morocco, Nigeria, and Tanzania). Even though caution is needed as a considerable number of papers published before 2000 were surely not reviewed, data suggest a tendency for research to increase in Asia and Africa, especially in secondary school, and in Europe, in primary school.

*Question 2: What school levels and samples were focused on in the research?*

Concerning school levels, 29,9% of the articles reviewed focused on primary school (grades 1 to 6), 50,7% on secondary school (grades 7 to 12), and the remaining 19,4% on both school levels. Most primary school studies were carried out in America and Europe. Also, most studies that focus on both school levels were carried out in Europe. Considering the two most recent periods, data indicate a growing interest in researching field-based activities at both the primary and secondary school levels. The size of the samples (or groups) involved in the research spanned a wide range of values, from 5 to 3110 students, uncovering secondary school students' conceptions of water pollution and primary school students' interest in animals and plants, respectively. Starting from the bottom, 25,4% of the studies used samples of 30 students maximum; 26,9% used samples of 31 to 60 students; 26,9% used samples of 61 to 250 students, and 19,4% used samples from 251 to 3110 students. One mixed-methods study does not mention the sample size. Studies focusing on conceptual learning tend to use smaller samples than those dealing with the affective dimension of learning, including opinions on field trips, fieldwork, or field activities, interest in science or a part of it, and motivation to learn science. Regarding changes in sample size over time, no pattern emerges when comparing the two most recent periods.

*Question 3: What are the aims of the research studies reviewed, and to what extent did their aims change over time and with school level?*

Most articles (51 out of 67 reviewed) focus on a single learning aim: science conceptual knowledge (e.g., earth resources, water erosion); procedural knowledge (e.g., observation, data collection or data analysis skills, problem solving abilities); epistemological knowledge (i.e., image of science, scientific methodology); attitudes towards science related to fieldwork; motivation to learn science (i.e., provide new and pleasant experiences and contact with nature to students); or opinions on outdoor science activities. The remaining 16 papers combine two or three of those areas. Additionally, data indicate that: 64,2% of the studies aimed at conceptual knowledge, with 41,8% dealing with it only; 13,5% focused on procedural knowledge, with 7,5% concentrating on it only; 22,4% dealt with attitudes, with 7,5% dealing solely with attitudes; 13,4% concentrated on motivation, with 10,4% addressing motivation alone; 9,0% focused on epistemological knowledge, with 6,0% dealing with it alone; and 3,0% focused on opinions alone. No study addressing solely the primary school level deals with procedural knowledge or epistemological knowledge. Most of the (few) epistemology-focused studies were conducted during the most recent period, which may indicate a tendency for this kind of study to increase. However, only future analysis may confirm or refute this tendency.

*Question 4: What science disciplines and themes were focused on in the research?*

General Science concentrates more than one third (38,8%) of the studies reviewed; they are followed by Biology (23,9%), Geology (13,4%), Chemistry (4,5%), Physics and STEM (3% each). A few articles (13,4%) report studies that include two or more science disciplines (e.g., Biology and Chemistry), which approach a theme (e.g., air quality) in a cross-disciplinary way. Primary school studies concentrate on general sciences and Biology themes, with no study dealing with Physics or Chemistry issues, as such. The number of studies per discipline is increasing over time, but those (few) focusing on STEM are recent and therefore a tendency can hardly be perceived. Concerning the science themes dealt with, primary school research concentrates on animals, plants, and ecosystems; in addition to those themes, secondary school research deals with themes like geological resources, plate tectonics, solar system, river formation, solar energy, and separation methods; studies addressing the primary and secondary school levels together, concentrated on biodiversity, plants, animals, ecosystems and geological resources.

*Question 5: What research paradigms were adopted in the research studies reviewed, and to what extent did their approaches change over time?*

The majority of the articles reviewed report studies (73,1%) that are quantitative, 19,4% are qualitative, and 7,5% are mixed-methods studies. Most of the studies reviewed (65 out of 67) aimed at assessing the efficacy of a pedagogical intervention on knowledge, attitudes, or motivation. The majority of the pedagogical intervention-based studies (n = 43) aim to promote conceptual knowledge. Of these 43 studies, 17 (39,5%) used a single-group pre-post-test design (i.e., no control group) and they prevailed in secondary school; 2 (4,7%) used a two-experimental groups pre-post-test design, and the remaining 24 (55,8%) used a one or two-experimental and control group designs. This variety of designs and the control and/or assessment weaknesses associated with single-group design can also be found in the studies reporting interventions on other domains. For instance, the majority (7 out of 9) of the procedure-targeted intervention studies are single-group studies, that is, have no comparison group. The same applies to epistemological (5 out of 6) and affective studies (13 out of 24). Concerning research designs, no clear tendency emerges from period comparisons. Additionally, 38 (56,8%) of the articles reviewed used inferential statistics to analyse the effect of the interventions under question; 16 (23,9%) used descriptive statistics, some of which (5, that is 7,5%) in conjunction with inferential statistics; 18 (26,9%) articles reviewed did not use any kind of statistics. Not using statistics is fine for qualitative studies, but it is a shortcoming for experimental studies, as it limits conclusions about the impact of interventions.

*Question 6: Types of field activities used in the studies reporting on field-based teaching interventions, and to what extent do they depend on time and school level?*

The 65 studies involving a pedagogical intervention asked students to carry out field activities covering all the kinds of activities described in the previous section, except the motivating ones. This applies to both school levels, even though training activities, focusing on observation and classification skills, were found in two primary school studies only. Guided observation activities are those used in more studies (27,7%), followed by inductive (20,0%), problem-solving (16,9%), and illustrative activities (15,4%). Hence, a few studies dealt with open problem-solving activities, but the majority used close (e.g., illustrative) or quite close (e.g., guided discovery and inductive) activities. There seems to be a tendency for illustrative activity-based studies to decrease over time, and problem-solving-based activity-based studies to increase. This may suggest a good change in field-based teaching practice, but it should be noted that changes in research contexts may take time to enter the classroom, if they succeed in doing so. About one-third of the 65 intervention studies do not mention any tool used in the field. Fourteen of the 65 pedagogical intervention-based studies used activities that draw on digital tools (e.g., soil and water sensors, digital weather instruments, including wind and sunlight recorders, and global positioning systems (GPS)), with 10 of them aiming to promote learning of conceptual knowledge, alone or together with other types of knowledge (e.g., procedural knowledge). The reduced number of studies using this kind of device makes it difficult to securely uncover a tendency, even though it seems that their number is increasing.

*Question 7: What and how were data collected in the reviewed studies?*

The studies reviewed used the inquiry, observation, and document analysis techniques, alone or combined. Most studies (94,0%) collected data through a questionnaire or knowledge test, alone (56,7%) or combined with other techniques (37,3%), and 41,8% through observation, alone (1,5%) or combined with other techniques (40,3%) (e.g., questionnaire). Document analysis (e.g., field reports, drawings, and filled-in worksheets) and interviews about tasks carried out in the field were scarcely used. No differences seem to exist between school levels and periods.

*Question 8: What was the impact of field-based teaching interventions, considering the diverse types of knowledge addressed?*

About field-based teaching interventions, all but two articles reporting intervention studies (65 out of 67) report positive effects of the experimental pedagogical approaches, which are based on more or less closed activities. Then, irrespective of the aim, school level, and period, field activities used in the experimental group(s) studies had a positive impact on the diverse kinds of students' learning. Almost all (41 out of 43) conceptual knowledge field-based intervention studies report positive effects of the experimental group, which are larger than those observed in the comparison group. Four of these articles report on more retention by experimental group students, as suggested by delayed post-test results. The remaining two conceptual knowledge intervention studies, centred on secondary school, report a negative effect or no effect on the acquisition of conceptual knowledge. This may be due to the features of the pedagogical materials supporting the experimental group intervention, which may have made students feel a lack of support. Positive impact of the field-based interventions is reported by all 9 studies focusing on learning of procedural (namely, science methodology skills), and 6 epistemological (namely, views on what is science and how it is done) knowledge. Positive changes in attitudes towards open space activities and favourable opinions about fieldwork are also reported by all 24 studies that involve field-based interventions, based on a wide range of sample sizes, and have somehow measured the impact on the affective dimension. Overall, the results of the field-based teaching interventions do not seem to depend on the school level, even though students' starting and intended ending points are naturally different.

## 10.6 Conclusion and Ways Forward

The set of 67 original research articles reviewed, focusing on the use of field activities in primary and secondary school science, deals with all the objectives that field activities may contribute to and which were referred to above. While a few of them report on attempts to make a diagnosis (e.g., how children appreciate nature) or description of a variable (e.g., what are students' opinions on field activities), the majority of the articles reviewed report (sometimes not only but also) on pedagogical interventions to promote a variety of knowledge types. About the consistency of the studies, some studies that do not include an intervention use very large samples, enabling generalizable conclusions; others use small samples, able to provide some insight into what was being studied. The studies focusing on pedagogical intervention(s) are more demanding in terms of support for their conclusions. A few of them used appropriate designs to conclude about the short and long-term efficacy of the intervention, even though they rely on convenience and reduced samples. Others, in addition to small convenience samples, did not use comparison groups and or a pre-/post-test evaluation, which weakens the reliability of their conclusions. Studies focusing on motivational activities may need to be reimagined so that motivation continues to be addressed, as it should (Dettweiler et al., 2015), and becomes a concern integrated with contextualized learning of content knowledge (Ayotte-Beaudet et al., 2023; Giamellaro et al., 2022) to better harness field trips.

The number of research studies reporting explicitly on the use of digital technologies is reduced, and these were seldom at the centre of the research. Besides, no study reported on the use of assistive technologies to make field activities more inclusive. It may happen that the research groups did not include students with special educational needs, or maybe they were disregarded from a research point of view. However, it seems very important to help teachers find the best ways to deal with diverse groups and to make field trips inclusive (Chasen et al., 2025). Digital technology may play a key role in helping to overcome this sort of challenge (Stagg et al., 2022).

From a pedagogic point of view, the large number of studies using close (e.g., guided observation) or quite close (e.g., inductive field activities) field activities limits the educational potential of field trips in terms of promoting students' learning how to learn abilities. These require students' engagement in open activities, like problem-solving activities, which are relevant for learning how to do science. Long-distance field trips may be hard to fit into the curriculum in such a way as to enable the best combination of content to be taught and the type of field activity to be performed. Even though this may explain the large amount of relatively close activities, research should concentrate on the diversity of knowledge that students may learn from open activities in an integrated way (Braund & Reiss, 2004; Remmen & Iversen, 2023). Short-distance field trips are easier to fit into the teaching sequence, and they could include more open activities for promoting students' learning and learning how to learn in a contextualized way.

About the intention to compare time periods and school levels, before 2000, few papers were available, which conditions the analysis of the evolution of the research. The same applies to the comparison between school levels, due to the reduced number of studies focusing on primary school. During the screening phase, some undue exclusion of articles may have happened. Despite these limitations, the review suggests that field-based activities may be a valuable educational tool, even though research focusing on activities with a higher degree of openness and integrating appropriate digital and assistive technologies is needed. Those studies should consider the key steps of field activities as well as their safety and ethical requirements. In addition, qualitative or mixed-methods studies (which include a qualitative dimension) should be carried out to allow a deep understanding of how open field activities and technology power each other and harness students' development of not only conceptual knowledge but also procedural and epistemological knowledge.

**Acknowledgements** Thanks are due to CIEd, Research Centre on Education of the Institute of Education of the University of Minho (UID/01661/2025), which funded this work through national funds of FCT/MCTES-PT, namely the Pluriannual Program for the Funding of R&D Units.

## Supplementary material

The list of articles reviewed and tables with data that support the findings section are available from https://doi.org/10.34622/datarepositorium/HUZQYG.

## References


Ayotte-Beaudet, J.-P., Chastenay, P., Beaudry, M.-C., L'Heureux, K., Giamellaro, M., Smith, J., Desjarlais, E., & Paquette, A. (2023). Exploring the impacts of contextualised outdoor science education on learning: The case of primary school students learning about ecosystem relationships. *Journal of Biological Education, 57*(2), 277–294. https://doi.org/10.1080/00219266.2021.1909634

Behrendt, M., & Franklin, T. (2014). A review of research on school field trips and their value in education. *International Journal of Environmental & Science Education, 9*, 235–245. https://doi.org/10.12973/ijese.2014.213a

Braund, M., & Reiss, M. (2004). The nature of learning science outside the classroom. In M. Braund & M. Reiss (Eds.), *Learning science outside the classroom* (pp. 1–12). Routledge.

Chasen, A., Tripp, H. A., & Borrego, M. (2025). Disability and postsecondary fieldwork experiences in the natural sciences: A systematic review. *Journal of Research in Science Teaching, 62*, 1006–1039. https://doi.org/10.1002/tea.21989

Compiani, M., & Carneiro, C. (1993). Os papéis didácticos das excursões geológicas [The didactic roles of geological excursions]. *Enseñanza de las Ciencias de la Tierra, 1*(2), 90–98. https://raco.cat/index.php/ECT/article/view/88098

Del Carmen, L. (1999). El estudio de los ecosistemas [The study of ecosystems]. *Alambique: Didáctica de las Ciencias Experimentales, 20*, 47–54.

Dettweiler, U., Ünlü, A., Lauterbach, G., Becker, C., & Gschrey, B. (2015). Investigating the motivational behavior of pupils during outdoor science teaching within self-determination theory. *Frontiers in Psychology, 6*(125). https://doi.org/10.3389/fpsyg.2015.00125

DeWitt, J., & Storksdieck, M. (2008). A short review of school field trips: Key findings from the past and implications for the future. *Visitor Studies, 11*(2), 181–197. https://doi.org/10.1080/10645570802355562

Dourado, L., & Leite, L. (2013). Field activities, science education and problem-solving. *Procedia Social and Behavioral Sciences, 106*, 1231–1241. https://doi.org/10.1016/j.sbspro.2013.12.138

Dummer, T., Cook, I., Parker, S., Barrett, G., & Hull, A. (2008). Promotion and assessing ´deep learning´ in geography fieldwork: An evaluation of reflective field diaries. *Journal of Geography in Higher Education, 32*(3), 459–479. https://doi.org/10.1080/03098260701728484

Falk, J., & Dierkering, L. (2000). *Learning from museums.* Altamira Press.

Garner, N., & Eilks, I. (2015). The expectations of teachers and students who visit a non-formal student chemistry laboratory. *Eurasia Journal of Mathematics, Science & Technology Education, 11*(5), 1197–1210. https://doi.org/10.12973/eurasia.2015.1415a

Giamellaro, M., L'Heureuz, K., Buxton, C., Beaudry, M.-C., Ayotte-Beaudet, J.-P., & Alajmi, T. (2022). Learning to teach science from a contextualized stance. In J. A. Luft & M. Gail Jones (Eds.), *Handbook of research on science teacher education* (pp. 439–451). Routledge. https://doi.org/10.4324/9781003098478

Gott, R., & Roberts, R. (2008, December). *Concepts of evidence and their role in open-ended practical investigations and scientific literacy; background to published papers.* University of Durham School of Education. https://cofev.webspace.durham.ac.uk/wp-content/uploads/sites/299/2022/05/Gott-Roberts-2008-Research-Report.pdf

Guilherme, E., Faria C., & Boaventura, C. (2016). Exploring marine ecosystems with elementary school Portuguese children: inquiry-based project activities focused on 'real-life' contexts. *Education 3–13, 44*(6), 715–726. https://doi.org/10.1080/03004279.2015.1007884

Hamidy, M., Errami, E., & Orion, N. (2024). Proposed geo-educational activities at the Sidi Bouzid geosite, Safi Province, Marrakech-Safi region, Morocco. *International Journal of Geoheritage and Parks, 12*, 209–222. https://doi.org/10.1016/j.ijgeop.2024.03.003

Henriques, M.-H., Tomaz, C., & Sá, A. (2012). The Arouca Geopark (Portugal) as an educational resource: A case study. *Episodes, 35*(4), 481–488. https://doi.org/10.18814/epiiugs/2012/v35i4/004

King, H., & Glackin, M. (2010). Supporting science learning in out-of-school contexts. In J. Osborne & J. Dillon (Eds.), *Good practice in science teaching: What research has to say* (pp. 259–273). Open University Press.

Kumpulainen, M., & Seppänen, M. (2022). Combining web of science and scopus datasets in citation based literature study. *Scientometrics, 127*, 5613–5631. https://doi.org/10.1007/s11192-022-04475-7

Lima, A., Vasconcelos, C., Félix, N., Barros, J., & Mendonça, A. (2010). Field trip activity in an ancient gold mine: Scientific literacy in informal education. *Public Understanding of Science, 19*(3), 322–334. https://doi.org/10.1177/0963662509104725

Lock, R. (2010). Biology fieldwork in schools and colleges in the UK: An analysis of empirical research from 1963 to 2009. *Journal of Biological Education, 44*(2), 58–64. https://doi.org/10.1080/00219266.2010.9656195

Mann J., Gray, T., Truong, S., Brymer, E., Passy, R., Ho, S., Sahlberg, P., Ward, K., Bentsen, P., Curry, C., & Cowper, R. (2022). Getting out of the classroom and into nature: A systematic review of nature-specific outdoor learning on school children's learning and development. *Frontiers in Public Health, 10*, Article 877058. https://doi.org/10.3389/fpubh.2022.877058

Moll, R. (2010). An amusement park physics competition. *Physics Education, 45*(4), 362–367. https://doi.org/10.1088/0031-9120/45/4/006

Orion, N., & Hofstein, A. (1994). Factors that influence learning during a scientific field trip in a natural environment. *Journal of Research in Science Teaching, 31*(10), 1097–1119. https://doi.org/10.1002/tea.3660311005

Orion, N. (1993). A model for the development and implementation of field trips as an integral part of the science curriculum. *School Science and Mathematics, 93*(6), 325–331. https://doi.org/10.1111/j.1949-8594.1993.tb12254.x

Ortega-Torres, E., & Moncholí-Pons, V. (2021). «We explain the Albufera»: Transforming a field trip into an interdisciplinary project. *Enseñanza De Las Ciencias, 39*(2), 241–252. https://doi.org/10.5565/rev/ensciencias.3241

Osborne, J. (2015). Practical work in science: Misunderstood and badly used? *School Science Review, 96*(357), 16–24.

Osborne, J., & Pimentel, D. (2023). Science education in an age of misinformation. *Science Education, 107*(3), 553–571. https://doi.org/10.1002/sce.21790

Page, M., Moher, D., Bossuyt, P., Boutron, I., Hoffmann, T., Mulrow, C., Shamseer, L., Tetzlaff, J., Akl, E., Brennan, S., Chou, R., Glanville, J., Grimshaw, J., Hróbjartsson, A., Lalu, M., Li, T., Loder, E., Mayo-Wilson, E., McDonald, S.,... & McKenzie, J. (2021). PRISMA 2020 explanation and elaboration: Updated guidance and exemplars for reporting systematic reviews. *BMJ, 372*(160), 1–36.https://doi.org/10.1136/bmj.n160

Pedrinaci, E., Sequeiros, L., & De La Torre, E. (1994). El trabajo de campo y el aprendizagem de la geología [Fieldwork and geology learning]. *Alambique Didáctica De Las Ciencias Experimentales, 2*, 37–45.

Pranckutė, R. (2021). Web of science (WoS) and scopus: The titans of bibliographic information in today's academic world. *Publications, 9*(12), 1–59. https://doi.org/10.3390/publications9010012

Reiss, M., & Braund, M. (2004). Practicalities and safety issues. In M. Braund & M. Reiss (Eds.), *Learning science outside the classroom* (pp. 13–17). Routledge.

Remmen, K., & Iversen, E. (2023). A scoping review of research on school-based outdoor education in the Nordic countries. *Journal of Adventure Education and Outdoor Learning, 23*(4), 433–451. https://doi.org/10.1080/14729679.2022.2027796

Rowe, S., & Humphries, S. (2004). The outdoor classroom. In M. Braund & M. Reiss (Eds.), *Learning science outside the classroom* (pp. 19–34). Routledge.

Skea, C., & Fulford, A. (2021). Releasing education into the wild: An education in, and of the outdoors. *Ethics and Education, 16*(1), 74–90. https://doi.org/10.1080/17449642.2020.1822612

Stagg, B., Dillon, J., & Maddison, J. (2022). Expanding the field: Using digital to diversify learning in outdoor science. *Disciplinary and Interdisciplinary Science Education Research, 4*(9). https://doi.org/10.1186/s43031-022-00047-0

Telford, A., Valentine, A., & Godby, S. (2024). The paradox of the 'sustainable fieldtrip'? Exploring the links between geography fieldtrips and environmental sustainability. *Journal of Geography in Higher Education, 48*(1), 115–132. https://doi.org/10.1080/03098265.2023.2190961

Wells, N., Myers, B., Todd, L., Barale, K., Gaolach, B., Ferenz, G., Aitken, M., Henderson, C., Jr., Tse, C., Pattison, K., Taylor, C., Connerly, L., Carson, J., Gensemer, A., Franz, N., & Falk, K. (2015). The effects of school gardens on children's science knowledge: A randomized controlled trial of low-income elementary schools. *International Journal of Science Education, 37*(17), 2858–2878. https://doi.org/10.1080/09500693.2015.1112048

Yemini, M., Engel, L., & Simon, A. (2025). Place-based education: A systematic review of literature. *Educational Review, 77*(2), 640–660. https://doi.org/10.1080/00131911.2023.2177260

# Chapter 11
# Problem-Based Practical Work in Real Contexts: Cultivating Scientific Literacy for Social Change and Youth Protagonism

**Pedro Reis**

## 11.1 Introduction

Practical work in science education encompasses a diverse array of pedagogical practices, from controlled verification experiments to open-ended exploratory activities (Oliveira & Bonito, 2023; Spaan et al., 2025). Traditional practical work typically reinforces theoretical concepts by guiding students through highly structured procedures that confirm expected results. While this form of practical work can develop basic laboratory skills and understanding of scientific concepts, it often fails to engage students meaningfully with science as a dynamic and socially embedded practice. In contrast, "Problem-Based Practical Work in Real Contexts" (PBPWRC) prioritizes student agency, contextual authenticity, and the integration of science with societal issues. Rather than following a script, students identify, investigate, and address real-world problems they find relevant, meaningful, and urgent (Bencze & Alsop, 2014; Hodson, 2020).

PBPWRC challenges the notion of science as an isolated discipline by embedding scientific inquiry within broader societal, environmental, and ethical contexts. Students are encouraged to address issues such as climate change, waste management, public health, and biodiversity loss-problems they encounter in their everyday lives. By designing and conducting investigations that respond to these concerns, students not only develop scientific knowledge and skills, but also build an understanding of science as a tool for social analysis and change (Bencze, 2017). This kind of practical work resonates with students' experiences, enhances motivation, and fosters a sense of ownership over their learning.

P. Reis (✉)
UIDEF - Institute of Education, University of Lisbon, Lisbon, Portugal
e-mail: preis@ie.ulisboa.pt



L. Leite et al. (eds.), *Practical Work in Science Education in the Digital Era*, Advances in Technology-Rich Science Education 2, https://doi.org/10.1007/978-3-032-13788-3_11

The transformative nature of PBPWRC also lies in its alignment with democratic educational values (Reis, 2020). By allowing students to co-construct the learning agenda, it fosters participation, critical dialogue, and mutual respect. This shift from transmissive to participatory models of teaching positions learners as knowledge producers rather than passive recipients. Moreover, it opens the door to collective inquiry and action, allowing science education to become a platform for civic engagement and ecological consciousness (Bencze & Carter, 2011). In doing so, PBPWRC not only responds to calls for more relevant and responsive science education but also contributes to students' development as informed, critical, and engaged citizens capable of acting upon the issues that shape their lives and futures.

PBPWRC integrates problem-based learning (PBL) methodologies with real-world scenarios to enhance students' practical skills and critical thinking abilities. This approach is particularly effective in bridging the gap between theoretical knowledge and practical application, as it encourages students to engage with real-life problems in a collaborative and reflective manner. PBPWRC is a versatile and effective pedagogical strategy that not only enhances students' practical skills but also fosters critical thinking, problem-solving, and the ability to apply knowledge in diverse real-world contexts across various disciplines (Bencze, 2017, 2025; Reis, 2025). This approach has demonstrated significant improvements in cultivating students as critically aware, socially responsible agents rather than passive recipients of scientific knowledge. It also fosters student agency: students are not only consumers of knowledge, but are encouraged to conduct their own research, explore possible interventions, design and implement actions, and evaluate the outcomes of those actions. This helps to build skills such as critical thinking, problem solving, inquiry, and ethical reasoning.

Digital technologies play a crucial role in amplifying the reach and effectiveness of PBPWRC (Bencze, 2017; Reis, 2025). They allow students to capture and analyse evidence through mobile sensors, photographs or coding tools, and to visualise results transparently in dashboards that can support public dialogues. They enable spatial thinking through mapping and geotagging, making interventions highly place-specific. They facilitate collaboration with external partners through shared digital platforms, and they provide opportunities for modelling and simulating scenarios before acting, which helps students reason about trade-offs and uncertainties. Digital media also expand the possibilities for public communication, with videos, podcasts or online campaigns extending the impact of students' work. Assessment practices benefit from e-portfolios that compile data, reflections, stakeholder feedback and impact indicators, while inclusive technologies such as translation or multimodal authoring tools broaden participation and ensure that diverse voices are heard.

## 11.2 Theoretical Foundations: Learning Theories Supporting PBPWRC

PBPWRC is grounded in several educational theories that emphasize the active, situated, and transformative nature of learning. Among these, constructivist learning theories provide a foundational perspective. Piaget (1972) emphasized that learners build knowledge through experiences that challenge existing cognitive structures, while Vygotsky (1978) highlighted the social and cultural dimensions of learning. In the context of PBPWRC, these ideas support the notion that meaningful learning arises when students engage in complex, real-world tasks in collaborative settings.

PBPWRC as a pedagogical approach aligns closely with constructivist theory. It centres learning on ill-defined problems that require inquiry, research, and solution development. In PBPWRC, problems are not invented scenarios, but authentic issues drawn from students' communities or global challenges. This enhances motivation and engagement while encouraging students to integrate multiple knowledge domains and develop higher-order thinking skills. PBPWRC is supported by theories of contextual learning, which suggest that knowledge is best acquired and retained when learned in the context in which it will be used, thus promoting deeper understanding and transferability of skills. Collaborative learning, which involves students working together to solve problems, complete tasks, or learn new concepts, is also integral to PBPWRC.

Situated learning theory (Lave & Wenger, 1991) reinforces the importance of context, emphasizing that learning is most effective when embedded in authentic social practices. Through PBPWRC, students engage in the practices of scientists, activists, and community members, enhancing the transferability and relevance of their learning.

Critical pedagogy (Freire, 1970) adds a vital dimension to the theoretical foundation of PBPWRC. It frames education as a political act and emphasizes the empowerment of learners through critical consciousness ("conscientização"). Freire's call for dialogic, problem-posing education aligns with the participatory ethos of PBPWRC, where students are not merely learning about science, but using it to interrogate and act upon the world. This approach positions students as agents of transformation, capable of challenging injustice and proposing alternatives grounded in ethical reasoning and empirical evidence.

Together, these theories underpin a model of science education that is inquiry-driven, socially situated, and action-oriented. They provide the pedagogical scaffolding necessary to design learning experiences that are intellectually rigorous, emotionally engaging, and civically empowering. By drawing from these perspectives, educators can create science learning environments that not only foster academic achievement but also promote social responsibility and ecological stewardship.

## 11.3 Developing Critical Thinking Through Real-World Inquiry

PBPWRC cultivates critical thinking by immersing students in authentic, complex, and socially relevant problems that demand thoughtful engagement. This approach is rooted in Problem-Based Learning (PBL), but it goes much forward, representing a paradigm shift in science education, as it will be explained further in this text. It emphasizes student-centred learning where students engage with complex, real-world problems that lack a single correct answer, thereby fostering critical thinking and problem-solving skills. This process requires students to formulate questions, evaluate evidence, consider multiple perspectives, and make informed judgments—all essential elements of critical thinking. In contrast to rote learning and standardized assessments that reward memorization, PBPWRC challenges students to grapple with ambiguity and uncertainty. The open-ended nature of real-world problems compels them to continuously revise their understanding and refine their approaches. As students design and carry out investigations, they must assess the quality and credibility of information sources, determine appropriate methodologies, and analyse results critically. This demands an understanding of the nature of evidence, the limits of scientific knowledge, and the ethical implications of research. Engaging in discussions about possible solutions to socio-environmental issues also strengthens their capacity for argumentation and reflective judgment. Students learn to identify biases, weigh trade-offs, and defend their positions based on data, values, and logical reasoning.

Through collaborative inquiry, students refine their communication and negotiation skills, learning to respect diverse viewpoints while maintaining a critical stance. Such practices build intellectual autonomy and resilience—key components of critical thinking.

Educational contexts that adopt PBPWRC also promote a classroom culture of inquiry, dialogue, and critique. Teachers serve as facilitators rather than knowledge gatekeepers, modelling critical questioning and providing scaffolding for deeper understanding. These dynamics create a supportive environment where students feel safe to take intellectual risks and challenge conventional ideas.

The development of critical thinking through PBPWRC is not only an academic goal but a democratic imperative. Informed citizens must be able to critically evaluate scientific claims presented in the media, understand the implications of technological developments, and participate in public discussions about science-related issues. By engaging students with real-world problems and positioning them as researchers and problem-solvers, PBPWRC empowers them to apply critical thinking beyond the classroom—in their communities, careers, and civic lives.

## 11.4 Promoting Scientific Literacy for Action

Scientific literacy for action extends beyond understanding scientific concepts; it involves the ability to use scientific knowledge to make decisions, engage in dialogue, and take responsible action on personal, local, and global issues. PBPWRC is suited to promote this expanded vision of scientific literacy, as it bridges classroom learning with societal application. In PBPWRC, students engage in the full spectrum of scientific processes—posing questions, developing hypotheses, collecting and analysing data, and drawing conclusions—in the service of investigating authentic problems (Bencze & Carter, 2011; Hodson, 2020). These investigations often stem from students' concerns about their environments and communities, such as pollution in a local river, the health effects of diet and lifestyle, or the impact of climate change. Through inquiry and experimentation, students connect scientific concepts to lived experiences, thereby deepening understanding and relevance.

Moreover, PBPWRC fosters communication skills critical to scientific literacy. Students learn to present findings through diverse media—reports, presentations, videos, posters, and digital storytelling—aimed at multiple audiences, from peers to policymakers. These communicative acts help students develop fluency in scientific language while encouraging them to tailor messages for real-world impact (Bencze, 2017).

A key dimension of scientific literacy for action is the ethical consideration of science's role in society. PBPWRC invites students to reflect on the values, risks, and uncertainties associated with scientific interventions (Reis et al., 2022). This reflection enhances their capacity for responsible decision-making and ethical reasoning. When students confront issues such as genetically modified organisms, renewable energy, or water access, they must weigh competing interests and values, often engaging in deliberative discussions that mirror democratic processes.

Importantly, PBPWRC situates scientific knowledge within interdisciplinary and socio-political contexts. By connecting science with history, geography, economics, and ethics, students understand that solutions to complex problems require integrated approaches. This holistic view nurtures systems thinking and prepares students to address the multifaceted challenges of the twenty-first century.

When students take informed action—whether by launching awareness campaigns, proposing sustainability initiatives, or collaborating with community partners—they experience science as a transformative force. Such experiences build confidence and illustrate the power of collective agency. In this way, PBPWRC enacts a vision of scientific literacy that is dynamic, participatory, and oriented toward social and environmental justice.

The promotion of scientific literacy for action through PBPWRC aligns with broader educational goals of preparing learners for active citizenship. It redefines science education as not only the pursuit of knowledge but the practice of engagement, advocacy, and meaningful intervention in the world.

PBPWRC holds significant educational potential in promoting scientific literacy for action by engaging students in authentic, context-driven learning experiences that bridge the gap between theoretical knowledge and real-world application. This approach encourages students to engage with local scientific or environmental issues, thereby fostering a deeper understanding of scientific concepts and their relevance to everyday life. Moreover, involving students in projects that connect science with their cultural and community experiences can make science more relevant and accessible, thereby supporting the development of scientific literacy that is grounded in students' lived experiences (Adler & Karam, 2024; Freire et al., 2013; Kowasch et al., 2021; Reis et al., 2020). Overall, PBPWRC not only enhances students' understanding of scientific concepts but also equips them with the skills and motivation needed to take informed action on scientific issues, thus promoting scientific literacy for action (Schusler & Krasny, 2015).

One of the most powerful contributions of PBPWRC is its ability to foster student agency and social protagonism. Agency refers to the capacity of individuals to act independently and make choices about their own lives and learning. Social protagonism, meanwhile, builds upon this concept by involving students in collective, purposeful action to influence social and environmental issues. Together, these concepts are central to cultivating informed, empowered, and active citizens (Schulz et al., 2018).

In the traditional classroom, students often follow pre-determined steps with limited scope for decision-making. PBPWRC shifts this dynamic by positioning students as co-designers of their learning experiences. They are encouraged to identify problems that matter to them and their communities, formulate research questions, and design investigations that aim not only to understand but to transform reality. This model nurtures a sense of ownership and commitment that goes far beyond compliance or performance for grades (Schusler & Krasny, 2015).

When students engage in real-world problems through scientific inquiry, they come to see themselves as capable of producing knowledge and effecting change. This transformative shift in identity—from passive learners to active change-makers—strengthens their confidence, autonomy, and motivation. Furthermore, they learn to navigate the complexities of civic engagement: negotiating goals, collaborating with peers and community members, and taking responsibility for ethical and practical implications of their actions (McInerney & Adshead, 2013).

Social protagonism emerges from this process when students take public stances, lead campaigns, and participate in collective actions. Projects such as 'We Act' and 'STEPWISE' offer concrete examples. In 'We Act', students used scientific and artistic methods to address socio-environmental issues like food insecurity, waste management, and environmental justice. They created podcasts, staged performances, and organised school-wide initiatives that connected learning with activism (Reis, 2025). Similarly, in STEPWISE, students investigated local problems and implemented context-sensitive interventions, often involving advocacy and policy dialogue (Bencze, 2017).

Such practices cultivate essential skills and dispositions for democratic life: critical consciousness, ethical reasoning, communication, collaboration, and leadership. Students learn to work across differences, navigate institutional structures, and persist in the face of setbacks—key features of resilient citizenship. They also discover the interdependence of knowledge and action: that science is not merely an abstract domain but a powerful tool for envisioning and shaping better futures (Kirk, 2024).

Ultimately, PBPWRC supports a vision of education that is emancipatory and future-oriented. It not only prepares students for exams and employment but equips them to participate meaningfully in the public sphere. By nurturing student agency and social protagonism, this approach transforms science education into a platform for democratic participation and socio-environmental transformation.

## 11.5 Challenges and Strategies: Addressing Barriers to Implementation

Despite the documented benefits of PBPWRC, its implementation faces several challenges, primarily due to the complexity and variability of real-world settings. Teachers often report systemic constraints, such as rigid curricula, high-stakes assessments, lack of time, and insufficient resources (Akuma & Callaghan, 2019; Reis, 2014). These constraints create pressure to prioritize content coverage and standardized testing over inquiry-based, socially engaged practices. Additionally, educators may feel unprepared or unsupported in facilitating open-ended investigations, managing interdisciplinary learning, or navigating controversial topics (Linhares & Reis, 2023a, 2023b).

Students, too, face challenges. Many are unaccustomed to autonomy in school settings and may initially struggle with the uncertainty and complexity of real-world problems (Marques & Reis, 2017; Reis et al., 2020, 2022). Others may feel disempowered by social or institutional dynamics, lacking the confidence or opportunity to see themselves as capable agents of change. These barriers are especially pronounced in under-resourced schools or contexts with limited access to supportive networks and partnerships.

To overcome these obstacles, a multi-pronged strategy is essential. At the systemic level, educational policies must create space for interdisciplinary and project-based learning. This can include integrating socio-scientific issues into national curricula, recognizing alternative assessment methods, and encouraging school-community partnerships. Flexibility in curriculum design allows educators to tailor projects to local contexts and student interests, enhancing relevance and engagement.

Professional development is equally critical. Teachers need ongoing training in inquiry-based pedagogy, facilitation of student agency, and the integration of science with social and ethical dimensions. Networks of practice—such as the communities established in "STEPWISE" and "We Act"—can provide spaces for collaboration, reflection, and mutual support. Mentoring programs and shared resources also help reduce the burden on individual educators (Bencze, 2017; Reis, 2014, 2025).

Scaffolding student learning is another key strategy. PBPWRC can be structured progressively, with increasing levels of complexity and independence. Teachers can begin with guided inquiries and gradually shift toward student-led projects. Clear expectations, supportive feedback, and access to resources empower students to navigate open-ended tasks. Embedding opportunities for reflection and discussion also helps students develop the metacognitive skills needed for sustained inquiry and action.

Partnerships with community organizations, NGOs, and local governments can enrich projects with real-world expertise and amplify their impact (Hadjichambis et al., 2020). These collaborations offer authenticity, visibility, and support, while helping students connect their work to broader societal discourses. Highlighting and celebrating student efforts through exhibitions, publications, and events further reinforces the value of their contributions and motivates continued engagement (Reis et al., 2022).

Ultimately, the successful implementation of PBPWRC requires a cultural shift in how science education is understood and practiced. It calls for a move from knowledge transmission to knowledge co-creation, from passive learning to active engagement, and from individual achievement to collective transformation. By addressing these challenges with vision and strategy, educators and institutions can create learning environments where all students have the opportunity to think critically, act responsibly, and lead change in their communities and beyond (Bencze, 2025).

## 11.6 Conclusion and Ways Forward

Problem-Based Practical Work in Real Contexts (PBPWRC) represents a paradigm shift in science education—one that moves beyond content transmission and laboratory routines toward the cultivation of critical, socially engaged learners capable of understanding and transforming the world around them. By empowering students to engage with real-life socio-scientific issues through inquiry and action, this approach helps develop critical thinking, scientific literacy, personal agency, and social protagonism. These competencies are not only essential for academic success but also for meaningful participation in democratic societies and in addressing global challenges such as climate change, inequality, and public health crises.

PBPWRC represents a significant development when compared to both traditional Problem-Based Learning (PBL) and conventional practical work. Whereas traditional practical work usually relies on pre-determined problems with highly structured procedures designed to verify known effects or to reinforce specific concepts and skills, and PBL tends to present carefully constructed scenarios where students analyse, research and propose solutions often within the confines of the classroom, PBPWRC is rooted in authenticity. Here, the problems are situated in real contexts, frequently emerging from the students' own environments and communities, and they are negotiated with external stakeholders. Instead of being artificial exercises, they are locally meaningful socio-scientific issues such as pollution, waste management or biodiversity loss, where the objective is not only to understand but also to transform the context.

The purpose of inquiry also shifts significantly. While traditional practical work seeks to consolidate disciplinary content and PBL aims at developing reasoning and integrative understanding, PBPWRC explicitly seeks learning for transformation. It equips students with the knowledge, skills and agency necessary to participate in civic life and to take action in relation to real problems. In this sense, the role of teachers changes from directors of laboratory procedures or facilitators of pre-set inquiry cycles into coaches and organisers of civic investigations. Students, in turn, become co-designers of projects, decision-makers and protagonists, exercising voice and agency alongside community partners.

Knowledge practices are also broadened. Instead of focusing solely on applying prescribed methods in controlled settings, students in PBPWRC engage in a full ecology of practices that combine scientific investigation with dialogue with stakeholders, ethical deliberation, policy literacy, campaigning and practical implementation. The expected outcomes mirror this expansion. Traditional practical work often culminates in a lab report and PBL in the quality of a proposed solution, but PBPWRC includes measurable community impact, such as changes in practices, public communication of results, or local policy engagement, in addition to disciplinary learning. Assessment thus becomes multidimensional, capturing data quality, argumentation, ethical reasoning, collaboration, implementation fidelity and impact, often through portfolios or public exhibitions.

The paradigm shift is also visible in how curriculum boundaries are redefined. Whereas practical work generally operates within a single subject and PBL aspires to interdisciplinarity but remains largely classroom-based, PBPWRC is transdisciplinary, place-based and deliberately permeable, occurring in community spaces such as rivers, markets, or municipal forums. This reflects a different view of science itself: rather than being portrayed as a neutral and apolitical method, as in traditional practical work, or a problem-solving tool, as in PBL, science in PBPWRC is treated as a socio-political practice where values, trade-offs and issues of justice are interrogated alongside empirical inquiry.

Epistemologically, it moves from "knowing that" towards "knowing and doing with others for change." Power relations are transformed as agendas are no longer set exclusively by teachers but co-constructed with students and communities. The locus of validity shifts from correctness in closed, well-defined tasks to fitness-for-purpose in open, complex contexts where effectiveness, ethics and equity matter. Accountability no longer focuses solely on individual performance but on collective responsibility and civic outcomes.

Taken together, these features show that PBPWRC is not simply PBL with an outdoor or practical dimension. It fundamentally redefines what counts as science learning, who controls it, where it takes place, and how success is judged. This is why it can be considered a paradigm shift in science education. In this light, PBPWRC can be understood as a genuine paradigm shift.

Projects like 'We Act' and 'STEPWISE' illustrate how theory can be translated into practice (Bencze, 2017; Reis, 2025). These projects offer robust frameworks for designing context-rich, participatory science education that is both transformative and sustainable. They show that when students are treated as citizens capable of producing knowledge and change, science education becomes a vehicle for empowerment and social justice. Their methodologies—blending scientific inquiry, creative expression, and digital media—highlight the value of interdisciplinary approaches and diverse modes of communication in engaging broader publics and fostering civic literacy.

However, these promising models also underscore the need for structural support and continued research. Future investigations should explore long-term outcomes of student activism in science education, the transferability of PBPWRC across diverse cultural and educational contexts, and strategies for assessing student learning in open-ended, action-oriented projects. Research could also examine the evolving role of teachers as facilitators of inquiry and agents of change themselves, particularly in systems that often prioritize standardization and accountability over innovation and responsiveness.

From a practical standpoint, professional development must be reimagined to equip educators with the pedagogical tools and socio-political sensibilities required for this work. Teacher education programs should include coursework and field experiences that engage preservice teachers with PBPWRC methodologies, critical pedagogy, and community-based science initiatives. Similarly, school leadership and policy must support curricular flexibility, interdisciplinary collaboration, and partnerships with external stakeholders.

Ultimately, the future of PBPWRC depends on a broader redefinition of science education—not as preparation for elite careers or standardized assessments, but as preparation for life in a complex, interconnected, and rapidly changing world. By prioritizing relevance, equity, and agency, this approach offers a compelling vision for science education that is not only intellectually rigorous but also ethically grounded and civically empowering. As global challenges intensify, the need for such a vision has never been greater. Researchers, educators, and policymakers must work together to ensure that science education becomes a foundation for social transformation and collective wellbeing in the digital era and beyond.

**Acknowledgements** This work was supported by National Funds through FCT-Portuguese Foundation for Science and Technology, I.P., under the scope of UIDEF—*Unidade de Investigação e Desenvolvimento em Educação e Formação*, UIDB/04107/2020, https://doi.org/10.54499/UIDB/04107/2020.

## References

Adler, I., & Karam, C. (2024). Djaji Mahsheye, Moghrabeye, and Labaneh: Making science relevant. *Journal of Research in Science Teaching, 61*(1), 103–136. https://doi.org/10.1002/tea.21866

Akuma, F. V., & Callaghan, R. (2019). A systematic review characterizing and clarifying intrinsic teaching challenges linked to inquiry-based practical work. *Journal of Research in Science Teaching, 56*(5), 619–648. https://doi.org/10.1002/tea.21516

Bencze, J. L. (2025). *Building networks for critical and altruistic science education: Seeking ubiquitous social justice and environmental vitality* (Vol. 63). Springer Nature.

Bencze, L. (2017). *Science and technology education promoting wellbeing for individuals, societies and environments (STEPWISE)* (Vol. 14). Springer. https://doi.org/10.1007/978-3-319-55505-8

Bencze, J., & Alsop, S. (2014). Activist science and technology education. *Springer*. https://doi.org/10.1007/978-94-007-4360-1_1

Bencze, L., & Carter, L. (2011). Globalizing students acting for the common good. *Journal of Research in Science Teaching, 48*(6), 648–669. https://doi.org/10.1002/tea.20419

Freire, P. (1970). *Pedagogy of the oppressed*. Continuum.

Freire, S., Faria, C., Galvão, C., & Reis, P. (2013). New curricular material for science classes: How do students evaluate it? *Research in Science Education*, 43, 163–178. https://link.springer.com/article/https://doi.org/10.1007/s11165-011-9247-0

Hadjichambis, A. Ch., Reis, P., Paraskeva-Hadjichambi, D., Činčera, J., Boeve-de Pauw, J., Gericke, N., & Knippels, M.-C. (2020). *Conceptualizing environmental citizenship for 21st century education*. Series "Environmental discourses in Science Education". Springer. https://doi.org/10.1007/978-3-030-20249-1

Hodson, D. (2020). Going beyond STS education: Building a curriculum for sociopolitical activism. *Canadian Journal of Science, Mathematics and Technology Education, 20*(4), 592–622. https://doi.org/10.1007/s42330-020-00114-6

Kirk, M. (2024). Student voice and agency for transformative change in matters that matter: Impactful inquiry in primary science. *The Australian Educational Researcher, 52*, 411–430. https://doi.org/10.1007/s13384-024-00721-4

Kowasch, M., Cruz, J., Reis, P., Gericke, N., & Kicker, K. (2021). Climate youth activism initiatives: Motivations and aims, and the potential to integrate climate activism into ESD and transformative learning. *Sustainability, 13*, Article 11581. https://doi.org/10.3390/su132111581

Lave, J., & Wenger, E. (1991). *Situated learning: Legitimate peripheral participation*. Cambridge University Press.

Linhares, E., & Reis, P. (2023a). Education for environmental citizenship and activism through the development of nature-based solutions with pre-service teachers. *Journal of Social Science Education, 22*(4). https://doi.org/10.11576/jsse-6498

Linhares, E., & Reis, P. (2023b). Education for environmental citizenship in pre-service teachers: Potentialities and limitations of a pedagogical approach applied at a distance. *Sustainability, 15*(21), Article 15411. https://doi.org/10.3390/su152115411

Marques, R., & Reis, P. (2017a). Producción y difusión de vídeos digitales sobre contaminación ambiental. Estudio de caso: Activismo colectivo basado en la investigación [Production and dissemination of digital videos on environmental pollution. Case study: Research-based collective activism]. *Revista Eureka sobre Enseñanza y Divulgación de las Ciencias, 14*(1), 215–226. http://hdl.handle.net/10498/18857

McInerney, C., & Adshead, M. (2013). Problem-based learning and civic engagement—shifting the focus of learning in public policy education. *Political Science & Politics, 46*(3), 630–636. https://doi.org/10.1017/S1049096513000620

Oliveira, H. & Bonito, J. (2023). Practical work in science education: A systematic literature review. *Frontiers in Education,* 8, Article1151641. https://doi.org/10.3389/feduc.2023.1151641

Piaget, J. (1972). *The psychology of the child*. Basic Books.

Reis, P. (2025). The Project We Act: Informed collective activism regarding real-life problems as part of an ecojustice network. In L. Bencze (Ed.), *Building networks for critical and altruistic science education* (Vol. 3, pp. 547–574). Springer. https://doi.org/10.1007/978-3-031-83837-8_23

Reis, P. (2020). Environmental citizenship & youth activism. In A. C. Hadjichambis, P. Reis, D. Paraskeva-Hadjichambi, J. Činčera, J. Boeve-de Pauw, N. Gericke, & M.-C. Knippels (Eds.), *Conceptualizing environmental citizenship for 21st century education* (pp. 139–148). Springer.

Reis, P. (2014). Promoting students' collective socio-scientific activism: Teacher's perspectives. In S. Alsop & L. Bencze (Eds.), *Activism in science and technology education* (pp. 547–574). Springer.

Reis, P., Baptista, M., Tinoca, L., & Linhares, E., et al. (2022). Educational potentialities of student-curated exhibitions on socioscientific issues: The students' perspective. In Y.-S. Hsu (Ed.), *Innovative approaches to socioscientific issues and sustainability education, learning sciences for higher education* (pp. 217–233). Springer.

Reis, P., Tinoca, L., Baptista, M., & Linhares, E. (2020). The impact of student-curated exhibitions about socio-scientific issues on students' perceptions regarding their competences and the science classes. *Sustainability, 12*(7), Article 2796. https://doi.org/10.3390/su12072796

Schulz, W., Ainley, J., Fraillon, J., Losito, B., Agrusti, G., & Friedman, T. (2018). *Becoming citizens in a changing world: IEA international civic and citizenship education study 2016 international report*. IEA and Springer.

Schusler, T. M., & Krasny, M. E. (2015). Science and democracy in youth environmental action—Learning "good" thinking. In M. P. Mueller & D. J. Tippins (Eds.), *EcoJustice, citizen science and youth activism—Situated tensions for science education* (pp. 363–384). Springer.

Spaan, W., Oostdam, R., Schuitema, J., & Pijls, M. (2025). Teaching thinking-back-and-forth in practical work: Results of an educational design study in secondary education. *Research in Science & Technological Education*. https://doi.org/10.1080/02635143.2024.2446795

Vygotsky, L. S. (1978). *Mind in society: The development of higher psychological processes*. Harvard University Press.

# Chapter 12
# Science Museums in Support of School Laboratory Work

**Jorge Martín-García, Ana S. Afonso, and Maria Eugenia Dies Álvarez**

## 12.1 Introduction

Laboratory work has been commonly regarded as a central and distinctive element of science education (Gericke et al., 2023). In this chapter, we consider laboratory work as a synthesis of diverse activities pursued through an integrated approach that seamlessly integrates theoretical and practical elements, where students, working either individually or in small groups, try to increase their comprehension of the world through experience. This definition, based on the work of different academics (e.g., Akuma & Callaghan, 2019; Ferreira & Morais, 2020), attempts to consider the significant heterogeneity of laboratory work and encompasses situations in which students are involved in:

– Activities with a hands-on approach, in which they handle laboratory equipment and materials while working with real objects or models,
– Activities that demand the mobilization of some competencies associated with scientific processes but, at the same time, the mobilization of their scientific knowledge and critical thinking with a minds-on approach, and
– Activities that include an empirical data collection or even interaction with secondary sources of data, which need to be analysed and made sense of.

Despite the recognised benefits attributed to laboratory work in science education (Gericke et al., 2023), dissatisfaction with the quantity and quality of school

J. Martín-García (✉) · M. E. Dies Álvarez
IUCA - University of Zaragoza, Zaragoza, Spain
e-mail: araujo@unizar.es

M. E. Dies Álvarez
e-mail: medies@unizar.es

A. S. Afonso
IE/CIEd - University of Minho, Minho, Portugal
e-mail: aafonso@ie.uminho.pt



L. Leite et al. (eds.), *Practical Work in Science Education in the Digital Era*, Advances in Technology-Rich Science Education 2, https://doi.org/10.1007/978-3-032-13788-3_12

laboratory work has persisted for years (Akuma & Callaghan, 2019; Itzek-Greulich et al., 2017). The challenges in implementing effective laboratory work stem from a series of interconnected problems involving a crowded, content-heavy curriculum that limits the available time for laboratory work (Cheung, 2011), inadequate and outdated laboratory infrastructure, large class sizes (Boesdorfer & Livermore, 2018), insufficient budgets for materials and equipment (Cheung, 2011), and teachers' difficulties in designing meaningful learning experiences through laboratory work (Chng, 2024).

In this context, it is necessary to explore new ways of expanding students' opportunities to actively engage in laboratory work. In this regard, non-formal settings can contribute (Itzek-Greulich et al., 2017; Martín-García & Dies Álvarez, 2022; Shaby et al., 2023). In fact, opportunities for laboratory work are frequently one of the objectives teachers delineate for students' participation in non-formal settings (Martín-García & Dies Álvarez, 2022). Among such settings, science museums are ones that each year host a large number of students on school field trips. To what extent, however, do they contribute to laboratory work?

This chapter explores the potential of science museums to act as facilitators of laboratory work in science education. To achieve this, we will examine various types of science museum learning facilities and propose a collaborative science museum-school model. This model is grounded in existing literature recommendations for the creation of an effective *third space* for learning, thereby enhancing the practical science experiences available to students.

## 12.2 Science Museums and Laboratory Work, is that Possible?

Science Museums were created to house collections of objects of scientific interest. In these institutions, staff members (e.g. scientists, curators, educators) hold specialized knowledge and skills in disciplinary knowledge, collections, and/or learning (Hooper-Greenhill, 2000). While collections of objects are the core of these settings, Dillon et al. (2016) add 'facilitation' as another dimension of the science museum. Facilitation transforms objects and collections, content and expertise into learning experiences, which are conveyed through face-to-face interactions (talks, tours, shows, workshops) or interactive exhibits, to name just a few (Dillon et al., 2016). In the new generation of science museums, there are different types of spaces, including exhibitions, hands-on areas, object areas, science labs, lecture rooms, or science shows.

Teachers value the contribution of science museums in enhancing students' learning of science that is part of the curriculum or that complements it in ways not possible in schools (Kisiel, 2005). Yet research shows that teachers often find it difficult to support learning during field trips (Lock & Glackin, 2009). Given the characteristics of science museums, and the fact that laboratory work is central to school science learning, can these institutions provide opportunities to enrich school laboratory work and foster deeper learning? To explore this potential role, it is important to examine the various types of museum learning facilities.

### 12.2.1 Interactive Exhibits, Objects and Digital Technology in the Exhibition Area

Interactive exhibits demand visitors to perform a particular action, and the exhibit then produces some form of response (Allen, 2014). They were originally established in science centres, but are increasingly being attached into conventional science museums. These exhibits are often placed alongside objects under the same theme, increasing the learning experience of both when explicit links between them are decoded by the audience (Gilbert & Afonso, 2015). Interactive exhibits can be grouped into exemplars of phenomena (e.g. Bernoulli effect), and into analogy-based exhibits which convey scientific principles (e.g. Black Hole, using a ball spinning in a funnel) (Stocklmayer & Gilbert, 2002).

Many interactive exhibits use large-scale equipment (e.g. prisms or pulleys), which resemble laboratory setups. In this sense, they provide opportunities to observe or predict phenomena under varying conditions (e.g. light refraction in different media), to engage with replicas of historical experiments, e.g. Newton's optical crucial experiment (Afonso et al., 2019), and even to experience a phenomenon in an embodied way, as simulated through a replica of a tornado (Stevens & Hall, 1997). Hence, interactive exhibits go beyond providing awareness of phenomena that are difficult to replicate in the classroom and increase motivation towards them (Falk et al., 2004). They may complement the outstanding objectives of school laboratory work, as defined by Leite and Dourado (2013). This potential complementarity depends on the exhibit's design, namely its degree of openness. While some exhibits require only a simple action (e.g. pushing a button), open-ended exhibits allow multiple manipulation of variables to investigate a phenomenon (e.g. shadow size). Open-ended exhibits play a central role in science museums as they foster scientific thinking, including raising questions, planning investigations, finding patterns in observation, using evidence critically, and reaching conclusions (Gutwill, 2008).

Research shows that conceptual gains from the free exploration of interactive exhibits tend to be low. In exhibits as exemplar of phenomena, difficulties arise in describing phenomena, and misconceptions are sometimes reinforced. In analogy-based exhibits, conceptual learning occurs when the source and target analogues are perceived and mapped (Afonso & Gilbert, 2007). Regarding thinking skills, students often formulate questions and make predictions, but they rarely follow a coherent line of investigation, draw conclusions, or make generalizations (Randol, 2005). While it is acknowledged that developing such skills can promote a bridge between school (namely laboratory work) and science museum, difficulties emerge regarding how to promote them for all students.

One factor that enhances learning from interactive exhibits is the dialogic interaction between students and science museum educators. In particular, storytelling and good analogies provide opportunities to enrich explanatory talk (Zhai & Dillon, 2014). Another successful approach is the "juicy questions game" (Gutwill & Allen, 2011). This is a collaborative approach in which students, initially supported by a science museum educator and later by an adult, explore an interactive exhibit, generate questions, and agree on one question to investigate. They then share and discuss their findings, reflecting on the process. However, care is needed when implementing guided inquiry in science museums. While guiding students to engage with exhibits in a specific way can enhance skills aligned with the school curriculum, this approach primarily benefits those who already identify as "science people" (Shaby & Vedder-Weiss, 2019). In particular, when science museums reproduce students' roles, interactions, and positions from school, those regarded as non-science learners in school tend to remain so in the science museum context. Consequently, field trips may compromise affective goals (e.g. identity development), thereby undermining the potential of science museums to leverage students with low science capital (Shaby & Vedder-Weiss, 2019).

Objects are at the core of science museums and are valued for their authenticity, which is understood in terms of their natural origin, long history, belonging to notable scientists (de Kluis et al., 2023). In contemporary science museums, objects are no longer organized as object-based exhibitions, which were once perceived as stable entities carrying intrinsic and fixed meaning (Pedretti & Iannini, 2020). The earlier approach reflected the Enlightenment model of science, characterised by reason, neutral rationality, and dispassion (Pedretti & Iannini, 2020). Currently, objects are reframed to convey meaning that go beyond their physical entity, which is shaped by the theme of the exhibition (Hooper-Greenhill, 2000). For example, laboratory collections are not merely milestones of famous scientists, they can reveal epistemic practices, institutional contexts, and cultural values, connecting scientific activity to its social and historical dimensions.

Since objects are not self-explanatory, students need support in identifying and interpreting them, which calls for activities that facilitate learning (Paris & Hapgood, 2002). These activities, termed object-based learning (Chatterjee & Hannan, 2015), often require collaborative dialogue with science museum educators. Melber (2003) reported that object-based learning focused on inquiry-based activities, that resemble the work of museum scientists (e.g. handling objects, critically examining them and discussing their characteristics), fascinates students. Achiam et al. (2016) revealed how object-based learning can prompt practical work in an object-based palaeontology programme for upper secondary students. In their programme, students took time to look closely at objects, compare them, ask questions and pursue their inquiries, which were initially supported by science museum educators. More recently, Yeh (2023) developed successful object-based learning in a Natural History Museum. In this approach, students did not touch the specimens, but engaged in collaborative conversation with a museum educator using a Predict-Observe-Explain framework.

Digital technologies, namely augmented reality (AR), hold promise in supporting object-based learning. Yoon et al. (2018) found that AR was most effective for revealing hidden information, but learning gains increased when it was complemented with additional scaffolding procedures. A recent study by Ga et al. (2024) showed that an AR dinosaur exhibit captured student interest but diverted their attention from the exhibit's intended purpose. Despite AR's promising educational value, few studies have analysed its impact on science museum learning (Ga et al., 2024). Moreover, visitors tend to prefer authentic objects over touchable replicas or computer simulations (Lindgren-Streicher & Reich, 2007). As Cooper (2019) stated, digital technologies should be used as a medium to support engagement, not as a replacement for objects and collections. Research on science museums focuses on the exhibition space, and without adequate guidance, students' learning gains during field trips are primarily affective. At least two factors can explain this trend: exhibition areas often overwhelm students with stimuli, causing faster fatigue (Shaby et al., 2023); and most field trips are stand-alone events, lasting only a few hours (Itzek-Greulich et al., 2017).

### *12.2.2 Science Shows*

Science shows can be used alone or as a complement to an exhibition. Demonstrations are key in science shows, conveying scientific ideas through laboratory activities or analogies of phenomena, which are presented with the help of physical props (Shaby et al., 2025). Demonstrations in science shows trace back to historical lecture demonstrations (e.g. Faraday's demonstrations at the Royal Institution in London), designed to engage the audience in science, but unlike their predecessors, they are not only presented by scientists (Sadler, 2021). Furthermore, science shows sometimes incorporate storylines and characters that can experience conflict over scientific issues (Austin & Sullivan, 2019). Science shows for school students in science museums, which include laboratory demonstrations, are linked to curriculum-related science areas, such as physics of sound (Fish et al., 2017) or pressure (Karim & Roslan, 2020), and are aimed at extending students' comprehension.

The few known studies on the educational value of science shows suggest that they can contribute to positive attributes towards science and support cognitive learning. Effective science shows use surprise demonstrations, which are organized around a specific theme, and engage the audience in dialogue (Austin & Sullivan, 2019). Regarding demonstrations that elicit surprise, Sadler (2021) noted that those that are novel, bizarre or counter-intuitive are best remembered. In terms of dialogue, the most effective interactions are unstructured and initiated by visitors, who often ask questions related to the content of the show (Shaby et al., 2025). Yet, evidence of the long-term impact of science shows remains scarce (.

### 12.2.3 Science Learning Laboratories, Makerspaces and 'Glass Museums'

Science learning laboratories are spaces in science museums where students can perform laboratory work under supervision (Bell et al., 2009). These spaces provide opportunities to engage in laboratory activities on themes that complement the school curriculum; the Cell Lab is one such example (Bell et al., 2009).

Few studies have evaluated the impact of science learning laboratories in science museums. The known studies suggest that informal science laboratories are effective in increasing motivation (Itzek-Greulich et al., 2017). Bencze and Lemelin (2001) reported that in a computer-based science and technology workshop supported by science museum educators, students developed inquiry skills, engaged in problem-solving and presented their reasoning to peers. Conversely, Shaby et al. (2023) mentioned that students often do not work collaboratively during these activities. Instead, their focus is on operating the equipment without engaging in scientific explanations.

In 'Maker' and 'Tinkerer' spaces in science museums, learners construct and deconstruct objects to explore science, technology, engineering, art, and mathematics (STEAM). It is the learner who, through design thinking and project-based learning, creates a problem and identifies challenges within the constraints of available materials and design (Gutwill et al., 2015). There is a consensus on the theoretical foundations of Maker and Tinkerer spaces in science museums, but little is known about learning outcomes from them, partly due to the challenges involved in evaluating the experiences they offer (Wardrip et al., 2024).

Science museums have been criticised for focusing on the products of science (e.g., objects, specimens, collections, and facts) while offering little representation of science-in-the-making. Particularly, science museums often do not reveal to visitors how objects are investigated or how knowledge is produced within these institutions (Wylie, 2020). To address this limitation, some science museums have developed 'glass museums', which incorporate a research laboratory within the museum. These spaces allow visitors to observe scientists at work, interact with them directly (e.g., Wylie, 2020) and, in some cases, even participate in a laboratory activity (Busch Chandran et al., 2020). Nevertheless, Wylie (2020) reported that visitors' understanding of what was happening in the lab was limited due to the few interpretative resources and interaction with staff was restricted.

In sum, there are several opportunities for students to engage in laboratory work in science museums or to enrich school-based laboratory experiences. Despite these opportunities, learning in science museums is complex and influenced by many factors beyond the design of physical spaces, namely the novelty of the experience, the quality of dialogue between museum educators and students, orientation within the science museum, social interactions, science capital, and the alignment of students', teachers', and museums' agendas (Rennie, 2014). While empirical research on students' learning in science museums has focused on exhibitions and cognitive learning outcomes, some studies have analysed the development of scientific skills. Nevertheless, a wider range of learning outcomes (e.g. affective, social and behavioural) requires further investigation.

## 12.3 Implications for the Practice: The Need to Build a Third Space

For years, it has been considered necessary to build bridges to promote closer collaboration between schools and providers of non-formal education activities (Rocard et al., 2007). Researchers such as Stocklmayer et al. (2010), Dunlop et al. (2019), Raaijmakers et al. (2021), and Toscano and Millar (2022) argue that there is a need in science education to create a *third space*. This is defined as a place where the spheres of formal science education (the first space) and non-formal programmes or activities (the second space) converge and merge. This approach provides a more complete and enriching way of learning science, allowing for the limitations of each separate educational domain to be overcome.

This *third space* should be understood as an interface that satisfies the needs of both sides, leverages their individual strengths to increase educational impact, and presents a unified image of scientific knowledge. Therefore, it is essential to strengthen and promote connections between non-formal education spaces and their educators and the school (McCauley et al., 2018; Rocard et al., 2007). This requires joint planning and cooperation between teachers and educators based on mutual respect and an understanding of each other's practices.

Based on this premise, authors such as Fallik et al. (2013) and Dunlop et al. (2019) have attempted to propose models of the *third space* in various contexts. In the science museum context, for example, Raaijmakers et al. (2021) have put forward a proposal focused on creating a *third space* by providing structure and resources to science museum educators and schoolteachers to cooperatively create learning situations in both school and science museum contexts. The authors highlight a series of recommendations derived from the literature, including:

- Establish equal partnerships between schools and science museums, based on mutual understanding of each other's practices (Fallik et al., 2013).
- Encourage teachers and science museum educators to become familiar with each other's workspaces, where students learn (Fallik et al., 2013).
- Plan and conduct post-visit classroom activities to reinforce the experience.

Other contexts in which successful *third spaces* have been described include collaborations between the school and scientists and university staff (e.g., Dunlop et al., 2019; McCauley et al., 2018) and school clubs that promote collaboration between the school and institutions such as universities, research centres, and museums.

This brings us to a critical question for science education: Can science museums play a role as educational agents in enhancing laboratory work? The preceding section presented diverse examples illustrating how science museums have endeavoured to offer visitors a clearer insight into and practical opportunities for engaging with laboratory work. It is evident, therefore, that science museums can contribute in various ways to promoting or facilitating such experiences. Specific examples can be found in seminal works by Falk et al. (2004), Gutwill (2008), Melber (2003), Yeh (2023), Shaby et al. (2025), and Busch Chandran et al. (2020), but also those conducted by Domenici (2022) or Franza and Patresi (2024).

However, this dynamic still has clear limitations if the experiences that take place in school are unrelated to those that take place in the science museum, despite the diversity of resources that science museums can provide to engage students in laboratory activities (Ferreira & Morais, 2020; Shaby et al., 2023). Therefore, if the intention is to promote sustained, effective, and high-quality laboratory work for students, a way forward is through the building of a *third space* in the direction taken by the proposals found in the literature (Dunlop et al., 2019; Fallik et al., 2013; Raaijmakers et al., 2021), which imply that non-formal education activities should be more than a once-off or occasional experience that happens in the science museum; they would be a reality integrated into the everyday workings of the formal science education sector.

Science museums are out-of-school contexts that are common school trip destinations, but attempting to convert them into a *third space* will involve a change in how science museums collaborate and relate to schools, moving towards a model of continuous presence. In this model of partnership, the science museum must become a travel companion, a permanently connected agent that provides challenges, collaborates on and promotes projects, and allows pupils to have more direct contact with the reality and culture of science.

This could involve science museums offering sustained project-based learning experiences, co-developed with teachers, that span several weeks or months, rather than single-day workshops (e.g. Xanthoudaki et al., 2007). Here, a greater interaction between teachers and museum staff could mutually benefit both parties. This interaction could lead to science museum offerings that are more aligned with school realities and better integrated into the school curriculum. By understanding the specific needs and limitations of teachers, science museums can tailor their programmes to be more relevant and impactful (Stocklmayer et al., 2010). In addition, this interaction facilitates a deepening of the disciplinary knowledge underlying the experiments, helping teachers to understand not only the 'how' but also the 'why' of scientific phenomena. This transfer of know-how and professional demeanour directly enriches teachers' pedagogical practice, enabling them to implement more authentic, safer, and meaningful activities, which ultimately results in a substantial improvement in students' experimental learning and understanding of science.

## 12.4 Conclusion and Ways Forward

In this chapter, we addressed the question of how science museums can contribute to promoting and improving laboratory work. Throughout this chapter, we have reviewed various types of science museum learning facilities, providing examples of how they can contribute to facilitating laboratory work. Following this review, the pressing need to foster a more sustained relationship between schools and science museums becomes apparent. Such a relationship should give rise to projects where the laboratory work undertaken is more closely connected with the content addressed in the classroom, thereby ensuring coherence and integration between practical laboratory engagement and theoretical classroom learning. Therefore, drawing inspiration from the efforts found in the literature to create a *third space*, this chapter advocates for a transformative model where science museums transition from being primarily venues for isolated visits to becoming active, year-round partners with schools.

That demands a shift to a collaboration model that is more sustained, equitable, and provides greater support for teachers. This mechanism should be more oriented towards positioning them as promoters of projects where laboratory work facilitates learning framed within authentic scientific practices. This involves science museums providing ongoing guidance, co-designing curriculum-aligned activities, lending essential materials and equipment, and offering targeted professional development to enhance teacher confidence and expertise. By embedding practical support directly within the school context, this collaborative model ensures thematic continuity, pedagogical suitability, and optimal utilisation of laboratory experiences. The model ensures that laboratory experiences are not just engaging, but also deeply integrated and systematically reinforced throughout the academic year. This empowers coherent and comprehensible lab work for students, where doing and thinking go hand in hand to produce meaningful learning.

**Acknowledgements** This work was supported by the research group "Grupo Beagle de Investigación en Didáctica de las Ciencias Experimentales" (S27_23R, Gobierno de Aragón, IUCA); by the Agencia Estatal de Investigación (PID2024-156735NA-I00); and by CIEd, Research Centre on Education of the Institute of Education of the University of Minho (UID/01661/2025), which funded this work through national funds of FCT/MCTES-PT, namely the Pluriannual Program for the Funding of R&D Units.

## References

Achiam, M., Simony, L., & Lindow, B. E. K. (2016). Objects prompt authentic scientific activities among learners in a museum programme. *International Journal of Science Education, 38*(6), 1012–1035. https://doi.org/10.1080/09500693.2016.1178869

Afonso, A. S., & Gilbert, J. K. (2007). Educational value of different types of exhibits in an interactive science and technology center. *Science Education, 91*(6), 967–987. https://doi.org/10.1002/sce.20220

Afonso, S., Afonso, A. S., & Rodrigues, F. (2019). Towards an effective use of language to explain light in the museum. *Science Education, 103*(4), 923–946. https://doi.org/10.1002/sce.21517

Akuma, F. V., & Callaghan, R. (2019). A systematic review characterizing and clarifying intrinsic teaching challenges linked to inquiry-based practical work. *Journal of Research in Science Teaching, 56*(5), 619–648. https://doi.org/10.1002/tea.21516

Allen, S. (2014). Interactive exhibits. In R. Gunstone (Ed.), *Encyclopedia of science education* (pp. 1–5). Springer. https://doi.org/10.1007/978-94-007-6165-0_296-4

Austin, R. P. S., & Sullivan, M. (2019). How are we performing? Evidence for the value of science shows. *International Journal of Science Education, Part B, 9*(1), 1–12. https://doi.org/10.1080/21548455.2018.1532620

Bell, P., Lewenstein, B., Shouse, A. W., & Feder, M. A. (2009). *Learning science in informal environments: People, places, and pursuits.* National Academies Press. https://doi.org/10.17226/12190

Bencze, J. L., & Lemelin, N. (2001). Doing science at a science centre: Enabling independent knowledge construction in the context of schools' museum visits. *Museum Management and Curatorship, 19*(2), 141–157. https://doi.org/10.1080/09647770100301902

Boesdorfer, S. B., & Livermore, R. A. (2018). Secondary school chemistry teacher's current use of laboratory activities and the impact of expense on their laboratory choices. *Chemistry Education Research and Practice, 19*(1), 135–148. https://doi.org/10.1039/c7rp00159b

Busch Chandran, K., Jarrett, K., & Wyss, J. M. (2020). Creating a sustainable partnership between a science center, university, and local school districts: A retrospective on over 20 years of successful programming and partnership. *The Journal of STEM Outreach, 3*(3), 1–15. https://doi.org/10.15695/jstem/v3i3.03

Chatterjee, H. J., & Hannan, L. (2015). *Engaging the senses: Object-based learning in higher education.* Routledge.

Cheung, D. (2011). Evaluating student attitudes toward chemistry lessons to enhance teaching in the secondary school. *Educación Química, 22*(2), 117–122. https://doi.org/10.1016/S0187-893X(18)30123-X

Chng, E. (2024). Designing computer vision support for science practical work: A qualitative investigation into the noticing practices and support preferences of science teachers. *Journal of Science Education and Technology, 33*(5), 718–728. https://doi.org/10.1007/s10956-024-10116-w

Cooper, C. (2019). You can handle it: 3D printing for museums. *Advances in Archaeological Practice, 7*(4), 443–447. https://doi.org/10.1017/aap.2019.39

de Kluis, T., Romp, S., & Land-Zandstra, A. M. (2023). Science museum educators' views on object-based learning: The perceived importance of authenticity and touch. *Public Understanding of Science, 33*(3), 325–342. https://doi.org/10.1177/09636625231202617

Dillon, J., DeWitt, J., Pegram, E., Irwin, B., Crowley, K., Haydon, R., King, H., Knutson, K., Veall, D., & Xanthoudaki, M. (2016). *A learning research agenda for natural history institutions.* Natural History Museum.

Domenici, V. (2022). STEAM project-based learning activities at the science museum as an effective training for future chemistry teachers. *Education Sciences, 12*(1), 30. https://doi.org/10.3390/educsci12010030

Dunlop, L., Clarke, L., & McKelvey-Martin, V. (2019). Free-choice learning in school science: A model for collaboration between formal and informal science educators. *International Journal of Science Education, Part B, 9*(1), 13–28. https://doi.org/10.1080/21548455.2018.1534023

Falk, J. H., Scott, C., Dierking, L., Rennie, L., & Jones, M. C. (2004). Interactives and visitor learning. *The Museum Journal, 47*(2), 171–198. https://doi.org/10.1111/j.2151-6952.2004.tb00116.x

Fallik, O., Rosenfeld, S., & Eylon, B.-S. (2013). School and out-of-school science: A model for bridging the gap. *Studies in Science Education, 49*(1), 69–91. https://doi.org/10.1080/03057267.2013.822166

Ferreira, S., & Morais, A. M. (2020). Practical work in science education: Study of different contexts of pedagogic practice. *Research in Science Education, 50*(4), 1547–1574. https://doi.org/10.1007/s11165-018-9743-6

Fish, D., Allie, S., Pelaez, N., & Anderson, T. (2017). A cross-cultural comparison of high school students' responses to a science centre show on the physics of sound in South Africa. *Public Understanding of Science, 26*(7), 806–814. https://doi.org/10.1177/0963662516642725

Franza, A., & Pratesi, G. (2024). Learning sciences from the past: Recovery, study, and cataloguing of a historical natural history school museum. *Education Sciences, 14*(1), 80. https://doi.org/10.3390/educsci14010080

Ga, S. H., Cha, H. J., & Yoon, H. G. (2024). How do students respond to the intended affordance of augmented reality dinosaur exhibits in a science museum? *Journal of Science Education and Technology, 33*, 687–698. https://doi.org/10.1007/s10956-024-10113-z

Gericke, N., Högström, P., & Wallin, J. (2023). A systematic review of research on laboratory work in secondary school. *Studies in Science Education, 59*(2), 245–285. https://doi.org/10.1080/03057267.2022.2090125

Gilbert, J. K., & Afonso, A. S. (2015). Lifelong learning: Approaches to increasing the understanding of chemistry by everybody. In J. García-Martínez & E. Serrano-Torregrosa (Eds.), *Chemistry education: Best practices, opportunities and trends* (pp. 123–148). Wiley-VCH. https://doi.org/10.1002/9783527679300.ch6

Gutwill, J. P. (2008). Challenging a common assumption of hands-on exhibits: How counterintuitive phenomena can undermine inquiry. *The Journal of Museum Education, 33*(2), 187–198. http://www.jstor.org/stable/40479643

Gutwill, J. P., & Allen, S. (2011). Deepening students' scientific inquiry skills during a science museum field trip. *Journal of the Learning Sciences, 21*(1), 130–181. https://doi.org/10.1080/10508406.2011.555938

Gutwill, J. P., Hido, N., & Sindorf, L. (2015). Research to practice: Observing learning in tinkering activities. *Curator, 58*(2), 151–168. https://doi.org/10.1111/cura.12105

Hooper-Greenhill, E. (2000). *Museums and the interpretation of visual culture* (1st ed.). Routledge. https://doi.org/10.4324/9781003124450

Itzek-Greulich, H., Flunger, B., Vollmer, C., Nagengast, B., Rehm, M., & Trautwein, U. (2017). Effectiveness of lab-work learning environments in and out of school: A cluster randomized study. *Contemporary Educational Psychology, 48*, 98–115. https://doi.org/10.1016/j.cedpsych.2016.09.005

Karim, N., & Roslan, R. (2020). The impact of interactive science shows on student's learning achievement on fire and pressure science concept for 9th grader in Brunei. *Journal Pendidikan IPA Indonesia, 9*(3), 294–308. https://doi.org/10.15294/jpii.v9i3.23684

Kisiel, J. (2005). Understanding elementary teacher motivations for science trips. *Science Education, 89*(1), 936–955. https://doi.org/10.1002/sce.20085

Leite, L., & Dourado, L. (2013). Laboratory activities, science education and problem-solving skills. *4th International Conference on New Horizons in Education, 106*, 1677–1686. https://doi.org/10.1016/j.sbspro.2013.12.190

Lindgren-Streicher, A., & Reich, C. A. (2007). Visitor usage of digital and physical artifacts in two museum programs. *Visitor Studies, 10*(2), 152–167. https://doi.org/10.1080/10645570701603666

Lock, R., & Glackin, M. (2009). Teaching out-of-classroom science: Implications from the initial teacher training experience. *School Science Review, 90*(333), 111–118.

Martín-García, J., & Dies Álvarez, M. E. (2022). An examination of teacher's goals for a school crystal growing competition: More than having fun. *International Journal of Science Education, 44*(6), 962–979. https://doi.org/10.1080/09500693.2022.2059121

McCauley, V., Martins Gomes, D., & Davison, K. G. (2018). Constructivism in the third space: Challenging pedagogical perceptions of science outreach and science education. *International Journal of Science Education, Part B, 8*(2), 115–134. https://doi.org/10.1080/21548455.2017.1409444

Melber, L. M. (2003). Partnerships in science learning: Museum outreach and elementary gifted education. *Gifted Child Quarterly, 47*(4), 251–258. https://doi.org/10.1177/001698620304700402
Paris, S. G., & Hapgood, S. E. (2002). Children learning with objects in informal learning environments. In S. G. Paris (Ed.), *Perspectives on object-centered learning in museums* (pp. 37–54). Lawrence Erlbaum Associates.
Pedretti, E., & Iannini, A. M. N. (2020). Towards fourth-generation science museums: Changing goals, changing roles. *Canadian Journal of Science, Mathematics and Technology Education, 20*, 700–714. https://doi.org/10.1007/s42330-020-00128-0
Raaijmakers, H., Mc Ewen, B., Walan, S., & Christenson, N. (2021). Developing museum-school partnerships: Art-based exploration of science issues in a third space. *International Journal of Science Education, 43*(17), 2746–2768. https://doi.org/10.1080/09500693.2021.1986646
Randol, S.M. (2005). *The nature of inquiry in science centers: Describing and assessing inquiry at exhibits.* [Unpublished doctoral dissertation]. University of California, Berkeley.
Rennie, L. J. (2014). Learning science outside of school. In S. K. Abell & N. G. Lederman (Eds.), *Handbook of research on science education* (Vol. II., pp. 120–144) Lawrence Erlbaum Associates.
Rocard, M., Csermely, P., Jorde, D., Lenzen, D., Walberg-Henriksson, H., & Hemmo, V. (2007). *Science education now: A renewed pedagogy for the future of Europe*. European Commission.
Sadler, W. J. (2021). Evaluating the short-term and long-term impact of an interactive science show. *Research for All, 5*(2), 399–419. https://doi.org/10.14324/RFA.05.2.14
Shaby, N., & Vedder-Weiss, D. (2019). Science identity trajectories throughout school visits to a science museum. *Journal of Research in Science Teaching, 57*(5), 733–764. https://doi.org/10.1002/tea.21608
Shaby, N., Assaraf, O.B.-Z., & Koch, N. P. (2023). Students' interactions during laboratory group activity in a science museum. *International Journal of Science and Mathematics Education, 22*(4), 703–720. https://doi.org/10.1007/s10763-023-10404-8
Shaby, N., Peleg, R., Coombs, I., & Hemming, J. (2025). "It was just amazing!" Unstructured interactions following a planetarium and science shows. *Journal of Museum Education*, 1–15. https://doi.org/10.1080/10598650.2025.2461846
Stevens, R., & Hall, R. (1997). Seeing tornado: How video traces mediate visitor understandings of (natural?) phenomena in a science museum. *Science Education, 81*(6), 735–747. https://doi.org/10.1002/(SICI)1098-237X(199711)81:6%3c735::AID-SCE9%3e3.0.CO;2-J
Stocklmayer, S. M., Rennie, L. J., & Gilbert, J. K. (2010). The roles of the formal and informal sectors in the provision of effective science education. *Studies in Science Education, 46*(1), 1–44. https://doi.org/10.1080/03057260903562284
Stocklmayer, S., & Gilbert, J. K. (2002). New experiences and old knowledge: Towards a model for the personal awareness of science and technology. *International Journal of Science Education, 24*(8), 835–858. https://doi.org/10.1080/09500690210126775
Toscano, M., & Millar, V. (2022). Trust in the third space of science education. *Science & Education, 31*(6), 1691–1708. https://doi.org/10.1007/s11191-022-00356-w
Wardrip, P. S., White, A., Bank, A., & Brahms, L. (2024). Making observations: Co-design of an observation tool for and with maker educators. *Curator: The Museum Journal, 67*(3), 711–724. https://doi.org/10.1111/cura.12614
Wylie, C. D. (2020). Glass-boxing science: Laboratory work on display in museums. *Science, Technology, & Human Values, 45*(4), 618–635. https://doi.org/10.1177/0162243919871101
Xanthoudaki, M., Tirelli, B., Cerutti, P., & Calcagnini, S. (2007). Museums for science education: Can we make the difference? The case of the EST project. *Journal of Science Communication, 6*(2), 1–10. https://doi.org/10.22323/2.06020202
Yeh, J. H. (2023). Developing natural history museum object-based inquiry for museum's group visitors. In P. G. Patrick (Ed.), *How people learn in informal science environments* (pp. 433–455). Springer. https://doi.org/10.1007/978-3-031-13291-9_21

Yoon, S. A., Anderson, E., Park, M., Elinich, K., & Lin, J. (2018). How augmented reality, textual, and collaborative scaffolds work synergistically to improve learning in a science museum. *Research in Science and Technological Education, 36*(3), 261–281. https://doi.org/10.1080/02635143.2017.1386645

Zhai, J., & Dillon, J. (2014). Communicating science to students: Investigating professional botanic garden educators' talk during guided school visits. *Journal of Research in Science Teaching, 51*(4), 407–429. https://doi.org/10.1002/tea.21143

# Part III
# Digital Practical Work in Science Education

# Chapter 13
# Enhancing Practical Work through 3D Animated Virtual Labs: Opportunities and Challenges in Biology Education

**Nor Asniza Ishak, Muaz Mohd Zaini Makhtar, and Rafiza Rosli**

## 13.1 Introduction

Laboratory activities play a vital role in science education as they enhance students' learning experiences (Aljuhani et al., 2018; Arbaugh & Benbunan-Fich, 2007). Through hands-on engagement, students gain deeper comprehension of theories and principles, leading to more meaningful learning outcomes (Bellou et al., 2022; Chiu & Linn, 2012). Scientific experiments are particularly effective in supporting students' understanding of abstract and challenging concepts (De Jong et al., 2013). Additionally, laboratory work develops essential scientific skills that prepare students to become competent learners (Aljuhani et al., 2018; Bellou et al., 2022). With technological advancements, real experiments can also be presented virtually, further supporting knowledge acquisition (Dori & Belcher, 2005; Faulconer & Gruss, 2018).

Practical work has long been a cornerstone of science education, especially in Biology, where laboratory investigations and experiments allow students to directly engage in scientific inquiry (García-Peñalvo et al., 2021; Rosli & Ishak, 2024). Traditionally, these activities have emphasized hands-on manipulation of specimens, equipment, and procedures. However, rapid advances in digital technologies, particularly three-dimensional (3D) animated virtual laboratories, are reshaping the way practical work is designed and delivered (Hodson, 2014; Jeno et al., 2020).

Virtual laboratories, or virtual labs, are technology-based simulations of traditional laboratories designed with a learner-centered approach (Jonassen, 1999; Karam & Warren, 2021). They typically combine simulations and animations (Makransky & Petersen, 2021), allowing learners to explore freely while connecting

N. A. Ishak (✉) · R. Rosli
School of Educational Studies, Universiti Sains Malaysia, Gelugor, Penang, Malaysia
e-mail: asnizaishak@usm.my

M. M. Z. Makhtar
School of Industrial Technology, Universiti Sains Malaysia, Gelugor, Penang, Malaysia
e-mail: muazzaini@usm.my



L. Leite et al. (eds.), *Practical Work in Science Education in the Digital Era*, Advances in Technology-Rich Science Education 2, https://doi.org/10.1007/978-3-032-13788-3_13

theoretical concepts with practical applications (Margot & Kettler, 2019; Ng & Chua, 2023). Depending on their design, virtual labs can range from simple 2D models to advanced 3D interactive simulations and immersive virtual reality (VR) environments that promote more engaging learning experiences (Potkonjak et al., 2016).

Definitions of virtual labs vary. Broadly, they are described as software that allows students to conduct experiments in a virtual space (De Jong et al., 2013; Raman et al., 2020). Through these platforms, students can perform virtual experiments and observe results that closely mirror real laboratory outcomes (Raman et al., 2020). Virtual labs are electronic systems supporting learning without requiring a physical laboratory (Makransky & Petersen, 2021) and serve as platforms for simulation and remote experimentation (Rosli & Ishak, 2022). Digital tools accessible via mobile devices or laptops allow students to achieve experimental objectives without physical apparatus or chemicals (Viitaharju et al., 2023). Despite definitional variation, the core concept remains consistent: virtual labs facilitate science learning.

Integrating virtual labs into education provides multiple benefits. Their design emphasizes contextual learning supported by simulations and animations, which fosters creativity and critical thinking (Makransky & Petersen, 2021; Smetana & Bell, 2012). They also encourage collaboration among students by enabling the exchange of scientific ideas and observations, thereby enhancing academic performance. For teachers, virtual labs enrich instructional quality by providing visual and immersive experiences that clarify complex concepts (Raman et al., 2020). This leads to more engaging, interactive, and effective lessons (Aljuhani et al., 2018; Dori & Belcher, 2005). Moreover, virtual experiments often use multimedia elements such as text, images, and animations based on real-life contexts, which improve comprehension. They save time by eliminating lengthy setup processes and offering immediate, repeatable results. Importantly, virtual labs support flexible, self-paced learning, allowing students to engage anytime and anywhere (Makransky & Petersen, 2021; Smetana & Bell, 2012).

The digital era presents unprecedented opportunities to improve the accessibility, scalability, and quality of practical work in Biology (García-Peñalvo et al., 2021). As schools and higher education institutions increasingly integrate digital technologies, virtual laboratories have emerged as a meaningful complement to traditional practical experiences. This manuscript examines the pedagogical rationale, opportunities, challenges, and future prospects of 3D animated virtual labs in Biology education, with a particular emphasis on the Malaysian curriculum.

## 13.2 Literature Review

Virtual laboratories have emerged as a valuable alternative or complement to traditional laboratories in science education (Hodson, 2014; Rosli & Ishak, 2022). They help address limitations such as lack of resources by providing flexible learning opportunities, immediate feedback, and the ability to repeat experiments (Rosli & Ishak, 2024). Importantly, virtual labs enable students to engage with costly, hazardous, or highly complex experiments that are otherwise difficult to conduct in real settings. As a result, students can practice essential skills such as manipulating equipment, collecting data, testing hypotheses, and reporting findings without being restricted by physical resources.

A major justification for integrating virtual labs is the abstract nature of many scientific concepts, which often leads to misconceptions (Rosli & Ishak, 2022; Smetana & Bell, 2012). To overcome this, effective learning materials should connect new knowledge with prior understanding. Virtual labs offer strong visualization tools that simplify and concretize abstract principles, enhancing comprehension (Rosli & Ishak, 2024). For instance, LabVIEW-based virtual modules help students better understand challenging concepts such as electromagnetism by providing simulations and visual representations. Such tools allow learners to link theoretical principles with observable scientific phenomena.

In terms of cognitive learning outcomes, achievement is often the most measured aspect when using virtual labs. Realistic simulations and visualizations allow students to explore concepts in ways similar to real laboratory settings, leading to better academic performance (Aljuhani et al., 2018). Studies show that 3D visualizations and multimedia features such as color, sound, and animation strengthen students' understanding of abstract topics (Rosli & Ishak, 2024). Instructional strategies such as inquiry-based learning, collaboration, scaffolding, feedback, and modality (Potkonjak et al., 2016) significantly improve learning outcomes. Combining visualization with structured instructional methods enhances achievement and fosters meaningful learning.

Beyond achievement, virtual labs strengthen students' conceptual understanding (Aljuhani et al., 2018). For example, PhET simulations improve secondary students' understanding of electric circuits, while animations enhance students' focus on conceptual aspects of experiments (Smetana & Bell, 2012). Integrating virtual and physical labs allows students to investigate otherwise inaccessible phenomena, deepening comprehension. Multimedia-based learning materials, including animations and simulations, also contribute significantly to conceptual development.

Regarding psychomotor learning outcomes, virtual labs are recognized as effective substitutes for physical labs in improving practical skills. Students can practice setting up experiments with minimal teacher support (Hofstein & Lunetta, 2004; Zacharia & Olympiou, 2011), while interactive simulations provide hands-on experiences (Kapici et al., 2020; Rutten et al., 2012). They enable learners to manipulate experimental variables and observe results visually, such as changes in color, mass, or temperature (De Jong et al., 2013; Smetana & Bell, 2012). Combining virtual labs (e.g., Olabs, vLAB) with physical labs enhances laboratory skills in areas like microscopy and microbiology (Rosli & Ishak, 2024; Tatli & Ayas, 2013). Hence, virtual labs develop experimental competencies and prepare students for real laboratory sessions (Hodson, 2014; Sotiriou & Bogner, 2021).

In the affective domain, virtual labs enhance motivation and foster positive attitudes toward science learning (Rosli & Ishak, 2022). They encourage independent and self-paced learning while providing interactive experiences that increase students' interest in science (Smetana & Bell, 2012). By making abstract concepts tangible, virtual labs create engaging learning environments that positively shape students' perceptions of science (Sotiriou & Bogner, 2021).

Globally, research on virtual laboratories has expanded significantly in the past two decades. Studies in Europe and North America demonstrate that virtual labs enhance inquiry learning, conceptual understanding, and engagement in STEM education (García-Peñalvo et al., 2021; Hodson, 2014). Blended formats combining physical and virtual labs yield the strongest learning outcomes (Kapici et al., 2020). Digital environments boost intrinsic motivation and sustain learners' interest (Jeno et al., 2020).

Regionally, in Asia, adoption of virtual labs has been growing, particularly in China, Korea, and Singapore, where researchers emphasize scalability, affordability, and continuity of science learning (Rosli & Ishak, 2024). Virtual labs have been successfully employed in physics, chemistry, and biology to address safety and logistical barriers. Their critical role in ensuring educational continuity during the COVID-19 pandemic has been highlighted (Bogusevschi et al., 2020).

In Malaysia, recent studies focus on curriculum alignment and local design. 3D virtual labs contextualized for the national Science curriculum directly address issues of accessibility and teacher readiness (Faulconer & Gruss, 2018) especially in biology. Systematic reviews confirm that virtual labs improve conceptual understanding, inclusivity, and engagement, supporting integration into Malaysian science education (Rosli & Ishak, 2024). Besides, in Malaysia, the secondary and pre-university Science curriculum is structured to progressively develop students' scientific understanding, skills, and values. Pedagogical approaches emphasize inquiry-based and project-based learning to nurture students' early engagement with scientific investigation.

At the upper secondary level (Form 4–5), the KSSM Science curriculum in biology advances students' understanding of life processes, physiology, inheritance, evolution, ecology, and the intersection of technology and society. Specific topics include cell structure and division, human and plant physiology (e.g., nutrition, respiration, transport, and excretion), reproduction, genetics, biodiversity, conservation, and biotechnology. Instructional strategies prioritize problem-based learning, STEM integration, and the use of digital tools such as simulations and virtual laboratories to enhance conceptual understanding.

Pre-university programs, such as STPM and Matriculation Biology, emphasize molecular and cellular biology, organismal biology, ecology, genetics, evolution, and applied biology. The curriculum is designed to foster research-based and laboratory-centered inquiry, critical thinking, and collaborative learning, preparing students for higher education and scientific careers. Across all levels, the curriculum integrates cross-cutting elements, including critical and creative thinking, communication, sustainability awareness, and ethical considerations, particularly in topics such as biotechnology and environmental management.

Table 13.1 provides a concise overview of the structure and focus of the Malaysian Biology curriculum at different educational stages, highlighting the alignment of pedagogical approaches with the content and learning objectives. This structured framework demonstrates the potential for integrating innovative instructional tools,

**Table 13.1** Overview of Malaysia Biology Curriculum (KSSM and Pre-University)

| Level | Core themes/Strands | Key topics | Pedagogical emphasis |
|---|---|---|---|
| Lower Secondary (Integrated Science, Form 1–3) | Scientific Skills, Life Processes, Interaction of Living Things, Sustainability of the Environment | Cell as the basic unit of life, nutrition, respiration, reproduction, ecosystem dynamics, environmental issues | Inquiry-based learning, introduction to scientific investigation, project-based learning |
| Upper Secondary Biology (KSSM Form 4–5) | Life Processes, Physiology, Inheritance, Evolution, Ecology, Technology & Society | Molecules of life, cell structure & division, physiology of humans & plants (nutrition, respiration, transport, excretion), reproduction & growth, genetics & inheritance, principles of evolution, biodiversity & conservation, biotechnology | Problem-based learning, STEM integration, use of ICT (e-learning, simulations), field studies |
| Pre-University (STPM/ Matriculation Biology) | Molecular & Cellular Biology, Organismal Biology, Ecology, Genetics, Evolution, Applied Biology | Biochemistry, molecular genetics, physiology of major organ systems, developmental biology, ecology, population biology, biotechnology & bioethics | Research-based learning, laboratory-based inquiry, critical thinking, collaborative projects |
| Cross-Cutting Elements (All Levels) | Scientific Skills & Values | Critical & creative thinking skills, communication, sustainability awareness, ethical issues in biology (e.g., GMOs, stem cells) | 21st-century skills, authentic assessment, project-based learning, service learning |

*Source* KSSM (Ministry of Education, Malaysia)

such as 3D animated virtual labs, to enhance students' conceptual understanding, motivation, and engagement in line with national curriculum goals.

## 13.3 Theoretical and Pedagogical Foundations

The adoption of virtual labs is informed by several educational theories. Constructivist theory emphasizes that learners actively construct knowledge through interaction and exploration, principles directly embodied in virtual labs (Jonassen, 1999).

Inquiry-based learning, which encourages students to investigate problems and formulate solutions, is supported by the interactive design of 3D simulations (Bellou et al., 2022; Kapici et al., 2020).

Cognitive load theory provides insights into how digital environments enhance learning by reducing extraneous load and emphasizing essential processes (Viitaharju et al., 2023). For instance, animations of digestion processes allow learners to visualize complex mechanisms that are invisible in traditional laboratory contexts (Hodson, 2014).

Biology is particularly well-suited for these innovations due to its reliance on visual representations of microscopic or abstract phenomena. Studies confirm that complex concepts such as molecular biology, human anatomy, and ecological systems are effectively taught using 3D animations (Dori & Belcher, 2005; Jeno et al., 2020). 3D virtual labs strengthen theory–practice integration in science classrooms, bridging the gap between abstract concepts and hands-on experiences (Rosli & Ishak, 2024).

## 13.4 Opportunities of 3D Animated Virtual Labs in Biology

The integration of 3D animated virtual laboratories in science especially biology education presents multiple pedagogical opportunities that can transform teaching and learning. By leveraging interactive simulations, these virtual labs allow students to engage with complex biological processes that are often difficult to observe directly in traditional classroom settings. Beyond mere visualization, learners can manipulate experimental variables, test hypotheses, and explore outcomes in a safe, repeatable, and controlled environment. Consequently, virtual labs not only support cognitive development but also enhance motivation, accessibility, inclusivity, and alignment with curriculum and STEM goals. The following sections elaborate on these key opportunities, highlighting how 3D virtual labs can enrich conceptual understanding, foster engagement, improve accessibility, and ensure curricular relevance in the Malaysian context.

### *13.4.1 Enhancing Conceptual Understanding*

3D virtual labs enable learners to visualize dynamic processes such as enzyme activity, photosynthesis, and cell division (Chen & Tsai, 2021; Kapici et al., 2020). Students can manipulate variables, test hypotheses, and observe outcomes beyond the limitations of physical labs, promoting deeper understanding and long-term retention of abstract concepts (De Jong et al., 2013). Virtual labs significantly enhance conceptual mastery across science domains (Rosli & Ishak, 2024).

### 13.4.2 Increasing Student Motivation and Engagement

Virtual labs foster immersive and often gamified experiences, which increase engagement and motivation (Dori & Belcher, 2005; Jeno et al., 2020; Makransky & Petersen, 2021). They reduce learner anxiety around experiments and allow self-paced exploration, thereby promoting autonomy and self-directed learning (Bogusevschi et al., 2020; Margot & Kettler, 2019). By providing interactive and visually stimulating environments, virtual labs help sustain curiosity, reduce cognitive barriers, and encourage students to take an active role in the learning process (Chiu & Linn, 2012; Ng & Chua, 2023). Furthermore, when integrated into collaborative tasks, these platforms promote teamwork and shared problem-solving, strengthening peer learning and engagement (Potkonjak et al., 2016; Rosli & Ishak, 2024).

### 13.4.3 Improving Accessibility and Inclusivity

Virtual labs democratize access to science education, particularly in schools with limited laboratory resources (Bellou et al., 2022). They provide individualized learning pathways that accommodate diverse student needs (Chen & Tsai, 2021). For students with disabilities, virtual labs offer flexible and inclusive environments (De Jong et al., 2013). Such innovations foster equity and broaden participation in science (Rosli & Ishak, 2024).

### 13.4.4 Curriculum Alignment and STEM Goals

Virtual labs can be carefully aligned with curriculum frameworks, ensuring they address critical competencies such as critical thinking, scientific reasoning, and digital literacy (Chen & Tsai, 2021; Rosli & Ishak, 2024). In Malaysia, where inquiry-based learning is central to the Biology curriculum, 3D virtual labs contextualize abstract theories into hands-on digital experiences (Rosli & Ishak, 2024).

## 13.5 Case Study: Malaysian Curriculum Virtual Lab

A focused case study of Malaysia's secondary Biology curriculum highlights the potential of localized virtual laboratory environments to enhance science learning and inquiry-based practice (Faulconer & Gruss, 2018). Within the national curriculum, one 3D animated module was specifically developed to address the topic of nutrition and the human digestive system, a concept traditionally taught through textbook readings, chalk-and-talk methods, and limited laboratory demonstrations due to resource and time constraints.

In the virtual lab, students are immersed in an interactive 3D environment where they can explore the anatomical structures of the digestive organs, manipulate models of the stomach and intestines, and visualize how food molecules are broken down by enzymes. Beyond observation, learners are encouraged to design and test hypotheses for example, by simulating how variations in pH or temperature influence enzymatic activity. This allows them to bridge theoretical knowledge with experimental inquiry, a connection often difficult to achieve in conventional classrooms where access to specialized laboratory facilities is limited.

### *13.5.1 Design Principles*

The design of the virtual lab was guided by four pedagogical and technological principles:

- *Interactivity*: Students actively manipulate variables (e.g., enzyme concentration, food types, temperature) and immediately observe outcomes in a safe, simulated environment. Such interaction not only replicates but often expands the scope of experiments beyond what is feasible in a physical school laboratory (Chen & Tsai, 2021; García-Peñalvo et al., 2021).
- *Curriculum alignment*: The content strictly adheres to Malaysia's Form Four and Form Five Biology syllabus, ensuring that the digital module directly supports national learning objectives. This alignment also allows teachers to integrate the virtual lab seamlessly into lesson plans without requiring significant adaptation (Rosli & Ishak, 2024).
- *Scaffolding*: The module incorporates built-in hints, step-by-step guides, and reflective prompts that support students as they engage in inquiry learning. For example, when students encounter difficulties in formulating hypotheses, the system provides guiding questions that scaffold the process, promoting deeper understanding rather than rote memorization (Bogusevschi et al., 2020; Hodson, 2014).
- *Assessment integration*: Quizzes, diagnostic checks, and short reflection tasks are embedded throughout the module. These formative assessments provide immediate feedback to students while simultaneously enabling teachers to track progress and identify misconceptions in real-time (Chen & Tsai, 2021).

### *13.5.2 Observed Benefits*

Pilot trials conducted with secondary school students revealed significant educational benefits. Learners demonstrated higher levels of engagement, spending longer periods interacting with the module compared to traditional textbook-based lessons. They also showed improved comprehension of digestion processes, particularly the dynamic role of enzymes, as evidenced by post-test gains (Faulconer & Gruss, 2018).

Teachers reported that the virtual lab environment stimulated more active questioning, peer discussion, and collaboration, suggesting that the digital tool fostered both cognitive and social dimensions of learning. These findings are consistent with prior studies, which observed sustained improvements in student motivation, curiosity, and achievement when virtual labs were integrated into classroom instruction across multiple schools (Rosli & Ishak, 2024).

Overall, the Malaysian case study illustrates how context-specific virtual labs, when thoughtfully aligned with curriculum, supported with scaffolds, and integrated with assessment, can serve as an effective bridge between abstract concepts and experiential learning in Biology education.

## 13.6 Challenges and Limitations in Virtual Labs for Biology Education

Despite the significant advantages of virtual laboratories, their effective implementation in Biology education is not without challenges. These limitations span technical, pedagogical, and student-related domains, and must be addressed to maximize the potential of digital learning environments. Understanding these challenges is critical for designing interventions, professional development programs, and blended learning models that ensure equitable and meaningful learning experiences.

### *13.6.1 Technical Challenges*

Virtual labs require stable ICT infrastructure, powerful hardware, and high-speed internet to function effectively (Jeno et al., 2020). In many contexts, especially in rural or underfunded schools, these requirements remain a significant barrier, contributing to the digital divide (Bogusevschi et al., 2020). For Biology education, this challenge is particularly critical as many biological simulations, such as cell processes, genetics experiments, or ecological modeling, are data-intensive and demand robust systems. A promising future direction is the development of lightweight, mobile-compatible platforms that can run efficiently with minimal bandwidth. Cloud-based solutions and open-source biology lab simulations could also democratize access, ensuring that students in resource-limited environments can still engage with authentic biological experimentation.

### 13.6.2 Pedagogical Challenges

The success of virtual labs in Biology depends heavily on teacher readiness and pedagogical integration. Without sufficient training in digital pedagogy, educators may adopt these technologies superficially, using them as add-ons rather than embedding them in inquiry-based, constructivist learning approaches (Aljuhani et al., 2018; Arbaugh & Benbunan-Fich, 2007). In Biology, this often results in students memorizing processes (e.g., mitosis stages or enzyme reactions) without truly engaging in scientific reasoning. To overcome this, professional development must equip teachers not only with technical skills but also with strategies for integrating simulations into problem-based and project-based learning (Rosli & Ishak, 2024). Future directions include designing Biology-specific professional learning communities where teachers co-develop lesson plans that blend physical and virtual lab experiences, fostering both conceptual and procedural understanding.

### 13.6.3 Student-Related Challenges

While virtual labs offer accessibility and interactivity, students may develop an over-reliance on simulations, neglecting the manual laboratory skills critical to Biology, such as microscopy techniques, dissection, or experimental data collection (Chen & Tsai, 2021). Furthermore, learners with limited digital literacy may struggle to navigate these platforms, risking disengagement and inequity. To mitigate this, blended learning models that balance physical and virtual labs are essential (De Jong et al., 2013). The future of Biology education could see scaffolded approaches, where virtual labs introduce concepts (e.g., molecular interactions or ecological dynamics), while physical labs reinforce hands-on competencies. Additionally, adaptive virtual platforms supported by AI can provide differentiated learning paths to support students with varying levels of digital fluency, ensuring equitable participation in Biology experiments.

Building on the understanding of these challenges, it becomes evident that overcoming technical, pedagogical, and student-related barriers is essential for the sustainable integration of virtual labs in Biology education. Future strategies must focus on blending physical and virtual laboratory experiences, leveraging emerging technologies such as AI and VR/AR, and strengthening systemic support through policy, infrastructure, and teacher professional development. By addressing these areas, educators can create enriched, inclusive, and effective learning environments that maximize the potential of virtual labs while preparing students for the demands of 21st-century science education.

## 13.7 Future Directions for Biology Education

To address the challenges and enhance the effectiveness of virtual labs in Biology education, several forward-looking strategies can be considered.

### 13.7.1 *Blended Practical Models*

Blended models that integrate both physical and virtual laboratories offer a balanced approach to teaching and learning in Biology. By combining hands-on experimentation with digital simulations, these models maximize the strengths of both modalities, enhancing students' achievement and conceptual understanding (Bellou et al., 2022; Chen & Tsai, 2021). Physical laboratories provide opportunities to develop essential manual and technical skills, while virtual labs allow for repetition, safe exploration of complex biological processes, and visualization of abstract phenomena that are difficult to replicate in real-world settings. Rosli and Ishak (2022) emphasize that blended practices represent the most sustainable strategy for long-term curriculum integration, as they ensure continuity of learning while addressing both cognitive and practical dimensions of Biology education.

### 13.7.2 *Integration of Artificial Intelligence and VR/AR*

The integration of advanced technologies, such as Artificial Intelligence (AI), augmented Reality (AR), and Virtual Reality (VR), is increasingly transforming the landscape of Biology education. AI can provide adaptive feedback in virtual labs, tailoring tasks and scaffolding to learner progress, which supports differentiated instruction and promotes mastery of complex biological concepts (Bogusevschi et al., 2020). When combined with AR and VR, these technologies create immersive environments that replicate authentic biological scenarios, such as molecular interactions, genetic manipulations, or ecological systems, offering learners a highly personalized and experiential form of learning (Jeno et al., 2020). Such innovations not only deepen understanding but also foster critical thinking, problem-solving, and scientific inquiry skills, which are central to effective Biology education.

### 13.7.3 *Policy, Infrastructure, and Teacher Professional Development*

For the successful integration of blended and technology-enhanced models in Biology education, systemic support remains essential. Policymakers must invest in robust digital infrastructure, ensure equitable access by providing subsidies for underfunded schools, and promote inclusive policies that reduce the digital divide (Aljuhani et al., 2018; De Jong et al., 2013). At the same time, teachers play a pivotal role in shaping the pedagogical effectiveness of these innovations. Professional development programs are therefore crucial to equip educators with the necessary digital pedagogy skills and confidence to integrate virtual and AI-enhanced labs into inquiry-driven instruction. Teacher education programs should also embed these approaches, preparing future educators to leverage digital tools and blended models effectively in their Biology classrooms (Rosli & Ishak, 2024).

## 13.8 Conclusion and Ways Forward

3D animated virtual labs offer transformative potential for Biology education by combining accessibility, scalability, and pedagogical effectiveness. They strengthen inquiry-based and constructivist learning, improve inclusivity, and align with STEM goals (Chen & Tsai, 2021; García-Peñalvo et al., 2021; Rosli & Ishak, 2024). By providing immersive simulations, interactive manipulations, and opportunities for hypothesis testing, these virtual labs bridge the gap between theoretical knowledge and practical application. Nevertheless, technical barriers, teacher readiness, and over-reliance on simulations remain challenges that require attention.

Moving forward, integrating global best practices with localized solutions, alongside strong policy support and professional development, will be key to ensuring that virtual labs realize their promise. Blended models, which combine physical and virtual labs, alongside emerging technologies such as AI, AR, and VR, provide rich, adaptable, and engaging learning environments for diverse student populations (Bogusevschi et al., 2020; Jeno et al., 2020; Rosli & Ishak, 2024). As Rosli and Ishak, (2024) argue, meaningful integration of virtual labs can elevate science education, preparing learners for the digital future while enhancing conceptual understanding, practical skills, and motivation.

**Acknowledgements** The authors would like to express their sincere gratitude to Ministry of Higher Education, Malaysia, Universiti Sains Malaysia and Changsha Daqin Network Technology for their generous financial support of this research under Prototype Research Grant Scheme (Ref. code PRGS/1/2025/SSI09/USM/02/1) and International Grant (Ref. code #2024/575/ET2661, Account R504-LR-GAL008-0000000991-0000).

## References

Aljuhani, K., Sonbul, H., Althaqafi, A., & Meccawy, M. (2018). Creating a virtual biology lab as a simulation-based learning environment for secondary education. *Smart Learning Environments, 5*(1), 16. https://doi.org/10.1186/s40561-018-0074-7

Arbaugh, J. B., & Benbunan-Fich, R. (2007). The importance of participant interaction in online environments. *Decision Support Systems, 43*(3), 853–865. https://doi.org/10.1016/j.dss.2006.12.013

Bellou, J., Papadakis, S., & Tselios, N. (2022). Blended laboratory education: Integrating virtual and hands-on experiences in biology teaching. *Education and Information Technologies, 27*(5), 6735–6754. https://doi.org/10.1007/s10639-021-10834-2

Bogusevschi, D., Muntean, C., & Muntean, G. M. (2020). Teaching and learning physics using 3D virtual learning environment: A case study of combined virtual reality and virtual laboratory in secondary school. *Journal of Computers in Mathematics and Science Teaching, 39*(1), 5–18. https://doi.org/10.70725/297454nsjryb

Chen, C. M., & Tsai, Y. H. (2021). Personalized intelligent tutoring and adaptive feedback mechanisms for virtual science labs. *Computers & Education, 164*, Article 104117. https://doi.org/10.1016/j.compedu.2020.104117

Chiu, M. H., & Linn, M. C. (2012). The role of self-monitoring in learning chemistry with dynamic visualizations. *Metacognition and Learning, 7*(2), 141–164. https://doi.org/10.1007/s11409-012-9080-4

De Jong, T., Linn, M. C., & Zacharia, Z. C. (2013). Physical and virtual laboratories in science and engineering education. *Science, 340*(6130), 305–308. https://doi.org/10.1126/science.1230579

Dori, Y. J., & Belcher, J. (2005). How does technology-enabled active learning affect undergraduate students' understanding of electromagnetism concepts? *The Journal of the Learning Sciences, 14*(2), 243–279. https://doi.org/10.1207/s15327809jls1402_3

Faulconer, E. K., & Gruss, A. B. (2018). A review to weigh the pros and cons of online, remote, and distance science laboratory experiences. *International Review of Research in Open and Distributed Learning, 19*(2), 156–168. https://doi.org/10.19173/irrodl.v19i2.3386

García-Peñalvo, F. J., Corell, A., Abella-García, V., & Grande, M. (2021). Online assessment in higher education in the time of COVID-19. *Education in the Knowledge Society, 22*, e25056. https://doi.org/10.14201/eks.25056

Hodson, D. (2014). Learning science, learning about science, doing science: Different goals demand different learning methods. *International Journal of Science Education, 36*(15), 2534–2553. https://doi.org/10.1080/09500693.2014.899722

Hofstein, A., & Lunetta, V. N. (2004). The laboratory in science education: Foundations for the twenty-first century. *Science Education, 88*(1), 28–54. https://doi.org/10.1002/sce.10106

Jeno, L. M., Adachi, P. J. C., Grytnes, J. A., Vandvik, V., & Deci, E. L. (2020). The effects of m-learning on motivation, achievement and well-being: A self-determination theory approach. *British Journal of Educational Technology, 51*(5), 1455–1473. https://doi.org/10.1111/bjet.12993

Jonassen, D. H. (1999). Designing constructivist learning environments. In C.M. Reigeluth (Ed.), *Instructional-design theories and models: A new paradigm of instructional theory* (Vol. II, pp. 215–239). Lawrence Erlbaum Associates.

Kapici, H. O., Akcay, H., & de Jong, T. (2020). Using hands-on and virtual laboratories alone or together—Which works better for acquiring knowledge and skills? *Journal of Science Education and Technology, 29*(6), 857–875. https://doi.org/10.1007/s10956-020-09850-0

Karam, R., & Warren, M. (2021). Virtual reality applications for science education: Towards authentic learning environments. *Journal of Science Education and Technology, 30*(1), 1–15. https://doi.org/10.1007/s10956-020-09877-1

Makransky, G., & Petersen G. B. (2021). Investigating the process of learning with desktop virtual reality: A structural equation modeling approach. *Computers & Education, 157*, Article 103963. https://doi.org/10.1016/j.compedu.2020.103963

Margot, K. C., & Kettler, T. (2019). Teachers' perception of STEM integration and education: A systematic literature review. *International Journal of STEM Education, 6*(1), 2. https://doi.org/10.1186/s40594-018-0151-2

Ng, P. K., & Chua, C. C. (2023). Exploring the effectiveness of virtual laboratories in science learning: A meta-analysis. *Journal of Science Education and Technology* (Online First). https://doi.org/10.1007/s10956-023-10002-x

Potkonjak, V., Gardner, M., Callaghan, V., Mattila, P., Guetl, C., Petrović, V., & Jovanović, K. (2016). Virtual laboratories for education in science, technology, and engineering: A review. *Computers & Education, 95*, 309–327. https://doi.org/10.1016/j.compedu.2016.02.002

Raman, A., Sabarish, K., & Pathmanathan, S. (2020). Collaborative virtual laboratory learning: Impact on science achievement and engagement. *Education and Information Technologies, 25*, 3153–3171. https://doi.org/10.1007/s10639-020-10121-0

Rosli, R., & Ishak, N. A. (2022). Virtual laboratories in secondary biology education: Review and perspectives. *Asian Journal of Science Education, 4*(2), 23–44.

Rosli, R., & Ishak, N. A. (2024). Integration of virtual labs in science education: A systematic literature review. *Jurnal Pendidikan Sains Dan Matematik Malaysia, 14*(1), 81–103.

Rutten, N., van Joolingen, W. R., & van der Veen, J. T. (2012). The learning effects of computer simulations in science education. *Computers & Education, 58*(1), 136–153. https://doi.org/10.1016/j.compedu.2011.07.017

Smetana, L. K., & Bell, R. L. (2012). Computer simulations to support science instruction and learning: A critical review of the literature. *International Journal of Science Education, 34*(9), 1337–1370. https://doi.org/10.1080/09500693.2011.605182

Sotiriou, S., & Bogner, F. X. (2021). Conceptualizing virtual labs for science learning: Implications for teacher education. *Education Sciences, 11*(4), 176. https://doi.org/10.3390/educsci11040176

Tatli, Z., & Ayas, A. (2013). Virtual chemistry laboratory: Effect of constructivist learning environment. *Turkish Online Journal of Educational Technology, 12*(1), 183–199.

Viitaharju, P., Nieminen, M., Linnera, J., Yliniemi, K., & Karttunen, A. J. (2023). Student experiences from virtual reality-based chemistry laboratory exercises. *Education for Chemical Engineers, 44*, 191–199. https://doi.org/10.1016/j.ece.2023.06.004

Zacharia, Z. C., & Olympiou, G. (2011). Physical and virtual manipulative experimentation: Combined effects on conceptual understanding in physics. *Learning and Instruction, 21*(3), 317–331. https://doi.org/10.1016/j.learninstruc.2010.03.001

**Nor Asniza Ishak** is an associate professor at the School of Educational Studies, Universiti Sains Malaysia (USM) and deputy director of the Institutional Planning and Strategic Centre, Universiti Sains Malaysia. She holds a master's in science education and a PhD in curriculum studies from USM. With over 23 years of teaching experience, her expertise includes STEM/Science education, curriculum studies, education for sustainable development, and innovative pedagogies supported by numerous international awards and research grants.

**Muaz Mohd Zaini Makhtar** is a senior lecturer at the School of Industrial Technology, Universiti Sains Malaysia (USM), and deputy director of the Centre for Innovation and Consultation, Universiti Sains Malaysia. He holds a master's and PhD in engineering from USM, and researches bioprocess technology, waste bioconversion, and microbial fuel cells for energy recovery. He has secured multiple national research grants and received international awards recognizing his contributions to sustainable scientific innovation.

**Rafiza Rosli** is a biology educator under the Ministry of Education in Malaysia. She holds a PhD in STEM education (Biology) from Universiti Sains Malaysia and a master's in Educational Administration from University Putra Malaysia. Her research interests include virtual laboratories, multimedia learning, and innovative pedagogies in STEM, promoting creative and technology-driven approaches to biology education.

# Chapter 14
# Simulations-Based Practical Work

**Joaquim Bernardino Lopes, Raimundo Nonato de Medeiros Jr, and Marco Duarte Naia**

## 14.1 Introduction

Practical work has a central role in science education, traditionally anchored in hands-on experimentation and laboratory work. In spite its recognized value the literature highlights some ambiguity of its goals, logistical constraints (e.g., resources, safety), and sometimes superficial engagement with scientific practices (Abrahams & Millar, 2008). On the other hand, recent studies identify the specificities of laboratory work and how difficult it is to develop without progressive and congruent curriculum development (Agustian, 2025).

Anyway, it seems sure that "what is important in physics learning is manipulation, physical or virtual, rather than physicality" (Zacharia & Olympiou, 2011, p. 328). Thus, there are conditions for computer simulations to be a way to extend, complement, or transform traditional forms of practical work. An epistemological analysis finds no reason why computer simulations should be considered epistemically inferior to traditional experimentation (Durán, 2021). On the other hand, computer

J. B. Lopes (✉) · R. N. de Medeiros Jr · M. D. Naia
University of Trás-Os-Montes and Alto Douro, Vila Real, Portugal
e-mail: blopes@utad.pt

R. N. de Medeiros Jr
e-mail: nonato.junior@escolar.ifrn.edu.br

M. D. Naia
e-mail: duarte@utad.pt

J. B. Lopes
CIDTFF, University of Aveiro, Aveiro, Portugal

R. N. de Medeiros Jr
Federal Institute for Education, Science and Technology of Rio Grande Do Norte, Natal, Brazil

M. D. Naia
CEMMPRE, University of Coimbra, Coimbra, Portugal



L. Leite et al. (eds.), *Practical Work in Science Education in the Digital Era*, Advances in Technology-Rich Science Education 2, https://doi.org/10.1007/978-3-032-13788-3_14

simulations can work and be considered as epistemic tools to support epistemic practices as means through which knowledge is produced in science classrooms requiring attention to material, pedagogical, and social factors, including students' degree of autonomy in their role as knowledge-producing agents (Koretsky et al., 2023; Lopes & Costa, 2021; McLaughlin & Farris, 2025).

Although there is significant research on computer simulations in science education (Çelik, 2022; de Jong et al., 2013; de Medeiros Jr et al., 2024; Rutten et al., 2012; Smetana & Bell, 2012) and some on instrumental orchestration (Drijvers et al., 2020; Trouche, 2004), the literature often lacks detailed theoretical and empirical syntheses of how simulations function differently in diverse practical configurations (e.g., virtual lab vs. field work) and how instrumental orchestration strategies can be tailored to each. On the other hand, the instrumental orchestration framework is underutilized in science education (Lopes & Costa, 2019, 2021) to make sense of this integration.

The research gap mentioned above manifests itself, in particular, in the following dimensions in which there is little work that integrates:

- The epistemic dimension of simulations-based practical work (Develaki, 2019; Greca et al., 2014; McLaughlin & Farris, 2025) to conceptualize computer simulations as epistemic tools.
- The nuances of instrumental orchestration configurations in different practical contexts (Hoyos et al., 2024).
- Concrete pedagogical scaffolding strategies linked to Instrumental orchestration dimensions to use computer simulations as an epistemic tool to promote epistemic practices (Lopes & Costa, 2019, 2021).
- Linking how instrumental orchestration enables epistemic practices.

Thus, the main objective of this chapter is: to conceptualize and exemplify the role of computer simulations as a central mediating tool in diverse configurations of practical work in science education, using the framework of instrumental orchestration to explain how these digital tools support students' epistemic practices in Physics and Chemistry classrooms.

In particular, the authors aim to:

- Enrich the concept of practical work in Science Education using simulations to obtain different configurations of practical work: Virtual Lab work, Simulations-based practical work, lab work, experimental work and field work.
- Highlight the role of orchestration—not just the simulation per se—in fostering epistemic practices.
- Present empirical evidence/examples of effective implementation in science classrooms.

In this chapter the authors: (a) explore five types of practical work and how simulations can mediate them (Sect. 14.2); (b) present the instrumental orchestration framework and its relevance to simulations-based work (Sect. 14.3); (c) provide classroom examples with documented learning outcomes (Sect. 14.4), and (d) discuss implications for teaching, learning, and future research (Sect. 14.5).

## 14.2 Contributions from Computer Simulations in Different Configurations of Work

By introducing computer simulations into the conceptualization made by Leite (2000) about practical work, we can obtain five configurations of practical work in which computer simulations are central or can be used: Virtual Lab work, Simulations-based practical work, lab work, experimental work and field work (Fig. 14.1).

We will attempt to provide a brief definition of each of the practical work configurations and explain the respective role that computer simulations can play.

*Virtual Lab work*: are activities that take place in an interactive environment for designing and performing simulated experiments, consisting of simulation programs, data files, and tools that operate on these objects. These environments use Information and Communication Technology that mimic physical laboratory settings, ensuring interactivity and manipulation. They can combine technologies such as virtual reality and 3D simulation. Students can interact with the virtual laboratory interface, simulate experiments, and adjust parameters to observe new results (Wahyudi et al., 2023). There are three types of Virtual Laboratory (Wahyudi et al., 2023, 2025): (a) measurement-based (cover a range of instrumentation controls that provide students with the opportunity to conduct tests and engage in measurement activities); (b) remote-triggered (access to real laboratory equipment that is remotely connected) (Fonseca et al., 2014); (c) Modelling-based/simulation (unrestricted by physical constraints such as location and time, and does not require the use of actual laboratory equipment). Computer simulations are used in combination with Virtual Laboratory types (a) and (b) and are central to type (c). In the case of type (c) virtual labs, numerical simulators should be considered. These are software programs

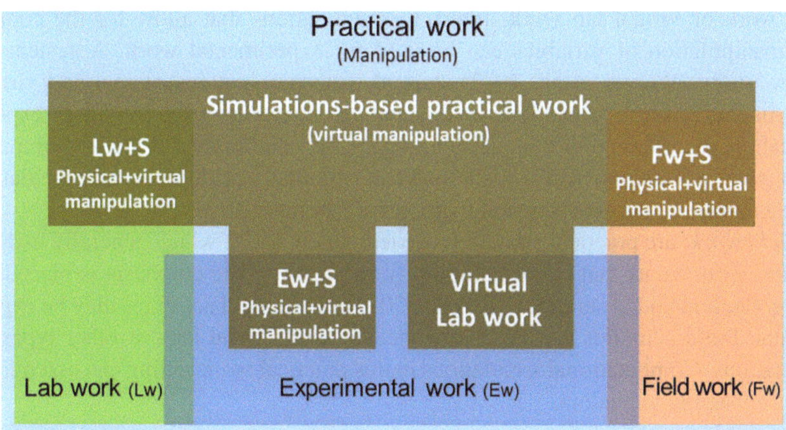


**Fig. 14.1** Computer simulations as pivot to obtain different configurations of practical work: Virtual Lab work, Simulations-based practical work, Lab work (Lw+S), Experimental work (Ew+S) and Field work (Fw+S) where S means simulation (extension of the framework developed by Leite[, 2000])

applied to the general scientific context—not specifically to teaching—developed to numerically solve mathematical models that describe different physical phenomena. These software programs are essential, as such models often present difficult or even non-existent analytical solutions.

*Simulations-based practical work*: activities that use a computer, which in turn operates a representation or model of an event, object, phenomenon or system (natural or artificial) (de Jong & van Joolingen, 1998). These activities consist of simulate dynamic systems of objects in a real or imagined world (Sahin, 2006). A simulation simplifies reality by omitting or changing detail (Sahin, 2006). There are essentially two classes of simulations (and within each class there are several types) (Sahin, 2006): (a) simulations that serve instructional purposes, i.e., without much intervention from students (e.g., animations, informative, demonstrative, or operational simulations); (b) simulations that allow the user to interact and build knowledge. Some computational simulations of type (b) allow experimental work to be carried out with the same degree of student agency as in a real or remote laboratory (Peirson, 2025). Computational simulations are central to this practical work configuration.

*Lab work*: are practical activities that require the use of laboratory materials, more or less conventional, and that must be carried out in a laboratory or equivalent (Leite, 2000). Some laboratory activities may be of the experimental type. Laboratory work allows for the development of six domains in the learning process, one of which (psychomotor domain) requires physicality (Agustian, 2025). However, several laboratory activities involving sophisticated equipment can be carried out in remote laboratories or virtual laboratories (Ogunsola-bandele & Kennepohl, 2024). Computational simulations in this practical work configuration can essentially act as epistemic mediators.

*Experimental work*: are practical activities that involve control and manipulation of variables (Leite, 2000). Some experimental activities can be laboratory work, field work or virtual lab work. Interactive simulations that allow for the control and manipulation of variables can be used for experimental work. A systematic review study (Wörner et al., 2022) showed that combinations of real and virtual experiments promote conceptual understanding better than a single type of experimentation, regardless of the sequence in which the combination is made. Computer simulations in this practical work setting can essentially act as epistemic mediators or as a substitute for experimental work in the laboratory.

*Field work*: are practical activities, carried out outdoors, which generally involve dealing with events that occur naturally (Leite, 2000). The emphasis is placed on data collection and observation (Peirson, 2025). Some field activities may be experimental. Field activities can be complemented with virtual labs or other types of simulations. Computational simulations in this practical work setting can essentially act as epistemic mediators.

## 14.3 Instrumental Orchestration as a Framework for the Simulations-Based Practical Work

There is still no comprehensive theoretical approach on how to take advantage of the use of artefacts, in general, and computer simulations, in particular, to learn science in formal or informal educational contexts, as the known theoretical approaches on the adoption of technology are not sufficient to account for how to take advantage of its use to learn science in epistemic terms (Drijvers et al., 2020; Euler et al., 2023).

To achieve this goal, we need to combine two theories: (a) instrumental genesis theory, originally proposed by Rabardel (1995), which concerns the process of learning using artefacts; (b) instrumental orchestration, originally proposed by Artigue (2002) and Guin and Trouche (2002), which concerns the teacher's role in designing and guiding tool use by students.

Instrumental genesis involves two complementary processes in a cycle of 'instrumentalization', when, based on prior knowledge, an individual appropriates and guides the use of the artefacts with which they interact, and may even modify them, and 'instrumentation', when the possibilities and restrictions of artefacts shape and influence the actions and knowledge constructed by the individual (Gueudet & Trouche, 2009; Gueudet et al., 2014). Instrumental genesis requires students to use artefacts with a mental framework, and teachers to resort to instrumental orchestration, which consists of a concert of tasks, teacher facilitation strategies and articulation with other artefacts (Gueudet & Trouche, 2009).

Instrumental orchestration is always understood as an intentional configuration of teaching, in accordance with learning objectives, to take advantage of the use of technology (Drijvers et al., 2020; Guin & Trouche, 2002). It refers to the deliberate configuration and exploitation of teaching resources, particularly technological ones, to scaffold conceptual development (Trouche, 2004) and was initially focused on mathematics teaching. This approach has been extended to science and technology education since 2019 (Lopes & Costa, 2019, 2021).

### *14.3.1 Components of Instrumental Orchestration*

Instrumental Orchestration is a systematic arrangement by an intentional agent to model the teaching process in a technological environment, involving artefacts and humans. It has three key elements (Drijvers et al., 2020): (a) Didactical Configurations, spatial and organisational arrangements of artefacts; (b) Exploitation Modes, how the teacher uses these configurations during the lesson, e.g., demo mode, student-centred exploration, collective discussion); (c) didactical performance to the model of didactic configuration and exploitation mode as a third layer (e.g., ad hoc decisions taken while teaching on how to actually perform in the chosen didactic configuration and exploitation mode). This theorisation led to the identification of several types of instrumental orchestration, underestimating the complexity of teaching practices.

**Table 14.1** Dimensions of instrumental orchestration (adapted from Lopes and Costa [2021])

| Dimensions of instrumental orchestration | Specification in each dimension of instrumental orchestration |
|---|---|
| Time | Duration of autonomous use of artefacts by students (for instrumental genesis) |
| | Articulation of artefacts throughout the curriculum |
| Teacher mediation | Preparation of the initial instrumental genesis |
| | Formulating the task as a challenge and what is actually required of students |
| | Teaching initiatives by the teacher (help, suggestions, relevant information, etc.) |
| | Connection between the use of artefacts and learning |
| | Degree of autonomy allowed to students |
| | Building a productive learning environment |
| Articulation of resources | Sequential articulation (use of one resource followed by another) |
| | Networking (concomitant use of various resources) |

Thus, it is necessary to abandon the quest to expand the repertoire of types of instrumental orchestration and focus efforts on characterising the most relevant aspects of teaching and learning in order to take advantage of the use of artefacts (Lopes & Costa, 2021). In this context, the same authors identified three important dimensions for instrumental orchestration: time, teacher mediation, and type of articulation between resources (Lopes & Costa, 2021).

Table 14.1 Specifies each of the dimensions of instrumental orchestration mentioned.

### *14.3.2 Applying the Framework to Simulations-Based Practical Work and Connecting Simulations to Epistemic Practices*

The guidelines on how to use the dimensions of instrumental orchestration in different practical work settings assume that students will develop epistemic practices (Lopes & Costa, 2021) in the context of inquiry-based work, supporting hypothesis formulation, model testing and refinement, interpretation of data, argumentation, and decision-making. (Duschl & Grandy, 2013; Louca & Zacharia, 2012). The issue of students' epistemic agency is very important in science teaching and learning (Miller et al., 2018; Zhang et al., 2022). Instrumental orchestration fosters students' epistemic agency insofar as the latter has as its fundamental dimension the teacher's facilitation strategies in the sense of transferring epistemic agency to the students (Lin & Puntambekar, 2025) and the former cannot be achieved without the same teacher facilitation strategies based on transforming simulations into epistemic tools for knowledge construction (Lopes & Costa, 2021). Computer simulations, when orchestrated effectively, can shift from visualisation aids to epistemic tools for reasoning, modelling, and sense-making. The models themselves can function as artefacts for scaling up the level of abstraction (Mashood & Chandrasekharan, 2025).

Thus, all practical work settings using computer simulations require the same type of requirements in terms of time and teacher mediation to ensure instrumental genesis by students and instrumental orchestration by teacher that guarantees the use of simulations as epistemic tools. These requirements for the dimensions of time and teacher mediation are (Lopes & Costa, 2021) as follows.

Time: (a) allow students to use the computer simulation autonomously for as long as necessary, as it is vital for the instrumental genesis of students; (b) provide for the simulation to be revisited at various points in the curriculum to confirm, reformulate or refine data, processes or ideas.

Teacher mediation: (a) provide time, simple tasks and initial technical support inside and/or outside the classroom for students to freely engage with computer simulation, allowing for initial instrumental genesis; (b) present a teaching guide in which the task is formulated as a challenge, indicating the desired outcome and the context in which the task is performed; (c) while students are using the simulation, the teacher must: (c1) take care and decide what teaching initiatives to take (help, suggestions, relevant information, etc.); (c2) ensure the connection between the use of simulation and learning; (c3) progressively increase the degree of autonomy granted to students; (c4) build a productive learning environment.

The dimension of resource articulation is what truly distinguishes the different practical work configurations. In some practical work settings, simulations can be used without any other resources. However, several studies show that the articulation of various resources is always more advantageous from the point of view of the learning it promotes (Lopes & Costa, 2021). This happens when real and virtual experiments are combined, regardless of the order (Wörner et al., 2022). When simulations are used in conjunction with other resources (other simulations, instructional texts, concept maps, graphics and videos, creation of a specific website, electronic testing with feedback, etc.), there are significant gains in student learning compared to the use of simulations without the combination of other resources (de Medeiros, 2025).

Let us examine how the articulation of resources can occur in each of the practical work settings. In any practical work setting, the articulation between different resources is achieved through a teaching guide.

a. Virtual lab work: The simulations can work as full substitutes for physical labs, without any other resources. However, the articulation of a virtual lab simulation can be done sequentially with experimental work with real equipment in the form of AB, BA or ABA (where A is virtual lab) and/or concomitantly with other simulations or other digital or physical resources.

b. Simulations-based practical work: Simulations are used to investigate systems, test hypotheses, or build models without any other resources. However, a simulation can be articulated sequentially with experimental work using real equipment or fieldwork in AB, BA or ABA form (where A is simulation) and/or concomitantly with other simulations or other digital or physical resources.
c. Lab work: Simulations augment the influence of traditional experiments (e.g., as predictive tools or for error analysis). Lab work can be articulated sequentially with virtual lab, or simulations in AB, BA or ABA form (where A is lab work) and/or concomitantly with simulations or other digital or physical resources.
d. Experimental work: Simulations are used for pre-lab, post-lab reflection, or conceptual abstraction. The articulation of experimental work can be done sequentially with virtual lab, fieldwork, or simulations in the form of AB, BA, ABA, or ABAB (where A is experimental work) and/or concomitantly with simulations, other digital or physical resources.
e. Field work: Simulations offer virtual replication or preparation for inaccessible environments. Field work can be coordinated sequentially with simulations in AB, BA, ABA or ABAB format (where A is field work) and/or concurrently with simulations or other digital or physical resources.

## 14.4 Classroom Scenarios of Simulations-Based Practical Work: Teaching by Instrumental Orchestration

In this section, all classroom scenarios of simulation-based practical work and results are taken from a training workshop reported in de Medeiros (2025).

The study reports on a training workshop that was responsible for triggering the creation of teaching guides for using PhET simulations in various practical work settings. The teaching guides were developed in two cycles, characterised by a gradual provision of information. In the first cycle, PhET project guidelines for developing inquiry activities, accessible to the general public, and a collection of teaching guides geared towards inquiry activities, sent to the PhET repository, were made available. In the second cycle, the theoretical framework underlying the proposal was addressed, in particular the instrumental orchestration presented above, as well as the initial results of the research, obtained from a systematic review of the literature and the study conducted on the PhET website. After completing these two cycles, the teachers implemented the activities orchestrated by the teaching guide they had developed in their classroom teaching practices and assessed their students' learning in a pre- and post-test scheme. Nine teachers (two men and seven women) participated in the workshop, involving 17 classes with a total of 340 students (200 male and 140 female) from grade 9 to grade 12. Eight teachers combined the simulations with other resources (in planning and implementation).

Activity 1. One teacher used the practical work configuration 'Field work' combining the use of the PhET simulation 'States of Matter' with the observation of everyday phenomena (wooden block, water, milk, balloon filled with air and coarse table salt) and also combining it with the use of videos. In two of the three classes in which the activity was implemented, learning outcomes improved in a statistically significant way.

Activity 2. One teacher used the PhET simulation 'Light Deflection' in conjunction with other resources (a cup of water, a button inside the cup, and a straw outside) in a practical 'field work' setting. In the only class where the activity was implemented, learning outcomes did not improve in a statistically significant way.

Activity 3. Three teachers used the practical work configuration 'Simulations-based practical work,' combining the use of the PhET simulation 'Build an Atom' with other resources (videos, texts, Google Classroom, and the adopted manual). In all seven classes, learning outcomes improved statistically significantly.

Activity 4. Two teachers used the practical work configuration 'Virtual lab work', combining the PhET simulation 'Circuit Building Kit' with electrical circuit assembly work in the laboratory that took place afterwards and, in both phases, combining it with other resources (graphing calculator, Google Classroom and smartphone for photos). In all three classes where the activity was implemented, learning outcomes improved statistically significantly.

Activity 5. A teacher used the PhET simulation 'Circuit Building Kit' without linking it to any other resource in a 'Simulations-based practical work' setting. In the only class where the activity was implemented, learning outcomes improved in a statistically significant way.

Activity 6. A teacher used the practical work setting "Experimental work" combining the use of the PhET simulation "Capacitor Lab" with the Experimental Activity "Building a Logarithmic Clock" in an AB sequence in one class and a BA sequence in another class, in each case in conjunction with other resources. In both classes where the activity was implemented, learning outcomes improved statistically significantly.

In summary, the use of computer simulations gave rise to four practical work configurations, and only one teacher did not combine the simulation with other resources. Of the 17 classes, learning outcomes improved statistically significantly in 15 classes.

### *14.4.1 An Example of a Teaching Guide*

Activity 3 uses the PhET simulation 'Build an Atom,' is intended for 8th grade students, and has the following structure:

a. Instructions for using the simulation;
b. Explanation of learning objectives;

c. Formulation of challenge task 1 ('Everything that exists is made up of atoms so small that it is impossible to see them under a microscope (…). Today we describe them as a set of a nucleus, formed by protons, neutrons and electrons that move around the nucleus.' A human being with a mass of 70 kg has about $70 \times 10^{26}$ atoms, most of which belong to the chemical elements' hydrogen, oxygen, carbon and nitrogen. What does the mass number of an atom depend on? What distinguishes the chemical elements present in the materials around us?);
d. Provides a link to further information on the problem, referring to Google Classroom with various resources;
e. Formulation of challenge task 2 (As you know, today it is possible to determine the age of Egyptian mummies! Carbon-14 is an unstable isotope that forms in the atmosphere through the action of cosmic rays and is incorporated into living organisms. When a living being dies, it stops incorporating carbon-14, which is transformed into other isotopes. By measuring the amount of carbon-14, it is possible to know the date of death of the organism. What characterises an isotope?);
f. Provides a link to a video about carbon dating;
g. Formulation of challenge task 3;
h. Asks students to propose a research path so that they can respond to the problems presented using the simulation and a predicted response for each problem;
i. Asks students to carry out the experiments with the simulation and record their observations in a table (a structured table is provided to be filled in);
j. Asks them to fill in the blanks to obtain scientifically correct sentences;
k. Asks them to compare the results obtained with the simulation with the predicted answers to each problem;
l. Proposes a game with the simulation.

Analysing the structure of this teaching guide, we can see that there are characteristics in the three dimensions of instrumental orchestration, namely:

a. In the time dimension: (a1) Students are given time to freely explore the simulation's features; (a2) The activity is structured around three problems; (a3) There are several opportunities to test knowledge using the simulation, increasing the time spent interacting with it;
b. In the teacher mediation dimension: (b1) Learning objectives are specified; (b2) Challenging tasks are formulated and contextualised in problematic situations (related to everyday life); (b3) Students are given a certain amount of autonomy to solve the problem; (b4) The simulation is fundamental to the task and allows students to operationalise all the transformations that can occur in an atom, transmuting it into atoms of different substances, thus becoming an epistemic tool for responding to problems and learning about the constitution of matter;
c. In the articulation dimension: There is articulation with additional resources, in a network.

## 14.5 Conclusion and Ways Forward

In conclusion, in this chapter we argued that computer simulations, when meaningfully orchestrated by teachers, can serve as powerful epistemic tools in science education. Their educational value is not intrinsic but emerges through the intentional design of teaching guides that foster epistemic practices such as modelling, reasoning, and hypothesis testing. In particular, we show how simulations support five configurations of practical work combining physical and virtual manipulation. In each of the five practical work configurations, computational simulations are orchestrated with other resources, and when instrumental orchestration is effective, computational simulations can move from being visualisation aids to epistemic tools for reasoning, experimentation, modelling, and understanding. In addition, we presented an instrumental orchestration framework with three dimensions (time, teacher mediation, and articulation between resources), which allows monitoring the actions of teachers and students so that the use of computer simulations, in any practical work configuration, promotes students' epistemic practices.

The ways forward in teaching practices and research have as central aspect the instrumental orchestration framework.

With regard to teaching practice, teachers need to see themselves as designers, as does teacher education. Teachers must be supported in developing instrumental fluency—the ability to configure, select, and exploit simulations intentionally through a teaching guide that promotes inquiry-based activities for students. Planning practical work in any of the five configurations that use computer simulations requires deliberation on the resources to be mobilised, the tasks to be performed by students, the patterns of interaction, and follow-up discussions. It is essential that the tasks are genuine challenges, that students have time for instrumental genesis, and that resources are well orchestrated. Schools should rethink practical work as a hybrid ecosystem, integrating physical, virtual, and simulation-based environments in coherent ways.

Regarding research directions we need empirical research on instrumental orchestration, in particular more classroom-based studies to understand how teachers orchestrate simulations in real-time and how this affects student learning, epistemic agency and scientific literacy in the short and long term. Other direction of the research should be on teacher professional development to know how to effectively prepare teachers to use simulation technologies with epistemic intentionality.

## References

Abrahams, I., & Millar, R. (2008). Does practical work really work? A study of the effectiveness of practical work as a teaching and learning method in school science. *International Journal of Science Education, 30*(14), 1945–1969. https://doi.org/10.1080/09500690701749305

Agustian, H. Y. (2025). Recent advances in laboratory education research. *Chemistry Teacher International, 7*(2), 217–224. https://doi.org/10.1515/cti-2024-0071

Artigue, M. (2002). Learning mathematics in a CAS environment: The genesis of a reflection about instrumentation and the dialectics between technical and conceptual work. *International Journal of Computers for Mathematical Learning, 7*, 245–274. https://doi.org/10.1023/A:1022103903080

Çelik, B. (2022). The effects of computer simulations on students' science process skills: Literature review. *Canadian Journal of Educational and Social Studies, 2*(1), 16–28. https://doi.org/10.53103/cjess.v2i1.17

de Jong, T., & van Joolingen, W. R. (1998). Scientific discovery learning with computer simulations of conceptual domains. *Review of Educational Research, 68*(2), 179–201. https://doi.org/10.3102/00346543068002179

de Jong, T., Linn, M. C., & Zacharia, Z. C. (2013). Physical and virtual laboratories in science and engineering education. *Science, 340*(6130), 305–308. https://doi.org/10.1126/science.1230579

de Medeiros Jr, R. N. (2025). *Contribuições das simulações interativas do PhET para o desenvolvimento profissional centrado nas práticas de ensino da física* [Contributions of PhET interactive simulations to professional development focused on physics teaching practices] (Doctoral dissertation). University of Trás-os-Montes and Alto Douro.

de Medeiros Jr, R. N., Naia, M. D., & Lopes, J. B. (2024). Simulações interativas do PhET nas práticas de ensino da física: uma meta-análise [PhET interactive simulations in physics teaching practices: a meta-analysis]. *Revista Brasileira de Ensino de Física, 46*, e20240186. https://doi.org/10.1590/1806-9126-RBEF-2024-0186

Develaki, M. (2019). Methodology and epistemology of computer simulations and implications for science education. *Journal of Science Education and Technology, 28*(4), 353–370. https://doi.org/10.1007/s10956-019-09772-0

Drijvers, P., Grauwin, S., & Trouche, L. (2020). When bibliometrics met mathematics education research: The case of instrumental orchestration. *ZDM Mathematics Education, 52*, 1455–1469. https://doi.org/10.1007/s11858-020-01169-3

Durán, J. M. (2021). A formal framework for computer simulations: Surveying the historical record and finding their philosophical roots. *Philosophy & Technology, 34*(1), 105–127. https://doi.org/10.1007/s13347-019-00388-1

Duschl, R. A., & Grandy, R. (2013). Two views about explicitly teaching nature of science. *Science & Education, 22*(9), 2109–2139. https://doi.org/10.1007/s11191-012-9539-4

Euler, E., Solvang, L., Gregorcic, B., & Haglund, J. (2023). Visualization and mathematization: How digital tools provide access to formal physics ideas. *International Handbook of Physics Education Research.* https://doi.org/10.1063/9780735425514

Fonseca, M., Maidana, N. L., Severino, E., Barros, S., Senhora, G., & Vanin, V. R. (2014). O laboratório virtual: uma atividade baseada em experimentos para o ensino de mecânica [Virtual laboratory: an experimental-based educational activity on mechanics]. *Revista Brasileira de Ensino de Física, 35*, 4503. https://doi.org/10.1590/S1806-11172013000400014

Greca, I. M., Seoane, E., & Arriassecq, I. (2014). Epistemological issues concerning computer simulations in science and their implications for science education. *Science & Education, 23*(4), 897–921. https://doi.org/10.1007/s11191-013-9673-7

Gueudet, G., Buteau, C., Mesa, V., & Misfeldt, M. (2014). Instrumental and documentational approaches: From technology use to documentation systems in university mathematics education. *Research in Mathematics Education, 16*(2), 139–155. https://doi.org/10.1080/14794802.2014.918349

Gueudet, G., & Trouche, L. (2009). Towards new documentation systems for mathematics teachers? *Educational Studies in Mathematics, 71*, 199–218. https://doi.org/10.1007/s10649-008-9159-8

Guin, D., & Trouche, L. (2002). Mastering by the teacher of the instrumental genesis in CAS environments: Necessity of instrumental orchestrations. *ZDM Mathematics Education, 34*(5), 204–211. https://doi.org/10.1007/BF02655823

Hoyos, V., Silverman, J., Carlsen, L. M., & Lee, A. (2024). Teachers' knowledge and capacity for using digital resources in mathematics education. In B. Pepin, G. Gueudet, & J. Choppin (Eds.),

*Handbook of digital resources in mathematics education* (pp. 851–881). Springer International Publishing. https://doi.org/10.1007/978-3-031-45667-1_50

Koretsky, M. D., Nefcy, E. J., Nolen, S. B., & Champagne, A. B. (2023). Connected epistemic practices in laboratory-based engineering design projects for large-course instruction. *Science Education, 107*(2), 510–550. https://doi.org/10.1002/sce.21769

Leite, L. (2000). O trabalho laboratorial e a avaliação das aprendizagens dos alunos [Laboratory work and assessment of students' learning]. In M. Sequeira, et al. (Eds.). *Trabalho prático e experimental na educação em ciências* [Practical and experimental work in science education] (pp. 91–108). University of Minho.

Lin, F., & Puntambekar, S. (2025). Epistemic scaffolding: Understanding and designing the support for epistemic growth in science. *International Journal of Science Education, 47*(6), 769–793. https://doi.org/10.1080/09500693.2024.2349973

Lopes, J. B., & Costa, C. (2019). Digital resources in science, mathematics and technology teaching—How to convert them into tools to learn. In *International Conference on Technology and Innovation in Learning, Teaching and Education* (pp. 243–255). Springer International Publishing. https://doi.org/10.1007/978-3-030-20954-4_18

Lopes, J. B., & Costa, C. (2021). Converting digital resources into epistemic tools enhancing STEM learning. In *International Conference on Technology and Innovation in Learning, Teaching and Education* (pp. 3–20). Springer International Publishing. https://doi.org/10.1007/978-3-030-73988-1_1

Louca, L. T., & Zacharia, Z. C. (2012). Modeling-based learning in science education: Cognitive, metacognitive, social, material and epistemological contributions. *Educational Review, 64*(4), 471–492. https://doi.org/10.1080/00131911.2011.628748

Mashood, K. K., & Chandrasekharan, S. (2025). The learning of modeling. In T. Knuuttila, N. Carrillo, & R. Koskinen (Eds.), *The Routledge handbook of philosophy of scientific modelling* (pp. 412–426). Routledge. https://doi.org/10.4324/9781003205647-35

McLaughlin, G., & Farris, A. V. (2025). Toward an epistemology of simulation: Preservice elementary teachers' perspectives on educational simulations and epistemic agency in science and engineering. *International Journal of Science and Mathematics Education, 1–29.* https://doi.org/10.1007/s10763-025-10572-9

Miller, E., Manz, E., Russ, R., Stroupe, D., & Berland, L. (2018). Addressing the epistemic elephant in the room: Epistemic agency and the next generation science standards. *Journal of Research in Science Teaching, 55*(7), 1053–1075. https://doi.org/10.1002/tea.21459

Ogunsola-bandele, M., & Kennepohl, D. (2024). Alternatives to traditional physical science laboratory experiences: a review. *West African Journal of Open and Flexible Learning, 12*(2), 165–200. https://wajofel.org/index.php/wajofel/article/view/159

Peirson, N. (2025). *What is truth? Lived experiences of science education: Explorations of real and virtual laboratories in Further Education* (Doctoral dissertation). The University of Sunderland.

Rabardel, P. (1995). *Les hommes et les technologies: Approche cognitive des instruments contemporains.* Armand Colin. https://hal.science/hal-01017462

Rutten, N., Van Joolingen, W. R., & Van Der Veen, J. T. (2012). The learning effects of computer simulations in science education. *Computers & Education, 58*(1), 136–153. https://doi.org/10.1016/j.compedu.2011.07.017

Sahin, S. (2006). Computer simulations in science education: Implications for distance education. *Turkish Online Journal of Distance Education, 7*(4), Article 12.

Smetana, L. K., & Bell, R. L. (2012). Computer simulations to support science instruction and learning: A critical review of the literature. *International Journal of Science Education, 34*(9), 1337–1370. https://doi.org/10.1080/09500693.2011.605182

Trouche, L. (2004). Managing the complexity of human/machine interactions in computerized learning environments: Guiding students' command process through instrumental orchestrations. *International Journal of Computers for Mathematical Learning, 9*(3), 281–307. https://doi.org/10.1007/s10758-004-3468-5

Wahyudi, M. N. A., Budiyanto, C. W., Widiastuti, I., Hatta, P., & bin Bakar, M. S. (2023). Understanding virtual laboratories in engineering education: A systematic literature review. *International Journal of Pedagogy and Teacher Education, 7*(2), 102–118. https://doi.org/10.20961/ijpte.v7i2.85271

Wahyudi, M. N. A., Budiyanto, C. W., Widiastuti, I., Hatta, P., & Bakar, M. (2025). Exploring students' perception of virtual laboratory adoption on an IoT course. *TEM Journal, 14*(2). https://doi.org/10.18421/TEM142-52

Wörner, S., Kuhn, J., & Scheiter, K. (2022). The best of two worlds: A systematic review on combining real and virtual experiments in science education. *Review of Educational Research, 92*(6), 911–952. https://doi.org/10.3102/00346543221079417

Zacharia, Z. C., & Olympiou, G. (2011). Physical versus virtual manipulative experimentation in physics learning. *Learning and Instruction, 21*(3), 317–331. https://doi.org/10.1016/j.learninstruc.2010.03.001

Zhang, J., Tian, Y., Yuan, G., & Tao, D. (2022). Epistemic agency for costructuring expansive knowledge-building practices. *Science Education, 106*(4), 890–923. https://doi.org/10.1002/sce.21717

**Joaquim Bernardino Lopes** is full professor at Universidade de Trás-os-Montes and Alto Douro, Portugal, a researcher at CIDTFF, Director of the doctoral program on Science and Technology Education Research, and editor of the APEduC Journal—Research and Practices in Science, Mathematics and Technology Education. He researches S&T Education, namely teaching practices, teacher education and professional development, artefacts as epistemic tools, and instrumental orchestration in an epistemic learning context.

**Raimundo Nonato de Medeiros Júnior** holds a degree in physics from the Federal University of Rio Grande do Norte in Brazil and is currently pursuing a PhD in applied physical sciences at the University of Trás-os-Montes and Alto Douro in Portugal. He is a full-time professor at the Federal Institute of Education, Science, and Technology of Rio Grande do Norte in the João Câmara campus.

**Marco Duarte Naia** holds a PhD in experimental physics from the University of Coimbra in Portugal. Currently, he is a full-time assistant professor at the Department of Physics of the University of Trás-os-Montes and Alto Douro, Portugal. He teaches fundamental physics and experimental physics, giving attention to the use of computer simulations, and investigating its usage in physics education.

# Chapter 15
# Modeling in the Digital Era: An Overview of Practical Studies on Computerized Model-Based Science Instruction

**Emil Eidin, Jonathan Bowers, and Peng He**

## 15.1 Introduction

The DNA double helix, Bohr's model of the atom, and Einstein's equation $E = mc^2$ are iconic scientific models that have permeated popular culture and helped shape public perceptions of science. Despite their widespread visibility, research shows that the public often has only a superficial understanding of these models and the scientific ideas they represent (Bodanis, 2005; Kemp, 2003). Indeed, science teachers worldwide continue to prioritize helping students understand these canonical representations (Haskel-Ittah & Yarden, 2022; Justi & Gilbert, 2000; Kapon, 2013).

However, research has shown that merely presenting canonical scientific models does little to shift students' naïve epistemological views about the nature of models, particularly their iterative, provisional, and constructed aspects (Gouvea & Passmore, 2017b; Harrison & Treagust, 2000). In response, alternative instructional approaches have emphasized student-generated models as a means for learners to make sense of phenomena (Hallström & Schönborn, 2019; Schwarz et al., 2009). Yet this shift introduces new challenges: students are often asked to model phenomena that are already well understood and are expected to produce models that align with predetermined standards by the end of a unit. These conditions raise concerns about the

E. Eidin (✉)
University of Wyoming, Laramie, WY, USA
e-mail: eeidin@uwyo.edu

J. Bowers
Michigan State University, East Lansing, MI, USA
e-mail: bowerj0824@lansingschools.net

P. He
Washington State University, Pullman, WA, USA
e-mail: peng.he@wsu.edul



L. Leite et al. (eds.), *Practical Work in Science Education in the Digital Era*, Advances in Technology-Rich Science Education 2, https://doi.org/10.1007/978-3-032-13788-3_15

authenticity of students' engagement and the degree of agency they are afforded in the modeling process (Guy-Gaytán et al., 2019). Our definition of authenticity is based on the works of Herrington and Parker (2013) and Hallström and Schönborn (2019), perceiving it *as* "a participation in practices and activities of professional scientists and technologists, or activities appropriate for, or corresponding closely to these" (Hallström & Schönborn, 2019, p. 3). Guy-Gaytán et al. (2019) coined the term "pseudo-modeling," describing it as a continuum of model-based instructional approaches with features that become increasingly disconnected from modeling in science. As those features are further from the scientific practice, the more inauthentic the teaching approach is.

While prior work on "pseudo-modeling" has focused on implementing authentic modeling practices, this discourse has not sufficiently addressed differences across modeling approaches and tools used at various grade levels. In the science education literature, two predominant types of modeling have emerged: paper–pencil models and computational models (Ke et al., 2021). Over the past few decades, growing evidence has highlighted the benefits of engaging students in different computational modeling approaches, including gains in content knowledge (Saba et al., 2023), mechanistic reasoning (Hsiao et al., 2019), systems thinking (Cole et al, 2025), computational thinking (Weintrop et al., 2016), and meta-modeling knowledge (Göhner et al., 2022).

In this chapter, we explore central issues concerning modeling in science education, with a particular focus on computational modeling approaches. Our emphasis on computational modeling is grounded in three key considerations. First, any discussion of authentic modeling implementation in the science classroom would be incomplete without acknowledging both the unique contributions and challenges posed by computational modeling. Second, the growing prominence of computational modeling in science education research over the past two decades necessitates a critical examination of its impact on teaching and learning. Third, the recent emergence of artificial intelligence (AI) technologies compels us to consider AI interfaces with computational modeling tools and its implications for science education.

We consider how different types of modeling interventions support authentic engagement in modeling practices, explore the affordances these approaches offer within specific educational contexts, and address the emerging role of Artificial Intelligence (AI) in shaping modeling implementation. Our goal is to surface current tensions, identify pressing questions, and outline future directions for supporting deeper and authentic modeling practices in science classrooms.

## 15.2 Making Modeling Authentic, Is It at All Possible?

Engaging students in constructing, evaluating, revising, and using models to make sense of phenomena (Schwartz, 2009), often referred to as modeling-based instruction, is intended to immerse them in modeling practices mirroring those of scientists, while considering constraints such as cognitive level, time, and resources. The claim for authenticity of this approach, is based on its alignment with core tenets of Nature of Science (NOS), including the tentativeness of scientific knowledge (through model revision), its empirical basis (by constructing models grounded in observation and real-world data), and the role of creativity (as many modeling environments allow for diverse forms of expression) (Tala & Vesterinen, 2015; Windschitl et al., 2020). However, the calls for modeling in science education- such as those advanced by the Next Generation Science Standards (NGSS)—often fall short in practice. Students are frequently asked to model phenomena whose mechanisms are already known, and the modeling process is typically expected to culminate in a standardized model that aligns with a canonical scientific representation (Gouvea & Passmore, 2017a; Guy-Gaytán et al., 2019).

Rinehart et al. (2016) proposed several design principles to support the implementation of authentic scientific modeling in classrooms. While we concur with their suggestion that the selection of phenomena be contingent on grade-appropriate evidence, existing scientific understanding, and practical considerations, we wish to emphasize that such actions limit the authenticity of the modeling experience. Therefore, like other researchers (Guy-Gaytán et al., 2019), we view authenticity as a spectrum. At one end lies inauthentic modeling, characterized by the rote memorization of canonical models (e.g., the DNA double helix), while at the other end is highly authentic modeling, where students construct models to explain phenomena with mechanisms that are not yet fully understood. In between, we consider a range of strategies offered to engage students in modeling practices sharing common features with those applied in scientific settings (Campbell et al., 2011; Rinehart et al., 2016; Windschitl et al., 2008a).

The authenticity of modeling implementation can be conceptualized in terms of how closely it aligns with the practices of professional scientists. For instance, consider the contrast between a classroom where the teacher guides the entire class in constructing a single target model of a phenomenon, and one in which students work in small groups to develop their models to explain the underlying mechanisms. In the latter scenario, students are also afforded opportunities to revise their models based on peer feedback and emerging evidence. Although both approaches engage students in model construction-reflecting an element of authentic practice-the first imposes consensus prematurely, leading to an epistemological misrepresentation of the role and development of models in science.

Such teacher-led convergence on a "correct" model may result in a corrective rather than exploratory modeling process (Guy-Gaytán et al., 2019; Nielsen & Nielsen, 2021).

While professional development for teachers is essential to support high-fidelity implementation of authentic modeling practices (Miller & Kastens, 2018), we argue that the choice of modeling tools and approaches also plays a critical role. Different modeling environments offer distinct affordances that can enhance authenticity by better supporting core modeling practices—namely, constructing, evaluating, revising, and using models to make sense of phenomena and solve problems.

## 15.3 Different Learning Environments and Their Affordances

We identify two prominent types of modeling approaches in the science education literature: paper–pencil models (often referred to as hand-drawn or representational models) and computational models. The overarching purpose of both type of modeling is to represent an underlying mechanism for a given phenomenon, encompassing both concrete and abstract features, as well as observable and unobservable elements (Chandrasekharan & Nersessian, 2015; Rachmatullah & Wiebe, 2022; Wilensky & Rand, 2015).

As the name implies, paper–pencil models involve visual representations—typically drawings supplemented by text or symbols. These models offer a high degree of expressive flexibility, enabling students to externalize their ideas in diverse ways. While there are general conventions for their construction—such as the use of legends to explain symbolic elements—the approach is relatively intuitive and requires minimal equipment, making it highly accessible in K-12 classrooms. These affordances contribute to the widespread use of paper–pencil models in science education, as reflected in the predominance of such models in the literature.

However, several limitations have been identified. First, research shows that students' initial use of paper–pencil modeling often results in descriptive representations, rather than mechanistic explanations. That is, rather than capturing underlying causal relationships—including unobservable variables—students tend to depict what is directly observed (Bielik et al., 2021). Second, while paper–pencil models can support engagement with Crosscutting Concepts (CCC's) such as cause and effect or thinking across scales, they offer limited support for reasoning about proportion and quantity, essential dimensions of CCC's. For example, paper–pencil models are limited in their ability to represent physical quantities such as mass, energy, and time, and therefore do not support reasoning about proportional relationships. As a result, an assessment of paper–pencil models tends to rely on qualitative interpretation, raising concerns about their alignment with authentic scientific practices, which often demand more rigorous or quantifiable criteria (Bielik et al, 2018; VanLehn et al., 2016). Third, the iterative nature of modeling is constrained by the medium itself. Revising a paper–pencil model often requires starting over on a new sheet, thus limiting the number of feasible iterations. Although a comprehensive review of iteration frequency is lacking, some prominent works suggest that typical implementations involve no more than three rounds of revision (Bowers & Eidin, 2024). Finally, paper–pencil models face challenges in representing temporal dynamics, such as changes over time, dynamic equilibria, or abrupt disruptions within a system. For example, they are not suited for representing even basic systems analysis, such as calculating the accumulation or depletion of a variable over time based on its inflows and outflows (Meadows, 2008; Eidin et al., 2023). The static nature of the paper–pencil medium hinders the depiction of these processes, which are often essential to modeling complex dynamic systems.

Over the past three decades, researchers have sought to address the limitations of paper–pencil models by developing school-appropriate computational modeling environments. Tools such as Stella (Mandinach, 1989), Model-It (Metcalf et al., 2000), and SageModeler (Bowers et al., 2022) adopted a system dynamics approach. Wilensky's work paved the way for the integration of agent-based modeling (Tisue & Wilensky, 2004; Wilensky & Resnick, 1999), and later block-based modeling implementation was researched (Fuhrmann et al., 2025; Vasconcelos & Kim, 2020), expanding the possibilities for learner-centered modeling in science education. Under the category of computational models, we include modeling environments that provide users with:

(a) a degree of agency and control over variables and their relationships;
(b) ability to run simulations that evaluate model behavior;
(c) ability to operate multiple model revisions (Shin et al., 2022).

Although showing promising results in supporting students in making sense of phenomena (Seoane et al., 2022), we do not include static or non-interactive computational simulations under our designation of computational modeling, as they do not support modeling aspects such as evaluation and revision.

The most prominent computational modeling tools thoroughly explored in the science education literature are Agent-based modeling (e.g. Net Logo), Systems Dynamics (e.g. SageModeler), and Block-based modeling (e.g. MoDa). While those modeling environments have different features and affordances (see Table 15.1), they all share the aforementioned features of computational modeling.

The ability to simulate models represents a key feature of authentic modeling practice, as simulations produce not only qualitative representations but also quantitative outputs- such as graphs displaying change in variables over time (Eidin et al., 2023). Furthermore, many contemporary modeling tools enable users to upload real-world or empirical datasets, allowing comparisons between model behavior and observed data (Bielik et al., 2018; Shin et al., 2022). This functionality strengthens the connection between model construction and model evaluation, prompting informed revision.

However, such features do not inherently promote increased meta-modeling knowledge—particularly in recognizing the limitations of models. Without explicit instructional attention, students may come to view computational models as more accurate or trustworthy than empirical data (Eidin et al., 2023). Conversely, with adequate scaffolding, computational modeling environments can foster students' meta-modeling awareness and deepen their understanding of models' epistemic role in science (Bielik et al., 2020).

Another common aspect of computational models is the potential to support students in applying Systems Thinking (ST) to make sense of phenomena and solve problems (Samon & Levy, 2020; Shin et al., 2022; Weintrop et al., 2016). ST refers to the ability to understand and analyze the components of a system, the relationships among those components, and the dynamic behaviors that emerge from their interactions over time. This cognitive skill enables learners to recognize patterns, feedback loops, and interdependencies within complex systems, making it essential for reasoning about scientific phenomena that span across multiple scales and

**Table 15.1** Comparison between various computational modeling tools

| Criterion | Description | Agent-Based Modeling (ABM) | Systems Dynamics Modeling (SDM) | Block-Based Modeling (BBM) |
|---|---|---|---|---|
| Platforms/Tools | Common tools used | NetLogo, AnyLogic, | Vensim, Stella, Insight Maker, SageModeler | Scratch, StarLogo, MoDa |
| Modeling paradigm | How the model represents systems and interactions | Bottom-up: behavior of individual agents | Top-down: aggregate feedback loops and stock-flow relationships | Visual rule-based logic, often hybrid (agent/system) |
| Entities modeled | Units being represented | Individual agents (e.g., particles, animals) | Aggregated variables (e.g., population, energy) | Components or modules (blocks) with predefined functions |
| Variables | How variables are set and used | Set per agent; can be heterogeneous (differentiates characteristics between agents) | Set globally or per compartment/stock; usually homogeneous (assumes an average of variables) | Set via block parameters and pre-defined inputs |
| Change over time visualization | How the model shows outputs over time | Graphs showing change over time | Graphs showing change over time | Graphs showing change over time |
| Interaction mechanism | How elements in the model affect one another | Direct interactions between agents; rules-based | Influences through cause and effect and feedback structures | Defined by connecting blocks through data/control flows |
| Simulation output | What the simulation typically visualizes or tracks | Individual behavior and emergent patterns | Trend lines over time (stocks, flows, feedback behavior) | State transitions, behavior trees, or system outputs depending on block setup |
| Visualization | How the model is typically displayed | Agents in space (e.g., grids, networks), often animated | Graphs of variables over time (stocks/flows), causal loop diagrams | Flowcharts, system diagrams, or animations depending on platform |

(continued)

**Table 15.1** (continued)

| Criterion | Description | Agent-Based Modeling (ABM) | Systems Dynamics Modeling (SDM) | Block-Based Modeling (BBM) |
|---|---|---|---|---|
| Context in which it was used in science education | Types of systems/problems it was used for in K-12 classrooms | Complex adaptive systems (formation of ant, predator pray interaction) | Population dynamics, climate models, economic systems | Introductory STEM education, control systems, simple programming environments |
| Limitations | Core challenges or constraints | Computationally intensive if includes programing; difficult to scale and validate | May oversimplify individual variation; hard to represent discrete agents | Limited in complexity; abstraction may obscure underlying mechanisms |

characterized by nonlinear processes. Recognized as crucial for supporting students' efforts in making sense of complex phenomena (Assaraf & Orion, 2005; Bielik et al., 2023; Yoon et al., 2018), ST has informed science curriculum design and education policy (KMK, 2005; NGSS Lead States, 2013). For example, *A Framework for K–12 Science Education* (NRC, 2012) identifies "systems and system models" as one of its CCC's, emphasizing ST as integral to scientific reasoning and instruction. Though not explicitly articulated, literature shows that these different modeling environments support certain ST aspects and apply to different types of phenomena. Next, we describe how features of prominent modeling approaches support different aspects of ST.

Agent-Based Models (ABMs) simulate the actions and interactions of autonomous, individual entities called agents within a defined environment. Each agent operates according to a set of rules defining its behavior and interactions with other agents. The model simulates the behavior of the system as emerging from agents' behavior and interactions over time. ABMs have shown much promise in shifting students' explanations of a phenomenon from a simple, centralized, cause-and-effect mechanism to a more complex and decentralized complex adaptive behaviors that arise from local interactions (Haensel et al., 2023; Levy & Wilensky, 2011). As such, ABM is useful for shifting students away from simplistic, linear cause-and-effect explanations of complex phenomena (Chi et al., 2012) and supports students in adopting counterintuitive mechanisms of emergence from what seems chaotic and arbitrary behavior of elements within a system to an observable pattern. ABM also addresses ST aspect of change of a system's behavior over time, allowing a concrete timeline representation showing how agents' interactions accumulate to observable system patterns over time (Hsiao et al., 2019; Railsback & Grimm, 2019).

Systems Dynamics Models (SDMs) aim to represent the behavior of complex systems over time. They utilize stocks, flows, and feedback loops as structural components to define the interconnective nature of a system's components. Time delays are coded into the modeling tool, impacting the model's simulation output (Sterman, 2002). Unlike agent-based models, SDM's can be thought of as a top-down approach, in which systems' components are assumed to represent an average behavior of a variable. A stock represents a variable that the model tracks for its accumulation and depletion over time (e.g., population), and flows represent the rate at which the stock increases or decreases over time (e.g., birth rate). The affordances of SDM have been shown to support students in applying two prominent aspects of ST, documented as significantly challenging: feedback and thinking in terms of change over time (Asaraf, 2005; Sweeney & Sterman, 2000). For example, Eidin et al. (2023) have shown that using SageModeler, an open-source systems dynamic modeling tool, improved students' thinking in terms of change over time regarding the evaporative cooling effect.

Block-based modeling (BBM) is a visual modeling approach in which users construct models by dragging, connecting, and configuring pre-defined "blocks" that represent conditions, variables, actions, processes, or system components. Each block typically corresponds to a specific function or rule, and the connections between blocks define the structure and behavior of the model. One of the major affordances of BBM is its rather intuitive user interface, requiring little to no background in computer programming languages (Fuhrmann et al., 2025; Weintrop, 2019). Recent years have shown a growing prominence of modeling tools that combine ABM and BBM approaches. For example, recently developed MoDa (Eloy et al., 2022), a tool that combines block-based pre-determined coding blocks to set the rules for agents and their environment. The outputs include an animated simulation of the agent's behavior and a graph showing change over time.

ABM and SDM were often framed as modeling tools that support the application of ST aspects and, as such, are considered tools that engage students in making sense of complex phenomena (Bowers et al., 2023; Eidin et al., 2023; Jacobson & Wilensky, 2006; Samon & Levy, 2020). Block-based modeling, however, was often framed as a tool for supporting students' Computational Thinking (CT) (Koray & Bilgin, 2023; Weintrop, 2019). CT was defined by Janet Wing as "the thought process involved in formulating problems and their solutions so that the solutions are represented in a form that can be effectively carried out by an information-processing agent" (Wing, 2011, p. 1). Nonetheless, such a definition says very little about how CT can be implemented in the science classroom. Shin et al. (2022) suggested a theoretical framework describing how ST and CT competencies are operationalized through the iterative process of computational modeling, allowing for better conceptualization of the overlapping aspects of both. Although this framework was developed in the context of SDM, the inherent synergies between CT and ST in Computational Modeling are also evident in ABM and BBM (Grover & Pea, 2013; Wientrop et al., 2016).

Despite the difference in the affordances of different computational modeling approaches, few works have tested how those can work in synergy. Bielik et al. (2021) implemented a curriculum with three types of modeling approaches to support student sense-making of the same phenomenon- the formation of an ant trail. The construction of paper-and-pencil, ABM, and SDM has shown a prominent gain in students' knowledge and understanding of the phenomenon, with each modeling approach supporting different aspects of ST.

## 15.4 Challenges of Supporting Teachers in Integrating Computational Models

Despite this work and calls for engaging students in various modeling approaches (Eilam & Reisfeld, 2017; Stroup & Wilensky, 2014), the feasibility and scalability of such implementations remain questionable. Teachers have mentioned various hurdles in integrating computational modeling into their science classrooms, mentioning time constraints and lack of comfort with adopting such tools (Ertmer & Ottenbreit-Leftwich, 2013; Sands et al., 2018; Vasconcelos & Kim, 2020). Moreover, Eidin et al. (2023) have documented teachers who have shown inconsistency in allowing their students to "tinker" with their models as they felt that time was pressing to move on to other Disciplinary Core Ideas (DCI's) that they are expected to cover. As the unit progressed, teachers were more inclined to guide students to construct the "right" model while cutting off time for additional model revisions.

These results correspond with teachers' product-oriented instruction characterized as perceiving modeling as an "artefact to communicate, explain, evaluate and facilitate students' understanding of the content knowledge" (Nielsen & Nielsen, 2019, p. S584) while ignoring aspects that familiarize students with modeling's iterative and collaborative nature, or using it as a predictive tool.

Bowers and Eidin (2024) recognized some of the challenging aspects in facilitating a computational modeling classroom discourse. Such aspects included the facilitation of a peer-reviewed discourse that will overcome well-documented challenges of creating a safe space for students to share their thinking and engage them in an authentic intellectual discourse to achieve a common goal (Sandoval et al., 2021; Weizman et al., 2008). Teachers' talking moves are imperative for facilitating modeling practice and discourse as they walk around groups and push students to reflect on their model's use to explain the mechanism of the overarching phenomenon impacts model evaluation and revision (Gray & Rogan-Klyve, 2018; Windschitl et al., 2008b). However, computational models add further complexity to such facilitation as teachers need to familiarize themselves with the tool, its affordances and limitations, and support students in expressing their ideas through the given rules of the modeling tool (Aljowaed & Alebaikan, 2018; Bowers & Eidin, 2024). In addition, there are technical computer problems teachers should be ready to address in a computational modeling environment (e.g., saving, importing data, glitches and bugs). Eidin and Fuhrmann (2025) found that, despite a two-week emphasis on modeling and computational modeling in particular within a science methods course, followed by a teaching residency, preservice teachers demonstrated a stronger inclination to implement paper–pencil models rather than the computational modeling tools introduced during their training.

While significant progress has been made in increasing student access to computational modeling (Fuhrmann et al., 2022) several of its inherent features and associated pedagogical demands continue to present challenges for classroom implementation. Notably, questions remain about the scalability of computational modeling in a typical school setting, especially given the substantial support, time, and resources required to adequately prepare teachers for its effective use. We believe that a more effective way to address these challenges is to move away from the haphazard dissemination of individual computational modeling tools. Instead, these tools should be introduced to teachers within an overarching framework. Such an approach would first help teachers recognize and become familiar with the shared features of various tools, classify them into distinct types, and build a shared understanding of how each type can best support specific DCIs and Crosscutting Concepts (CCCs).

## 15.5 Students' Challenges When Using Computational Models

The attempts to design as authentic a modeling experience as possible for students revealed some challenges they face while constructing computational models. We identify three primary categories: technical, cognitive, and epistemological.

*Technical challenges* arise from navigating the user interface and operational logic of the modeling environments. Each tool presents a learning curve for understanding basic functionalities, such as running simulations, recording results, or importing real-world data. Additionally, users must learn which operations are invalid and produce error messages. For example, in BBM, certain blocks cannot be connected due to predefined syntax rules, while in SDM, improper linking (e.g., failing to define a causal relationship between a stock and an auxiliary variable) may result in an unusable model. These constraints require time and experience before students can effectively communicate their conceptual understanding through the model's interface.

*Cognitive challenges* stem from the required abstraction involved in representing complex phenomena through the computational modeling tool's interface. Students often struggle with the inherent abstraction of scientific concepts, such as how microscopic particle behavior produces macroscopic effects, or how a single genetic mutation can lead to systemic changes in an organism. These challenges are compounded in contexts that involve complexity, interconnectivity, and emergent properties (Chi et al., 2012; Perkins & Grotzer, 2005). Computational modeling adds yet another level of abstraction: translating conceptual elements into symbolic representations interpretable by a computer. While research demonstrates that computational modeling can enhance students' conceptual understanding and disciplinary knowledge, this benefit typically requires substantial scaffolding and instructional support (Magana et al., 2024).

*Epistemological challenges* relate to students' meta-modeling knowledge and their beliefs about the nature and role of models in science. Students' conceptions of what a model is and what it is meant to accomplish influence their modeling practices (Sins et al., 2009). Common difficulties include viewing models as exact representations of reality (Eidin et al., 2023) overestimating the accuracy of computational outputs and undervaluing empirical data- leading to resistance toward model revision (Eidin et al., 2023) and interpreting models as simply "right" or "wrong" without considering factors such as explanatory power, communicative clarity, or alignment with scientific consensus.

## 15.6 Modeling Assessment

We contend that there remains a degree of ambiguity in defining what constitutes modeling assessment in science education. There is a growing international consensus among researchers regarding the nature and purpose of modeling in science classrooms, with overlapping modeling practices definitions and goals across national contexts (Böschl et al., 2023; Chiu & Lin, 2019). Moreover, some works suggest modeling competence framework, to support the assessment of students' modeling process and product (Chiu & Lin, 2019; Gilbert & Justi, 2016). However, it is still unclear how the modeling process and product should function as an assessment tool for evaluating students' conceptual understanding. For instance, it is not evident to what extent summative three-dimensional (3D) assessments, (assessments aimed to evaluate students' competency in applying scientific and engineering practices, CCC's and DCI's) that prompt students to construct a model to explain a phenomenon aligned with the broader goals of modeling practice. Often, such tasks involve students producing a static representation- typically a paper–pencil model or a computer-generated drawing (He et al., 2023)—without engaging in other essential aspects of the modeling process, such as evaluating the model, revising it based on feedback or evidence, or applying it to new contexts. In these cases, the modeling task risks becoming an inauthentic school exercise focused on producing a final product that conforms to canonical scientific explanations (Guy-Gaytán et al., 2019).

While these tasks can offer valuable insights into students' thinking and provide alternative ways for students to represent their knowledge, we caution that in an assessment-driven educational climate, there is a risk that the full modeling process will be reduced to a singular focus on model construction. This shift may undermine the epistemic and pedagogical value of modeling by sidelining the iterative and exploratory dimensions that are central to authentic scientific practices (Guy-Gaytán, 2019; Hallström & Schönborn, 2019).

Moreover, as the field is progressing towards AI automated assessment of paper–pencil and computational models (Zhai et al., 2022), we urge the community to think not only about the opportunities AI technology might have but also about the possible ramifications of using it. We identify four primary concerns regarding the use of AI to assess student-generated models. The first is that AI may reduce the authenticity of the modeling practice by focusing solely on the output rather than the process, thereby marginalizing other essential components of the modeling process, such as evaluation, revision, and the iterative refinement of ideas. This risks further exacerbating the concern mentioned for reducing modeling to a static output rather than a dynamic practice.

The second concern involves the potential for misinterpretation and inaccurate evaluation of students' models. AI systems may incorrectly assess a model's quality based on superficial features or fail to recognize the intent behind certain representations (Zhai, 2024). A related issue-possibly a subcategory of the second concern-is the tendency to evaluate models based solely on structural criteria. For example, an AI system might detect the presence of a feedback loop in a systems model and mark it as complete or accurate based on structural patterns indicating a feedback mechanism, despite a possible conceptual flaw of a mechanism or causal relationships. Such assessments, while structurally informed, may overlook the contextual coherence and scientific validity of the model's internal logic.

The third concern relates to potential biases arising from the training data on which AI-based assessments rely. Suppose the AI is primarily trained on canonical scientific models. In that case, it may evaluate student models based on a narrow set of criteria, emphasizing structural correctness over explanatory usefulness or contextual relevance. This can reinforce the misunderstanding that models are either "right" or "wrong", rather than tools for prediction, sense-making, and communication.

The fourth concern involves the erosion of student and teacher agency. Increased reliance on AI tools for model evaluation may diminish teachers' scrutiny of students' modeling processes and interpretations. This, in turn, could signal to students that the AI's evaluation is superior to their own reasoning, and lead them to construct AI-approved models. Moreover, such trends risk undermining key epistemological dimensions of modeling, including iterative comparison with real-world data and collaborative peer review and discussion for reaching a consensus.

## 15.7 Conclusion and Ways Forward

Despite the concerns mentioned, the field of computational models can gain from integrating AI into its interface and supporting students and teachers in some of the challenges mentioned. AI individual support systems can help reduce the learning curve of using computational models by automatically training programs, suggestions ways for communicating ideas, and guide students through the modeling process. AI can also assist in the iterative cycle of model revision by providing teachers with insights and analyses of students' modeling progress. These supports have the potential to make computational modeling more accessible and scalable across schools.

Beyond embedding AI into existing computational modeling tools, teachers can utilize publicly available AI tools. For example, teachers can support students in practicing their meta-modeling knowledge model evaluation skills by asking an AI agent to create a visual model of a phenomenon and ask students to evaluate the generated model. Teachers can also ask students to compare AI evaluation with peer evaluation of a model, to reveal some of AI's limitations. Such simple activities have the potential to emphasize epistemic aspects of modeling and address some of the challenges regarding these aspects.

To conclude this chapter has outlined both challenges and opportunities for advancing modeling, particularly computational modeling. We call for further research into the use of different computational models, with attention to the alignment between scientific contexts, disciplinary concepts, and the most appropriate modeling tools. We also argue that framing the various modeling approaches under a single, overarching concept of "computational modeling" can help teachers develop a simpler and more coherent understanding, potentially leading to greater adoption.

All indications point to the growing use of AI in science education and within computational models. We hope that the concerns and opportunities mentioned in this chapter will raise awareness among the community and encourage informed integration of AI in the context of scientific modeling moving forward.

## References

Aljowaed, M., & Alebaikan, R. A. (2018). Training needs for computer teachers to use and teach computational thinking skills. *International Journal for Research in Education, 42*(3), 237–284.

Assaraf, O. B. Z., & Orion, N. (2005). Development of system thinking skills in the context of earth system education. *Journal of Research in Science Teaching: THe Official Journal of the National Association for Research in Science Teaching, 42*(5), 518–560.

Bielik, T., Delen, I., Krell, M., & Assaraf, O. B. Z. (2023). Characterising the literature on the teaching and learning of system thinking and complexity in STEM education: A bibliometric analysis and research synthesis. *Journal for STEM Education Research, 6*(2), 199–231. https://doi.org/10.1007/s41979-023-00087-9

Bielik, T., Fonio, E., Feinerman, O., Duncan, R. G., & Levy, S. T. (2021). Working together: Integrating computational modeling approaches to investigate complex phenomena [Journal Articles; Reports—Research]. *Journal of Science Education and Technology, 30*(1), 40–57. https://doi.org/10.1007/s10956-020-09869-x

Bielik, T., Opitz, S. T., & Novak, A. M. (2018). Supporting students in building and using models: Development on the quality and complexity dimensions. *Education Sciences, 8*(3), 149. https://doi.org/10.3390/educsci8030149

Bielik, T., Stephens, L., Damelin, D., & Krajcik, J. S. (2020). Designing technology environments to support system modeling competence. In *Towards a competence-based view on models and modeling in science education* (pp. 275–290). Springer.

Bodanis, D. (2005). *E= $mc^2$: A biography of the world's most famous equation.* Bloomsbury Publishing.

Böschl, F., Forbes, C., & Lange-Schubert, K. (2023). Investigating scientific modeling practices in U.S. and German elementary science classrooms: A comparative, cross-national video study. *Science Education, 107*(2), 368–400. https://doi.org/10.1002/sce.21780

Bowers, J., Damelin, D., Eidin, E., & McIntyre, C. (2022). Keeping cool with SageModeler. *The Science Teacher, 89*(4), 18–25. https://doi.org/10.1080/00368555.2022.12293680

Bowers, J., & Eidin, E. (2024). Analyzing students' systems thinking in-situ through screencasts in the context of computational modeling: A case study. *Disciplinary and Interdisciplinary Science Education Research, 6*(1), 24.

Bowers, J., Eidin, E., Stephens, L., & Brennan, L. (2023). Examining student testing and debugging within a computational systems modeling context. *Journal of Science Education and Technology, 32*(4), 607–628. https://doi.org/10.1007/s10956-023-10049-w

Campbell, T., Zhang, D., & Neilson, D. (2011). Model based inquiry in the high school physics classroom: An exploratory study of implementation and outcomes. *Journal of Science Education and Technology, 20*(3), 258–269. https://doi.org/10.1007/s10956-010-9251-6

Chandrasekharan, S., & Nersessian, N. J. (2015). Building cognition: The construction of computational representations for scientific discovery. *Cognitive Science, 39*(8), 1727–1763. https://doi.org/10.1111/cogs.12203

Chi, M. T., Roscoe, R. D., Slotta, J. D., Roy, M., & Chase, C. C. (2012). Misconceived causal explanations for emergent processes. *Cognitive Science, 36*(1), 1–61. https://doi.org/10.1111/j.1551-6709.2011.01207.x

Chiu, M.-H., & Lin, J.-W. (2019). Modeling competence in science education. *Disciplinary and Interdisciplinary Science Education Research, 1*(1), 12. https://doi.org/10.1186/s43031-019-0012-y

Cole, L. B., Justice, J., O'Brien, D., Aman, J., Kim, J. B., Akturk, A., & Zangori, L. (2025). Can green building science support systems thinking for energy education? *Sustainability, 17*(15), 7008. https://doi.org/10.3390/su17157008

Eidin, E., Bielik, T., Touitou, I., Bowers, J., Damelin, D., Mcintyre, C., & Krajcik, J. (2023). Thinking in terms of change over time: Opportunities and challenges of using system dynamics models. *Journal of Science Education and Technology, 33*, 1–28. https://doi.org/10.1007/s10956-023-10047-y

Eidin, E., & Fuhrmann, T. (2025). Preservice teachers' engagement with systems dynamic, block-based, and paper-pencil models. In *Proceedings of the 19th International Conference of the Learning Sciences-ICLS 2025* (pp. 2840–2842). International Society of the Learning Sciences.

Eilam, B., & Reisfeld, D. (2017). A curriculum unit for promoting complex system thinking: The case of combined system dynamics and agent based models for population growth. *Journal of Advances in Education Research,2*(2), 39–60. https://doi.org/10.22606/jaer.2017.22001

Eloy, A., Wolf, J., Fuhrmann, T., Bumbacher, E., Wilkerson, M., & Blikstein, P. (2022). MoDa: Designing an integrated platform for computational modeling & data analysis for sustained investigations in science classrooms. In *Proceedings of the 2022 Annual Meeting of the International Society for the Learning Sciences (ISLS 2022).*

Ertmer, P. A., & Ottenbreit-Leftwich, A. (2013). Removing obstacles to the pedagogical changes required by Jonassen's vision of authentic technology-enabled learning. *Computers & Education, 64*, 175–182. https://doi.org/10.1016/j.compedu.2012.10.008

Fuhrmann, T., Levy, S. T., Wilensky, U., Bumbacher, E., Blikstein, P., Saba, J., Langbeheim, E., Hel-Or, H., Fernandez, C., & de Deus Lopes, R. (2022). Developing accessible and sustainable computational modeling tools in learning science: What is next? In *Proceedings of the 16th International Conference of the Learning Sciences-ICLS 2022* (pp. 1715–1722).

Fuhrmann, T., Rosenbaum, L., Wagh, A., Eloy, A., Wolf, J., Blikstein, P., & Wilkerson, M. (2025). Right but wrong: How students' mechanistic reasoning and conceptual understandings shift when designing agent-based models using data. *Science Education, 109*(1), 3–26. https://doi.org/10.1002/sce.21890

Gilbert, J. K., & Justi, R. (2016). Learning scientific concepts from modelling-based teaching. In J. K. Gilbert & R. Justi (Eds.), *Modelling-based teaching in science education* (pp. 81–95). Springer.

Gouvea, J., & Passmore, C. (2017a). 'Models of' versus 'models for.' *Science & Education, 26*(1), 49–63. https://doi.org/10.1007/s11191-017-9884-4

Gouvea, J., & Passmore, C. (2017b). 'Models of' versus 'Models for' toward an agent-based conception of modeling in the science classroom. *Science & Education, 26*(1), 49–63. https://doi.org/10.1007/s11191-017-9884-4

Göhner, M. F., Bielik, T., & Krell, M. (2022). Investigating the dimensions of modeling competence among preservice science teachers: Meta-modeling knowledge, modeling practice, and modeling product. *Journal of Research in Science Teaching, 59*(8), 1354–1387. https://doi.org/10.1002/tea.21759

Gray, R., & Rogan-Klyve, A. (2018). Talking modelling: Examining secondary science teachers' modelling-related talk during a model-based inquiry unit. *International Journal of Science Education, 40*(11), 1345–1366. https://doi.org/10.1080/09500693.2018.1479547

Grover, S., & Pea, R. (2013). Computational thinking in K-12: A review of the state of the field. *Educational Researcher, 42*(1), 38–43. https://doi.org/10.3102/0013189X12463051

Guy-Gaytán, C., Gouvea, J. S., Griesemer, C., & Passmore, C. (2019). Tensions between learning models and engaging in modeling: Exploring implications for science classrooms. *Science & Education, 28*(8), 843–864. https://doi.org/10.1007/s11191-019-00064-y

Haensel, M., Schmitt, T. M., & Bogenreuther, J. (2023). Teaching the modeling of human-environment systems: Acknowledging complexity with an agent-based model. *Journal of Science Education and Technology, 32*(2), 256–266. https://doi.org/10.1007/s10956-022-10022-z

Hallström, J., & Schönborn, K. J. (2019). Models and modelling for authentic STEM education: Reinforcing the argument. *International Journal of STEM Education, 6*(1), 22. https://doi.org/10.1186/s40594-019-0178-z

Harrison, A. G., & Treagust, D. F. (2000). A typology of school science models. *International Journal of Science Education, 22*(9), 1011–1026. https://doi.org/10.1080/095006900416884

Haskel-Ittah, M., & Yarden, A. (2022). *Genetics education: Current challenges and possible solutions*. Springer.

He, P., Chen, I. C., Touitou, I., Bartz, K., Schneider, B., & Krajcik, J. (2023). Predicting student science achievement using post-unit assessment performances in a coherent high school chemistry project-based learning system. *Journal of Research in Science Teaching, 60*(4), 724–760. https://doi.org/10.1002/tea.21815

Herrington, J., & Parker, J. (2013). Emerging technologies as cognitive tools for authentic learning. *British Journal of Educational Technology, 44*(4), 607–615.

Hsiao, L., Lee, I., & Klopfer, E. (2019). Making sense of models: How teachers use agent-based modeling to advance mechanistic reasoning. *British Journal of Educational Technology, 50*(5), 2203–2216. https://doi.org/10.1111/bjet.12844

Jacobson, M. J., & Wilensky, U. (2006). Complex systems in education: Scientific and educational importance and implications for the learning sciences. *The Journal of the Learning Sciences, 15*(1), 11–34. https://doi.org/10.1207/s15327809jls1501_4

Justi, R., & Gilbert, J. (2000). History and philosophy of science through models: Some challenges in the case of the atom. *International Journal of Science Education, 22*(9), 993–1009. https://doi.org/10.1080/095006900416875

Kapon, S. (2013). Constructing conceptual meaning from a popular scientific paper—The case of $E= mc^2$. *Physics Education, 48*(1), 90. https://doi.org/10.1088/0031-9120/48/1/90

Ke, L., Sadler, T. D., Zangori, L., & Friedrichsen, P. J. (2021). Developing and using multiple models to promote scientific literacy in the context of socio-scientific issues. *Science & Education, 30*(3), 589–607. https://doi.org/10.1007/s11191-021-00206-1

Kemp, M. (2003). The Mona Lisa of modern science. *Nature, 421*(6921), 416–420. https://doi.org/10.1038/nature01403

Koray, A., & Bilgin, E. (2023). The effect of block coding (scratch) activities integrated into the 5E learning model in science teaching on students' computational thinking skills and programming self-efficacy. *Science Insights Education Frontiers, 18*(1), 2825–2845. https://doi.org/10.15354/sief.23.or410

Levy, S. T., & Wilensky, U. (2011). Mining students' inquiry actions for understanding of complex systems. *Computers & Education, 56*(3), 556–573. https://doi.org/10.1016/j.compedu.2010.09.015

Li, T., Haudek, K., & Krajcik, J. (2025). Utilizing deep learning AI to analyze scientific models: Overcoming challenges. *Journal of Science Education and Technology, 34*, 866–887. https://doi.org/10.1007/s10956-025-10217-0

Magana, A. J., Arigye, J., Udosen, A., Lyon, J. A., Joshi, P., & Pienaar, E. (2024). Scaffolded team-based computational modeling and simulation projects for promoting representational competence and regulatory skills. *International Journal of STEM Education, 11*(1), 34. https://doi.org/10.1186/s40594-024-00494-3

Mandinach, E. B. (1989). Model-building and the use of computer simulation of dynamic systems. *Journal of Educational Computing Research, 5*(2), 221–243. https://doi.org/10.2190/7W4F-XY0H-L6FH-39R8

Meadows, D. H. (2008). *Thinking in systems: A primer*. Chelsea Green Publishing.

Metcalf, S. J., Krajcik, J., & Soloway, E. (2000). Model-It: A design retrospective. In M. J. Jacobson & R. B. Kozma (Eds.), *Innovations in science and mathematics education* (pp. 77–115). Routledge. https://doi.org/10.4324/9781410602671

Miller, A. R., & Kastens, K. A. (2018). Investigating the impacts of targeted professional development around models and modeling on teachers' instructional practice and student learning. *Journal of Research in Science Teaching, 55*(5), 641–663. https://doi.org/10.1002/tea.21434

National Research Council. (2012). *A framework for K-12 science education: Practices, crosscutting concepts, and core ideas*. The National Academies Press.

NGSS Lead States. (2013). *Next generation science standards: For states, by states*. The National Academies Press.

Nielsen, S. S., & Nielsen, J. A. (2021). A competence-oriented approach to models and modelling in lower secondary science education: Practices and rationales among Danish teachers. *Research in Science Education, 51*(Suppl 2), 565–593.

Perkins, D. N., & Grotzer, T. A. (2005). Dimensions of causal understanding: The role of complex causal models in students' understanding of science. *Studies in Science Education, 41*(1), 117–165. https://doi.org/10.1080/03057260508560216

Rachmatullah, A., & Wiebe, E. N. (2022). Building a computational model of food webs: Impacts on middle school students' computational and systems thinking skills. *Journal of Research in Science Teaching, 59*(4), 585–618. https://doi.org/10.1002/tea.21738

Railsback, S. F., & Grimm, V. (2019). *Agent-based and individual-based modeling: A practical introduction*. Princeton University Press.

Rinehart, R. W., Duncan, R. G., Chinn, C. A., & Atkins, T. A. (2016). Critical design decisions for successful model-based inquiry in science classrooms. *International Journal of Designs for Learning,7*(2), 17–40. https://doi.org/10.14434/ijdl.v7i2.20137

Saba, J., Hel-Or, H., & Levy, S. T. (2023). Promoting learning transfer in science through a complexity approach and computational modeling. *Instructional Science, 51*(3), 475–507. https://doi.org/10.1007/s11251-023-09624-w

Samon, S., & Levy, S. T. (2020). Interactions between reasoning about complex systems and conceptual understanding in learning chemistry. *Journal of Research in Science Teaching, 57*(1), 58–86. https://doi.org/10.1002/tea.21585

Sandoval, W. A., Kawasaki, J., & Clark, H. F. (2021). Characterizing science classroom discourse across scales. *Research in Science Education, 51*(1), 35–49. https://doi.org/10.1007/s11165-020-09953-7

Sands, P., Yadav, A., & Good, J. (2018). Computational thinking in K-12: In-service teacher perceptions of computational thinking. In M. Khine (Ed.), *Computational thinking in the STEM disciplines: Foundations and research highlights* (pp. 151–164). Springer. https://doi.org/10.1007/978-3-319-93566-9_8

Schwarz, C. V., Reiser, B. J., Davis, E. A., Kenyon, L., Achér, A., Fortus, D., Shwartz, Y., Hug, B., & Krajcik, J. (2009). Developing a learning progression for scientific modeling: Making scientific

modeling accessible and meaningful for learners. *Journal of Research in Science Teaching, 46*(6), 632–654. https://doi.org/10.1002/tea.20311
Seoane, M. E., Greca, I. M., & Arriassecq, I. (2022). Epistemological aspects of computational simulations and their approach through educational simulations in high school. *SIMULATION, 98*(2), 87–102. https://doi.org/10.1177/0037549720930084
Shin, N., Bowers, J., Roderick, S., McIntyre, C., Stephens, A. L., Eidin, E., Krajcik, J., & Damelin, D. (2022). A framework for supporting systems thinking and computational thinking through constructing models. *Instructional Science, 50*(6), 933–960. https://doi.org/10.1007/s11251-022-09590-9
Sins, P. H., Savelsbergh, E. R., van Joolingen, W. R., & van Hout-Wolters, B. H. (2009). The relation between students' epistemological understanding of computer models and their cognitive processing on a modelling task. *International Journal of Science Education, 31*(9), 1205–1229. https://doi.org/10.1080/09500690802192181
Standing Conference of the Ministers of Education and Cultural Affairs of the Länder in the Federal Republic of Germany. (2020). *Bildungsstandards im Fach Biologie für die Allgemeine Hochschulreife*. Wolters Kluwer.
Sterman, J. (2002). *System dynamics: Systems thinking and modeling for a complex world*. http://hdl.handle.net/1721.1/102741
Stroup, W. M., & Wilensky, U. (2014). On the embedded complementarity of agent-based and aggregate reasoning in students' developing understanding of dynamic systems. *Technology, Knowledge and Learning, 19*(1), 19–52. https://doi.org/10.1007/s10758-014-9218-4
Sweeney, L. B., & Sterman, J. D. (2000). Bathtub dynamics: Initial results of a systems thinking inventory. *System Dynamics Review: The Journal of the System Dynamics Society, 16*(4), 249–286.
Tala, S., & Vesterinen, V.-M. (2015). Nature of science contextualized: Studying nature of science with scientists. *Science & Education, 24*(4), 435–457. https://doi.org/10.1007/s11191-014-9738-2
Tisue, S., & Wilensky, U. (2004). *Netlogo: A simple environment for modeling complexity. International conference on complex systems*. Retrieved from October 2, 2025, from http://www.ccl.sesp.northwestern.edu/papers/netlogo-iccs2004.pdf
VanLehn, K., Chung, G., Grover, S., Madni, A., & Wetzel, J. (2016). Learning science by constructing models: Can dragoon increase learning without increasing the time required? *International Journal of Artificial Intelligence in Education, 26*(4), 1033–1068. https://doi.org/10.1007/s40593-015-0093-5
Vasconcelos, L., & Kim, C. (2020). Preparing preservice teachers to use block-based coding in scientific modeling lessons. *Instructional Science, 48*(6), 765–797. https://doi.org/10.1007/s11251-020-09527-0
Weintrop, D. (2019). Block-based programming in computer science education. *Communications of the ACM, 62*(8), 22–25. https://doi.org/10.1145/3341221
Weintrop, D., Beheshti, E., Horn, M., Orton, K., Jona, K., Trouille, L., & Wilensky, U. (2016). Defining computational thinking for mathematics and science classrooms. *Journal of Science Education and Technology, 25*(1), 127–147. https://doi.org/10.1007/s10956-015-9581-5
Weizman, A., Shwartz, Y., & Fortus, D. (2008). The driving question board. *The Science Teacher, 75*(8), 33.
Wilensky, U., & Rand, W. (2015). *An introduction to agent-based modeling: Modeling natural, social, and engineered complex systems with NetLogo*. Mit Press.
Wilensky, U., & Resnick, M. (1999). Thinking in levels: A dynamic systems approach to making sense of the world. *Journal of Science Education and Technology, 8*(1), 3–19. https://doi.org/10.1023/A:1009421303064
Windschitl, M., Thompson, J., & Braaten, M. (2008a). Beyond the scientific method: Model-based inquiry as a new paradigm of preference for school science investigations. *Science Education, 92*(5), 941–967. https://doi.org/10.1002/sce.20259

Windschitl, M., Thompson, J., & Braaten, M. (2008b). How novice science teachers appropriate epistemic discourses around model-based inquiry for use in classrooms. *Cognition and Instruction, 26*(3), 310–378. https://doi.org/10.1080/07370000802177193
Windschitl, M., Thompson, J., & Braaten, M. (2020). *Ambitious science teaching*. Harvard Education Press.
Wing, J. (2011). Research notebook: Computational thinking—What and why. *The Link Magazine, 6*, 20–23.
Yoon, S. A., Goh, S.-E., & Park, M. (2018). Teaching and learning about complex systems in K-12 science education: A review of empirical studies 1995–2015. *Review of Educational Research, 88*(2), 285–325. https://doi.org/10.3102/0034654317746090
Zhai, X. (2024). AI and machine learning for next generation science assessments. *Machine Learning, Natural Language Processing, and Psychometrics*, 201. https://doi.org/10.48550/arXiv.2405.06660
Zhai, X., He, P., & Krajcik, J. (2022). Applying machine learning to automatically assess scientific models. *Journal of Research in Science Teaching, 59*(10), 1765–1794. https://doi.org/10.1002/tea.21773

**Emil Eidin, PhD,** is an assistant professor of secondary science education at the School of Teacher Education at the University of Wyoming, USA. A former high school chemistry teacher and college lecturer, his research focuses on harnessing a systems thinking approach to support teachers and students in making informed decisions about SSI as well as exploring strategies to integrate AI technologies into the science classroom in an informed manner.

**Jonathan Bowers** is an expert in systems thinking and computational modeling, publishing multiple high-profile articles in this subject area. He has previously worked with CREATE for STEM (Michigan State University) and the Concord Consortium on designing, implementing, and revising curricular units focused on integrating SageModeler and computational thinking into high school classrooms. Currently, he works for Lansing Public Schools, USA, advancing the goals of three-dimensional learning at Eastern High School.

**Peng He** is an assistant professor of science education, co-director of the Center of AI Research & Education (CAIRE) and XR Developmental Lab at Washington State University, USA. His research focuses on supporting science teaching and learning with innovative technologies (e.g., AI, robotics), including adaptive learning progression, AI tools for teaching and learning, and teacher AI literacy.

# Chapter 16
# Augmented Reality in Education: Didactic and Philosophical Implications for Science Education

**José-Antonio Marín-Marín, Óscar Bedoya-Cano, and Jesús López-Belmonte**

## 16.1 Introduction

This paper addresses augmented reality in science education from two perspectives: philosophical and pedagogical. From the first perspective, it traces the emergence of the concept as a form of virtuality and a possibility of occurrence (Lévy, 1999), closer to hypertext as a means of representation and projection of reality in the construction of hyperobjects. It examines its genesis in the image, its subsequent projection into three-dimensionality, and the technological features required for its practical implementation.

The second perspective explores the fields of Science, Technology, Engineering, and Mathematics (STEM), focusing on the incorporation of augmented reality into pedagogical practices as a tool for science education, particularly by harnessing its advantages in simulating and visualising objects that are otherwise difficult to access. In this regard, the potential of augmented reality to strengthen science education is presented, highlighting its versatility as a multi-purpose tool.

Methodologically, a review was conducted in the Web of Science database to identify classroom experiences in each STEM discipline since the emergence of the concept in the final decade of the twentieth century. The review also considers the types of pedagogical strategies in which such interventions have been applied, and

J.-A. Marín-Marín · J. López-Belmonte (✉)
University of Granada, Granada, Spain
e-mail: jesuslopez@ugr.es

J.-A. Marín-Marín
e-mail: jmarin@ugr.es

Ó. Bedoya-Cano
Pontifical Bolivarian University, Medellín, Colombia
e-mail: oscar.bedoya@upb.edu.co



L. Leite et al. (eds.), *Practical Work in Science Education in the Digital Era*, Advances in Technology-Rich Science Education 2, https://doi.org/10.1007/978-3-032-13788-3_16

it concludes with a discussion of the future possibilities of augmented reality and its implications for science education.

## 16.2 Augmented Reality in Science Education: From Hypertext to Hyperobject

Hypertextuality incorporates non-sequential forms of reading in which the reader determines the structure of the path. This reinforces a constructivist idea whereby the reader assigns meaning to objects of knowledge (Silva et al., 2023). The sequence of ideas and their conception generate a different effect, since the reading of texts produces an alternative relationship, in contrast to traditional forms that follow a system-suggested order (Landow, 1995). Furthermore, in the interpretation of a non-linear text, regarding the order of exploration, the reader may reinterpret in multiple ways the ideas proposed by the author within the digital content.

Hypertext employs words as connecting nodes that enable the linking of different textual contents across independent spaces, for example, navigation between web pages through hyperlinks, while preserving and enhancing network-based technology as a form of psychological tool (Kozulin, 2002; Rueda, 2003). This relationship between different contents highlights the associative component as a fundamental feature in the nature of understanding the world and in the theories that culminate in constructivist approaches.

This associative perspective has developed exponentially, as hypertextual networks connect represented objects through computation; in this case, words act as activators. In terms of virtual spaces, this phenomenon is recognised as deterritorialisation (Lévy, 1999), or, in other words, disembedding (Giddens, 1997). Hyperlinks, in this sense, provide a window for exploration towards other contents, freeing places and specific moments so that they may be experienced at will, much like the release of an idea that escapes into another material space or through thought itself.

In augmented reality, objects incorporate visual or image-based zones that function as the connecting nodes of hypertexts. In this way, they allow the reader—who interacts within systems of visual immersion—to establish associations with digital content that transcend the text and become interconnected objects. These, in turn, enhance associative and visual capacity, thereby taking shape as hyperobjects.

### *16.2.1 The Image: A Cognitive Connection*

Human beings perceive environmental signals through their senses. Aristotle's assertion, there is nothing in the mind that has not first passed through the senses, anticipates the way in which the transducers of electrical impulses act as catalysts, enabling the brain to interpret events in the surrounding world (Beaty & Kenett, 2023). Consequently, the representation of a physical object through an image fulfils the same function as the word in its intention to signify.

The image, understood as the representation of something absent (Aparici, 2009), makes possible the projection, reconstruction, and transfer of an object from one place to another. In this sense, it can be linked to the general nature of the text through its semiotic component as a visual sign. This perspective considers the representation of images of physical objects in the world to argue that, like words in a text, images may serve as both a starting point and a connecting device for associating content, in a manner similar to hypertextuality, by means of the technological supports designed for this purpose.

### *16.2.2 From Image to Hypertext in Augmented Reality*

Augmented reality emerges as an alternative mode of relating content through technological supports, grounded in the visual perception of the world. It represents a distinct approach compared with virtual reality, in which the user is fully isolated from the physical perception of their environment (Azuma, 1997). By contrast, AR overlays additional layers of information onto physical objects through technological devices. This occurs along the so-called reality-virtuality continuum, with a fully virtual environment at one extreme and physical reality at the other (Milgram & Kishino, 1994). A range of devices has been developed to support virtuality. Its pioneer, Ivan Sutherland, introduced as early as 1968 a computer-assisted visualisation system using a head-mounted display. Since the 1990s, technological infrastructures and digital content for AR have been designed to interact in real time with physical objects, projecting them simultaneously through visualisation devices that preserve the user's connection to their environment.

Augmented reality is enabled by technological supports such as mobile devices, immersive glass, and internet connectivity, which together allow the fusion of real and virtual domains (physical objects combined with deterritorialised digital texts). Two notable modalities can be identified: annotated augmented reality, involving the superimposition of text onto physical objects; and integrated augmented reality, which entails the construction of virtual objects via visualisation systems (Nijholt, 2022). To provide a clearer understanding, the ongoing transitions along the reality-virtuality continuum are illustrated (see Fig. 16.1).

From a philosophical perspective, augmented reality is conceived as a system that combines scenarios from the physical world with superimposed digital information (Sinclair, 2004). This philosophical stance highlights the relationship between hypertextuality and AR, pointing out a key similarity: while hypertext connects content through links, AR associates physical objects with digital content. When observing a physical object, it may be linked to the construction of meaning as formed in a mental image or, alternatively, processed as information, much like static texts presented alongside images in an advertisement (Díaz, 2011).

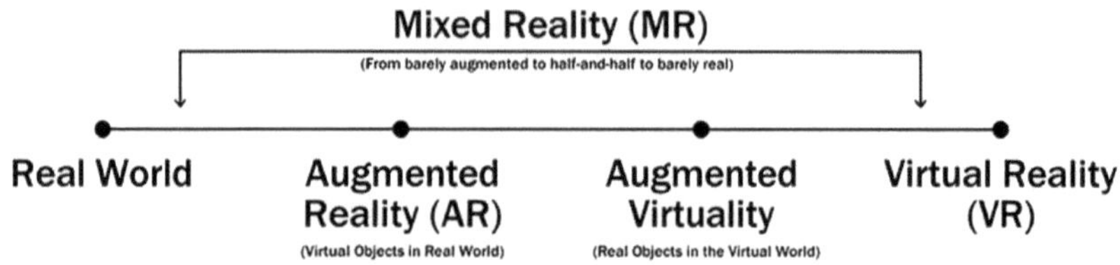


**Fig. 16.1** Reality vs. virtual reality (Milgram & Kishino, 1994, p. 1321)

Augmented reality may be understood as an evolution of perception and access to knowledge, as it constitutes a form of three-dimensional hypertext embedded in the physical world, transforming the traditional relationship between the learner and the object of knowledge (Mystakidis et al., 2022). Within the philosophical discourse on augmented reality, issues such as the phenomenology of augmented experiences, the metaphysics of virtual objects, and the ethical concerns raised by these technologies are explored (Turner, 2022). Classical phenomenology underscores the role of the body in perception and knowledge construction, in contrast to traditional views that separated mind and body, insofar as they depend on the corporeal involvement of the subject with the world (Stalheim & Somby, 2024). In this regard, augmented reality functions as an amplifier of such embodied experiences.

Augmented reality generates what some describe as a relation of immersion or fusion between human being, technology, and world (Nijholt, 2022). This extension of physical reality entails a rupture with the constructivist division between subject, mediation, and object. In immersive augmented reality, the object of knowledge (for instance, a virtual model of a molecule or an organism) coexists in the same space as the subject and responds to their actions, thereby blurring the boundary between observer and observed phenomenon. Technological mediation thus introduces a new epistemic framework in which knowing means coexisting and interacting with tangible representations of knowledge, challenging the traditional paradigm of observing phenomena from the outside. Just as conventional hypertext connects concepts through links on a page, AR directly links scientific ideas to virtual objects situated in physical space. Students can literally touch or move around them, constructing their own mental images (Ocanto, 2009) and exploring scientific concepts in a multisensory manner.

In the past decade (2010–2019), a standardised structure was developed, similar to Hypertext Markup Language (HTML), known as Augmented Reality Markup Language (ARML) (Liao, 2019). This standard is based on the interconnection and organisation of objects and their associated digital content. In this way, augmented reality system developers and manufacturers have access to an interoperable framework that facilitates the creation of object-linked content, ensuring that digital resources for science education can be reproduced without interpretative limitations across diverse devices.

Augmented reality technologies and visualisation elements must meet the following requirements for effective implementation:

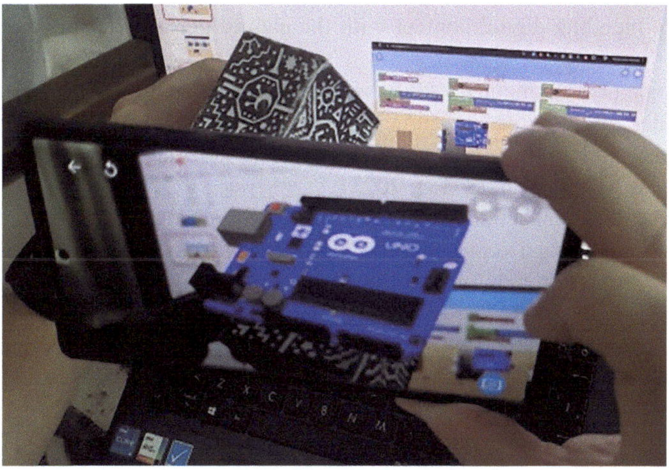

**Fig. 16.2** Application of augmented reality in computational thinking and electrical circuits with Merge Cube and Arduino system (Bedoya-Cano et al., 2025)

1. A device activated by location or by the identification of a physical object. Portability of the device is essential, as it must be operable across multiple contexts without being restricted to a specific space or time. For instance, trigger codes such as Quick Response (QR) and their three-dimensional variations are used (see Fig. 16.2).
2. The production of digital content associated with the physical object. Once the object has been identified, it must have associated digital content to be presented through its hypertextual link. At this point, hypertextuality becomes connected with content construction, using the physical object as a hypertextual node.
3. A technological support that enables the simultaneous display of the physical object and its associated information. As with activation through location or identification, the technological support must allow the physical object to be visualised simultaneously with the corresponding multimedia content.

### *16.2.3 Science Education and Augmented Reality*

Although augmented reality became popular through advances in computing and visualisation devices, its first conceptual introduction took place in the aviation industry in 1992, with an instructional purpose: supporting the assembly of the Boeing 747 through a visual field enhancement device known as the *Heads-up Display Set* (HUDset) (Caudell & Mizell, 1992). Subsequently, interest grew within the educational community in exploring an alternative way of combining computational processes with the physical environment (Feiner, 2002), particularly as digital screens, mobile devices, and glasses incorporating technologies capable of seamlessly blending digital content with the natural perception of objects became available.

Pedagogically, augmented reality is deployed as part of multimodal learning; it constitutes an additional resource that can be employed as a didactic strategy, bringing the physical object closer while creating associated hypertextual content (Mehler et al., 2018; Wang & Tan, 2023).

There are different modes of approaching the learning object: textual, oral, or two-dimensional visual representation, such as in books and blackboards; three-dimensional representation on two-dimensional screens; and, more recently, three-dimensional projection. In STEM areas, augmented reality has been incorporated into science education since the 1990s through pedagogical mediation. A literature review conducted in *Web of Science* reveals that the earliest practical publications in these disciplines, natural sciences (Cabascango, 2023; Fearn & Hook, 2023), technology (Kim et al., 1999), engineering (Arapis et al., 1997), and mathematics (Kaufmann & Schmalstieg, 2003), pioneered classroom experiences. However, other branches of the social sciences, such as history and geography, also adopted augmented reality in projects aimed at historical and cultural preservation (Bekele et al., 2018; Komianos et al., 2024; Skublewska-Paszkowska et al., 2022).

At different stages of science education—conceptualisation, practice, laboratory work, experiments, and fieldwork—augmented reality emerges as a tool that contributes to the construction of new learning experiences. For this reason, a literature review was conducted to examine how AR has been applied across different educational practices in STEM.

The literature review of academic publications indexed in Web of Science provides an overview of each STEM field. Using the search criteria "Augmented Reality and Science", "Augmented Reality and Education Technology", "Augmented Reality and Engineering", and "Augmented Reality and Mathematics", an exponential growth trend is observed from the 1990s through to 2024 (see Fig. 16.3). The fields with the highest number of publications are technology and science, whereas mathematics reports comparatively fewer experiences per year.

In the field of science education, new possibilities emerge when applying the principles of augmented reality to interactive books based on hypertextual links and the representation of abstract objects, making augmented reality and its variants valuable tools for pedagogical mediation. Augmented reality has a strong didactic component for educational experiences (Tzima et al., 2019), as evidenced by several initiatives developed in the first half of the previous decade, during which augmented reality experienced significant academic and educational expansion through classroom applications in the representation of objects (Bacca et al., 2014; Hernández et al., 2012; Ierache et al., 2014; Leiva & Moreno, 2015; Saidin et al., 2015; Telefónica, 2011). However, a critical perspective on augmented reality calls for appropriate orientation, since it transforms traditional epistemology.

Ensuring the quality and objectivity of digitally added information, as well as maintaining students' attention amidst multiple layers of superimposed content, represents a collateral challenge resulting from the incorporation of this technology. A critical epistemology applied to educational augmented reality would require

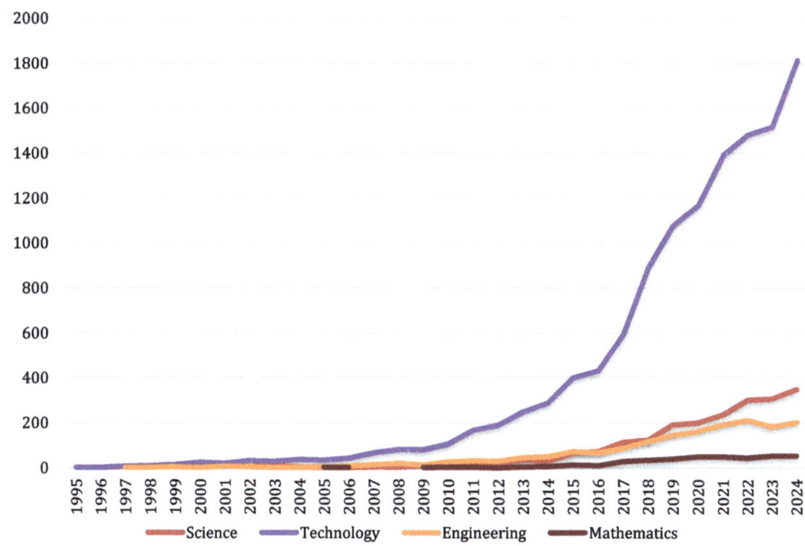


**Fig. 16.3** Augmented reality and publications in Web of Science (1995–2024)

examining these issues, ensuring that the integration of augmented reality does not reproduce extracurricular biases.

When conceived as a three-dimensional hypertext, augmented reality transforms the relationship between the learner and the object of knowledge. Instead of accessing knowledge solely through books or flat screens, where information remains separated from the physical world, the learner encounters information embedded in objects and places (Smart, 2012). Likewise, student motivation increases in augmented reality-based learning environments (Cabero-Almenara & Roig-Vila, 2019; Gómez et al., 2020), particularly when associated with specific teaching strategies (Diegmann et al., 2015).

Its integration into strategies traditionally considered conventional, such as books enhanced with digital supports, now enables three-dimensional representations over textual content through QR codes (AlNajdi, 2022) or other hyperobject activation mechanisms. Discovery-based learning is enhanced by embedding digital information into a physical object, thereby enriching the student's experience (Wahyu & Rusnilawati, 2025; Wen et al., 2023). Similarly, augmented reality can be applied to projections for the structural modification of physical objects, as in instructional processes of assembly and connection—an approach increasingly used in architecture, civil engineering, and construction.

The development of skills requiring interaction with physical objects, such as the maintenance of mechanical parts and assembly tasks, can be strengthened through related instructional content. Furthermore, game-based learning also incorporates augmented reality to enrich interaction and provide alternative access to learning objects (Lampropoulos et al., 2023). Through achievements, tokens, and rewards, augmented reality thus offers students a different and more engaging learning experience.

## 16.3 Discussion

Augmented reality in science education has implications that extend beyond the transformation of laboratory practices. It provides an alternative form of interaction with objects of knowledge, in which simulations characteristic of STEM areas are represented didactically. Nevertheless, several considerations must be addressed: teachers' digital competences, the risks of technological overuse and dependency, and students' overexposure to screens, all of which generate dynamics that cannot be overlooked (Beaty & Kenett, 2023). The technological novelty of pedagogical mediation requires the inclusion of reflective moments within classroom practice, in which both the opportunities offered by augmented reality and the challenges of its implementation are critically examined (Komianos et al., 2024).

The current trend in science education points towards the metaverse as an evolution of AR, alongside the incorporation of artificial intelligence into teaching and learning environments. This demonstrates how technological convergence redefines pedagogical practice. In other words, emerging elements and trends complement and combine with one another, producing not only practical applications but also a reconfiguration of the very concepts of learning (Wahyu & Rusnilawati, 2025).

The main technological limitation does not lie in the acquisition of devices. This access barrier is not the central obstacle, since augmented reality applications typically require only the basic features available in most mobile devices, rather than fully immersive equipment such as virtual reality headsets. Thus, the key challenge is not technical but pedagogical: it concerns the teacher's ability to design and implement practical learning experiences through the effective use of their digital competences (Bedoya-Cano et al., 2025).

## 16.4 Conclusion and Ways Forward

Digital content is adapting to new forms of presentation and is being rethought in relation to the world of physical objects, as technological media transform. With these new possibilities, hypertexts can be presented not only through digital media, but physical objects can also serve as hypertextual nodes thanks to technological media. In this way, hypertexts are linked depending on the digital content designed for physical objects, supported by technological media, thus offering great possibilities in the field of education, where hypertextual relationships are fundamental to the construction of knowledge.

Transcending a text that was tied to its support, to a deterritorialised text without borders, accessible in any virtual form thanks to digital devices, has generated a new, enriched space, almost saturated in terms of content and ideas, much closer to the associative form of thought, and which seems not to stop there and seeks to replicate itself in other types of visual forms, coded in the same way, taking advantage of the capabilities of the digital.

Science education finds new epistemological ways to represent difficult-to-access elements. The natural physical environment can be enriched with simulations that allow students to interact with the objects of study at different scales. This creates a sensorial advantage over the abstract capacity necessary to understand some scientific concepts.

## References

AlNajdi, S. (2022). The effectiveness of using augmented reality (AR) to enhance student performance: Using quick response (QR) codes in student textbooks in the Saudi education system. *Educational Technology Research and Development, 70*(3), 1105–1124. https://doi.org/10.1007/s11423-022-10100-4

Aparici, R. (2009). *La imagen: Análisis y representación de la realidad* [The image: Analysis and representation of reality]. Gedisa Editorial.

Arapis, C., Gibbs, S., & Breiteneder, C. (1997). Real-time segmentation of video on a multiprocessor platform. *Parallel Computing, 23*(12), 1777–1792. https://doi.org/10.1016/S0167-8191(97)00087-2

Azuma, R. (1997). A survey of augmented reality. *Presence: Teleoperators and Virtual Environments, 6*(4), 355–385. https://doi.org/10.1162/pres.1997.6.4.355

Bacca, J., Baldiris, S., Fabregat, R., Graf, S., & Kinshuk. (2014). Augmented reality trends in education: A systematic review of research and applications. *Educational Technology & Society, 17*(4), 133–149. https://www.jstor.org/stable/jeductechsoci.17.4.133

Beaty, R., & Kenett, Y. (2023). Associative thinking at the core of creativity. *Trends in Cognitive Sciences, 27*(7), 671–683. https://doi.org/10.1016/j.tics.2023.04.004

Bedoya-Cano, Ó., Marín-Marín, J.-A., Sánchez-Amate, J.-J., & López-Belmonte, J. (2025). The metaverse in STEM education: Interactive assessment with electronic circuits and Arduino mediated by artificial intelligence. *IJERI: International Journal of Educational Research and Innovation*, (24). https://doi.org/10.46661/ijeri.11217

Bekele, M., Pierdicca, R., Frontoni, E., Malinverni, E., & Gain, J. (2018). A survey of augmented, virtual, and mixed reality for cultural heritage. *Journal on Computing and Cultural Heritage, 11*(2), Article 7. https://doi.org/10.1145/3145534

Cabascango, G. (2023). El uso de la realidad aumentada en la enseñanza de ciencias: Un enfoque integrador en educación secundaria [He use of augmented reality in science teaching: An integrative approach in secondary education]. *Revista Científica Kosmos, 2*(1), 39–50. https://doi.org/10.62943/rck.v2n1.2023.43

Cabero-Almenara, J., & Roig-Vila, R. (2019). The motivation of technological scenarios in augmented reality (AR): Results of different experiments. *Applied Sciences, 9*(14), Article 2907. https://doi.org/10.3390/app9142907

Caudell, T., & Mizell, D. (1992). Augmented reality: An application of heads-up display technology to manual manufacturing processes. In *Proceedings of the Twenty-Fifth Hawaii International Conference on System Sciences* (pp. 659–669). IEEE. https://doi.org/10.1109/HICSS.1992.183317

Díaz, Á. (2011). La imagen metafórica publicitaria [The metaphorical advertising image]. *Cuadernos de Lingüística Hispánica, 18*, 27–42. https://www.redalyc.org/articulo.oa?id=322227526003

Diegmann, P., Schmidt-Kraepelin, M., Van den Eyden, S., & Basten, D. (2015). Benefits of augmented reality in education environments: A systematic literature review. In O. Thomas & F. Teuteberg (Eds.), *12th International Conference on Wirtschaftsinformatik* (pp. 1542–1556). https://www.scirp.org/reference/referencespapers?referenceid=2552062

Fearn, W., & Hook, J. (2023). A service design thinking approach: What are the barriers and opportunities of using augmented reality for primary science education? *Journal of Technology and Science Education, 13*(1), 329–351. https://doi.org/10.3926/jotse.1394

Feiner, S. K. (2002). Augmented reality: A new way of seeing. *Scientific American, 286*(4), 48–55. http://www.jstor.org/stable/26059641

Giddens, A. (1997). *Modernidad e identidad del yo: El yo y la sociedad en la época contemporánea* [Modernity and identity of the self: The self and the society in contemporary times]. Península.

Gómez, G., Rodríguez, C., & Marín, J. (2020). La trascendencia de la realidad aumentada en la motivación estudiantil: Una revisión sistemática y meta-análisis [The impact of augmented reality in student motivation: A systematic review and meta-analysis]. *Alteridad, 15*(1), 36–46. https://doi.org/10.17163/alt.v15n1.2020.03

Hernández, J., Pennesi, M., Sobrino, D., & Vázquez, A. (2012). *Tendencias emergentes en educación con TIC* [Emerging trends in education with ICT]. Espiral.

Ierache, J., Igarza, S., Mangiarua, N., Becerra, M., Bevacqua, S., Verdicchio, N., Ortiz, F., Sanz, D., Duarte, N., & Sena, M. (2014). Herramienta de realidad aumentada para facilitar la enseñanza en contextos educativos mediante el uso de las TICs. *Revista Latinoamericana de Ingeniería de Software, 2*(6), 365–368. https://doi.org/10.18294/relais.2014.365-3688

Kaufmann, H., & Schmalstieg, D. (2003). Mathematics and geometry education with collaborative augmented reality. *Computers & Graphics, 27*(3), 339–345. https://doi.org/10.1016/S0097-8493(03)00028-1

Kim, H., Kim, J., & Park, S. (1999). A bird's-eye view system using augmented reality. In *Proceedings of the 32nd Annual Simulation Symposium, California, USA* (pp. 126–131). IEEE. https://doi.org/10.1109/SIMSYM.1999.766463

Komianos, V., Tsipis, A., & Kontopanagou, K. (2024). Introducing digitized cultural heritage to wider audiences by employing virtual and augmented reality experiences: The case of the v-Corfu project. *Technologies, 12*(10), Article 196. https://doi.org/10.3390/technologies12100196

Kozulin, A. (2002). Sociocultural theory and the mediated learning experience. *School Psychology International, 23*(1), 7–35. https://doi.org/10.1177/0143034302023001729

Lampropoulos, G., Keramopoulos, E., Diamantaras, K., & Evangelidis, G. (2023). Integrating augmented reality, gamification, and serious games in computer science education. *Education Sciences, 13*(6), Article 618. https://doi.org/10.3390/educsci13060618

Landow, G. (1995). *Hipertexto: La convergencia de la teoría crítica contemporánea y la tecnología* [Hypertext: The convergence of contemporary critical theory and technology]. Paidós.

Leiva, J., & Moreno, N. (2015). Tecnologías de geolocalización y realidad aumentada en contextos educativos: Experiencias y herramientas didácticas [Geolocation and augmented reality technologies in educational contexts: Experiences and teaching tools]. *Revista DIM: Didáctica, Innovación y Multimedia, 11*(31), 1–18. https://dialnet.unirioja.es/servlet/articulo?codigo=5081652

Lévy, P. (1999). *¿Qué es lo virtual?* Paidós.

Liao, T. (2019). Standards and their (recurring) stories: How augmented reality markup language was built on stories of past standards. *Science, Technology, & Human Values, 45*(4), 712–737. https://doi.org/10.1177/0162243919867417

Mehler, A., Spiekermann, C., Abrami, G., & Jostock, M. (2018). VAnnotatoR: A framework for generating multimodal hypertexts. In *HT 18: Proceedings of the 29th on Hypertext and Social Media* (pp. 150–154). ACM. https://doi.org/10.1145/3209542.3209572

Milgram, P., & Kishino, F. (1994). Taxonomy of mixed reality visual displays. *IEICE Transactions on Information and Systems, E77*, 1321–1329. https://doi.org/10.1.1.102.4646

Mystakidis, S., Christopoulos, A., & Pellas, N. (2022). A systematic mapping review of augmented reality applications to support STEM learning in higher education. *Education and Information Technologies, 27*(2), 1883–1927. https://doi.org/10.1007/s10639-021-10682-1
Nijholt, A. (2022). Virtual worlds: A new open access journal of virtual reality, augmented and mixed reality technologies, and their uses. *Virtual Worlds, 1*(1), 18–19. https://doi.org/10.3390/virtualworlds1010002
Ocanto, S. (2009). La creación de imágenes mentales y su implicación en la comprensión, el aprendizaje y la transferencia [The creation of mental images and its implication in the understanding, the learning and the transference]. *Sapiens. Revista Universitaria de Investigación, 10*, 243–253. http://www.redalyc.org/articulo.oa?id=41021266013
Rueda, R. (2003). *Para una pedagogía del hipertexto: Una teoría entre la deconstrucción y la complejidad* [For a hypertext pedagogy: A theory between deconstruction and complexity] (Doctoral dissertation). University of Islas Baleares.
Saidin, N., Abd, N., & Yahaya, N. (2015). A review of research on augmented reality in education: Advantages and applications. *International Education Studies, 8*(13), 1–8. https://doi.org/10.5539/ies.v8n13p1
Silva, M., Bermúdez, K., & Caro, K. (2023). Effect of an augmented reality app on academic achievement, motivation, and technology acceptance of university students of a chemistry course. *Computers & Education: X Reality, 2*, Article 100022. https://doi.org/10.1016/j.cexr.2023.100022
Sinclair, P. (2004). *Integrating hypermedia techniques with augmented reality environments.*
Skublewska-Paszkowska, M., Milosz, M., Powroznik, P., & Lukasik, E. (2022). 3D technologies for intangible cultural heritage preservation: Literature review for selected databases. *Heritage Science, 10*(3), 1–24. https://doi.org/10.1186/s40494-021-00633-x
Smart, P. (2012). The web-extended mind. *Metaphilosophy, 43*(4), 446–463. https://doi.org/10.1111/j.1467-9973.2012.01756.x
Stalheim, O. R., & Somby, H. M. (2024). An embodied perspective on an augmented reality game in school: Pupil's bodily experience toward learning. *Smart Learning Environments, 11*(24), 1–21. https://doi.org/10.1186/s40561-024-00308-7
Telefónica, F. (2011). *Realidad aumentada* [Augmented reality]. Fundación Telefónica.
Turner, C. (2022). Augmented reality, augmented epistemology, and the real-world web. *Philosophy & Technology, 35*(19), 1–28. https://doi.org/10.1007/s13347-022-00496-5
Tzima, S., Styliaras, G., & Bassounas, A. (2019). Augmented reality applications in education: Teachers' point of view. *Education Sciences, 9*(2), 1–18. https://doi.org/10.3390/educsci9020099
Wahyu, A., & Rusnilawati, R. (2025). Augmented reality-based discovery learning: An effective strategy to improve critical thinking in sciences for Grade IV. *Lectura: Jurnal Pendidikan, 16*(1), 293–306. https://doi.org/10.31849/lectura.v16i1.25775
Wang, T., & Tan, L. (2023). The conceptualisation of user-app interactivity in augmented reality-mediated learning: Implications for literacy education. *Sustainability, 15*(14), Article 10949. https://doi.org/10.3390/su151410949
Wen, Y., Wu, L., He, S., Ern Ng, N., Teo, B., Looi, C., & Cai, Y. (2023). Integrating augmented reality into inquiry-based learning approach in primary science classrooms. *Educational Technology Research and Development, 71*(4), 1631–1651. https://doi.org/10.1007/s11423-023-10235-y

**José-Antonio Marín-Marín** holds a PhD in pedagogy and is an associate professor at the University of Granada, Department of Didactics and School Organization, Spain. He coordinates the Official Master's Degree in Secondary Education Teaching at the same university and leads the EducaTech-XXI Research Group, Education and Technology for the 21st Century (SEJ-666).

**Óscar Bedoya-Cano** is an electronic engineer, holds a master's degree in education, and is a PhD candidate at the Pontifical Bolivarian University, Colombia. His academic work focuses on the relationship between technology, pedagogy, and computational thinking, with a particular interest in teacher education and the educational use of digital innovation.

**Jesús López-Belmonte** holds a PhD in educational sciences. He is a professor in the Department of Didactics and School Organization at the University of Granada, Spain. He is a member of the EducaTech XXI research group (SEJ-666). He is the author of several scientific publications. He has completed several research and training visits in Europe and Latin America. He has also participated in various research and teaching innovation projects.

# Chapter 17
# Immersive Virtual Reality for Science Practicals: Opportunities, Challenges, and Pedagogical Considerations

**Henry Matovu and Mihye Won**

## 17.1 Introduction

Science practicals have long been recognised as a key component of effective science education (Gericke et al., 2023). These activities typically include laboratory experiments, investigative tasks, and fieldwork extending beyond passive observation to active manipulation and inquiry. In learning, practicals bring scientific concepts to life through tangible, sensory-rich experiences (vision, audio, touch, and smell). Practicals also immerse students in the epistemic habits of scientists, such as developing hypotheses, designing experiments, or constructing evidence-based arguments (Gericke et al., 2023). Traditionally, science practicals are conducted in specially organised physical spaces (or labs). However, growing enrolments in science classes, calls for greener and inclusive labs (Karpudewan, 2024), and the sudden move to remote teaching during the COVID-19 pandemic (Atchia & Rumjaun, 2023) have prompted educators to reimagine how practical learning can be delivered.

Immersive Virtual Reality (VR) technology has emerged as a compelling avenue for extending the reach and flexibility of science practicals. VR uses head-mounted display units to fully immerse learners in interactive, three-dimensional environments that simulate laboratory or field settings (Radianti et al., 2020). Within these virtual environments, students can interact with digital elements using motion-tracking devices to observe the consequences of their actions through visual, auditory, and sometimes tactile feedback. This vivid multi-sensory and embodied experience enhances learners' sense of presence and agency which are crucial for sustaining students' motivation and enhancing understanding (Makransky & Petersen, 2021).

H. Matovu (✉)
School of Chemistry, The University of Sydney, Sydney, Australia
e-mail: henry.matovu@sydney.edu.au

M. Won
Faculty of Education, Monash University, Melbourne, Australia
e-mail: mihye.won@monash.edu



L. Leite et al. (eds.), *Practical Work in Science Education in the Digital Era*, Advances in Technology-Rich Science Education 2, https://doi.org/10.1007/978-3-032-13788-3_17

When designed with sound pedagogical principles, VR technology can productively support scientific experimentation and practical learning.

With affordable standalone VR headsets now entering schools and universities (Radianti et al., 2020), many science educators have started integrating VR into their classes. Despite the growing interest, the integration of VR into science practicals has yielded mixed results in terms of learning outcomes and students' perceptions (Matovu et al., 2023; Reeves et al., 2021). This trend raises important questions about when, how, and why VR works in practical science education. Previous reviews have examined the implementation of virtual environments in science practicals broadly, including desktop-based applications (Potkonjak et al., 2016; Reeves & Crippen, 2021). These syntheses have not addressed the unique opportunities and challenges presented by immersive VR due to its distinct nature of visualisation and interaction.

The purpose of this chapter is to synthesise the existing literature and critically evaluate the benefits, challenges, and future directions of adopting immersive VR to teach practicals in the field of science education. The key research questions being addressed in the chapter are:

- Why do educators adopt immersive VR in science practicals?
- How effective is immersive VR-based learning for science practicals?
- What are the barriers to the adoption of immersive VR-based science practicals?

While the term "science practicals" can be interpreted broadly, in this chapter, we define "science practicals" as hands-on activities in which students directly engage with real or virtual objects, equipment, apparatus, and phenomena to develop experimental skills, practice scientific reasoning, or apply theoretical knowledge through direct interaction. In this context, we include studies (from journal articles) that describe VR-enhanced activities involving students' active, inquiry-driven interactions with real or virtual objects. We exclude studies limited to passive experiences (e.g., watching VR simulations), activities that model abstract concepts, or science-related virtual field trips, unless these activities involve active, inquiry-driven interactions with the material world. Given the emerging nature of the field, common themes from the literature have been organised and illustrated with examples using a narrative approach, rather than a systematic protocol. This analysis aims to provide interpretive insights into when and how immersive VR can be most effectively leveraged to maximize learning benefits in science practicals.

## 17.2 Rationale for Immersive VR in Science Practicals

The existing literature suggests that VR is not merely a high-tech (yet inferior) substitute for physical science laboratories but a medium with unique pedagogical benefits. Educators are increasingly leveraging VR to: (i) extend real-world fieldwork to otherwise inaccessible environments, (ii) remove safety, cost, and ethical barriers associated with experimentation, (iii) foster unconstrained experimentation for learning, (iv) boost students' confidence ahead of real-life labs and, (v) scale high-quality practicals across diverse learner populations. The reasons for the applications tend to vary depending on the discipline as illustrated below.

*Extending the reach of field-based learning*: In disciplines like geology, VR enables students to explore terrains that are otherwise inaccessible, such as deep marine environments and tectonic plate boundaries (Harknett et al., 2022; Métois et al., 2021). For instance, Métois et al. (2021) used high-resolution digital outcrops in VR to enable undergraduate students explore the submarine Roseau fault, located 1200 m underwater. Students observed and measured fault lines, and sediment layers—phenomena that are beyond reach in traditional field trips. Similarly, Harknett et al. (2022) used VR field trips to simulate large-scale tectonic processes. By offering pseudo-aerial perspectives, undergraduate students could 'fly' around a tectonic plate boundary, pause the subduction processes, and measure fault offsets with a virtual tape. These immersive environments provided an opportunity for learners to practice field-based tasks such as, observation, measurement, and mapping of structural features in these obscured places.

*Cost, safety, and ethical advantages*: VR offers a practical solution to cost, safety, and ethical barriers that often limit the integration of science practicals in learning. In geology, for example, VR can enable place-based learning without the logistical and accessibility challenges of organising physical field trips. In an undergraduate-level geology unit, Klippel et al. (2019) replicated real-world field trips using interactive 360° imagery, photos, and 3D models of field outcrops with audio guides. Students who used VR reported higher satisfaction levels and achieved improved lab grades compared to those who engaged in traditional field trips.

In biology and chemistry, VR simulations can be used to safely conduct ethically sensitive activities or hazardous experiments. These include animal or human dissections (Reeves et al., 2024; Shine et al., 2025), toxic chemical handling, or high-risk procedures (e.g., Broyer et al., 2021; Chitra et al., 2024). For example, in microbiology, Chitra et al. (2024) used VR to simulate six core undergraduate experiments, such as Gram staining and aseptic transfers, in an undergraduate-level unit. About 70% of students reported being confident in completing the labs without the fear of being exposed to biohazards (microbes).

*Unconstrained student experimentation in realistic environments*: VR creates immersive, realistic-looking environments where students can experiment with ideas, make mistakes, try alternative strategies, and refine their understanding through iteration. For instance, in physics, Canright and Brahmia (2024) used VR to immerse first and second-year undergraduate students into a simulated universe governed by fictitious physical laws. Students designed and performed experiments on the force laws governing interactions between virtual charged particles without pre-programmed "right answers". This activity design tasked students to generate, test, and refine their own scientific models, shifting students' approaches toward a more expert-like view of experimentation; not using labs to simply prove science concepts, but to engage in scientific thought processes and experimentation.

*Enhancing emotional readiness for science practicals:* The possibility of making experimental mistakes without penalties has made VR a valuable resource for boosting students' confidence in preparation for real-life labs. For instance, as a pre-lab resource in an undergraduate-level unit, VR greatly reduced students' anxiety and boosted their perceived self-efficacy in setting up organic chemistry reactions in fume hoods (Gungor et al., 2022). In another study, students reported that self-pacing in VR reduced performance anxiety stemming from peer comparison in real-world chemistry labs (Reeves et al., 2021). Some students even felt that VR tutors were non-judgemental and expressed willingness to complete experiments with the guidance of virtual tutors (Reeves et al., 2021). This work suggests that VR environments can create emotionally safe spaces and encourage more resilient, risk-taking, and exploratory learning experiences.

*Scalability and inclusivity*: Scalable VR labs broaden access to science practicals, democratising hands-on learning and narrowing equity gaps. For instance, VR platforms can offer high-quality laboratory experiences to institutions without access to costly infrastructure. To address cost-related barriers to accessing high performance liquid chromatography (HPLC) instruments, Taylor et al. (2024) developed an interactive simulation that allowed undergraduate students to explore and operate virtual HPLC systems. The authors found that VR enhanced undergraduate students' confidence and enjoyment of practical work and perceived understanding of HPLC instruments. Similarly, Ahmed et al. (2023) demonstrated that low-cost VR simulations could help students in an underprivileged high school to attend and productively complete biology labs.

VR also promotes inclusion for remote learners and those with disabilities (e.g., Lu et al., 2021; Williams et al., 2022). For instance, during the pandemic, Williams et al. (2022) developed first-person-perspective VR environments to provide a learning opportunity for remote learners enrolled in an undergraduate organic chemistry lab program. In another study, Lu et al. (2021) designed and implemented a system that allowed undergraduate students—especially those with mobility limitations—to engage in authentic experimental chemistry experiences through a fully immersive virtual environment connected to intelligent robotic systems. The system's design supported multiple modes of interaction (auto, custom, and manual), ensuring flexibility for learners with varying needs and preferences.

## 17.3 Effectiveness of Immersive VR-Based Practicals

VR has been hailed for its ability to enhance student engagement and motivation in learning, but there are mixed reports on its learning benefits (Matovu et al., 2023). In science practicals, research points to at least three interrelated factors that collectively shape the effectiveness of VR: intended outcomes, students' prior knowledge, and design and implementation of VR.

*Intended outcomes:* VR has been shown to enhance students' abilities to engage in authentic scientific inquiry (Canright & Brahmia, 2024), and their self-efficacy in conducting experiments (e.g., Petersen et al., 2022). Procedural knowledge benefits of VR-based practicals are also comparable to or exceed those from alternative approaches, such as in-person practicals (e.g., Dunnagan et al., 2020; Hu-Au & Okita, 2021; Klippel et al., 2019; Lu et al., 2021). By supporting limitless interactions with virtual elements, VR allows students to collect data, change parameters and observe the consequences of their actions, or draw patterns through scientific experimentation (Canright & Brahmia, 2024; Klippel et al., 2019; Lu et al., 2021).

However, compared to traditional face-to-face science practicals, VR is less effective in supporting fine motor, hands-on skills, such as pipetting or titration (Broyer et al., 2021; Petersen et al., 2022; Reeves et al., 2021). Technical and psychological considerations could explain this. The lack of tactile feedback in handling laboratory apparatus and the inability of VR controllers to accurately replicate real-world motor movements reduce the perceived authenticity and functional fidelity of VR labs (Chu et al., 2024; Klippel et al., 2019; Petersen et al., 2022). Haptic feedback also enables students to perceive the physical properties (e.g., texture and weight) of virtual objects (Huang et al., 2025). Without this feedback, students would find it impossible to judge the resistance of a pipette plunger or the weight of a scalpel, potentially impacting the usefulness of VR (e.g., Chitra et al., 2024; Chu et al., 2024). In addition, the risk-free nature of VR-based learning may inadvertently encourage more reckless behaviours in VR (where the consequences of errors are diminished) than in real labs (Hu-Au & Okita, 2021).

*Students' prior knowledge*: Prior experience –with subject content and VR technology – greatly impacts how students learn through VR. Studies indicate that students with more background knowledge benefit more from VR and tend to perceive VR as a useful alternative to physical laboratory-based learning (Reeves et al., 2024). Similarly, undergraduate students with prior exposure to VR tended to benefit more from the technology (Klippel et al., 2019). Students with higher prior knowledge can focus on content while those new to the content tend to focus on navigating the new technology to complete the assigned tasks (Reeves et al., 2021). The studies suggest that understanding students' prior experiences is critical when planning and designing VR activities for optimal learning experience. For instance, additional scaffolding may be necessary to support students with lower prior knowledge and limited prior VR experiences to overcome initial usability challenges.

*Design and implementation choices:* Generally, the integration of interactive, multimedia elements in VR environments, such as audio explanations and virtual rulers, has been well received by undergraduate students (Klippel et al., 2019). In addition, the students' ability to control the pace of learning and experiment in a risk-free environment in VR is valued to support diverse learning experiences (Chitra et al., 2024). However, feedback from undergraduate students regarding self-directed learning in VR is mixed (Canright & Brahmia, 2024; Reeves et al., 2021). In VR-based experiments designed to train students in conducting basic chemistry laboratory techniques (such as the use of a micropipette, titration, and melting point determination), some students without sufficient prior content knowledge appreciated the step-by-step procedural guidance in VR environments (Reeves et al., 2021). However, others found the step-by-step design restrictive and frustrating (Reeves et al., 2021). In a separate VR-based physics experiment that allowed undergraduate students to explore fictitious particles freely, all participants reported that the creative freedom VR offered was both compelling and enjoyable (Canright & Brahmia, 2024).

Findings from the existing studies also suggest that VR practicals could work best as a supplement to but not a replacement of traditional laboratory experiences (Carroll et al., 2025; Reeves & Crippen, 2021; Shine et al., 2025; Yang et al., 2024). For instance, when VR was used as a replacement of teacher-led pre-lab instruction in chemistry and biology, students' learning outcomes were lower than when VR was integrated with conventional teaching methods (Beatrice et al., 2024). Nevertheless, students who used VR only often reported higher levels of perceived learning (Petersen et al., 2022), suggesting that VR may boost confidence in certain contexts, but students may require more support to develop practical competencies fully.

## 17.4 Challenges and Barriers to Overcome in VR-Based Science Practicals

Despite its promise, the adoption of VR in science practicals still faces challenges. Below we highlight the common challenges identified in recent studies that should be addressed to realise the full educational potential of VR-based practicals:

*Technical and infrastructure barriers:* Developing customised VR experiences is still challenging due to technical and budgetary constraints among science educators. Despite supporting embodied movements, the absence of haptic feedback and tactile interactions in VR (e.g., holding a rock, a pipette, or a dissection kit) is still noticeable (Chitra et al., 2024; Chu et al., 2024). Other technical challenges constrain the full replication of physical practical experiences. Students often express difficulties in discerning fine details in virtual environments due to low-resolution graphics (Klippel et al., 2019). Usability issues, such as the inability to take notes during VR experiences (Métois et al., 2021) or repetitive actions caused by system glitches also tend to add frustration and lower students' overall enjoyment of VR-based practicals (Broyer et al., 2021; Carroll et al., 2025; Reeves et al., 2024; Wang et al., 2025; Wilkerson et al., 2022).

*Pedagogical design challenges:* When educators cannot develop in-house VR applications and must rely on off-the-shelf applications, it becomes challenging to incorporate pedagogically sound principles in the applications, or to meaningfully align VR content with specific course goals (Matovu et al., 2023). It also becomes challenging to balance exploratory, discovery learning with the necessary guidance and social interactions.

Current VR implementations often lack sufficient personalisation to accommodate diverse learning preferences and experiences. Some students find the step-by-step instructions tedious, while others struggle without structured guidance (Reeves et al., 2021; Williams et al., 2022). Yet, many VR applications (e.g., Carroll et al., 2025;

Chu et al., 2024) follow prescriptive, step-by-step designs that constrain inquiry-based learning, limiting one of VR's key affordances: risk-free, problem-solving exploration.

In addition, most VR applications support individual learning experiences, limiting learning opportunities through peer-to-peer or student-instructor interactions (Klippel et al., 2019; Reeves & Crippen, 2021). Some educators have experimented with asymmetric set-ups where one student uses a headset while another student or a teacher used a 2D screen (De Lorenzis et al., 2024; Métois et al., 2021). Peer-to-peer discussion was found to enhance students' recall of procedural aspects as assessed using a written pre-post-tests, but uninterrupted individual practice in VR allowed students to concentrate on the learning tasks and complete the hands-on aspects better (De Lorenzis et al., 2024). A few educators have also started experimenting with designs in which multiple students share both physical and virtual spaces (Métois et al., 2021). However, facilitating student interactions as well as exploration during such virtual excursions is still challenging.

*Teacher professional development*: For innovative learning technologies like VR to be productively employed in teaching, educators must themselves be adequately supported to effectively integrate the technology into their lesson plans and to assess learning outcomes meaningfully. Traditional assessment methods may not align well with the immersive and interactive nature of VR tasks, making it difficult to measure learning outcomes or provide timely feedback. Currently, science educators do not feel adequately prepared to adopt VR in their teaching (Bower et al., 2020), while developing meaningful assessments for VR-based science practicals also remains a challenge (Matovu et al., 2023). These gaps highlight the need for professional development initiatives, such as collaborations between educational VR researchers and science educators, focusing on both the pedagogical and technical aspects of VR-based learning.

## 17.5 Future Directions for Innovation of Immersive VR-Based Practicals

The existing literature suggests that the transformative potential of VR lies in its capacity to reshape how students engage with scientific inquiry, experimentation, and collaborative learning with VR. As VR gains traction in science practicals, future innovation should move beyond simply adopting the technology for its novelty. Below, we identify key directions that could be explored to maximise the pedagogical potential of VR-based science practicals.

*Reimagining the place of VR in science practicals:* With relatively few well-documented success stories of VR implementation, a central question of how VR should be positioned within science practicals remains. Should VR serve primarily as a preparatory tool for physical labs, offering risk-free rehearsal of procedures? Should it act as a complementary environment, enabling experiences that are impractical or unsafe in physical settings? Or might it replace certain forms of science practicals altogether?

In responding to these questions, educators, researchers, and designers should move beyond attempts to simply replicate traditional labs in virtual environments. Instead, they should capitalise on the unique benefits of VR—such as iterative experimentation, reconstruction of hazardous or hard-to-reach sites, time manipulation, and atomic or molecular-level scaling—that could enable new forms of experimentation and conceptual understanding. For instance, iterative experiment design in VR could allow students to quickly modify variables and observe outcomes, promoting hypothesis testing and metacognitive reflection in ways that are often constrained in physical laboratories (Canright & Brahmia, 2024). Educators should also engage in design-based research approaches to explore how VR can best be integrated into learning.

*Pedagogical innovations:* To further leverage the educational potential of immersive VR, pedagogical models must evolve in tandem with technology developments. A compelling example of a fruitful pedagogical innovation is designing VR applications that encourage students to learn science by thinking and acting like scientists (Canright & Brahmia, 2024). In their design, Canright and Brahmia (2024) challenged students to develop and test their own models and generate new explanations through open-ended inquiry in a fictitious environment. Students reported increased self-efficacy, more expert-like views about experimentation, and sustained engagement. The study demonstrated that pedagogically-rich VR environments can support authentic scientific practices beyond the novelty effect.

Another promising area is the continued development of VR environments that support real-time peer interaction and educator involvement. VR environments are also well-positioned to incorporate adaptive scaffolding to support individualised learning experiences. By integrating real-time guidance, intelligent prompts, and personalised learning trajectories based on learner input, VR can cater to individual student needs while maintaining high levels of cognitive challenge (Song et al., 2024). Such adaptive systems hold the potential to reduce learner frustration and support novice learners through complex experimental procedures.

*Integration with emerging technologies:* The next frontier of using VR in science practicals could also involve hybridising VR with other emerging technologies, such as mixed reality and artificial intelligence. Mixed reality, which overlays virtual elements onto real-world environments, can blend the representational power of digital augmentation with the tactile engagement of physical tools—currently missing in VR. For instance, Jiang et al. (2024) used a combination of VR with physical pipettes to teach pipette calibration skills in a second-year quantitative chemistry course. Students who used the mixed reality setup completed the skill faster than those who used the VR-only mode. These findings suggested that the tactile authenticity of lab tool manipulation provided in augmented set-up helped students navigate the lab experience more intuitively.

At the same time, generative Artificial Intelligence (AI) presents opportunities to embed intelligent virtual tutors that respond dynamically to student behaviours and provide real-time feedback (Song et al., 2024; Taylor et al., 2025), adjust task difficulty, and suggest new investigative directions. Such AI-enhanced systems can foster self-directed learning in VR-based practicals while maintaining alignment with curricular goals. For instance, AI can gauge a student's familiarity with the VR environment or track their conceptual and skill acquisition over time and use the data to personalise learning trajectories; novice users can receive scaffolded instructions and simpler manipulations, whereas experienced students are challenged with more complex, open-ended tasks (Li et al., 2024). By dynamically calibrating task difficulty and content based on prior knowledge, these VR-AI integrated systems can enhance content accessibility and deepen engagement in science practicals.

## 17.6 Conclusion and Ways Forward

The purpose of this chapter was to critically synthesise the existing literature on the benefits, challenges, and future directions of adopting immersive VR in science practicals. The review of the literature found that immersive VR is a maturing medium capable of redefining how students' access, practice, interrogate, and apply scientific ideas and skills. VR is attractive for extending fieldwork practices to hard-to-access areas, minimising safety-cost-ethical constraints, and broadening access to high-quality practicals. However, the learning benefits are contingent on thoughtful design. Affective and procedural knowledge gains from VR flourish when iterative experimentation is foregrounded, but tactile competencies still demand complementary physical experience. VR benefits are further moderated by students' prior knowledge, system usability, and level of guidance provided, reminding us that technology developments alone cannot guarantee learning effectiveness.

To ensure effective VR-based practical experiences, the agenda should shift from simply designing digital replicas of traditional laboratories to designing experiences that exploit the distinctive affordances of VR, such as iterative manipulation of digital environments or collaborative problem-solving in shared spaces. In addition, integrating mixed-reality interfaces promises to re-introduce haptic authenticity in VR-based practicals, while AI-driven tutors can help personalise guidance in practicals at scale. Advancing this agenda demands design-based research that weaves together pedagogy, technical innovation, and assessment scholarship.

This chapter identifies broad trends in VR adoption and offers thematic insights in this rapidly developing landscape. However, due to the narrative review approach taken, some relevant studies may have been overlooked. Moreover, the chapter discusses the synthesis of findings across all educational contexts without tailoring to specific science learning contexts. Different science disciplines and student levels may differ significantly in how immersive VR is adopted and in the kinds of learning benefits realised. More context-specific reviews may be required to provide more targeted guidance for VR use in science education practicals.

## References


Ahmed, T., Chowdhury, A. U. N., & Mozumder, Z. H. (2023). BIO-VR: Design and implementation of virtual reality-based simulated biology laboratory using Google cardboard with an emphasis on virtual education. In S. Smys, K. A. Kamel, & R. Palanisamy (Eds.), *Inventive computation and information technologies. Lecture notes in networks and systems* (Vol. 563, pp. 867–883). Springer Nature. https://doi.org/10.1007/978-981-19-7402-1_62

Atchia, S. M. C., & Rumjaun, A. (2023). The real and virtual science laboratories. In B. Akpan, B. Cavas, & T. Kennedy (Eds.), *Contemporary issues in science and technology education. Contemporary trends and issues in science education* (Vol. 56, pp. 113–127). Springer Nature. https://doi.org/10.1007/978-3-031-24259-5_9

Beatrice, P., Grimaldi, A., Bonometti, S., Caruso, E., Bracale, M., & Montagnoli, A. (2024). Adding immersive virtual reality laboratory simulations to traditional teaching methods enhances biotechnology learning outcomes. *Frontiers in Education, 9*, Article 1354526. https://doi.org/10.3389/feduc.2024.1354526

Bower, M., DeWitt, D., & Lai, J. W. M. (2020). Reasons associated with preservice teachers' intention to use immersive virtual reality in education. *British Journal of Educational Technology, 51*(6), 2215–2233. https://doi.org/10.1111/bjet.13009

Broyer, R. M., Miller, K., Ramachandran, S., Fu, S., Howell, K., & Cutchin, S. (2021). Using virtual reality to demonstrate glove hygiene in introductory chemistry laboratories. *Journal of Chemical Education, 98*(1), 224–229. https://doi.org/10.1021/acs.jchemed.0c00137

Canright, J. P., & Brahmia, W. S. (2024). Modeling novel physics in virtual reality labs: An affective analysis of student learning. *Physical Review Physics Education Research, 20*(1), Article 010146. https://doi.org/10.1103/PhysRevPhysEducRes.20.010146

Carroll, J. S., Najafi, H., & Steiner, M. (2025). Evaluating the effectiveness of virtual laboratory simulations for graduate-level training in genetic methodologies. *Biochemistry and Molecular Biology Education, Online First*. https://doi.org/10.1002/bmb.21898

Chitra, E., Mubin, S. A., Nadarajah, V. D., Se, W. P., Sow, C. F., Er, H. M., Mitra, N. K., Thiruchelvam, V., & Davamani, F. (2024). A 3-D interactive microbiology laboratory via virtual reality for enhancing practical skills. *Scientific Reports, 14*(1), Article 12809. https://doi.org/10.1038/s41598-024-63601-y

Chu, C. E., Sian Wee Cheong, G., Mishra, A., Wen, Y., Huei Leo, C., Yeo, D. J., & Cheong, K. H. (2024). Enhancing biology laboratory learning: student perceptions of performing heart dissection with Virtual Reality. *IEEE Access, 12*, 76682–76691. https://doi.org/10.1109/ACCESS.2024.3394713

De Lorenzis, F., Visconti, A., Restivo, S., Mazzini, F., Esposito, S., Garofalo, S. F., Marmo, L., Fino, D., & Lamberti, F. (2024). Combining virtual reality with asymmetric collaborative learning: a case study in chemistry education. *Smart Learning Environments, 11*(1), Article 43. https://doi.org/10.1186/s40561-024-00331-8

Dunnagan, C. L., Dannenberg, D. A., Cuales, M. P., Earnest, A. D., Gurnsey, R. M., & Gallardo-Williams, M. T. (2020). Production and evaluation of a realistic immersive virtual reality organic chemistry laboratory experience: Infrared spectroscopy. *Journal of Chemical Education, 97*(1), 258–262. https://doi.org/10.1021/acs.jchemed.9b00705

Gericke, N., Per, H., & Wallin, J. (2023). A systematic review of research on laboratory work in secondary school. *Studies in Science Education, 59*(2), 245–285. https://doi.org/10.1080/03057267.2022.2090125

Gungor, A., Kool, D., Lee, M., Avraamidou, L., Eisink, N., Albada, B., van der Kolk, K., Tromp, M., & Bitter, J. H. (2022). The use of virtual reality in a chemistry lab and its impact on students' self-efficacy, interest, self-concept and laboratory anxiety. *Eurasia Journal of Mathematics, Science and Technology Education, 18*(3), em2090. https://doi.org/10.29333/ejmste/11814

Harknett, J., Whitworth, M., Rust, D., Krokos, M., Kearl, M., Tibaldi, A., Bonali, F. L., Van Wyk de Vries, B., Antoniou, V., Nomikou, P., Reitano, D., Falsaperla, S., Vitello, F., & Becciani, U. (2022). The use of immersive virtual reality for teaching fieldwork skills in complex structural terrains. *Journal of Structural Geology*, *163*, Article 104681. https://doi.org/10.1016/j.jsg.2022.104681

Hu-Au, E., & Okita, S. (2021). Exploring differences in student learning and behavior between real-life and virtual Reality chemistry laboratories. *Journal of Science Education and Technology, 30*(6), 862–876. https://doi.org/10.1007/s10956-021-09925-0

Huang, L., Garbaya, S., & Khelladi, S. (2025). Haptic interaction with fluids in virtual environment: An analytical review. *Virtual Reality*, *29*(2), Article 62. https://doi.org/10.1007/s10055-025-01120-5

Jiang, Y., Hamadani, K., Ng, K., Ahmadinia, A., Aquino, A., Palacio, R., Huang, J., Macshane, J., & Hadaegh, A. (2024). Developing scalable hands-on virtual and mixed-reality science labs. *Virtual Reality*, *28*(4), Article 173. https://doi.org/10.1007/s10055-024-01062-4

Karpudewan, M. (2024). Uniting through green chemistry lab a inclusive, responsive pedagogical approach. Experiences from Malaysia. *Sustainable Chemistry and Pharmacy*, *37*, Article 101394. https://doi.org/10.1016/j.scp.2023.101394

Klippel, A., Zhao, J., Jackson, K. L., La Femina, P., Stubbs, C., Wetzel, R., Blair, J., Wallgrün, J. O., & Oprean, D. (2019). Transforming earth science education through immersive experiences: Delivering on a long held promise. *Journal of Educational Computing Research, 57*(7), 1745–1771. https://doi.org/10.1177/0735633119854025

Li, H., Xu, T., Zhang, C., Chen, E., Liang, J., Fan, X., Li, H., Tang, J., & Wen, Q. (2024). Bringing generative AI to adaptive learning in education, 1–14. arXiv:2402.14601. https://arxiv.org/pdf/2402.14601

Lu, Y., Xu, Y., & Zhu, X. (2021). Designing and implementing VR2E2C, a virtual reality remote education for experimental chemistry system. *Journal of Chemical Education, 98*(8), 2720–2725. https://doi.org/10.1021/acs.jchemed.1c00439

Makransky, G., & Petersen, G. B. (2021). The cognitive affective model of immersive learning (CAMIL): A theoretical research-based model of learning in immersive Virtual Reality. *Educational Psychology Review, 33*(3), 937–958. https://doi.org/10.1007/s10648-020-09586-2

Matovu, H., Ungu, D. A. K., Won, M., Tsai, C.-C., Treagust, D. F., Mocerino, M., & Tasker, R. (2023). Immersive virtual reality for science learning: Design, implementation, and evaluation. *Studies in Science Education, 59*(2), 205–244. https://doi.org/10.1080/03057267.2022.2082680

Métois, M., Martelat, J. E., Billant, J., Andreani, M., Escartín, J., Leclerc, F., & the, I. t. (2021). Deep oceanic submarine fieldwork with undergraduate students: an immersive experience with the Minerve software. *Solid Earth, 12*(12), 2789–2802. https://doi.org/10.5194/se-12-2789-2021

Petersen, G. B., Klingenberg, S., & Makransky, G. (2022). Pipetting in virtual reality can predict real-life pipetting performance. *Technology, Mind, and Behavior, 3*(3), 1–16. https://doi.org/10.1037/tmb0000076

Potkonjak, V., Gardner, M., Callaghan, V., Mattila, P., Guetl, C., Petrović, V. M., & Jovanović, K. (2016). Virtual laboratories for education in science, technology, and engineering: A review. *Computers & Education, 95*, 309–327. https://doi.org/10.1016/j.compedu.2016.02.002

Radianti, J., Majchrzak, T. A., Fromm, J., & Wohlgenannt, I. (2020). A systematic review of immersive virtual reality applications for higher education: design elements, lessons learned, and research agenda. *Computers & Education*, *147*, Article 103778. https://doi.org/10.1016/j.compedu.2019.103778

Reeves, S. M., & Crippen, K. J. (2021). Virtual laboratories in undergraduate science and engineering courses: A systematic review, 2009–2019. *Journal of Science Education and Technology, 30*(1), 16–30. https://doi.org/10.1007/s10956-020-09866-0

Reeves, S. M., Bolch, C. A., Bex Ii, R. T., & Crippen, K. J. (2024). Learner experience and motivational beliefs for a VR lab in advanced undergraduate biology. *Interactive Learning Environments, 32*(10), 7156–7169. https://doi.org/10.1080/10494820.2024.2307601

Reeves, S. M., Crippen, K. J., & McCray, E. D. (2021). The varied experience of undergraduate students learning chemistry in virtual reality laboratories. *Computers & Education*, *175*, Article 104320. https://doi.org/10.1016/j.compedu.2021.104320

Shine, S., Warznie, J., Zhou, G., & Zubek, J. (2025). From virtual to reality: Evaluating student attitudes through VR dissection preparations: A use case. *Advances in Physiology Education, 49*(3), 611–620. https://doi.org/10.1152/advan.00170.2024

Song, Y., Wu, K., & Ding, J. (2024). Developing an immersive game-based learning platform with generative artificial intelligence and virtual reality technologies - "LearningverseVR". *Computers & Education: X Reality, 4*, Article 100069. https://doi.org/10.1016/j.cexr.2024.100069

Taylor, M., Abdullah, N. B., Al-Dargazelli, A., Montaner, M. B., Kareem, F., Locks, A., Cao, Z., Bowles, B., Sarraf, J. C., Fajinmi, T., Muwaffak, Z., Beckwith, C., Parkinson, G. N., Waller, Z. A. E., Szulc, B. R., & Hilton, S. T. (2024). Breaking the access to education barrier: Enhancing HPLC learning with virtual reality. *Journal of Chemical Education, 101*(10), 4093–4101. https://doi.org/10.1021/acs.jchemed.4c00540

Taylor, M. V., Muwaffak, Z., Penny, M. R., Szulc, B. R., Brown, S., Merritt, A., & Hilton, S. T. (2025). Optimising digital twin laboratories with conversational AIs: Enhancing immersive training and simulation through virtual reality. *Digital Discovery, 4*(5), 1134–1141. https://doi.org/10.1039/D4DD00330F

Wang, X., Jordan, E., & Lundstrom, M. (2025). Design and evaluation of immersive virtual reality (VR) labs in an online undergraduate physics course. *American Journal of Distance Education*, 1–16. https://doi.org/10.1080/08923647.2025.2472447

Wilkerson, M., Maldonado, V., Sivaraman, S., Rao, R. R., & Elsaadany, M. (2022). Incorporating immersive learning into biomedical engineering laboratories using virtual reality. *Journal of Biological Engineering, 16*(1), Article 20. https://doi.org/10.1186/s13036-022-00300-0

Williams, N. D., Gallardo-Williams, M. T., Griffith, E. H., & Bretz, S. L. (2022). Investigating meaningful learning in virtual reality organic chemistry laboratories. *Journal of Chemical Education, 99*(2), 1100–1105. https://doi.org/10.1021/acs.jchemed.1c00476

Yang, C., Zhang, J., Hu, Y., Yang, X., Chen, M., Shan, M., & Li, L. (2024). The impact of virtual reality on practical skills for students in science and engineering education: a meta-analysis. *International Journal of STEM Education, 11*(1), Article 28. https://doi.org/10.1186/s40594-024-00487-2

**Henry Matovu** is a lecturer in the School of Chemistry at the University of Sydney, Australia. His research interests include laboratory-based learning and the use of emerging technologies, such as interactive and collaborative immersive virtual reality, to support students' visualisation and understanding of abstract chemistry concepts.

**Mihye Won** is an associate professor of science education at the Faculty of Education, Monash University, Australia. Her research focuses on building constructive learning environments through student-generated representations and collaborative immersive virtual reality.

# Chapter 18
# A Systematic Review of Research on the Integration of Real and Virtual Laboratories for Inquiry-Based Learning

**Umesh Ramnarain**

## 18.1 Introduction

Inquiry-based learning (IBL) is widely recognized as a powerful pedagogical approach in science education in fostering active engagement, critical thinking, and deeper conceptual understanding (Crawford, 2014; Pedaste et al., 2015). IBL affords learners the opportunity to engage in scientific practices such as questioning, experimenting, and drawing evidence-based conclusions (Hmelo-Silver et al., 2007). Hands-on experiences whether physical, virtual, or a combination of both are pivotal in enabling such engagement, as they allow students to manipulate variables, observe phenomena, and test hypotheses in controlled settings (Hofstein & Lunetta, 2004).

Recent advances in educational technology have expanded the scope of laboratory-based learning beyond the confines of traditional physical spaces. Virtual laboratories provide interactive, computer-simulated environments that allow for the visualization of abstract scientific concepts, and the safe repetition of experiments (Dunn & Ramnarain, 2020; Rutten et al., 2012). Furthermore, they provide a means by which constraints related to time, cost, and accessibility can be overcome (Deriba et al., 2024). There is growing interest in hybrid laboratory environments that integrate real and virtual experiences. Despite promising findings, questions remain about the optimal design, sequencing, and implementation of hybrid laboratories for IBL. Variations in pedagogical approaches, teacher facilitation, and technological infrastructure influence the extent to which these environments can promote deep learning (Haberbosch et al., 2025). This chapter presents a systematic review of research on the integration of real and virtual laboratories for IBL, critically analysing how such hybrid environments have been designed, enacted, and assessed across diverse educational contexts. It aims to synthesize evidence on the pedagogical affordances

U. Ramnarain (✉)
University of Johannesburg, Johannesburg, South Africa
e-mail: uramnarain@uj.ac.za



L. Leite et al. (eds.), *Practical Work in Science Education in the Digital Era*, Advances in Technology-Rich Science Education 2, https://doi.org/10.1007/978-3-032-13788-3_18

and challenges of integration, the roles of teachers and learners, and the implications for promoting 21st-century competencies in science education.

## 18.2 Inquiry-Based Learning and Virtual Learning Experiences

Inquiry-Based Learning (IBL) is a learner-centred pedagogical approach that emphasizes the active construction of knowledge through the processes of questioning, investigating, and drawing evidence-based conclusions (Pedaste et al., 2015). Rooted in constructivist and socio-constructivist learning theories, IBL positions students as active participants in their learning, fostering deep engagement with scientific concepts and practices (Hmelo-Silver et al., 2007). It is recognized as a key strategy for promoting higher-order thinking skills such as critical thinking, creativity and problem-solving that are essential for success in Science, Technology, Engineering, and Mathematics (STEM) fields (Bybee, 2014; NRC, 2012). Within science education, IBL aligns closely with the authentic practices of scientists, encouraging learners to design experiments, collect and analyse data, and develop conceptual understanding through iterative cycles of exploration and reflection (Furtak et al., 2012). The inquiry framework as described by Pedaste et al. (2015) typically involves the phases of orientation, conceptualisation, investigation, conclusion, and discussion, supported by scaffolding that balances student autonomy with teacher guidance.

Virtual labs have emerged as a powerful educational technology capable of transforming inquiry-based learning (IBL) environments by providing immersive, interactive, and context-rich experiences. These labs enable students to actively engage in the processes of exploration, experimentation, and evidence-based reasoning in ways that closely simulate authentic scientific practices (Radianti et al., 2020). Through high levels of sensory immersion and interactivity, virtual experiences can situate learners within complex, otherwise inaccessible environments, ranging from submicroscopic molecular structures to large-scale ecosystems, thus enhancing both engagement and conceptual understanding (Makransky & Mayer, 2022).

A growing body of empirical research highlights the potential of virtual labs to enhance learning outcomes in IBL settings. For example, Stojšić et al. (2022) found that virtual physics experiments improved students' conceptual understanding and problem-solving skills compared to traditional laboratory work. Similarly, Ramnarain and Moosa (2017) demonstrated that virtual science investigations fostered higher levels of inquiry engagement among learners in under-resourced South African schools. Beyond cognitive gains, virtual labs can promote affective outcomes, such as increased motivation, self-efficacy, and collaborative behaviours, which are critical to sustaining inquiry processes (Makransky et al., 2020).

Moreover, virtual labs can address long-standing challenges in implementing IBL, such as limited laboratory resources, safety concerns, and time constraints. Virtual environments eliminate the risk of accidents associated with experimentation involving hazardous materials, and reduce dependency on costly or fragile equipment (Radianti et al., 2020). This is particularly valuable in low-resource contexts, where physical laboratory infrastructure is lacking (Ramnarain & Moosa, 2017).

Despite these advantages, the integration of virtual labs into IBL is not without challenges. High development costs, technical requirements, and the need for teacher professional development can hinder effective adoption (Parong & Mayer, 2018). Furthermore, while virtual labs can provide rich inquiry opportunities, its effectiveness depends on pedagogical design that needs to be structured with clear inquiry goals and scaffolding to prevent cognitive overload (Makransky & Mayer, 2022). Hybrid models that combine virtual with physical laboratory work are increasingly seen as a way to leverage the complementary affordances of different modalities for deeper learning (Stojšić et al., 2022). These considerations underscore the importance of examining how virtual labs, when integrated with physical labs is used in practice to support inquiry-based learning. This is an issue at the heart of this systematic review which critically analyses research on the integration of real and virtual laboratories to optimize learning outcomes.

The questions that are addressed in this systematic review are:

- What pedagogical sequencing has been used that combines real and virtual laboratories for IBL?
- What are the reported affordances of integrating real and virtual laboratories for IBL?
- What are the reported challenges of integrating real and virtual laboratories for IBL?
- What roles do teachers play in IBL within hybrid laboratory environments?

## 18.3 Methodology

### *18.3.1 Search Strategy*

A search strategy was followed to identify relevant literature on the integration of real and virtual labs for inquiry-based learning. The search strategy involved using keywords that included 'inquiry-based learning' AND 'hybrid laboratory' OR 'hybrid practical work' OR 'real and virtual labs' The search was conducted from ERIC (Education Resources Information Center), Education Source (EBSCO) and Scopus databases for the period 2010 to 2025.

### *18.3.2 Selection*

The selection criteria were based on the Preferred Reporting Items for Systematic Reviews and Meta-Analyses (PRISMA) statement (Moher et al., 2009).

Papers reviewed in this study were selected based on the following inclusion criteria:

- Empirical studies relating to science education;
- Studies published in referenced or peer-reviewed articles and written in the English language;
- Studies published from January 2010 to July 2025.

Exclusion criteria included:

- Publications in books, conferences, and grey literature (opinion pieces, technical reports, blogs, dissertations and theses, etc.);
- Not focused on science education;
- Not published in reputable (i.e. peer-reviewed) sources;
- Not published in English.

### *18.3.3 Search Process*

Following the PRISMA guidelines, an initial comprehensive database search yielded 101 publications relevant to real and virtual labs in science education. After removing duplicate records, 78 publications remained for further screening. Titles and abstracts were then assessed against the inclusion and exclusion criteria established for the review. This screening process led to the exclusion of publications that did not meet the eligibility requirements, such as those without connections to science education. After a full-text review and eligibility assessment, a total of 18 articles were retained for final inclusion in the systematic review.

The selected articles reviewed in this study are provided in Table 18.1.

## 18.4 Results

The results of the review of articles are presented according to the focus of each research question formulated.

### *18.4.1 Pedagogical Sequencing that Combines Real and Virtual Laboratories for IBL*

Based on the reviewed articles, several pedagogical sequencing strategies were employed when combining real and virtual laboratories for inquiry-based learning (IBL). These can be synthesized as follows:

*Blended laboratories (Integrated Use)*

**Table 18.1** Capture of articles meeting selection criteria

| No | Year | Author/s | Title | Journal | Context | Virtual lab software used |
|---|---|---|---|---|---|---|
| 1 | 2013 | Ajredini, F., Izairi, N., & Zajkov, O | Real experiments versus PhET simulations for better high-school students' understanding of electrostatic charging | European Journal of Physics Education | This research investigated the influence of computer simulations (virtual experiments) and real experiments on the conceptual understanding of electrical charging amongst 10th grade students of three gymnasiums in Macedonia | PhET |
| 2 | 2012 | Chini, J.J., Madsen, A., Gire, E., Rebello, N.S., & Puntambek, S | Exploration of factors that affect the comparative effectiveness of physical and virtual manipulatives in an undergraduate laboratory | Physics Education Research | This research explores how university-level non-science majors' understanding of physics concepts related to pulleys was supported by experimentation with real pulleys and a computer simulation of pulleys | Custom computer simulation of pulleys |
| 3 | 2014 | Darrah, M., Humbert, R., Finstein, J., Simon, M., & Hopkins, J | Are virtual labs as effective as hands-on labs for undergraduate physics? A comparative study at two major universities | *Journal of* Science Education and Technology | This study conducted at two major U.S. universities, compared the effectiveness of virtual labs and traditional hands-on labs in supporting undergraduate physics learning, both when virtual labs replace and when they supplement physical labs | Virtual Physics Lab |
| 4 | 2010 | Farrokhnia, M., & Esmailpour, A | A study on the impact of real, virtual and comprehensive experimenting on students' conceptual understanding of DC electric circuits and their skills in undergraduate electricity laboratory | *Procedia—Social and Behavioral Sciences* | This research at an Iranian university investigated the impact of real-only, virtual-only, and integrated (comprehensive) laboratory experiences on students' conceptual understanding of DC circuits and on their ability to carry out practical circuit-related tasks | Virtual circuit builder |

(continued)

**Table 18.1** (continued)

| No | Year | Author/s | Title | Journal | Context | Virtual lab software used |
|---|---|---|---|---|---|---|
| 5 | 2023 | Flegr, S., Kuhn, J., & Scheiter, K | When the whole is greater than the sum of its parts: Combining real and virtual experiments in science education | Computers & Education | This study that focused on German secondary school physics lessons on optics, investigated whether combining real and virtual experiments enhanced conceptual understanding compared to real-only instruction and whether the sequence of integration mattered | PhET and Go-Lab |
| 6 | 2019 | Gamo, J | Assessing a virtual laboratory in optics as a complement to on-site teaching | IEEE Transactions on education | This study assesses the learning experience of engineering students taking a virtual lab as a complement to on-site, traditional lab | OPTILAB |
| 7 | 2018 | Gumilar, S., Ismail, A., Budiman, D.M., & Siswanto, S | Inquiry instructional model infused blended experiment: helping students enhance critical thinking skills | Journal of Physics | This study examined the implementation of an inquiry instructional model infused blended experiment to enhance critical thinking skills of high school Indonesian students | Not specified |
| 8 | 2025 | Haberbosch, M., Deiters, M., & Schaal, S | Combining virtual and hands-on lab work in a blended learning approach on molecular biology methods and lab safety for lower secondary education students | Education Sciences | This study of German lower secondary biology classes examined whether combining virtual and real labs improves students' conceptual knowledge and practical experimental skills compared to using either alone | Desktop virtual simulation of molecular biology |

(continued)

**Table 18.1** (continued)

| No | Year | Author/s | Title | Journal | Context | Virtual lab software used |
|---|---|---|---|---|---|---|
| 9 | 2023 | Hurtado-Bermúdez, S., & Romero-Abrio, A | The effects of combining virtual laboratory and advanced technology research laboratory on university students' conceptual understanding of electron microscopy | Interactive Learning Environments | The aim of this study was to analyse the impact of combinations of virtual laboratory and a research laboratory on 4th year Spanish Forensic Physics students' learning process, and to examine the effect on their interest towards the subject | My Scope™ SEM Simulator |
| 10 | 2019 | Kapici, H. O., Akcay, H., & de Jong, T | Using hands-on and virtual laboratories alone or together—Which works better for acquiring knowledge and skills? | Journal of Science Education and Technology | In Turkish middle school science classes, this study investigated the effects of combining physical and virtual labs on students' academic achievement and attitudes toward science | Go-Labs |
| 11 | 2023 | Lestari, D. P., Supahar, Paidi, & Suwarjo, H | Effect of science virtual laboratory combination with demonstration methods on lower-secondary school students' scientific literacy ability in a science course | Education and Information Technologies | Conducted with Indonesian lower-secondary students, this quasi-experimental study compared virtual lab-plus-demonstration, virtual-only, and demonstration-only approaches in enhancing scientific literacy | Custom-developed science virtual laboratory system |
| 12 | 2023 | Mihret, A. T., Laja, T., & Tesema, A. G | Effectiveness of blended physics laboratory experimentation on pre-service physics teachers' understanding of the nature of science | Pedagogical Research | This study examined the effectiveness of blended physics laboratory experimentation on Ethiopian pre-service physics teachers' (PSPTs') understanding of the nature of science (NOS) during an electricity and magnetism laboratory course | Custom-developed virtual simulation environment |

(continued)

**Table 18.1** (continued)

| No | Year | Author/s | Title | Journal | Context | Virtual lab software used |
|---|---|---|---|---|---|---|
| 13 | 2011 | Olympiou, G., & Zacharia, Z. C | Blending physical and virtual manipulatives: An effort to improve students' conceptual understanding through science laboratory experimentation | Science Education | This study aimed to investigate the effect of experimenting with physical manipulatives (PM), virtual manipulatives (VM), and a blended combination of PM and VM on Cyprian undergraduate students' understanding of concepts in the domain of Light and Color | Custom computer-based simulations |
| 14 | 2020 | Shivam, S., & Wagoner, K | How well do remote labs work? A case study at Princeton University | arXiv | In an introductory physics course at Princeton during COVID-19, this study compared recorded, applet-based, and at-home remote labs to in-person labs to evaluate impacts on learning and student engagement | PhET |
| 15 | 2024 | Tokatlidis, C., Tselegkaridis, S., Rapti, S., Sapounidis, T., & Papakostas, D | Hands-on and virtual laboratories in electronic circuits Learning-Knowledge and skills acquisition | Information | This study explored the effect of the hands-on and virtual laboratories on 152 undergraduate Greek students' conceptual knowledge, inquiry, and measurement skills acquisition in the domain of operational amplifiers (op-amps) circuit learning | TINA-TI v9 |
| 16 | 2014 | Toth, E.E., Ludvico, L.R., & Morrow, B.L | Blended inquiry with hands-on and virtual laboratories: the role of perceptual features during knowledge construction | Interactive Learning Environments | This study examined the characteristics of virtual and hands-on inquiry environments for the development of blended learning in bionanotechnology at a university in the United States | 2D virtual laboratory |

(continued)

**Table 18.1** (continued)

| No | Year | Author/s | Title | Journal | Context | Virtual lab software used |
|---|---|---|---|---|---|---|
| 17 | 2018 | Wang, T-L, & Tseng, Y-K | The comparative effectiveness of physical, virtual, and virtual-physical manipulatives on third-grade students' science achievement and conceptual understanding of evaporation and condensation | Int J of Sci and Math Educ | This study investigated the relative effectiveness of experimenting with physical manipulatives alone, virtual manipulatives alone, and virtual preceding physical manipulatives (combination environment) on third-grade Taiwanese students' science achievement and conceptual understanding in the domain of state changes of water, focusing on the concepts of evaporation and condensation | Custom molecular simulation-based virtual manipulatives |
| 18 | 2014 | Zacharia, Z.C., & de Jong, T | The effects on students' conceptual understanding of electric circuits of introducing Virtual Manipulatives Within a Physical Manipulatives-Oriented Curriculum | Cognition and Instruction | This study investigated whether Virtual Manipulatives (VM) within a Physical Manipulatives (PM)- oriented curriculum affect conceptual understanding of electric circuits and related experimentation among processes among undergraduates in a preservice elementary school teacher program in Cyprus | Custom-built virtual manipulatives |

In the study by Gumilar et al. (2019) both real and virtual experiments we infused into an inquiry instructional model for high school physics as blended activities within the same lesson. This approach led to the highest gains in critical thinking compared to real-only or virtual-only inquiry sessions, emphasizing the complementary strengths of both methods. Hurtado-Bermúdez and Romero-Abrio (2023) report on a pedagogical model that integrated virtual simulations with advanced technology lab visits in a 4th year forensic physics course. This combination yielded significantly higher conceptual understanding than either alone. Haberbosch et al. (2025) implemented a full blended learning cycle that desktop virtual simulations with hands-on labs in molecular biology of German lower secondary classes. Similarly, in the study by Farrokhnia and Esmailpour (2010), university physics students switched between real and virtual labs to benefit from the strengths of each format.

*Virtual-Then-Real (V → R) Sequencing*

This sequencing introduces students first to virtual environments, typically to lower cognitive load, allow experimentation with minimal risk, and support conceptual understanding. It is then followed by real lab experiences to reinforce procedural skills and deepen understanding.

In the study by Gamo (2019), engineering study engaged with OPTILAB simulations as preparatory tools before real lab work. This model was found effective in helping students anticipate results and reduce errors, thereby maximizing real lab productivity and learning gain. Toth et al. (2012) implemented a blended inquiry sequence in which students first engaged with a 2D virtual lab to familiarize themselves with experimental variables and data interpretation. This was followed by hands-on investigation to enhance procedural and perceptual understanding. Zacharia and de Jong (2014) found that students who used virtual manipulatives before physical ones for complex electric circuits developed stronger conceptual models. The virtual lab made abstract phenomena (e.g., current flow) visible, facilitating better transfer to hands-on settings. Olympiou and Zacharia (2012) showed that students benefited from virtual pre-training before hands-on experiments in optics, especially for visualizing light paths and correcting misconceptions prior to physically engaging with equipment.

*Real-Then-Virtual (R → V) Sequencing*

In some cases, studies prioritized initial real-world engagement to ground student learning in tangible experiences, before using virtual tools to consolidate knowledge or explore advanced variations. Ajredini et al. (2014) started with hands-on lab work, followed by virtual simulations that enabled repeated trials and visualization of otherwise invisible processes, supporting hypothesis refinement and iterative inquiry. Darrah et al. (2014) reported a study where students engaged in real labs followed by virtual simulations as review or reinforcement activities.

### 18.4.2 *Affordances of Integrating Real and Virtual Laboratories for IBL*

The studies emphasize that real and virtual laboratories have distinct, complementary affordances, and their integration enhances learning more than either format alone. Real Labs offer tactile, hands-on experiences that foster practical, psychomotor, and procedural skills (Haberbosch et al., 2025; Kapici et al., 2019), while virtual labs enable visualization of abstract concepts, provide safe and repeatable experimentation, reduce cognitive load, and support students in hypothesis testing and data analysis (Chini et al., 2012; Farrokhnia & Esmailpour, 2010; Flegr et al., 2023). However, it was found that combining real and virtual labs was more effective that either of them.

One affordance that was extensively reported was that blending the two labs fostered students' conceptual understanding. Dual formats help students move between the abstract and concrete, enhancing conceptual coherence (Darrah et al., 2014; Hurtado-Bermúdez & Romero-Abrio, 2023). Farrokhnia and Esmailpour (2010) showed that the combination of real and virtual labs allowed students to better visualize abstract concepts, which enhanced they conceptual understanding of DC circuits. Virtual labs in combination with real labs allow students to visualize abstract or unobservable phenomena, such as electric current flow or wave propagation, which supports deeper conceptual understanding (Zacharia & de Jong, 2014). Virtual labs when blended with real labs often reduce cognitive load by abstracting and simplifying complex systems, enabling students to focus on core concepts without being overwhelmed by procedural details (Toth et al., 2012).

The articles did not explicitly address how the features of the software tools might have influenced the findings.

### 18.4.3 *Challenges in Integrating Real and Virtual Laboratories for IBL*

One of the major challenges flagged by the studies was that teachers were not adequately prepared to optimally leverage in their instruction the benefits of integrating real and virtual labs. Flegr et al. (2023) and Gumilar et al. (2019) emphasized that the success of hybrid labs depends heavily on effective instructional design and teacher facilitation. Teachers may not be adequately prepared to blend real and virtual lab elements effectively for IBL, limiting the potential benefits. Teachers often lack adequate training in designing and implementing blended lab experiences as discussed by Muilwijk and Lazonder (2023) and Lestari et al. (2023) in their respective studies. They remark that without professional development or robust instructional guidelines, the integration of real and virtual labs may become superficial or ineffective. Similarly, Wang and Tseng (2018) stress the importance of professional development. Olympiou and Zacharia (2012) stress the importance of aligning physical (PM) and virtual manipulatives (VM) with their unique affordances. They highlight major challenge is the lack of a guiding framework to help instructors combine real and virtual labs in a pedagogically meaningful way.

Another challenge revealed that was related to poor pedagogical sequencing of real and virtual labs was cognitive overload for students. Kapici et al. (2019) note that using hands-on and virtual labs in combination can improve inquiry skills and conceptual knowledge, but students may struggle if the sequence is not optimally designed. Cognitive overload can occur when students are required to switch between lab types without adequate scaffolding or conceptual coherence. Studies by Wang and Tseng (2018), and Olympiou and Zacharia (2012) reveal that without proper scaffolding, learners, especially at the lower level of schooling may struggle to integrate knowledge from physical and virtual contexts. This can lead to persistent misconceptions and cognitive overload, especially when switching between macroscopic physical setups and microscopic virtual representations.

A further challenge was resource and infrastructure constraints. Darrah et al. (2014) observed that while virtual labs can reduce costs and setup time, many institutions still struggle with maintaining both real and virtual environments due to budget and staffing constraints. Farrokhnia and Esmailpour (2010) highlight that unequal access to technology can limit the equitable implementation of virtual components. Shivam and Wagoner (2020) refer to equity issues in remote or virtual labs, where students' access to internet connectivity, devices, and physical space affects the consistency and fairness of lab-based learning opportunities.

These findings suggest that teacher professional development should be explicitly structured to address these challenges. Such support should go beyond tool training to include collaborative design workshops where teachers learn to develop parallel and complementary tasks across modalities. Practice-based professional development could also model how to provide adaptive scaffolding, offering step-by-step guidance at the start of hybrid investigations and gradually fading support as students gain independence. Furthermore, professional development should incorporate troubleshooting and contingency planning for infrastructure and access constraints. Anchoring teacher professional learning in these specific challenges will better equip teachers to leverage the complementary affordances of real and virtual laboratories to support inquiry-based learning.

### *18.4.4 Roles of the Teacher in Inquiry-Based Learning Within Hybrid Laboratory Environments*

One of the most important teacher responsibilities in hybrid IBL is designing and sequencing the learning experience. Gumilar et al. (2019) illustrate how teachers deliberately structured the inquiry cycle by sequencing real and virtual experiments to leverage their complementary strengths. For example, a real experiment might be used to introduce tangible manipulation and observation, while a virtual simulation could extend the investigation by enabling repeated trials or the manipulation of otherwise inaccessible variables. This sequencing ensures that each modality plays a distinct and reinforcing role in the inquiry process. Similarly, Hurtado-Bermúdez and Romero-Abrio (2023) designed parallel activities in both real and virtual lab settings so that students could work with equivalent problems and tasks across modalities. This alignment allowed for meaningful transfer of learning from one environment to the other.

Throughout the inquiry process, teachers act as scaffolders of thinking and skill development. In real sessions, as shown in Hurtado-Bermúdez and Romero-Abrio (2023), instructors modelled laboratory practices from sample preparation to instrument operation, thus making tacit procedural knowledge visible. In virtual sessions, they provided structured, step-by-step assignments that mirrored the real lab tasks, enabling students to manipulate variables in a risk-free yet authentic environment. Olympiou and Zacharia (2012) found that teachers' structured guidance during both real and virtual experiments was crucial for helping students make conceptual connections between the two environments. Teachers actively prompted students to interpret data, reflect on experimental outcomes, and reconcile differences between real and simulated results. Similarly, Muilwijk and Lazonder (2023) emphasise the role of adaptive scaffolding, where teachers adjust their level of guidance in response to student progress by offering more direct support during initial trials and gradually reducing it as learners gain independence in handling both real and virtual apparatus.

## 18.5 Discussion

The synthesis of findings across the reviewed studies demonstrates that integrating real and virtual laboratories in IBL environments offers substantial pedagogical value, but its success hinges on thoughtful design, sequencing, and teacher facilitation. Hybrid approaches consistently outperformed single-format instruction in promoting conceptual understanding, procedural competence, and inquiry skills. This aligns with prior research showing that combining the tactile authenticity of real labs with the visualization and flexibility of virtual environments leverages their complementary affordances (Hofstein & Lunetta, 2004; Makransky & Mayer, 2022).

A key insight from the review is the centrality of sequencing strategies in determining learning outcomes. Virtual-first approaches often reduced cognitive load, addressed safety concerns, and prepared students for more efficient real-lab engagement. Conversely, real-first approaches grounded learning in tangible experiences before using virtual tools for deeper exploration and consolidation. Integrated, within-session blends also proved effective for fostering critical thinking and supporting iterative inquiry, especially when carefully aligned to inquiry phases (Pedaste et al., 2015). However, poorly designed transitions between modalities risked cognitive overload, particularly when abstract virtual representations were not sufficiently scaffolded.

Teacher roles emerged as a critical determinant of hybrid lab success. Teachers functioned as designers of the inquiry pathway, aligning lab formats to learning objectives, and as scaffolders who mediated conceptual connections between environments. Adaptive guidance that provides structure early and gradually releasing responsibility helped students transfer skills and understanding across modalities. Yet, several studies highlighted gaps in teacher preparation, noting that without targeted professional development, hybrid lab use can become superficial and fail to exploit each modality's strengths.

The review also underscores contextual challenges. Resource constraints, unequal access to technology, and infrastructure limitations can hinder equitable adoption, especially in under-resourced settings. Additionally, the lack of established pedagogical frameworks for integrating real and virtual labs leaves many teachers to rely on trial-and-error approaches. Addressing these barriers will require sustained investment in teacher training, institutional support, and research into context-sensitive integration models.

## 18.6 Conclusion and Ways Forward

Overall, the evidence suggests that hybrid laboratory environments when well-sequenced, purposefully scaffolded, and supported by trained teachers can significantly enhance the depth and quality of inquiry-based learning. Future research should focus on refining design frameworks, exploring long-term impacts on 21st-century competencies, and developing scalable professional development programs to ensure that the benefits of hybrid labs are realized across diverse educational contexts.

While this review affirms the educational potential of hybrid laboratory environments, several gaps in the literature point to promising avenues for further investigation. There is a need for robust pedagogical frameworks that guide the sequencing, scaffolding, and assessment of real and virtual laboratories within inquiry-based learning. Future research should develop and validate such models, ensuring they are adaptable to diverse curricular contexts and resource settings. Most existing research focuses on short-term learning gains. Longitudinal studies could provide insights into the sustained effects of hybrid labs on students' conceptual retention, inquiry skills, and transfer of knowledge to novel contexts.

Further studies should examine how different professional development approaches such as mentoring, collaborative design, and practice-based workshops can equip teachers to effectively facilitate hybrid IBL. The findings of the reviewed studies point to promising practices such as designing parallel activities across real and virtual settings, using step-by-step scaffolded tasks to guide data interpretation and reflection, and providing adaptive facilitation that fades support as students gain independence. Future research could build on these practices by developing and testing design frameworks that teachers can use to better align real and virtual inquiry tasks, and by investigating how different scaffolding approaches influence students' conceptual gains and inquiry processes. Understanding how teacher expertise evolves over time will inform scalable training programs. Given disparities in infrastructure and technology access, particularly in under-resourced settings, future research should explore cost-effective and context-sensitive strategies for implementing hybrid labs. This includes investigating how mobile, low-cost, or open-source virtual tools can be paired with existing real lab resources. By addressing these areas, future research can move the field toward evidence-based, scalable, and equitable models for integrating real and virtual laboratories in inquiry-based science education.

While the review highlights that teacher roles are pivotal in hybrid laboratory environments, there is limited research on how to equip teachers to enact these roles effectively. The findings of the reviewed studies point to promising practices—such as designing parallel activities across real and virtual settings, using step-by-step scaffolded tasks to guide data interpretation and reflection, and providing adaptive facilitation that fades support as students gain independence. Future research could build on these practices by developing and testing design frameworks that teachers can use to align real and virtual inquiry tasks, and by investigating how different scaffolding approaches influence students' conceptual gains and inquiry processes. Longitudinal and design-based research could also examine how teachers' hybrid-lab facilitation skills evolve over time, and which professional development models (e.g., collaborative design teams, mentoring, or practice-based workshops) most effectively build these skills. Such studies would contribute to scalable, evidence-informed models of teacher preparation for hybrid inquiry-based laboratories.

## References

Ajredini, F., Izairi, N., & Zajkov, O. (2014). Real experiments versus PhET simulations for better high-school students' understanding of electrostatic charging. *European Journal of Physics Education, 5*(1), 59–69. https://files.eric.ed.gov/fulltext/EJ1051517.pdf

Bybee, R. W. (2014). *The BSCS 5E instructional model: Creating teachable moments.* NSTA Press. https://my.nsta.org/resource/100197/the-bscs-5e-instructional-model-creating-teachable-moments-e-book

Crawford, B. A. (2014). From inquiry to scientific practices in the science classroom. In N. Lederman & S. Abell (Eds.), *Handbook of research on science education* (Vol. II, pp. 529–556). Routledge. https://doi.org/10.4324/9780203097267

Chini, M., Madsen, A., Gire, E., Rebello, N. S., & Puntambekar, S. (2012). Exploration of factors that affect the comparative effectiveness of physical and virtual manipulatives in an undergraduate laboratory. *Physical Review Special Topics - Physics Education Research, 8*(1), Article 010113. https://doi.org/10.1103/PhysRevSTPER.8.010113

Darrah, M., Humbert, R., Finstein, J., Simon, M., & Hopkins, J. (2014). Are virtual labs as effective as hands-on labs for undergraduate physics? A comparative study at two major universities. *Journal of Science Education and Technology, 23*(6), 803–814. https://doi.org/10.1007/s10956-014-9513-9

Deriba, F. G., Saqr, M., & Tukiainen, M. (2024). Assessment of accessibility in virtual laboratories: A systematic review. *Frontiers in Education, 9*, 1351711. https://doi.org/10.3389/feduc.2024.1351711

Dunn, J., & Ramnarain, U. (2020). The effect of simulation-supported inquiry on South African natural sciences learners' understanding of atomic and molecular structures. *Education Sciences, 10*(10), 280. https://doi.org/10.3390/educsci10100280

Farrokhnia, M., & Esmailpour, A. (2010). A study on the impact of real, virtual and comprehensive experimenting on students' conceptual understanding of DC electric circuits and their skills in the procedures of experiments. *Procedia - Social and Behavioral Sciences, 2*(2), 5474–5479. https://doi.org/10.1016/j.sbspro.2010.03.888

Flegr, J., Kuhn, J., & Scheiter, K. (2023). When the whole is greater than the sum of its parts: Combining real and virtual experiments for in-depth knowledge acquisition. *Computers & Education, 197*, 1–15. https://doi.org/10.1016/j.compedu.2023.104745

Furtak, E. M., Seidel, T., Iverson, H., & Briggs, D. C. (2012). Experimental and quasi-experimental studies of inquiry-based science teaching: A meta-analysis. *Review of Educational Research, 82*(3), 300–329. https://doi.org/10.3102/0034654312457206

Gamo, J. (2019). Assessing a virtual laboratory in optics as a complement to on-site teaching. *IEEE Transactions on Education, 62*(2), 119–126. https://doi.org/10.1109/TE.2018.2871617

Gumilar, S., Ismail, A., Budiman, D. M., & Siswanto, S. (2019). Inquiry instructional model infused blended experiment: Helping students enhance critical thinking skills. *Journal of Physics: Conference Series, 1157, Article* 032009. https://doi.org/10.1088/1742-6596/1157/3/032009

Haberbosch, A., Deiters, J., & Schaal, S. (2025). Combining virtual and hands-on labwork in a blended learning approach on molecular biology methods and lab safety for lower secondary education students. *Education Sciences, 15*(2), 123. https://doi.org/10.3390/educsci15020123

Hofstein, A., & Lunetta, V. N. (2004). The laboratory in science education: Foundations for the twenty-first century. *Science Education, 88*(1), 28–54. https://doi.org/10.1002/sce.10106

Hmelo-Silver, C. E., Duncan, R. G., & Chinn, C. A. (2007). Scaffolding and achievement in problem-based and inquiry learning: A response to Kirschner, Sweller, and Clark (2006). *Educational Psychologist, 42*(2), 99–107. https://doi.org/10.1080/00461520701263368

Hurtado-Bermúdez, S., & Romero-Abrio, A. (2023). Combining virtual laboratory and an advanced technology research laboratory for the study of electron microscopy. *Learning Environments, 31*(2), 1126–1141. https://doi.org/10.1080/10494820.2020.1821716

Kapici, H. O., Akcay, H., & de Jong, T. (2019). Using hands-on and virtual laboratories alone or together - Which works better for acquiring knowledge and skills? *Journal of Science Education and Technology, 28*, 231–250. https://doi.org/10.1007/s10956-018-9762-0

Lestari, D. P., Supahar, Paidi, & Suwarjo. (2023). Effect of science virtual laboratory combination with demonstration methods on lower-secondary school students' scientific literacy ability in a science course. *Education and Information Technologies, 28,* 16153–16175. https://doi.org/10.1007/s10639-023-11857-8

Makransky, G., Terkildsen, T. S., & Mayer, R. E. (2019). Adding immersive virtual reality to a science lab simulation causes more presence but less learning. *Learning and Instruction, 60*, 225–236. https://doi.org/10.1016/j.learninstruc.2017.12.007

Makransky, G., Petersen, G. B., & Klingenberg, S. (2020). Can an immersive virtual reality simulation increase students' interest and career aspirations in science? *British Journal of Educational Technology, 51*(6), 2079–2097. https://doi.org/10.1111/bjet.12954

Makransky, G., & Mayer, R. E. (2022). Benefits of taking a virtual field trip in immersive virtual reality: Evidence for the immersion principle in multimedia learning. *Education Psychology Review, 34*, 1771–1798. https://doi.org/10.1007/s10648-022-09675-4

Mihret, Z., Alemu, M., & Assefa, S. (2023). Effectiveness of blended physics laboratory experimentation on pre-service physics teachers' understanding of the nature of science. *Pedagogical Research, 8*(1), Article em0144. https://doi.org/10.29333/pr/12607

Moher D., Liberati A., Tetzlaff J., & Altman D. G. (2009). The PRISMA group (2009) Preferred reporting items for systematic reviews and meta-analyses: The PRISMA statement. *PLoS Med, 6*(7), Article e1000097. https://doi.org/10.1371/journal.pmed.1000097

Muilwijk, N., & Lazonder, A. W. (2023). Learning from physical and virtual investigation in school science: A meta-analysis. *Frontiers in Education, 8*, Article 1163024. https://doi.org/10.3389/feduc.2023.1163024

National Research Council. (2012). *A framework for K-12 science education: Practices, crosscutting concepts, and core ideas*. National Academies Press. https://doi.org/10.17226/13165

Olympiou, G., & Zacharia, Z. C. (2012). Blending physical and virtual manipulatives: An effort to improve students' conceptual understanding through science laboratory experimentation. *Science Education, 96*(1), 21–47. https://doi.org/10.1002/sce.20463

Parong, J., & Mayer, R. E. (2018). Learning science in immersive virtual reality. *Journal of Educational Psychology, 110*(6), 785–797. https://doi.org/10.1037/edu0000241

Pedaste, M., Mäeots, M., Siiman, L. A., de Jong, T., van Riesen, S. A. N., Kamp, E. T., Manoli, C. C., Zacharia, Z. C., & Tsourlidaki, E. (2015). Phases of inquiry-based learning: Definitions and the inquiry cycle. *Educational Research Review, 14*, 47–61. https://doi.org/10.1016/j.edurev.2015.02.003

Radianti, J., Majchrzak, T. A., Fromm, J., & Wohlgenannt, I. (2020). A systematic review of immersive virtual reality applications for higher education: Design elements, lessons learned, and research agenda. *Computers & Education, 147*, Article 103778. https://doi.org/10.1016/j.compedu.2019.103778

Ramnarain, U., & Moosa, S. (2017). The use of simulations in correcting electricity misconceptions of grade 10 South African physical sciences learners. *International Journal of Innovation in Science and Mathematics Education, 25*(5), 1–20.

Rutten, N., van Joolingen, W. R., & van der Veen, J. T. (2012). Learning effects of computer simulations in science education. *Computers & Education, 58*(1), 136–153. https://doi.org/10.1016/j.compedu.2011.07.017

Shivam, S., & Wagoner, K. (2020). *How well do remote labs work? A case study at Princeton University*. arXiv:2008.04499. https://doi.org/10.48550/arXiv.2008.04499

Stojšić et al. (2022). Students' acceptance of mobile augmented reality applications in primary and secondary biology education. *International Journal of Cognitive Research in Science, Engineering and Education, 10*(3), 129–138.

Tokatlidis, I., Tselegkaridis, S., Rapti, E., Sapounidis, T., & Papakostas, G. A. (2024). Hands-on and virtual laboratories in electronic circuits learning - Knowledge and skills acquisition. *Information, 15*, 672. https://doi.org/10.3390/info15110672

Toth, E. E., Ludvico, L. R., & Morrow, B. L. (2012). Blended inquiry with hands-on and virtual laboratories: The role of perceptual features during knowledge construction. *Interactive Learning Environments, 22*(5), 614–630. https://doi.org/10.1080/10494820.2012.693102

Wang, T.-L., & Tseng, Y.-K. (2018). The comparative effectiveness of physical, virtual, and virtual-physical manipulatives on third-grade students' science achievement and conceptual understanding of evaporation and condensation. *International Journal of Science and Mathematics Education, 16*(2), 203–219. https://doi.org/10.1007/s10763-016-9774-2

Zacharia, Z. C., & de Jong, T. (2014). The effects on students' conceptual understanding of electric circuits of introducing virtual manipulatives within a physical manipulatives-oriented curriculum. *Cognition and Instruction, 32*(2), 101–158. https://doi.org/10.1080/07370008.2014.887083

**Umesh Ramnarain** is a professor of science education, and Director of the Centre for Advanced Learning Technologies in Science, Technology, Engineering, Arts and Mathematics (CALT-STEAM) research centre at the University of Johannesburg, South Africa. His main research interest is in inquiry-based science education, with a particular focus on its uptake in disadvantaged educational settings. He is associate editor of the Research in Science Education journal.

# Part IV
# Inclusion Through Practical Work

# Chapter 19
# Promoting Inclusion in Science Education through Low-Cost Practical Work

**Katerina Salta, Katerina Paschalidou, and Dionysios Koulougliotis**

## 19.1 Introduction

Social justice in education has been an important issue which is under active debate in the research literature (Francis et al., 2017). Ensuring inclusion of all learners in the educational process is a matter that "speaks directly to the issue of social justice" (Opertti et al., 2014). More specifically, in science education one of the main barriers to achieving inclusion is related to the elevated cost of conventional equipment and materials which are required for the execution of practical work. For this purpose, the possibility of producing and/or using low-cost equipment and materials by the educators or even the students themselves has been examined (Akuma & Callaghan, 2016).

The aim of this chapter is to examine how inclusion in science education may be promoted via low-cost methods in practical work by critical examination of indicative research literature and by concentrating on the case of chemistry as a subject where practical work plays a central role.

K. Salta (✉) · K. Paschalidou
Department of Chemistry, National and Kapodistrian University of Athens, Athens, Greece
e-mail: ksalta@chem.uoa.gr

K. Paschalidou
e-mail: kpaschalidou@chem.uoa.gr

D. Koulougliotis
Department of Environment, Ionian University, Zakynthos, Greece
e-mail: dkoul@ionio.gr



L. Leite et al. (eds.), *Practical Work in Science Education in the Digital Era*, Advances in Technology-Rich Science Education 2, https://doi.org/10.1007/978-3-032-13788-3_19

## 19.2 Inclusion in Science Education

As the *inclusion* concept has evolved during the last 70 years (Opertti et al., 2014), various definitions of inclusion have been reported in literature, indicating different perspectives of both inclusion and inclusive education (Ainscow et al., 2006). According to this evolution, four core ideas to conceptualize the inclusive education were developed. The rights-based approach of educational practice was the initial idea. Then, inclusive education was focused on improving the learning conditions of students with special needs. In the next step, inclusive education was widening its scope to all marginalized children. In our days, inclusive education aims to improve the entire education system to provide quality "education for all" (Opertti et al., 2014). Education for all is one of the six ways of thinking about the concept of inclusion that Ainscow et al. (2006) described as increasing access to, and participation within education for all, within the children's local communities. The concept of inclusive education includes together with school all the other forms of education within communities (e.g., lifelong learning, outreach programs), ascribing to the schools the responsibility to support the education of communities (Ainscow et al., 2006).

During the twenty-first century, inclusive education has increasingly adopted the sociocultural views of learning (Thomas, 2013). According to these views, learning depends on the context and culture for learning, therefore, students' failure at school is not due to learning disabilities, but to a set of factors which relate to acceptance and inclusion. Furthermore, the sociocultural views of learning explain the role of schools as a significant source of variation in students' performance "in poor and unequal countries" (Thomas, 2013). Thus, inclusive education must be based on knowledge regarding the harmful effects of "inequality, relative poverty and contrastive judgment in schools" and its educational practices should also heavily rely on the knowledge about the benefits of "communities of learning and social capital" (Thomas, 2013).

Within the current core idea, inclusive education focuses on both equity that encompasses access, processes and outcomes, and quality that concerns achievement of effective learning outcomes (Opertti et al., 2014). Therefore, promoting inclusion entails changes in the content, approaches, structures, and strategies of the education system (Horna-Saldaña & Canaleta, 2024).

Barton (1998) has reported that science for all students, although it is an egalitarian view of science education, has proved difficult to make feasible for all students, especially those of low-income or coming from low-income countries. These countries often run up against serious economic challenges, including poverty and limited access to education resulting in limited chances to improve the quality of their citizens' lives (Murtadlo et al., 2025).

## 19.3 Practical Work

*Practical work* refers to any science activity where students directly observe and/or manipulate the objects and materials of study. The term 'practical work' is a wider term than 'laboratory work' because it includes the students' observation and manipulation activities that take place in either a school laboratory or an out-of-school setting (e.g., student's home or outreach project). Teacher demonstrations are also included in practical work when students actively participate (Millar, 2010).

Most importantly, in practical work, the learning processes of data collection and interpretation, as well as the inquiry activities that students conduct, retain the characteristics of the tasks executed in "traditional" laboratory work. Practical work is crucial in science education; thus, it covers a huge part of science education literature. Although there is a thorough discussion about the types of learning, the teaching activities and the content knowledge regarding practical work, there seems to be limited concern regarding the difficulties of realizing practical work in science education in countries and in contexts with low economic resources (Oliveira & Bonito, 2023).

## 19.4 Low-Cost Practical Work

Practical work should be engaging all students, but its implementation in resource-constrained settings faces challenges. Especially, high-quality science education requires a variety of instructional resources, (e.g., lab equipment) and their acquiring and maintaining can create challenges, particularly for low-income countries. It has been reported that many schools, in less developed countries but even in industrialized ones, lack essential conventional equipment and materials for practical work (e.g., Ens et al., 2012) because either this equipment is expensive, or science education budgets have decreased (e.g., Set & Kita, 2014).

Additionally, chemistry teachers have identified many barriers to implementing practical work such as tight teaching material budgets, insufficient dedicated rooms regarding equipped material and chemicals (material resources), and large student numbers (Bögge & Lühken, 2021). On the other hand, the academic and future success of students positively relates to material resources and technology (Barrett et al., 2019). Various methods have been used around the world to address these challenges. These methods include using micro-scale experiments (Davila-Diaz, 2024), as well as development of "low-cost practical work" (di Fuccia et al., 2012).

The approach of "low-cost practical work" is referred to inexpensive, readily available materials (like repurposed everyday objects or medical supplies), which fulfill the functions of standard lab equipment and may be supported by common digital equipment. Most of the materials that have been used in the production of science low-cost equipment are initially intended for other purposes and are modified for use in practical work in science education. The low-cost practical work is an approach that has been used in science education for many years (e.g., DeWitt, 1936).

Based on this approach, science teachers or students produce equipment/apparatus and physical models from everyday materials. Subsequently, this equipment together with household materials as reagents, is used in practical work in science classrooms. Moreover, the low-cost practical work is environmentally friendly and provides a connection of science education to the everyday experiences of students (Kyle, 2006).

Unlike micro-scale experiments that primarily focus on minimizing chemical usage (Davila-Diaz, 2024), but usually require the use of specialized lab equipment, low-cost materials aim to make science education accessible and affordable for schools with limited resources and therefore they may promote social inclusion in science.

## 19.5 Low-Cost Practical Work in Chemistry Education

Low-cost practical work methods related to the inclusion of students from countries and/or contexts with limited economic resources in chemistry-related practicals can be categorized with the criterion of the mode via which the cost is lowered. Thus, it may be considered that the cost can be reduced via two main modes: (a) via construction of laboratory equipment/apparatus or simple physical models ("Lowering cost by construction") and (b) via the use of low-cost materials, chemicals or supplies during the conduction of practical work ("Lowering cost during use"). Subsequently, we will present these two categories of low-cost practical work methods via reference to representative works from the research literature. At this point, we note that among the examined indicative publications only two (Mundy & Nokeri, 2024; Tseng et al., 2024) involve investigation of the impact of the corresponding teaching approach.

### *19.5.1 Lowering Cost by Construction*

The development of low-cost laboratory equipment or apparatus has long been recognized (DeWitt, 1936; Jansen-van Vuuren et al., 2013; Srivastava et al., 1985). In addition to the equipment/apparatus the construction of self-created physical models can also be used for cost reduction purposes. These two sub-categories will be presented separately by giving emphasis to the following characteristics: relevant chemistry topic, educational level, type of equipment/apparatus/model, person carrying out the construction, place of use. The low-cost is mainly attained via the use of items which can be recyclable or bought in a standard supermarket.

## 19.6 Low-Cost Laboratory Equipment/Apparatus

The researchers and chemistry educators who are involved in the publications that will be discussed originate from all over the world. This fact is indicative of the worldwide importance of the need for developing low-cost laboratory equipment and/or apparatus in order to enhance inclusion in chemistry education. More specifically, the publications originate from Asia (Guidote et al., 2015; Jarujamrus et al., 2024; Set & Kita, 2014; Supasorn et al., 2021; Suzuki, 1995; Wakabayashi et al., 1998), North and South America (Campanizzi et al., 1999; Cesin-AbouAtme et al., 2021; Dávila-Díaz & Díaz-Vázquez, 2024; Dovidio & Harakas, 2024; Hamer et al., 2021; Turano et al., 2015), Europe (Grønneberg et al., 2006; O'Donoghue et al., 2024), and Oceania (Ng et al., 2020).

Different types of equipment/apparatus may be constructed: CD-ROM spectroscope (Wakabayashi et al., 1998), UV sensor (Dávila-Díaz & Díaz-Vázquez, 2024), apparatus for distillation (Campanizzi et al., 1999), water electrolysis (Suzuki, 1995), measurement of conductivity (Set & Kita, 2014) or fluorescence (Hamer et al., 2021), batteries-galvanic cells (Dovidio & Harakas, 2024; Grønneberg et al., 2006; O'Donoghue et al., 2024), syringe kits for gas collection (Jarujamrus et al., 2024; Ng et al., 2020; Supasorn et al., 2021), magnetic stirrers (Guidote et al., 2015), and spatulas (Turano et al., 2015).

The equipment/apparatus is, in several cases, constructed by the teacher so that it can be used by the students in the laboratory or for demonstration experiments (Campanizzi et al., 1999; Dávila-Díaz & Díaz-Vázquez, 2024; Guidote et al., 2015; Jarujamrus et al., 2024; Supasorn et al., 2021; Suzuki, 1995; Wakabayashi et al., 1998). Alternatively, the construction is administered by the students under the supervision of the teacher in the laboratory (Cesin-AbouAtme et al., 2021; Grønneberg et al., 2006; Ng et al., 2020; Set & Kita, 2014; Turano et al., 2015), or by the students themselves for experimentation at home (Cesin-AbouAtme et al., 2021; Hamer et al., 2021; Turano et al., 2015) or for outreach activities (Dovidio & Harakas, 2024; O'Donoghue et al., 2024).

Regarding the educational level of the different practical applications referred to in the selected publications, it is noted that all of them are suitable for secondary education. Some of them are also suitable for primary education (Campanizzi et al., 1999; Dávila-Díaz & Díaz-Vázquez, 2024; Wakabayashi et al., 1998), tertiary education (Cesin-AbouAtme et al., 2021; Guidote et al., 2015; Hamer et al., 2021; Ng et al., 2020; Supasorn et al., 2021) or for the general public during outreach activities (Dávila-Díaz & Díaz-Vázquez, 2024; Turano et al., 2015).

The chemical topics that are taught via this type of practical work are mainly related to different areas of physical chemistry such as electrochemistry/electrolysis (Cesin-AbouAtme et al., 2021; Dovidio & Harakas, 2024; Grønneberg et al., 2006; O'Donoghue et al., 2024; Suzuki, 1995), chemical kinetics (Jarujamrus et al., 2024; Supasorn et al., 2021), conductivity (Set & Kita, 2014), spectroscopy and fluorescence (Dávila-Díaz & Díaz-Vázquez, 2024; Hamer et al., 2021; Wakabayashi et al., 1998), photochromism and polymerization (Dávila-Díaz & Díaz-Vázquez, 2024), physical processes like distillation, evaporation and condensation (Campanizzi et al., 1999). In addition, we find applications related to food science (Set & Kita, 2014) and monitoring of chemical reactions (Ng et al., 2020).

## 19.7 Self-created Physical Models

Physical models can be very useful for teaching chemical concepts (Justi & Gilbert, 2002), particularly related to chemical structure and molecular properties such as lattice structure, packing arrangements and charge transfer. They can be constructed with inexpensive and easily accessible materials like, for example pom pons (Cady, 1997), waste plastic caps and wooden sticks or steel wires (Zhuo & Liang, 2023), simple electronic components (LEDs) and plastic tubes (Horikoshi et al., 2023). The construction of this model provides a good opportunity for hands on learning and teamwork and can be performed by the students themselves in the laboratory with teacher supervision. The accompanying teaching activities have been or may be applied in tertiary (Cady, 1997; Horikoshi et al., 2023; Zhuo & Liang, 2023) as well as in secondary education (Zhuo & Liang, 2023) with suitable modifications.

### *19.7.1 Lowering Cost During Use*

The reduction of the cost of chemistry practical work may also be achieved via the use of inexpensive materials which are easily accessible in everyday life. They are mostly used as chemical reagents either directly or after some process (for example boiling) but they may have other uses as well (e.g., the use of a kitchen balance as equipment). They can be materials which are either commonly available worldwide or which are indigenous in local communities. These two subcategories will be presented separately by giving emphasis to the following characteristics: relevant chemistry topic, educational level, type of material, place of use.

## 19.8 Low-Cost Materials/Chemicals/Supplies

As seen above for the case of low-cost equipment/apparatus, the researchers and chemistry educators who are involved in the publications to be discussed are also active worldwide. More specifically, the publications (19 in total) originate from North and South America (Baldwin et al., 2010; García-Zavala, 2023; García-Zavala et al., 2025 Johnson et al., 2022; Kuntzleman et al., 2021; Naimi et al., 2024; Nguyen et al., 2023; Palmer et al., 2018; Trujillo, 2005; Valdez et al., 2024; Wright, 2003), Asia (Duangpummet et al., 2022; Im et al., 2023; Kanda et al., 1995; Tseng et al.,

2024), and Europe (Binder & Lämmle, 2019; Costa et al., 2025; Moutsakis et al., 2025; Parkes et al., 2016).

The types of material used in the 19 publications belong mainly to two categories: (a) natural extracts from plant species or juices like anthocyanins, lycopene and β-carotene met in 7 publications (Binder & Lämmle, 2019; Costa et al., 2025; García-Zavala, 2023; Im et al., 2023; Kanda et al., 1995; Moutsakis et al., 2025; Naimi et al., 2024) and (b) whole household items which may or may not food based such as juices, vegetables, oils, ice cream, baking soda, salt, sugar, food dyes, yeast, iron tablets, digital kitchen balance, candles, bleach, acetone, cotton and so on (the remaining 12 publications).

Regarding the educational level of the different practical applications referred to in the selected publications, it is noted that they are suitable for secondary education except for four (García-Zavala et al., 2025; Johnson et al., 2022; Nguyen et al., 2023; Palmer et al., 2018) which are proposed for tertiary education. Some practical applications are proposed for either secondary or tertiary education (García-Zavala, 2023; Im et al., 2023; Kuntzleman et al., 2021), while a few are suggested for both secondary and primary education (Naimi et al., 2024; Valdez et al., 2024; Wright, 2003). Finally, some practical applications are also deemed suitable for the general public during outreach activities (Palmer et al., 2018; Parkes et al., 2016; Valdez et al., 2024).

Regarding the locality where the described practical work can be administered, almost half of the publications (9 out of 19) suggest that the students may also undertake the activities at home (Duangpummet et al., 2022; García-Zavala, 2023; García-Zavala et al., 2025; Im et al., 2023; Johnson et al., 2022; Kuntzleman et al., 2021; Moutsakis et al., 2025; Naimi et al., 2024; Nguyen et al., 2023).

Finally, the chemical topics which are dealt with in the selected publications are related to the following areas: acid–base chemistry (García-Zavala, 2023; Im et al., 2023; Kanda et al., 1995; Kuntzleman et al., 2021; Moutsakis et al., 2025; Naimi et al., 2024; Nguyen et al., 2023), chemical analysis like chromatography and elemental analysis (Baldwin et al., 2010; Binder & Lämmle, 2019; Costa et al., 2025; García-Zavala et al., 2025; Tseng et al., 2024), solution properties and chemistry (Costa et al., 2025; Duangpummet et al., 2022; Johnson et al., 2022), chemical reaction monitoring and kinetics (Costa et al., 2025; Palmer et al., 2018; Trujillo, 2005; Wright, 2003), electrochemistry (Parkes et al., 2016), magnetism (Palmer et al., 2018) and polymer chemistry/waste management (Valdez et al., 2024).

## 19.9 Materials from Local Communities

The incorporation of indigenous knowledge into science and technology promotes inclusion in terms of justice and respect for the values of indigenous peoples (Zidny et al., 2023). More generally, the use of locally available materials (for example plants, fruits, vegetables etc.) in the chemistry laboratory may increase students' learning motivation and involvement and promote meaningful learning (Mundy & Nokeri,

2024). In addition, the materials from local communities are usually more easily available at low-cost. Four characteristic examples from the international literature concern the cases of the use of the Japanese eggplant (Kanda et al., 1995), the native to southern Mexico festive plant poinsettia (García-Zavala, 2023), local food products of South Africa like brandy and pineapple beer (Mundy & Nokeri, 2024) and the application of anthocyanins in Peking opera facial masks and paper cutting (Xu et al., 2025). The practical work may be applied either mainly in secondary (Kanda et al., 1995; Xu et al., 2025) or tertiary education (Mundy & Nokeri, 2024) and in some cases in both (García-Zavala, 2023). Some activities may also be administered at home (García-Zavala, 2023; Xu et al., 2025). The chemistry topics that are dealt with in the four selected publications are related to acid–base chemistry (García-Zavala, 2023; Kanda et al., 1995; Xu et al., 2025), electrolysis (Xu et al., 2025), physical properties of matter and separation (Mundy & Nokeri, 2024) and isomerism/tautomerism (García-Zavala, 2023).

## 19.10 Conclusion and Ways Forward

Low-cost practical work can contribute significantly to the promotion of social justice in science education via fostering inclusion of students originating from countries or social contexts which are less favored economically. In this work, a cohesive presentation of specific examples of low-cost practical work for the case of chemistry education was attempted, via careful examination of indicative research literature. The selected publications originated from researchers and educators residing worldwide and they were categorized by considering two main modes of cost reduction: via construction and during use. Analysis of the publications gave rise to several conclusions as follows. Low-cost practical work may be applied at all educational levels (primary, secondary, tertiary) as well as for the execution of outreach activities suitable for the general public. In some cases, the practical work is performed mainly by the teacher (mainly in the case of equipment construction) but most of the time the active involvement of students is the preferred choice. The low-cost equipment and materials are mostly household-based, a fact which helps students more easily conceive the strong connection between chemistry and everyday life phenomena. In the case of usage of materials originating from the local community or culture, a more holistic teaching approach to chemistry can be adopted which may more effectively foster students' cognitive as well as affective skills and competences. Finally, there is a multitude of chemistry topics related mostly to introductory general, inorganic, physical as well as organic chemistry that can be approached didactically via the use of low-cost equipment and materials.

Based on the current work, it may be deduced that there is a need for continuing support and dissemination of methods for incorporating low-cost equipment and materials in chemistry (and in general science education). In addition, the research community should intensify its efforts for assessing aspects regarding the effectiveness of the application of low-cost practical work in the teaching practice. Finally, since the adoption of low-cost practical work is accompanied by several challenges such as time constraints and teachers' insufficient practical skills, pedagogical knowledge, motivation or even training, the education community must intensify its efforts for their alleviation.

## References


Ainscow, M., Booth, T., & Dyson, A. (2006). *Improving schools, developing inclusion*. Routledge.

Akuma, F. V., & Callaghan, R. (2016). Framework for reducing teaching challenges relating to improvisation of science education equipment and materials in schools. *Eurasia Journal of Mathematics, Science and Technology Education, 12*(10), 2697–2717. https://doi.org/10.12973/eurasia.2016.1305a

Baldwin, B. W., Hasbrouck, S., Smith, J., & Kuntzleman, T. S. (2010). Classroom activity connections: Demonstrating various flame tests using common household materials. *Journal of Chemical Education, 87*(8), 790–792. https://doi.org/10.1021/ed100296s

Barrett, P., Treves, A., Shmis, T., Ambasz, D., & Ustinova, M. (2019). *The impact of school infrastructure on learning: A synthesis of evidence*. International Development in Focus. World Bank Group.

Barton, A. C. (1998). Reframing "science for all" through the politics of poverty. *Educational Policy, 12*(5), 525–541. https://doi.org/10.1177/0895904898012005004

Binder, R., & Lämmle, C. (2019). Reversed phase high performance thin layer chromatography of aqueous samples in student laboratories using the example of anthocyanin patterns from flower petals. *Journal of Chemical Education, 96*(9), 1922–1927. https://doi.org/10.1021/acs.jchemed.8b00900

Bögge, L., & Lühken, A. (2021). Zur schülerwahrnehmung von low-cost-experimentiermaterialien. Die destillation im kleinmaßstab [On students' perception of low-cost experimental equipment. Distillation in small-scale]. *CHEMKON*, *28*(5), 196–202. https://doi.org/10.1002/ckon.202000010

Cady, S. G. (1997). Use of pom pons to illustrate cubic crystal structures. *Journal of Chemical Education, 74*(7), 794–795. https://doi.org/10.1021/ed074p794

Campanizzi, D. R. D., Mason, B., & Hermann, C. K. F. (1999). Distillation apparatuses using household items. *Journal of Chemical Education, 76*(8), 1079–1080. https://doi.org/10.1021/ed076p1079

Cesin-AbouAtme, T., Lopez-Almeida, C. G., Molina-Labastida, G., & Ibanez, J. G. (2021). Light-emitting diodes as voltage generators: Demonstrating the fuel cell principle with low-cost, magnetically enhanced, homemade solar electrolysis. *Journal of Chemical Education, 98*(9), 3045–3049. https://doi.org/10.1021/acs.jchemed.1c00093

Costa, P. A., Sampaio, C. I., & Dias, A. M. (2025). Lycopene and beta-carotene as colorful and versatile tools to illustrate chromatography, colorimetry, alkene-reactions and health-benefits. *Journal of Chemical Education, 102*(2), 829–838. https://doi.org/10.1021/acs.jchemed.4c01261

Davila-Diaz, K. (2024). On-the-go lab for aqueous reactions demonstrations: Activities at the microscale. *Journal of Chemical Education, 101*(5), 2149–2155. https://doi.org/10.1021/acs.jchemed.3c01317

Dávila-Díaz, K., & Díaz-Vázquez, L. M. (2024). Alginate-based UV sensor: A simple and inexpensive tool for educational purposes. *Journal of Chemical Education, 101*(8), 3596–3602. https://doi.org/10.1021/acs.jchemed.4c00291

DeWitt, C. B. (1936). The household comes to the laboratory. *Journal of Chemical Education, 13*(12), 577–580. https://doi.org/10.1021/ed013p577

di Fuccia, D., Witteck, T., Markic, S., & Eilks, I. (2012). Trends in practical work in German science education. *Eurasia Journal of Mathematics, Science and Technology Education, 8*(1), 59–72. https://doi.org/10.12973/eurasia.2012.817a

Dovidio, L., & Harakas, G. N. (2024). A new design of the air-aluminum battery, optimized for STEM activities. *Journal of Chemical Education, 101*(4), 1747–1749. https://doi.org/10.1021/acs.jchemed.3c01341

Duangpummet, P., Yodyingyong, S., & Chenprakhon, P. (2022). Fruit puzzle: An inquiry-based activity to investigate high school students' understanding of the relationship between the concepts of density and concentration of solution at the submicroscopic level. *Journal of Chemical Education, 99*(12), 4175–4180. https://doi.org/10.1021/acs.jchemed.2c00161

Ens, S., Olson, A., Dudley, C., Ross, N., Siddiqi, A., Umoh, K., & Schneegurt, M. A. (2012). Inexpensive and safe DNA gel electrophoresis using household materials. *Biochemistry and Molecular Biology Education, 40*(3), 198–203. https://doi.org/10.1002/bmb.20596

Francis, B., Mills, M., & Lupton, R. (2017). Towards social justice in education: Contradictions and dilemmas. *Journal of Education Policy, 32*(4), 414–431. https://doi.org/10.1080/02680939.2016.1276218

García-Zavala, A. (2023). Demonstrations with poinsettia extract indicator: Acid–base chemistry at home during winter holidays. *Journal of Chemical Education, 100*(9), 3731–3738. https://doi.org/10.1021/acs.jchemed.3c00113

García-Zavala, A., Jiménez, C. C., Martínez-Bourget, D., & Romero-Ávila, M. (2025). Exploring thin-layer and column chromatography fundamentals via experiential learning with simple and affordable materials. *Journal of Chemical Education, 102*(5), 2181–2189. https://doi.org/10.1021/acs.jchemed.4c00957

Grønneberg, T., Kvittingen, L., & Eggen, P. O. (2006). Small-scale and low-cost galvanic cells. *Journal of Chemical Education, 83*(8), 1201–1203. https://doi.org/10.1021/ed083p1201

Guidote, A. M., Jr., Pacot, G. M. M., & Cabacungan, P. M. (2015). Low-cost magnetic stirrer from recycled computer parts with optional hot plate. *Journal of Chemical Education, 92*(1), 102–105. https://doi.org/10.1021/ed500153r

Hamer, M., Beraldi, A. M., Gomez, S. G., Ortega, F., Onna, D., & Hamer, M. (2021). Glowing-in-the-screen: Teaching fluorescence with a homemade accessible setup. *Journal of Chemical Education, 98*(8), 2625–2631. https://doi.org/10.1021/acs.jchemed.1c00328

Horikoshi, R., Shirotani, D., Nakanishi-Masuno, T., & Shioyama, H. (2023). Structural models of TTF and TCNQ based on electronic components linked by plastic tubes. *Journal of Chemical Education, 100*(8), 3089–3092. https://doi.org/10.1021/acs.jchemed.3c00065

Horna-Saldaña, C., & Canaleta, X. (2024). Application of universal design for learning and digital fabrication in the creation of a tool for inclusive teaching of the ordering of chemical elements. *Journal of Chemical Education, 101*(12), 5261–5271. https://doi.org/10.1021/acs.jchemed.4c00679

Im, R., Iwayama, T., & Osa, M. (2023). Development of acid-base indicators from natural pigments in agar gel. *Journal of Chemical Education, 100*(12), 4707–4713. https://doi.org/10.1021/acs.jchemed.3c00131

Jansen-van Vuuren, R. D., Buchanan, M. S., & McKenzie, R. H. (2013). Connecting resources for tertiary chemical education with scientists and students in developing countries. *Journal of Chemical Education, 90*(10), 1325–1332. https://doi.org/10.1021/ed3006105

Jarujamrus, P., Chairam, S., & Supasorn, S. (2024). Portable syringe kit demonstration of gas generating reactions for upper secondary school chemistry. *Chemistry Teacher International*, in press. https://doi.org/10.1515/cti-2024-0090

Johnson, T. R., Shaffer, T. A., Holland, L. A., Veltri, L. M., Lucas, J. A., Elshamy, Y. S., & Rutto, P. K. (2022). A low-cost and simple demonstration of freezing point depression and colligative properties with common salts and ice cream. *Journal of Chemical Education, 99*(10), 3590–3594. https://doi.org/10.1021/acs.jchemed.2c00626

Justi, R., & Gilbert, J. (2002). Models and modelling in chemical education. In J. K. Gilbert, O. De Jong, R. Justi, D.F. Treagust & J. H. Van Driel (Eds.), *Chemical education: Towards research-based practice* (pp. 47–69). Springer. https://doi.org/10.1007/0-306-47977-X_3

Kanda, N., Asano, T., Itoh, T., & Onoda, M. (1995). Preparing "chameleon balls" from natural plants: Simple handmade pH indicator and teaching material for chemical equilibrium. *Journal of Chemical Education, 72*(12), 1131. https://doi.org/10.1021/ed072p1131

Kuntzleman, T. S., Corts, S., & Schmidt, A. (2021). At-home titration: Magnesium hydroxide in milk of magnesia using an inexpensive digital balance and natural food dye as indicators. *Journal of Chemical Education, 98*(8), 2592–2595. https://doi.org/10.1021/acs.jchemed.1c00612

Kyle, W. C. (2006). The road from Rio to Johannesburg: Where are the footpaths to/from science education. *International Journal of Science and Mathematics Education, 4*, 1–18. https://doi.org/10.1007/s10763-005-0856-9

Millar, R. (2010). Practical work. In J. Osborne & J. Dillon (Eds.), *Good practice in science teaching: What research has to say* (pp. 108–134). Open University Press.

Moutsakis, G., Paschalidou, K., & Salta, K. (2025). Chemistry laboratory experiments focusing on students' engagement in scientific practices and central ideas of chemical practices. *Chemistry Teacher International, 7*(1), 173–182. https://doi.org/10.1515/cti-2024-0070

Mundy, C. E., & Nokeri, B. K. (2024). Investigating the effects of a context-based laboratory exercise for meaningful learning. *Journal of Chemical Education, 101*(8), 3118–3125. https://doi.org/10.1021/acs.jchemed.3c01260

Murtadlo, M., Albana, H., Nisa, Y. F., Izazy, N. Q., Sumiati, N. T., Dewi, M. S., Abdullah, H., & Henry, C. (2025). Inclusive education in Africa: Transforming higher education in low-income countries. *Scientific African, 28*, Article e02708. https://doi.org/10.1016/j.sciaf.2025.e02708

Naimi, W., Vinnacombe-Willson, G. A., Saldana, S., Ronduen, L., Domjan, H., & Chiang, N. (2024). Teaching acid-base fundamentals and introducing pH using butterfly pea flower tea. *Journal of Chemical Education, 101*(3), 1373–1378. https://doi.org/10.1021/acs.jchemed.3c01058

Ng, E. M. W., Song, Z., Yau, C. D., Liew, O. W., & Ng, T. W. (2020). Safe handling of gas generating experiments using disposable plastic syringes. *Journal of Chemical Education, 98*(1), 237–242. https://doi.org/10.1021/acs.jchemed.0c00102

Nguyen, J., Akdeniz, A., Anfuso, C. L., & Morris, J. D. (2023). A simple at home titration: Quantifying citric acid in lemon juice with baking sodaand Mentos. *Jurnal of Chemical Education, 100*(2). 739–744.

O'Donoghue, J., Doménech, N. G., Tully, D., McGoldrick, N., McArdle, F., Connolly, M., Otway, D. J., Daly, W., Keeny, L., & Horgan, M. (2024). Do you want to make a battery? Insights from the development and evaluation of a chemistry public engagement activity. *Journal of Chemical Education, 101*(11), 5089–5096. https://doi.org/10.1021/acs.jchemed.4c01123

Oliveira, H., & Bonito, J. (2023) Practical work in science education: A systematic literature review. *Frontiers in Education, 8*, Article 1151641. https://doi.org/10.3389/feduc.2023.1151641

Opertti, R., Walker, Z., & Zhang, Y. (2014). Inclusive education: From targeting groups and schools to achieving quality education as the core of EFA. In L. Florian (Ed.), *The SAGE handbook of special education* (pp. 149–169). Sage. https://www.torrossa.com/en/resources/an/5018795

Palmer, M. J., Martinez, K. A., Gadgil, M. G., & Campbell, D. J. (2018). Demonstrations of magnetism and oxidation by combustion of iron supplement tablets. *Journal of Chemical Education, 95*(3), 423–427. https://doi.org/10.1021/acs.jchemed.7b00475

Parkes, M. A., Chen, T., Wu, B., Yufit, V., & Offer, G. J. (2016). "Can" you really make a battery out of that? *Journal of Chemical Education, 93*(4), 681–686. https://doi.org/10.1021/acs.jchemed.5b00496

Set, S., & Kita, M. (2014). Development of a handmade conductivity measurement apparatus and application to vegetables and fruits. *Journal of Chemical Education, 91*(6), 892–897. https://doi.org/10.1021/ed400611q

Srivastava, P. K., Sane, K., Sane, K. V., & Waddington, D. J. (1985). Development of low-cost, locally produced laboratory equipment: An international initiative. *Journal of Chemical Education, 62*(5), 428–430. https://doi.org/10.1021/ed062p428

Supasorn, S., Jarujamrus, P., Chairam, S., & Amatatongchai, M. (2021). Portable syringe-vial kit of gas-generating reactions for easy demonstration of chemical reaction rate. *Journal of Physics: Conference Series, 1806*, Article 012175. https://doi.org/10.1088/1742-6596/1806/1/012175

Suzuki, C. (1995). A new low-cost apparatus for electrolysis of water. *Journal of Chemical Education, 72*(10), 912–913. https://doi.org/10.1021/ed072p912

Turano, M. A., Lobuono, C., & Kirschenbaum, L. J. (2015). Coffee stirrers and drinking straws as disposable spatulas. *Journal of Chemical Education, 92*(1), 197–198. https://doi.org/10.1021/ed5006322

Thomas, G. (2013). A review of thinking and research about inclusive education policy, with suggestions for a new kind of inclusive thinking. *British Educational Research Journal, 39*, 473–490. https://doi.org/10.1080/01411926.2011.652070

Trujillo, C. A. (2005). A modified demonstration of the catalytic decomposition of hydrogen peroxide. *Journal of Chemical Education, 82*(6), 855. https://doi.org/10.1021/ed082p855

Tseng, Y. J., Hong, Z. R., & Lin, H. S. (2024). Exploring high school students' chemical explanatory levels of thin-layer chromatography through reflective inquiry. *Journal of Chemical Education, 101*(9), 3635–3642. https://doi.org/10.1021/acs.jchemed.4c00084

Valdez, S., Dunn, C. B., Hullum, M. D., Harper, E., & Qiang, Z. (2024). Reprocessable Networks from vegetable oils, salts, and food acids: A green polymer outreach demonstration for middle school students. *Journal of Chemical Education, 101*(7), 2947–2953. https://doi.org/10.1021/acs.jchemed.3c01258

Wakabayashi, F., Hamada, K., & Sone, K. (1998). CD-ROM spectroscope: A simple and inexpensive tool for classroom demonstrations on chemical spectroscopy. *Journal of Chemical Education, 75*(12), 1569–1570. https://doi.org/10.1021/ed075p1569

Wright, S. W. (2003). Where there's fire there's... *Journal of Chemical Education, 80*(10), Article 1160A. https://doi.org/10.1021/ed080p1160A

Xu, W., Huang, D., Li, J., Dai, Y., Feng, X., Pang, A., Lin, L., Chen, M., Liu, W., & Zhong, J. (2025). A home chemistry experiment based on Chinese traditional culture: Anthocyanins in Peking opera facial makeup and paper-cutting. *Journal of Chemical Education, 102*(1), 372–378. https://doi.org/10.1021/acs.jchemed.4c01117

Zhuo, J., & Liang, H. (2023). Reusing waste plastic caps to build inexpensive and easily changeable crystal structure models. *Journal of Chemical Education, 100*(7), 2793–2801. https://doi.org/10.1021/acs.jchemed.3c00007

Zidny, R., Sjöström, J., & Eilks, I. (2023). Indigenous knowledge and science and technology education. In B. Akpan, B. Cavas & T. Kennedy (Eds.), *Contemporary issues in science and technology education* (pp. 165–179). Springer. https://doi.org/10.1007/978-3-031-24259-5_12

**Katerina Salta** Ph.D., is a teaching staff member of the Department of Chemistry of the National and Kapodistrian University of Athens, Greece. Her published work is mainly related to the affective and cognitive domains of science/chemistry education. Her current research focuses on social and epistemic dimensions of science/chemistry teaching and learning.

**Katerina Paschalidou** Ph.D., is a teaching staff member of the Department of Chemistry of the National and Kapodistrian University of Athens, Greece. Her published work is in the field of organic chemistry and chemistry education research. Her current research focuses on designing new teaching and learning methods and integrating inquiry and green chemistry.

**Dionysios Koulougliotis** is full professor at the Department of Environment of Ionian University in Zakynthos, Greece. He conducted postgraduate studies in biophysical chemistry. Since 2010 he has been active in science/chemistry education research. His current research interests focus on the affective and cognitive domains in science education and education for sustainable development.

# Chapter 20
# Designing Sustainable and Inclusive Physics Laboratories for the Digital Age

**Mustafa Şahin Bülbül**

## 20.1 Introduction

STEM laboratories serve not only as environments for experimental learning but also as gateways through which students engage with science in a tangible, inquiry-driven manner. However, when these spaces are not equally accessible or pedagogically inclusive, they inadvertently reinforce educational disparities rather than reduce them (Mutch-Jones et al., 2012). Traditional laboratory designs are often developed with an "average" student in mind, overlooking the wide spectrum of physical, sensory, and cognitive diversity among learners. This oversight results in the systematic exclusion of students who do not conform to normative assumptions about mobility, perception, or communication. Ensuring equity in lab-based learning is not a supplementary consideration—it is foundational to the broader mission of inclusive education. Inclusive laboratory practices democratize science by making experimental learning experiences meaningful and accessible for all. For underrepresented and marginalized groups, accessible labs become critical sites of identity formation and academic resilience (Sukhai et al., 2014). Therefore, in physics, chemistry, biology, and engineering labs alike, both structural access and pedagogical accommodation must be addressed in tandem.

UNESCO and other international bodies have consistently highlighted the importance of equity and inclusion in 21st-century education systems. The United Nations Sustainable Development Goal 4 (Quality Education) explicitly frames inclusive and equitable STEM education as a fundamental human right. Within this framework, science laboratories are not merely instructional spaces but instruments of social justice, where systemic inequities can be addressed or perpetuated. Inclusion in STEM must extend beyond physical or cognitive disability to encompass diverse socioeconomic, linguistic, and cultural backgrounds. Inclusive laboratories

M. Ş. Bülbül (✉)
Kafkas University, Kars, Turkey
e-mail: msahinbulbul@gmail.com



L. Leite et al. (eds.), *Practical Work in Science Education in the Digital Era*, Advances in Technology-Rich Science Education 2, https://doi.org/10.1007/978-3-032-13788-3_20

are thus not only about accessibility—they are also about agency, representation, and belonging. Educators must be trained not just in universal design principles but also in adaptive pedagogy, culturally responsive teaching, and digital accessibility (Bülbül, 2016). Institutional frameworks should align with these goals by implementing global accessibility standards and promoting open access resources. When inclusivity is systemically embedded, labs transform into equitable ecosystems that welcome difference as a strength rather than a limitation (Sukhai & Mohler, 2016).

Physical barriers remain one of the most visible yet least addressed challenges in laboratory-based STEM education. Laboratory stations that are too high, inaccessible equipment layouts, and lack of assistive technologies can turn a hands-on learning experience into an exclusionary ordeal for students with disabilities. Such barriers not only limit participation but can profoundly affect students' sense of scientific identity, belonging, and career aspirations (Deane & Burrill, 2024). Inaccessible lab environments send implicit messages about who belongs in science, and who does not. Addressing these inequities requires a shift toward universal design thinking—where accessibility is designed into the system from the outset rather than retrofitted later. When accessibility becomes an integral design principle, it benefits all students—not just those with visible disabilities (Persson et al., 2015). For instance, adjustable lab benches, voice-activated tools, and tactile learning aids can support multiple modes of engagement for all learners. When thoughtfully implemented, inclusive designs enhance the overall safety, flexibility, and efficacy of laboratory instruction. For instance, incorporating blindfolded participation during teacher training experiments not only enhances pre-service teachers' empathy toward individuals with visual impairments, but also fosters a deeper sense of communicative responsibility among their peers (Bülbül, 2013b). When a blindfolded student struggles to follow the procedure and expresses confusion, it prompts others to reflect on the clarity, precision, and inclusiveness of their explanations. This experiential strategy cultivates both awareness of diverse learning needs and the pedagogical skill of articulating scientific processes in an accessible and supportive manner.

Acknowledging that student diversity is the norm and the 'average' student is a myth is the fundamental starting point for inclusive lab design. Universal Design principles advocate for designing from the outset to anticipate the varied needs of all students, rather than making 'special accommodations' for disabled individuals later (Wilson, 2017). This approach extends from architectural adjustments (adjustable benches, wide pathways) to pedagogical strategies (multiple means of representation, multiple means of action & expression). For instance, presenting an experiment protocol not only in written text but also with visual diagrams and audio narration enhances understanding not just for a student with visual impairment or dyslexia, but for all students with diverse learning styles. Inclusive design is built on the principle of 'including' rather than 'fixing'.

While physical access is essential, digital inequity has emerged as a significant barrier in modern STEM education, particularly as laboratories integrate online and hybrid learning formats (Tate & Warschauer, 2022). Students' unequal access to devices, internet connectivity, and accessible learning software can silently exclude them from meaningful participation. This digital divide is often compounded by a lack of institutional support, leaving students to navigate complex systems on their own. The promise of digital innovation—through simulations, virtual labs, and augmented reality—will only be fulfilled if these tools are equitably distributed and pedagogically integrated. Digital transformation must be accompanied by a parallel commitment to digital justice (Donoghue, 2017). Open-source platforms, multilingual content, and screen-reader-compatible interfaces are crucial for leveling the playing field. Furthermore, faculty development in digital inclusion strategies is as important as infrastructure investment. True equity in STEM labs requires a dual focus: physical redesign and digital reimagination (Heimerdinger, 2019).

## 20.2 Universal Design Principles for Inclusive STEM Learning Spaces

The Universal Design for Learning (UDL) framework offers a powerful foundation for reimagining STEM laboratories as inclusive spaces for all learners. UDL emphasizes the proactive design of learning environments that accommodate learner variability from the outset, rather than reactively adapting them for students with specific needs (Rao et al., 2016). In the context of laboratory-based science education, UDL requires the integration of multiple means of engagement, representation, and expression into both physical and instructional design. This means rethinking traditional lab practices to ensure that all students can participate meaningfully in experimentation and inquiry. For example, providing audio instructions alongside written protocols, or offering lab simulations for students with mobility limitations, aligns with UDL's principle of flexible access. Rather than targeting disability alone, UDL approaches variability as the norm—not the exception—in the classroom (Wilson, 2017). Visually impaired students can use scientific process skills and conduct experiments with universally designed laboratory instruments such as audible pulse meters and audible thermometers (Bülbül, 2013a). This shift in mindset has profound implications for how laboratories are built, equipped, and used. Institutions that embed UDL principles into lab design send a clear message: inclusion is not optional, but essential for excellence in science education.

The architectural and environmental design of laboratory spaces plays a pivotal role in supporting or hindering inclusive participation. Physical space should not merely facilitate movement but actively encourage equitable interaction with laboratory tools, peers, and instructors. This requires flexible workstations, adjustable-height tables, and wheelchair-accessible sinks and storage areas. Visual signage should be complemented with tactile and auditory cues, especially for students with sensory differences. Ventilation systems must consider both safety and comfort, especially for students with respiratory or sensory sensitivities. Lighting must be adequate but not overwhelming, and acoustic conditions must minimize echo and distraction. Inclusive laboratory environments go beyond compliance with disability regulations—they reflect a deep pedagogical commitment to equity. For example; drawing graphics with wires and pins on a corkboard when needed allows not only the visually impaired but also the sighted students to learn by touching, that is, using an additional sense. In this way, learning is both multisensory, interesting and memorable (Bülbül, 2013c). These designs support not only students with disabilities, but also enhance usability and comfort for all learners, including those managing temporary injuries, chronic conditions, or anxiety in crowded spaces. There is no need for high-tech laboratories for inclusive activities. Sometimes, even a correct and understandable description of samples collected simultaneously from nature with a paper microscope (foldscope) is enough. Stones, leaves, branches and soil in nature can be sufficient tactile materials for drawing graphics (Bülbül, 2024).

Balancing safety and accessibility in laboratory environments requires careful and intentional design decisions. Laboratories are, by nature, settings that involve potential risks—chemical spills, sharp instruments, electrical devices—making safety a non-negotiable priority. However, safety protocols must be designed inclusively to avoid inadvertently excluding students with disabilities (Sawyer & Bright, 2014). For example, emergency exit paths should be wide and clearly marked for wheelchair users, while safety instructions must be available in multiple formats, including large print, Braille, or audio. Eye-wash stations and fire extinguishers should be placed at accessible heights, and lab coats or protective gear must accommodate diverse body types and mobility needs. Institutions must provide training for both students and instructors on inclusive safety practices. Inclusive safety is not a contradiction in terms—it is a redefinition of what safety means in an educational context committed to equity (Kozleski et al., 2014). Balancing risk management with universal access creates a lab culture that prioritizes care without compromise.

The integration of UDL and physical design principles into laboratory spaces must be understood as a long-term, systemic commitment. Inclusive laboratory design is not a one-time architectural project but a continuous process of assessment, feedback, and iterative improvement (Bülbül, 2018). This involves regular consultation with students, including those with disabilities, to identify unforeseen barriers and evaluate the effectiveness of current accommodations. Feedback loops should inform infrastructure upgrades, equipment procurement, and even curriculum planning. Collaboration with occupational therapists, accessibility experts, and inclusive design professionals can enrich the process. Moreover, institutions must develop internal policies that require all new lab constructions or renovations to adhere to universal design benchmarks. Sustainability in inclusion is achieved through policy integration, not ad hoc fixes. A truly inclusive laboratory is not simply a space—it is a culture embedded in the values, practices, and everyday routines of STEM education.

## 20.3 Planning Inclusive and Collaborative Laboratory Activities

Differentiating laboratory instruction based on student ability levels is essential for creating equitable and inclusive STEM learning environments. Differentiated experiment planning provides multiple pathways for students to access, engage with, and demonstrate understanding of scientific concepts. Instead of offering a one-size-fits-all approach, educators design experiments that align with students' cognitive, sensory, and motor profiles. For example, one student may investigate light refraction using augmented reality, while another performs the same experiment with tangible optical materials. It is also possible to make simulations tactile for visually impaired students (Bülbül et al., 2013d). Offering diverse versions of the same conceptual task enriches the learning process while ensuring accessibility. This approach benefits not only students with disabilities but also supports a broader range of learners who process information differently. Differentiation includes variations in process, content, and product—empowering all students to contribute meaningfully to scientific inquiry. Ultimately, such inclusive planning deepens student engagement and fosters ownership of learning.

Collaborative laboratory work enhances both social interaction and academic engagement, making it a key strategy in inclusive science education. Well-structured group activities can mitigate individual disparities and create environments where every student has a valued role. Peer support within lab groups provides scaffolding for students who may need assistance, whether due to a disability or other learning differences. Role distribution—such as assigning positions like material handler, data recorder, or observer—ensures shared responsibility and balanced participation. These roles can be rotated to encourage skill development and prevent hierarchies. When implemented thoughtfully, collaborative strategies promote mutual respect, responsibility, and scientific discourse. Diversity within groups enhances problem-solving by integrating multiple perspectives. The most powerful factor that makes a classical laboratory accessible is the learning atmosphere. In learning environments where collaboration is strong, I have witnessed peers supporting each other with a common language. While explaining concave and convex lenses, visually impaired middle school students would take turns holding them. While I was explaining, I was also recording their conversations and when I examined them later, I noticed that they were producing new and understandable terms among themselves, such as "fat lens". The emotional benefits of peer interaction, such as increased confidence and motivation, are especially critical for students who may feel marginalized in traditional settings.

Physics laboratories offer a rich context for applying differentiated and collaborative teaching methods, especially in experiments related to light, force, and electricity. In a light experiment, for instance, students might choose between using real lenses and mirrors, or exploring simulations on tablets with screen readers. Force-related activities could combine hands-on use of spring scales with digital modeling of vectors and motion graphs. These blended approaches demonstrate how inclusive practices and technology can converge to support deeper conceptual understanding. In electricity labs, students with visual impairments may use tactile circuit boards with Braille labels, while others engage with color-coded interactive kits. Because physics often involves abstract concepts, the use of multimodal representation—such as visual diagrams, tactile models, and auditory feedback—can dramatically enhance comprehension. The integration of concrete materials, simulations, and data visualization narrows the learning gap between students of varying abilities. Moreover, there is no need to use high technology. For example, it is possible to make drawings on optics tactile by using a corkboard, some string and pins (Bülbül et al., 2015). As a result, learners transform from passive observers into active participants in the scientific process.

Inclusive laboratory environments that combine differentiated instruction and collaborative learning not only promote cognitive growth but also foster emotional and social development. A student's sense of belonging in the lab is as important as their academic performance (Li & Singh, 2023). When learners feel they are welcomed, valued, and capable, they are more likely to persist in science fields. In group-based labs, students develop both scientific reasoning and interpersonal skills through peer learning and shared accountability. Empathy, cooperation, and adaptability emerge as integral components of scientific exploration. These learning experiences transcend disciplinary boundaries, equipping students with transferable life skills. Inclusivity through differentiation and collaboration enables both individual achievement and collective success. Thus, planning physics labs with these principles in mind is not merely pedagogical best practice—it is an ethical imperative for 21st-century science education.

## 20.4 Designing Accessible and Multimodal Learning Materials

Tactile materials, Braille labeling, and audio guidance are essential components of inclusive STEM laboratory environments (Koehler & Picard, 2024). These elements play a particularly crucial role for students with visual impairments, enabling them to physically engage with scientific content. Tactile models—such as 3D representations of molecules, circuit elements, or vector forces—help translate abstract ideas into sensory-rich experiences. Braille labeling allows for independent and safe identification of lab tools, chemicals, and devices. This becomes especially important in physics and chemistry labs, where misidentification may pose serious safety risks. Audio guidance systems, such as step-by-step experiment narration, not only support visually impaired students but also benefit learners with attention or processing differences. Multisensory learning materials increase both accessibility and conceptual retention, offering a richer educational experience for all students (Di Fuccio et al., 2025). Incorporating these tools into mainstream settings normalizes accessibility and fosters a more inclusive lab culture.

The accessibility of digital learning materials is a critical, yet often overlooked, factor in modern STEM education. Being digital does not guarantee accessibility; in fact, many digital resources exclude learners with disabilities due to poor design (Zhang et al., 2020). Screen reader compatibility, captioned videos, keyboard navigation, and simplified interfaces are all necessary features for making content accessible to diverse learners. The Web Content Accessibility Guidelines (WCAG) provide a robust framework for ensuring usability across abilities. In STEM fields—where materials often include graphs, equations, and simulations—these standards must be applied rigorously and early in the design phase. Collaborating with accessibility experts, special educators, and users from the disability community can significantly improve the inclusivity of digital tools. Accessible digital materials not only enhance equity but also promote universal usability, benefiting all learners regardless of ability. It is possible to do graphic and motion studies by sending only the audio file to remote areas by the simplest means; what is important is to establish the smart accessibility structure correctly rather than the level of technology used (Bülbül & Sahyun, 2010). Institutional commitment to these standards reflects a broader investment in inclusive pedagogy. Having a colleague help a visually impaired researcher by reading is also an accessible method, but the fact that the researcher can collect and analyze data alone shows the accessibility of that laboratory.

Redesigning STEM kits like Foldscope, Arduino, and sensor-based tools represents a transformative approach to equitable education. These tools demonstrate that accessibility need not be expensive or complex. Foldscope, a paper-based microscope, offers a low-cost, lightweight, and adaptable alternative to traditional microscopes—ideal for both under-resourced settings and adaptation for sensory differences. Arduino microcontrollers, being open-source and programmable, can be modified to include audio feedback, vibration signals, or large-button interfaces to accommodate various needs. Sensors that capture light, temperature, or motion data can be paired with tactile displays or sonified outputs, making scientific data accessible in real-time for students who cannot rely on visual cues. Customizable and hackable kits encourage learner agency, allowing students to tailor tools to their own preferences and needs Mäkelä & Vellonen, 2018; Bülbül, 2017). This process fosters deeper engagement, technical fluency, and a sense of ownership in scientific inquiry. The shift from passive use to active modification of tools aligns with the ethos of inclusive and inquiry-based learning.

Accessible materials are not just functional aids—they are powerful pedagogical instruments that signal the values of equity and belonging (Tate & Warschauer, 2022). When students see that materials are designed with their needs in mind, they feel acknowledged, empowered, and more motivated to engage. Accessibility improves not only usability but also the depth and personalization of learning. Materials that support multiple representations (e.g., visual, tactile, auditory) allow students to build conceptual understanding through varied pathways. This supports students with disabilities and also benefits multilingual learners, neurodivergent students, and others with diverse learning profiles. Designing for accessibility is not a technical add-on; it is an ethical responsibility and a pedagogical imperative (Bülbül, 2017). The way teachers introduce, scaffold, and reflect on the use of these materials determines how effective they will be. Thus, material design must be approached through the combined lenses of engineering, education, and equity.

## 20.5 The Role of Digital Tools in Expanding Laboratory Access

Digital technologies can reshape STEM laboratories to accommodate a wide range of learner needs. Audio description, screen-reader support and haptic cues may open experiments to students with visual impairments; captioning, sign language and clear visual signaling can assist those who are deaf or hard of hearing. Voice or remote control can lower physical barriers, and step-by-step guides with low-stimulation interfaces may help neurodivergent learners. Used in this way, technology often moves practice beyond one-size-fits-all routines toward multisensory, adaptable workflows that build autonomy.

Personalization works best when learners can approach the same experiment through multiple routes—hands-on manipulation, narrated video, interactive simulation or data visualization—and when interfaces allow simple adjustments such as text size, contrast and language. Where appropriate, light-touch automation can suggest pacing or resources without removing learner choice. Assessment can follow the same logic: oral or video responses, interactive diagrams and brief AR tasks capture understanding in varied formats, while timely feedback helps learners notice and correct misconceptions. In this model, evaluation emphasizes how ideas are processed and applied, not only final answers.

Lasting impact tends to come from pedagogy rather than tools alone. Teachers who map activities to students' cognitive and sensory profiles can offer parallel pathways toward common goals: one learner uses traditional instruments, another collects data with a tablet, a third dictates observations. Such flexibility keeps inquiry central and positions inclusion as an ordinary feature of laboratory work. At the same time, institutional routines that treat accessibility—both physical and digital—as core infrastructure make these designs easier to sustain.

Virtual and remotely accessible laboratories can widen participation when time, location or mobility make in-person work difficult. Live video, remote sensors and cloud visualizations let learners observe, control and analyze experiments from elsewhere, turning the lab into a distributed service rather than a fixed room. Immersive tools, including interactive simulations and AR, can also make abstract ideas more tangible while providing safe spaces to rehearse complex or hazardous procedures.

Simulations are particularly useful in physics because they allow rapid hypothesis testing, controlled variation and immediate feedback. Repeated, low-risk iteration helps consolidate concepts and can reduce material, safety and cost constraints. Still, their value depends on thoughtful integration: aligning tasks with clear objectives, framing activities with preparation and prediction, and closing with reflection and sense-making. The aim is not to bolt technology onto existing lessons, but to let it deepen the quality of scientific thinking and experience.

## 20.6 Building Digital Infrastructure for Sustainability and Scalability

The sustainability and scalability of digital accessibility in education require robust infrastructure at school and institutional levels. Beyond device distribution, planning must address bandwidth, software compatibility, power, and maintenance. For STEM laboratories, high-speed internet, synchronized access systems, and reliable data storage are essential. Equity in hardware distribution across urban and rural contexts directly affects fair access, making investment in digital infrastructure as critical as physical upgrades. Accordingly, policymakers should treat digitalization not only as a technological initiative but as a pedagogical priority.

Equally vital are platform design, data security, and accessible content delivery. Open-source learning platforms offer cost-effective, customizable options; accessible LMSs should support screen readers, keyboard navigation, high-contrast visuals, and flexible layouts while meeting stringent data-protection standards. Because platforms shape how learning occurs—not merely how content is delivered—accessibility and security must be built in and supported through intuitive interfaces and adaptive functionality.

Human capacity underpins sustainability. Teachers need preparation not just to operate tools but to integrate them meaningfully into learning goals, supported by ongoing professional development and institutional backing. Students likewise require digital literacy to use simulations, analyze data, operate remote systems, and adapt tools. Short orientation sessions, accessible user guides, and peer-learning models foster independence—especially for students with disabilities. Misconceptions unique to certain profiles illustrate the stakes: while a sighted student observes a ball's motion throughout life, a blind student may hold the misconception that "the ball speeds up as it rises and slows down as it falls" (Bülbül et al., 2015); well-designed digital and tactile supports can pre-empt such errors.

Finally, sustainable infrastructure balances technology, policy, and pedagogy through long-term, human-centered strategies. Ministries, universities, and local authorities should coordinate system-wide plans—needs assessments, budgeting, support, and evaluation—while ensuring backups, updates, and technical support. Visionary, not quick-fix, planning secures present access and future equity.

## 20.7 Case Study: An Inclusive Physics Laboratory in Practice

In 2024, a pilot project was launched at a public university in Kars, Turkey, to redesign a physics laboratory based on accessibility and inclusion principles. The redesign process integrated Universal Design for Learning (UDL) principles, addressing physical, digital, and pedagogical accessibility in a holistic manner. The physical layout of the lab was modified to include height-adjustable tables for wheelchair users, widened pathways, and voice-activated experimental apparatuses. Laboratory materials were labeled in Braille, and experiment protocols were prepared in both printed and screen-reader-compatible digital formats. The aim of the project was not merely to accommodate students with disabilities, but to create a multi-modal, inclusive learning environment for all learners. The redesign of the physics lab demonstrated that accessibility is not a burden but an opportunity for educational innovation. The initiative was also designed to serve as a model for other institutions in similar geographic or infrastructural conditions.

Several classical physics experiments were adapted into accessible formats and implemented in the redesigned lab. For example, force and equilibrium experiments were enhanced with sensors that provided audio feedback, allowing visually impaired students to independently interpret measurement results. Light reflection and refraction experiments were conducted using augmented reality (AR) glasses and tablet applications, while students with hearing impairments received instructions in written and graphical formats. Basic electrical circuit experiments were structured with audio instructions, LED indicators, and tactile layouts to support students with cognitive differences. In addition, data analysis software was optimized with keyboard navigation options and color-blind-friendly visual themes. Each experiment was not simply retrofitted with assistive tools but was restructured through a pedagogical lens to ensure meaningful engagement. This allowed all students to participate not just as observers, but as active experimenters and scientific thinkers. Sometimes experiments are designed using a large number of simple circuit elements such as resistors, voltmeters, and switches, while other times, students with mental disabilities can focus on building a simple circuit using fewer circuit elements. Pluggable circuit elements make it easier to understand the circuit itself without having to learn the working mechanism inside it (Fig. 20.1).

During the implementation phase, 18 students participated in the lab sessions, and extensive feedback and observational data were collected. Eighty-nine percent of students reported feeling more engaged, confident, and included compared to their previous lab experiences. One visually impaired student described the audio-enabled force sensors as "the first time I truly experienced physics independently." A hearing-impaired student showed remarkable conceptual understanding during AR-assisted light experiments, which was described by instructors as a "deepening of conceptual reasoning." Student-generated video reports revealed not only their grasp of the scientific content but also their ability to analyze challenges encountered during the experiments. In the inclusive setting, students demonstrated notable growth in

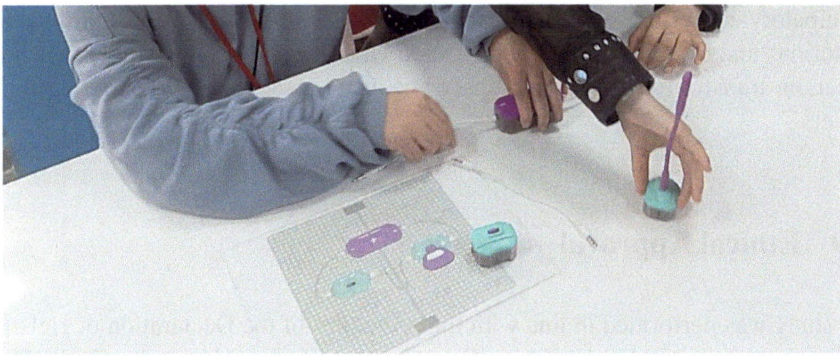

**Fig. 20.1** Students with mental disabilities are using pluggable circuits elements

scientific reasoning, problem-solving, and self-efficacy. These outcomes indicate that learning is shaped not just by curriculum, but by the quality and design of the learning environment. Student feedback highlighted the dual benefits for both learners and instructors in fostering a more inclusive educational culture.

This case illustrates that accessibility in science education should be understood not only as a matter of individual rights but also as a marker of educational quality and social equity. Redesigning physics laboratories with inclusive principles can establish new standards for universities in underserved or rural regions. The Kars model shows that impactful change can be achieved through low-cost adaptations, strategic planning, and genuine teacher-student collaboration. Furthermore, this framework can be adapted to other disciplines, including engineering, biology, and chemistry. The results of the project will be published on an open-access platform to enable replication and scaling by other institutions. An inclusive laboratory is not defined by its technology alone, but by its human-centered design and pedagogical vision. Such initiatives do more than improve physics education—they redefine who science is for. Ultimately, accessibility must become not just a policy or a feature, but a sustainable educational culture.

## 20.8 Conclusion and Ways Forward

The transformation of STEM laboratories into inclusive spaces requires more than technical adjustments; it demands cultural change, systemic policy alignment, and interdisciplinary collaboration. Sustainable progress depends on embedding inclusivity into educational policy, teacher training, and R&D initiatives, while promoting open-source, accessible digital tools. Engineers, educators, designers, and students must co-create laboratories that reflect equity, usability, and pedagogical rigor. Most importantly, students—particularly those with disabilities—must be actively involved in design, evaluation, and feedback processes, ensuring that their lived experiences guide improvements. Inclusive labs are not static projects but evolving, participatory ecosystems where accessibility becomes a shared responsibility across institutions and disciplines. Ultimately, such laboratories redefine equity in science education, transforming them into spaces of justice, agency, and belonging for all learners.

## 20.9 Ethical Approval

This study was performed in line with the principles of the Declaration of Helsinki. Approval was granted by the Ethics Committee of Kafkas University (23.01.2025/66).

## References


Bülbül, M. Ş. (2013a). A description of a blind student's science process skills through health physics. *European Journal of Physics Education, 4*(2), 6–13.

Bülbül, M. Ş. (2013b). Blindfold experiments: Seven principles for inclusive classes. *Latin American Journal of Physics Education, 7*(2), 265.

Bülbül, M. Ş. (2013c). Görme engelli öğrenciler ile grafik çalışırken nasıl bir materyal kullanılmalıdır? [What kind of material should be used when working with visually impaired students?] *Fen Bilimleri Öğretimi Dergisi, 1*(1), 1–11.

Bülbül, M. Ş., Demirtas, D., Garip, B.,& Oktay, O. (2013d, March, 14–15). *Re-simulating: Physics simulations for blind students*. New Perspectives in Science Education Conference, PIXEL, Italy.

Bülbül, M. Ş. (2016). Görme engelli öğrenciyi fizikçi yapan fonksiyon [The function that makes visually impaired student physicist]. *Alan Eğitimi Araştırmaları Dergisi, 2*(1), 17–26.

Bülbül, M. Ş. (2017). A universal design for robotics education. *Journal of Science Education for Students with Disabilities, 20*(1), 16–19. https://doi.org/10.14448/jsesd.08.0005

Bülbül, M. Ş. (2018). From academician's office to physics lab for students with special needs. In M. Koomen, S. Kahn, C. L. Atchison & T. A. Wild (Eds.), *Towards inclusion of all learners through science teacher education* (pp. 141–150), Brill. https://doi.org/10.1163/9789004368422_016

Bülbül, M. Ş. (2024). Graph literacy activities in natural outdoor learning environments for students with visual impairments: Natural graphs. In *STEAM Research Group in Turkey*. Zenodo. https://doi.org/10.5281/zenodo.14602479

Bülbül, M. Ş, & Sahyun, S. C. (2010). A distance experiment with a blind partner. *Balkan Physics Letters, 18*, 165–170.

Bülbül, M. Ş., Garip, B., & Özdemir, Ö. F. (2015). Using a force concept ınventory test with visually impaired and blind students. *European Journal of Physics Education, 6*(3), 20–31. https://doi.org/10.20308/ejpe.09583

Deane, K., & Burrill, B. (2024). Step up and be an ally - accessible labs benefit everyone. *The Biochemist, 46*(4), 23–24. https://doi.org/10.1042/bio_2024_138

Di Fuccio, R., Ponticorvo, M., Nadim, M. A., & Limone, P. (2025). Exploring the effect of digital and multisensory educational materials on retention in primary school using tangible user interfaces. *Interactive Learning Environments, 33*(4), 2928–2938. https://doi.org/10.1080/10494820.2024.2427277

Donoghue, J. (2017). The rise of digital justice: Courtroom technology, public participation and access to justice. *The Modern Law Review, 80*(6), 995–1025. https://doi.org/10.1111/1468-2230.12300

Heimerdinger, F. (2019). *Student voice in a participatory design process focused on developing equity-based STEM teaching practices* [Doctoral dissertation, Portland State University].

Koehler, K. E., & Picard, K. M. (2024). Making STEM accessible for students with visual ımpairments: Implications for practice. *Teachıng Exceptional Children.* https://doi.org/10.1177/00400599241231211

Kozleski, E., Artiles, A., & Waitoller, F. (2014). Equity in inclusive education: A cultural-historical comparative perspective. In L. Florian (Ed.), *The SAGE handbook of special education* (2nd ed., pp. 231–249). SAGE. https://doi.org/10.4135/9781446282236.n16

Li, Y., & Singh, C. (2023). Sense of belonging is an important predictor of introductory physics students' academic performance. *Physical Review Physics Education Research, 19*(2), Article 020137. https://doi.org/10.1103/PhysRevPhysEducRes.19.020137

Mäkelä, S., & Vellonen, V. (2018). Designing for appropriation: A DIY kit as an educator's tool in special education schools. *International Journal of Human-Computer Studies, 118*, 14–23. https://doi.org/10.1016/j.ijhcs.2018.05.004

Mutch-Jones, K., Puttick, G., & Minner, D. (2012). Lesson study for accessible science: Building expertise to improve practice in inclusive science classrooms. *Journal of Research in Science Teaching, 49*(8), 1012–1034. https://doi.org/10.1002/tea.21034

Persson, H., Åhman, H., Yngling, A. A., & Gulliksen, J. (2015). Universal design, inclusive design, accessible design, design for all: Different concepts - one goal? On the concept of accessibility - historical, methodological and philosophical aspects. *Universal Access in the Information Society, 14*, 505–526. https://doi.org/10.1007/s10209-014-0358-z

Rao, K., Currie-Rubin, R., & Logli, C. (2016). *UDL and inclusive practices in IB schools worldwide.* CAST Professional Learning.

Sawyer, A., & Bright, K. (2014). *The access manual: Designing, auditing and managing inclusive built environments.* Wiley.

Sukhai, M. A., Mohler, C. E., Doyle, T., Carson, E., Nieder, C., Levy-Pinto, D., Duffett, E., & Smith, F. (2014). *Creating an accessible science laboratory environment for students with disabilities.* Council of Ontario Universities.

Sukhai, M. A., & Mohler, C. E. (2016). *Creating a culture of accessibility in the sciences.* Academic Press.

Tate, T., & Warschauer, M. (2022). Equity in online learning. *Educational Psychologist, 57*(3), 192–206. https://doi.org/10.1080/00461520.2022.2062597

Wilson, J. D. (2017). Reimagining disability and inclusive education through universal design for learning. *Disability Studies Quarterly, 37*(2). https://doi.org/10.18061/dsq.v37i2.5417

Zhang, X., Tlili, A., Nascimbeni, F., Burgos, D., Huang, R., Chang, T.-W., Jemni, M., & Koutheair Khribi, M. (2020). Accessibility within open educational resources and practices for disabled learners: A systematic literature review. *Smart Learning Environments, 7*, 1–19. https://doi.org/10.1186/s40561-019-0113-2

**Mustafa Şahin Bülbül** is the head of the department of special education, Kafkas University, Turkey. Holding undergraduate degrees in physics, sociology, and philosophy, he earned his PhD in physics education at METU, focusing on teaching physics to students with visual impairments. He also teaches physics at the Science and Art Centre (BİLSEM) and serves as director of the STEAM Research Group. His research centres on gifted education and inclusive STEAM learning.

# Chapter 21
# Inclusive STEM Education for Students who are Blind or Low Vision: From the Classroom into the Science Laboratory

**Cary A. Supalo and Jasodhara Bhattacharya**

## 21.1 Introduction

Persons who are Blind or have Low Vision (BLV) have traditionally been marginalized by their non-BLV counterparts for centuries. Numerous factors contribute to this phenomenon. The most common is the lack of vision itself, compounded by the stereotypes and misconceptions that the non-BLV community ascribe to this disability (Jernigan, 1965). People who are sighted may reflect on their prior experiences with representations of people with BLV they have encountered in their daily lives, or may have seen on television or in a book that they have read (Wunder, 2017). The accuracy, or lack thereof, of the portrayal of these characters with BLV can have a significant impact on expectations ascribed to a person with BLV who they may encounter in a science laboratory working scenario.

These attitudinal barriers affect people with disabilities broadly, limiting equitable access to STEM learning and work due to low expectations and assumptions about their abilities perpetuated through media and personal encounters. Since people with disabilities in general struggle to find equitable science work, when the opportunity for securing practical work arises, individuals with disabilities often demonstrate high levels of loyalty in their roles (Andreassi, 2018). For persons with BLV, this underscores a key advantage: reallocating the cost for resources which the person with BLV may forgo, like reserved parking spots in favor of rideshares or public transit, to individualized accommodations and supports, such as Text-To-Speech (TTS) screen reader software that the person with BLV may need to do their job. These tools

C. A. Supalo (✉) · J. Bhattacharya
Independence Science, West Lafayette, IN, USA
e-mail: csupalo@independencescience.com

J. Bhattacharya
e-mail: jbat@independencescience.com



L. Leite et al. (eds.), *Practical Work in Science Education in the Digital Era*, Advances in Technology-Rich Science Education 2, https://doi.org/10.1007/978-3-032-13788-3_21

may enable access to emails, word processing, spreadsheets, library databases, and internet searches (Albright, 2023), typically via annual licensing fees to keep the screen reader product current with the newest features and compatibility with other commercially available products.

Mobility aids further empower independence: white canes, dog guides, or a combination of both, facilitate commuting to and from the workplace, and onsite navigation (Siler, 2024). Dog guides require brief daily breaks in accessible relief areas, with owners handling cleanup. While harnessed, they are working animals and not pets! It is good decorum to ask the dog guide owner if you can pet the animal then simply do it without asking.

This chapter delves into dismantling such barriers across STEM education and employment. From navigating visual-centric gatekeepers in the classrooms and labs of the STEM pipeline through non-ambiguous verbal strategies, born-accessible materials, and tactile tools like the SALS probe, to fostering an inclusive STEM ecosystem through workplace interventions such as customized Access Technology (AT), orientation and mobility (O&M) training, and emerging innovations like RTDs and AI, we chart pathways to the authentic and full inclusion of persons with BLV. By embracing universal design, the STEM ecosystem can evolve to where BLV contributions illuminate innovation on merit alone.

## 21.2 Students with BLV in the Science Classroom Context

In the current STEM classroom context, students with BLV face many challenges to quality education (Rashid & Chehadeh, 2023). Amongst these is access to core content. Effective student learning requires active engagement with diverse information sources, including teachers, textbooks, peers, and emerging hybrid approaches that integrate digital tools. Yet the inherently visual nature of STEM—rich with diagrams, graphs, and dynamic models—may exclude students with BLV, changing these conduits of knowledge into gatekeepers.

In the STEM lecture instruction communication ecosystem, educators use a breadth of strategies to convey concepts and facilitate active learning: these include verbal explanations supported by visual presentation tools (e.g. slides, chalkboards, animations, simulations), interactive discourse, and non-verbal cues. However, approaches that rely on the visual modality can create barriers to learning for students with BLV. For example, verbal instructions that presuppose a shared visual frame of reference—such as pointing to a molecule on a whiteboard or alluding to a graph on a slide—without providing equivalent descriptive language, can lead to an in-the-moment fragmented understanding of spatial or relational concepts like molecular bonding or data trends.

This verbal communication needs to be non-ambiguous (Isaacson et al., 2010): for example, if educators do not verbally indicate when specific quantities and fractions begin and end in an algebraic expression, it introduces significant ambiguity and leads to confusion for the student with BLV (Frankel et al., 2016). Absent or ambiguous verbal equivalents can heighten cognitive demand and foster reliance on peers for clarification rather than independent learning. In disciplines like chemistry or physics, where visuals depict dynamic processes (e.g., particle motion or circuit diagrams), this ambiguity can perpetuate misconceptions, with BLV students missing the incidental learning opportunities that sighted peers gain from such visual aids.

Textbooks are another traditional conduit that complement lectures: they offer structured access to canonical knowledge, serving as a resource for background information or evidence-based claims. STEM textbooks abound with diagrams, graphs, chemical structures, and mathematical notations: the inaccessibility of such visual elements can be a barrier which hinders comprehension, denying students with BLV equitable entry to this foundational resource. To mitigate these barriers, content must be available in alternative formats which encompass a range of options that can be tailored to the BLV student's individual preferences and technological needs: some common formats used by students with BLV are hard copy braille, large print, and digital formats that can be read using a text-to-speech screen reader in conjunction with a refreshable braille device, such as e-texts with embedded graphic descriptions, and tagged PDFs.

Electronic files, whether textbooks or other lecture content, need to be rendered such that they are consumable by the student with BLV. If educators or education support staff render files in accessible file types like Microsoft Word (DOC), Excel (XLS) for data tables, rich text (RTF), ASCII text (TXT), or scalable vector graphics (SVG), these integrate seamlessly with text-to-speech and braille output devices. Conversely, scanned images of PDF files and handwritten content tend not to be accessible by a person with BLV using standard AT tools: these types of documents require extensive remediation to make the content accessible, incurring significant time investment and cost (Schmitt-Koopman et al., 2025). Therefore, educators who produce their own content should endeavor to make their materials "born-accessible", to be usable by students with BLV. As access technology evolves, a commitment to such accessibility standards can support timely, high-fidelity conversions, benefiting all learners through enhanced descriptive clarity.

Further, graphical STEM content, whether two- or three-dimensional, whether drawn or illustrated on a slide deck, should be made available in a raised line drawing or tactile graphics format. These images can be made into tactile forms using freehand drawn with a raised line drawing tool either during or prior to the class session. When a teacher refers to a pre-drawn image, it can simply be referenced by indicating verbally a figure number that can be indicated by drawing a certain number of vertical lines in one top corner of the raised line drawing to indicate a figure number (Supalo & Kennedy, 2014).

Overall, it is good practice to discuss with the student with BLV their individualized learning needs for accessing the full range of lecture content. Preferences for auditory, tactile, or hybrid modalities can vary. Therefore, initiating open pre-class conversations with the students with BLV can help instructors tailor accommodation to individual needs. Additionally, recognizing that it is difficult for teachers to maintain perfect non-ambiguous speaking during STEM lectures: encouraging the student with BLV to ask clarifying questions in real-time will support self-advocacy in addressing ambiguities, while refining instructors' verbal precision over time; this may yield universal benefits like reduced cognitive load for all.

## 21.3 Students with BLV in the Teaching Laboratory

Hands-on laboratory learning is a key element of STEM education: it brings together theoretical knowledge with skills like problem-solving, and enhances scientific literacy through experimentation and practical application. In disciplines like chemistry, manipulating materials, observing real-time reactions, and troubleshooting variables foster deeper understanding and scientific literacy beyond what lectures or simulations alone can achieve. Studies on the benefits of inclusive laboratory learning practices underscore the ethical imperative for universal design to dismantle ableist structures in science education (Boyd et al., 2023); however, the visual-centric designs of hands-on laboratory learning activities can exclude students with BLV from these formative experiences (Supalo et al., 2006).

Students with BLV can participate actively in laboratory learning situations (Miner et al., 2001), the extent of which is shaped by a constellation of interconnected factors including having good spatial awareness, effective verbal communication skills, and access to adapted curriculum in the student's preferred reading medium within the laboratory classroom. Collectively such a constellation of factors can mitigate exclusionary barriers (Miner et al., 2001): spatial navigation skills, which may be honed through orientation and mobility training, enables safe interaction with equipment; effective verbal skills support collaborative inquiry, reducing isolation in group settings (Supalo et al., 2014); accessible protocols and curriculum boost BLV students' engagement and learning (Prasertpong et al., 2023).

Accessible equipment is a cornerstone of independent laboratory work in STEM labs, allowing BLV students to participate actively in independent data collection tasks and engage in real-time experimentation (Supalo & Isaacson, 2014). Accessible lab equipment may consist of text-to-speech scientific data loggers or lab equipment with display screens that offer enlargement output support, with or without high contrast features (Supalo et al., 2014). The Submersible Audio Light Sensor (SALS), originally prototyped by Supalo et al. (2006), and commercialized by the American Printing House for the Blind (APH) based in Louisville, Kentucky, United States, is one example of an accessible science probe. This submersible sensor connects via Bluetooth to a smart phone or tablet, and can convert ambient light into an audible tone; it can be used to detect the formation of precipitates, evolution of bubbles, or slight to major color changes in test tubes in a qualitative manner including conducting acid–base titrations, where audio outputs are converted to spoken frequencies to allow for a numerical comparison. Note that the quantitative readings are dependent on ambient light and can be altered by shadows and the amount of light in the laboratory based on time of day and quality of available light (APH, 2022; Hoffmann, 2019; Supalo et al., 2006). Complementary technologies, such as tactile rulers and audible detectors, further universalize access. Emerging critiques, however, stress the need for cost-effective, scalable solutions to bridge implementation gaps, positioning these tools as vital for equitable STEM inquiry (Miner et al., 2001).

In addition to measuring tools, it is also important for students with BLV to have an individualized way of recording scientific observations and performing calculations, to foster autonomy in their lab work and minimize reliance on sighted assistance (Wild, 2025). These can encompass digital and tactile tools. For example, students may opt to record laboratory observations on computers or tablet devices; students who are braille readers may choose hard-copy braille note-taking with a slate and stylus or Perkins Braille Writer. Students may also prefer a specific accessible talking and/or audible scientific and/or graphing calculator: searching the Internet for "talking scientific calculators" show both web-based and physical solutions. Adequate time for orientation is needed ahead of the laboratory for the student with BLV to learn how to use the calculator effectively; further this personally selected calculator should be allowed on assessments and evaluations, thus requiring the STEM teacher to be familiar with the capabilities of the selected calculator in their course (Wild, 2025).

Laboratory safety is also a common concern for science teachers for all their students. To mitigate marginalization of students with BLV, we can implement proactive, layered strategies to position safety beyond an issue of mere compliance to an inclusive norm. Students with BLV should be required to observe the same safety videos and other training materials provided to their sighted peers; this should be coupled with an individualized and more hands-on safety instruction session prior to the start of class sessions which familiarizes the students with BLV with the layout, all exits, the location of safety showers and fire extinguishers, the location and use of disposal containers for safe disposal of glass, other materials and chemical waste (Canadian Chemical Institute, 2023). Disposal containers should be properly labeled in Braille or large print.

If a service animal is used in the laboratory environment, commercially available lab coats, foot protection, and eye protection for the animal is recommended; without personal protective equipment for the service animal, it is recommended they are placed in an adjacent room to the learning laboratory when possible, and other travel tools considered for working within the lab. Note that we are not prohibiting the use of the dog guide in the laboratory; rather we acknowledge the safety and welfare of the service animal, and the need for a properly trained service animal within a lab environment. With appropriate equipment and strategies, a dog guide can be used effectively in the teaching laboratory, and discussions of how this can be accomplished should occur prior to the lab course commencing (Miner et al., 2001).

As STEM labs in K-12 and higher education evolve, emerging technologies like AI-driven haptic tools and 3D-printed models promise transformative access for BLV students, building on foundational adaptations (Supalo et al., 2014). Recent innovations, such as oral tactile visualization via edible models from the Shaw Lab in Baylor, demonstrate equivalent accuracy to visual methods (Baumer et al., 2021). Partnerships like Independence Science's 2023 collaboration with ViewPlus have yielded refreshable Braille data collectors, enhancing real-time analysis (Purdue University, 2023). With widespread adoption of voice-activated interfaces now possible, this trajectory positions adaptive labs as catalysts for a broader and more inclusive inquiry.

## 21.4 Emerging Access Technology (AT) Tools

For decades, the AT industry has traditionally made only incremental improvements in the capabilities and quality of AT products used by the BLV, with a few notable exceptions. Similarly, one-line refreshable devices have been performing similar functions for nearly a half century: with these devices, students with BLV have had access to one line of braille text, which is effectively rendered. Graphical representations are also rendered one line at a time, limiting their presentation to a very part to whole approach: this constraint impacts effective access to STEM content, hindering the comprehensive visualization of non-linear mathematical expressions, graphical text including graphs and two-dimensional tables, and complex data.

These historical limitations of one-line braille devices have been surmounted by a recent transformative evolution: the multi-line refreshable tactile display (RTD). These displays feature multiple lines of braille, which vary in the number of lines and braille cells per line, enabling superior graphical representation and features like zooming in and out, scrolling, and real-time access to tabular, graphical, and data visualization tools. Though their low resolution still pales in comparison to the high-res LCD screens of the sighted world, this radical disruption holds immense innovation potential, representing the first major advancement in the industry in decades. These customizable multi-line displays may someday interface with research-grade laboratory equipment, further dismantling barriers to STEM participation in classrooms and employment!

As RTD technology evolves, text-to-speech applications are beginning to integrate support for multi-line braille and dynamic tactile graphics, with standards being collaboratively defined by AT companies and the BLV community. The developments in these emerging software and standards will extend to guidelines for authoring digital content, ensuring enhanced end-user experiences and greater accessibility for BLV individuals using RTDs in professional and educational settings. As RTD technology matures, both in terms of breadth of capabilities and breadth of use by the BLV, and its adoption within the BLV community expands, more advanced products will emerge at a commercial price-point that can reach a wider audience (Supalo et al., 2025).

Looking ahead, RTDs hold promise to make research-grade laboratory equipment inclusive of the BLV, by providing a viable vehicle for accessible STEM content to assist with data visualization and analysis tasks. To realize this potential, we hope that manufacturers of classroom and research tools commit to working with the BLV community to ensure their products are inclusively designed, and incorporate multimodal presentation of scientific data. Such a paradigm shift, where the design of laboratory equipment and other technical products includes accessibility considerations, will not only broaden the customer base and ensure compliance with emerging accessibility requirements, but also foster a more equitable landscape for scientific innovation.

## 21.5 Persons with BLV Pursuing Employment in the STEM Enterprise

The STEM professions have traditionally been lucrative for people that have proven themselves to have the skillsets for these more technical professions. The same should also be true for people with BLV (McDonnall et al., 2023). Yet the dearth of persons with BLV in the STEM professions and broader STEM enterprise points to multi-faceted barriers to inclusion such as: workforce entry and retention barriers such as discriminatory hiring, unaddressed accommodations, and mentorship gaps (McDonnall & Crudden, 2018; McDonnall et al., 2022); attitudinal barriers characterized by ableism and low expectations (McDonnall et al., 2022); and structural barriers in workplace infrastructure (Moon et al., 2012) such as inaccessible tools which directly impede BLV access to STEM content and activities.

Even where the workforce entry barriers are overcome, hiring managers may struggle with the concept of how a person with BLV can safely and competently make meaningful contributions on a project team in a STEM context. Such doubts about safety and competence in teams can be addressed through specific strategies for successful employment and broader workplace contributions, such as the employee's effective communication and technology proficiency, and the employer's environmental adaptations.

The employee with BLV needs to develop essential communication and technical skills for success in the workplace: they should be able to articulate their scientific data results to a diverse audience, which includes the sighted community; they should also develop proficiency using the required AT tools for their work. Alongside the employee's foundational skills in the STEM workplace, the employer should provide proper orientation to the physical workspace. Employees with BLV do not know innately how to get around an unfamiliar workspace; they need sufficient orientation time, including to hazards unmitigated and mitigated. Common mitigations include proper braille and large print signs, and installing barriers to prevent accidents with overhead obstructions such as a jutting staircase. Employees with BLV may wish to access external support in addition to this initial orientation, working with vocational rehabilitation agencies which provide certified orientation and mobility (O&M) support services; further, these services may offer job coaching and assistance with AT purchases for employers with documented limited financial resources required to purchase AT for the employment of the person with BLV.

## 21.6 Access Technology and Laboratory Tools for the STEM Enterprise

Beyond the foundational supports of O&M and initial AT acquisition, employers can foster not just workplace entry but ongoing independence and innovation for people with BLV, notwithstanding the oft-inaccessible demands of STEM labs. Prospective employers can proactively initiate discussion early in the hiring or onboarding process to align on the prospective employee with BLV's favored reading format, such as braille or large print, where production methods are well-documented online for straightforward implementation. Furthermore, collaborating on the types of raised line drawings that best suit the individual's needs will guide the acquisition of appropriate equipment, bridging preferences directly to practical tools like braille embossers—which operate like printers but output braille and often include graphics modes for direct tactile production. Other effective options include thermoform paper, a heat-sensitive plastic ideal for duplicating graphics from a braille paper master (with caution to avoid heat sources like heaters that could damage the material), and freehand tools using a stylus on plastic sheets over a drawing board for custom raised lines; numerous additional methods are extensively documented and searchable online (Maurer, 2001). In addition to tailoring media and tactile graphics for core workflow tasks, the conversation can extend to their preferred screen reader, these personalized insights serving as the foundation for selecting and acquiring the right AT for potentially seamless software compatibility.

While general AT lays the groundwork, targeted adaptations may ensure not only entry into the workplace, but sustained contribution to scientific innovation. Most research-grade lab equipment lacks native accessibility for people with BLV, especially when it involves unique, profession-specific software that is not in general use by the BLV community. Such unique applications may not be accessible out-of-the-box. However, depending on the screen-reader preference of the employee with BLV, Text-To-Speech (TTS) screen readers may allow customizable scripts that can offer an accessibility infrastructure for these specialized tools. Documentation for developing such scripts is publicly available; however, creating the scripts may be within the internal domain and in the purview of internal IT support, or may be contracted out to third party vendors that offer text-to-speech screen reader support services.

For hands-on laboratory tasks, specific adaptations promote independence and precision. When it comes to scientists with BLV, measuring liquids is challenging with traditional visual aids like graduated cylinders or columns; an alternative is using automatic pipettors which can be made accessible by setting and understanding the range of the device. For pipettors relying on tactile feedback, individuals require training to correspond each click with exact measurements for reliable task execution. Computer- or lab-automation automatic pipettors, despite their frequent dependence on mouse-driven graphical user interfaces that limit BLV usability, can be made keyboard-accessible through text-to-speech scripting languages (Graybill et al., 2008). Similarly, RTD devices may be leveraged to allow a person with BLV the opportunity to set their own experimental parameters, and the outputs from these types of experiments may *also* be rendered on the RTD.

## 21.7 Current Benefits and Limitations of Artificial Intelligence for Users with BLV

Artificial Intelligence (AI) tools have permeated everyday life, influencing tasks from literature searches to mundane decisions like comparing prices on household chemicals. At their core, these tools generate responses that are dependent on user prompts, refining their outputs over time as they process a growing volume of queries across diverse topics. The responses mirror the dominant user base. As a result, the evolution of AI is biased towards sighted audiences, under-representing the needs of a low-incidence group like people with BLV.

However, the BLV user benefits from the integration of AI into AT solutions. AI tools such as smart speakers and phones have been used by people with BLV for more than a decade now. Smart speakers can handle scientific queries, such as calculating distances between locations, retrieving atomic weights from the periodic table, or defining complex terms. In more recent times, TTS screen readers have incorporated AI solutions to offer low-level descriptions of images and other visual content, providing a meaningful entry point for interpreting everyday visuals without fully-sighted assistance.

Current benefits are tempered by limitations that underscore AI's current immaturity in nuanced recognition and functionality: AI responses struggle with the depth required for educational or professional use. Looking ahead, however, AI's maturation holds transformative promise for elevating BLV inclusion in scientific workflows. As this technology advances, alt-text for scientific visuals will evolve into more precise, user-ready descriptions; scientific referencing will sharpen, with improved accuracy in citing literature as adoption widens. There is no clear timeline for how long this will take, but the scientific community will determine when its reliability on such tasks is ready.

Further, AI tools will be used in data visualization tasks. For example, outputs on laboratory instrumentation can be scaled into a form that can be rendered in a more accessible format. This capability currently does not exist as a seamless solution: data visualization from an IR spectrophotometer, for example, requires visual outputs to be modified and scaled manually for viewing on an RTD device; this allows a person with BLV to examine and identify key peaks illustrated on a specific spectrum. More software development is necessary by the AT industry to ensure more functionality with RTD devices that can become more streamlined, thus providing more access to scientific data. Ultimately, AI could revolutionize data access by automating the rendering of lab outputs to accessible RTD formats—a progression that not only addresses today's shortcomings but paves the way for equitable, independent engagement with STEM knowledge.

## 21.8 Conclusion and Ways Forward

As refreshable tactile displays (RTDs), AI, and other innovative AT mature and gain widespread adoption within the BLV community, their seamless integration into the STEM pipeline and ecosystem—from classrooms to labs and workplaces—will redefine inclusion. STEM faculty can further this by building on foundational adaptations to deliver lectures and materials in "born-accessible" formats compatible with TTS and RTD scaffolds, enabling real-time engagement and reducing cognitive barriers for BLV students.

The science education technology industry can also follow suit, drawing inspiration from tools like the SALS probe to mainstream tactile markings on glassware, audible data loggers probes, and instrumentation and software with RTD/TTS interfaces to foster truly hands-on lab experiences. Equally vital is ensuring electronic scientific literature and databases align with common BLV solutions, unlocking pathways to the STEM workforce. Here, AI emerges as a powerhouse for data visualization and querying, but only if biases favoring sighted users are actively minimized through diverse training data.

Ultimately, the AT sector should accelerate commercialization of STEM-specific products as demand surges, driven by shifting expectations in both the BLV community when it comes to STEM, and the scientific community when it comes to learners and workers with BLV. As the expectations in the scientific community shift in a more positive direction as they pertain to the capabilities and contributions of people with BLV to make meaningful and potentially significant contributions in the scientific community, only then will a truly inclusive STEM workforce be achieved. By valuing contributions from persons with BLV for their ingenuity—not their rarity—we pave the way to normalize an equitable STEM landscape. Together, we can imagine a future where persons with disabilities are commonplace in the STEM community, with their contributions standing out not for overcoming "limitations," but rather for illuminating new frontiers in science.

## References

Albright, R. (2023, July 13). *How to create an inclusive workplace for blind employees*. https://Www.Inclusionhub.Com/Articles/How-to-Create-an-Inclusive-Workplace-for-Blind-Employees

Andreassi, D. G. (2018, December 6). *Disabled brings skills, loyalty, problem-solving and more to workforce*. https://www.corpmagazine.com/industry/human-resources/disabled-bring-skills-loyalty-problem-solving-and-more-to-workforce/

APH. (2022). *SALS*. Retrieved September 13, 2025, from https://www.aph.org/product/submersible-audio-light-sensor-sals/

Baumer, K. M., Lopez, J. J., Naidu, S. V., Rajendran, S., Iglesias, M. A., Carleton, K. M., Eisenmann, C. J., Carter, L. R., & Shaw, B. F. (2021). Visualizing 3D imagery by mouth using candy-like models. *Science Advances, 7*(22). https://doi.org/10.1126/sciadv.abh0691

Boyd, L., Lunsford, M., & Supalo, C. (2023). Designing the biology classroom & lab to support blind & visually impaired students. *The American Biology Teacher, 85*(1), 4–9. https://doi.org/10.1525/abt.2023.85.1.4

Canadian Chemical Institute. (2023). *Checklist for making science labs accessible for students with disabilities*. https://www.cheminst.ca/wp-content/uploads/2023/10/Checklist-for-Making-Science-Labs-Accessible-for-Students-with-Disabilities.pdf

Frankel, L., Brownstein, B., Soiffer, N., & Hansen, E. (2016). *Development and initial evaluation of the ClearSpeak style for automated speaking of algebra* (ETS Research Report No. RR-16-2). https://doi.org/10.1002/ets2.12103

Graybill, C. M., Supalo, C. A., Mallouk, T. E., Amorosi, C., & Rankel, L. (2008). Low-cost laboratory adaptations for precollege students who are blind or visually impaired. *Journal of Chemical Education, 85*(2), 243. https://doi.org/10.1021/ed085p243

Hoffmann, R. (2019). The SALS app: Making chemistry accessible with iOS devices *Journal of Science Education for Students with Disabilities, 22*(1), 4. https://jsesd.github.io/web-articles/JSESD-Volume-22/1096-Hoffman/

Isaacson, M., Schleppenbach, D., & Lloyd, L. L. (2010). Increasing STEM accessibility in students with print disabilities through MathSpeak. *Journal of Science Education for Students with Disabilities, 14*(1), 25–32.

Jernigan, K. (1965, July 6). *Blindness: Concepts and misconceptions*. https://Nfb.Org/Sites/Default/Files/Images/Nfb/Publications/Convent/Blndnesc.Htm

Maurer, M. (2001). *If blindness comes*. National Federation of the Blind.

McDonnall, M. C., & Crudden, A. (2018). Predictors of employer attitudes toward blind employees, revisited. *Journal of Vocational Rehabilitation, 48*(2), 221–231. https://doi.org/10.3233/JVR-180933

McDonnall, M. C., Sui, Z., & Fick, F. (2022). Beyond employment rates: Continuity of employment for people with visual impairments. *Journal of Visual Impairment & Blindness, 116*(2), 275–280. https://doi.org/10.1177/0145482X221091827

McDonnall, M. C., Sui, Z., & Fick, F. (2023). How degree major and demographic factors influence employment and earnings for college graduates with visual impairments. *Journal of Visual Impairment & Blindness, 117*(4), 363–374. https://doi.org/10.1177/0034355223 1187587

Miner, D. L., Nieman, R., Swanson, A. B., & Woods, M. E. (2001). *Teaching chemistry to students with disabilities: A manual for high schools, colleges, and graduate programs* (4th ed.). American Chemical Society Committee on Chemists with Disabilities.

Moon, N. W., Todd, R. L., Morton, D. L., & Ivey, A. (2012). *Accommodating students with disabilities in science, technology, engineering, and mathematics (STEM)*. Georgia Institute of Technology, Center for Assistive Technology and Environmental Access.

Prasertpong, P., Charmondusit, K., Taecharungroj, V., Rawang, W., Suwan, S., & Woraphong, S. (2023). Factors influencing science and environmental education learning of blind students: A case of primary school for the blind in Thailand. *International Journal of Education and Practice, 11*(2), 295–307. https://doi.org/10.18488/61.v11i2.3338

Purdue University. (2023). *Independence science and viewplus technologies partner to create science access tools*. https://www.purdue.edu/newsroom/2023/Q2/independence-science-and-viewplus-technologies-partner-to-create-science-access-tools-for-students-who-experience-blindness-or-low-vision

Rashid, I., & Chehadeh, D. (2023). Adaptation of chemistry experiments for middle school blind or visually impaired students. *Journal of Chemical Education, 100*, 2262–2268. https://doi.org/10.1021/acs.jchemed.3c00016

Schmitt-Koopman, F. M., Huang, E. M., Hutter, H.-P., & Darvishy, A. (2025). Towards More accessible scientific pdfs for people with visual impairments: Step-by-step remediation to improve tag accuracy. In *Proceedings of the 2025 CHI conference on human factors in computing systems* (pp. 1–16). https://doi.org/10.1145/3706598.3713084

Siler, B. E. (2024, March 12). *Service animals in the office? Here's how employers should handle accommodation requests*. https://Www.Adamsandreese.Com/Insights/Best-Practices-for-Employers-Service-Animal-Accommodation-Requests
Supalo, C. A., & Kennedy, S. H. (2014). Using commercially available techniques to make organic chemistry representations tactile and more accessible to students with blindness or low vision. *Journal of Chemical Education, 91*(10), 1745–1747. https://doi.org/10.1021/ed4005936
Supalo, C. A., Kreuter, R. A., & Mallouk, T. E. (2006). Seeing chemistry through sound: A Submersible audible light sensor for observing chemical reactions for students who are blind or visually impaired. *Assistive Technology Outcomes and Benefits, 3*(1), 110–116. https://www.learntechlib.org/p/55599/
Supalo, C., Schmid, S., Motion, A., Katselas, A., Bhattacharya, J., & Wilkinson, S. (2025). Refreshable tactile displays: A paradigm shift in chemistry learning for blind students who read Braille. *Journal of Chemical Education* (submitted and accepted with minor revision as of Oct 2025).
Supalo, C. A., Isaacson, M. D., & Lombardi, M. V. (2014). Making hands-on science learning accessible for students who are blind or have low vision. *Journal of Chemical Education, 91*(2), 195–199. https://doi.org/10.1021/ed3000765
Wild, T. (2025). *Accessible science for students with visual impairments*. APH Press, American Printing House for the Blind. https://books.google.com/books?id=elvw0AEACAAJ
Wunder, G. (2017, January). Breaking the ceiling for blind actors. *Braille Monitor, 60*(1), 13–13.

**Cary A. Supalo PhD** is a USA blind chemist, entrepreneur, inventor, strategist, and science accessibility expert. He founded Independence Science, an access technology and consulting firm that promotes hands-on science laboratory learning experiences for the blind. He also founded the annual ISLAND (Inclusion in Science Learning a New Direction) conference on disability and STEM, which convenes researchers, educators, and others interested in making STEM inclusive of the disabled. He works at the intersection of STEM education, access technology, and evaluation and assessment.

**Jasodhara Bhattacharya** a 15-year veteran of teaching STEM in public high schools in North America, led the Global Citizenship Education Working Group, co-convened by the Brookings Institution, UNESCO, and UN Secretary General Ban-Ki-Moon, to define a metric for measuring SDG 4.7.1. Her interest in making lifewide and lifelong learning in STEM accessible for all is motivated by improving the well-being of people and the planet.

# Chapter 22
# Unlocking Young Children's Potential with STEAM: Special Online Programme for Gifted Young Children

**Manabu Sumida**

## 22.1 Introduction

Recent research findings systematically argue that the scope and sequence of early childhood education should be considered not only in relation to age but also in terms of children's individuality and their social and cultural contexts. Nevertheless, the concepts of 'early childhood' and 'early years' remain predominantly framed in age-based terms. Consequently, discussions focusing on young children's learning abilities, or on the social and cultural demands placed upon them as citizens to mitigate disadvantage, remain limited.

Economist and Nobel laureate James Heckman has contributed significantly to this debate through his theory on the relationship between educational investment returns and age. Heckman views infancy and early childhood as uniquely valuable periods for investing in human capital, arguing that this foundational stage of learning is critical for later outcomes. His work has profoundly influenced global early childhood education reforms, promoting a shift towards more grassroots-oriented approaches to education and policy (OECD, 2006).

Gifted children exist irrespective of age or region; however, young children are particularly vulnerable to oversight. Therefore, this chapter focuses on STEAM (science, technology, engineering, art, and mathematics) education programmes designed for gifted young learners. It highlights an innovative initiative, established at Ehime University in Japan in 2010, that targets 'young children with gifts and talents'—a rare endeavour globally (Sumida, 2015). The definition of 'gifted and talented children' differs across countries and continues to evolve, reflecting both diverse research findings and cultural contexts (e.g., Dai & Chen, 2013). Although IQ is employed as one criterion in certain settings, giftedness is widely recognised

M. Sumida (✉)
Faculty of Education, Centre for Gifted Education and Talent Development, Ehime University, Matsuyama, Japan
e-mail: sumida.manabu.mm@ehime-u.ac.jp



L. Leite et al. (eds.), *Practical Work in Science Education in the Digital Era*, Advances in Technology-Rich Science Education 2, https://doi.org/10.1007/978-3-032-13788-3_22

as emerging and developing through the interaction between innate potential and the environment. Accordingly, comprehensive and multifaceted processes of evaluation and assessment are strongly recommended. Early childhood STEAM education has the potential to unlock children's abilities by transcending existing educational frameworks, simultaneously expanding learning opportunities and driving innovative transformation.

However, significant barriers persist. These include reductive assumptions, such as 'young children are primitive and passive'; beliefs that 'young children can only learn within the confines of the classroom or home'; and enduring stereotypes surrounding discipline and gender (for example, that 'science is for boys'). Furthermore, highly capable and motivated gifted children may experience intellectual isolation, develop a fear of failure, or become prone to perfectionism. An interdisciplinary approach to STEAM offers a constructive response to these challenges. By integrating diverse domains, educators can create environments that foster a wide range of personalities and talents. Breaking down entrenched barriers not only unlocks the potential of children who will shape the future but also compels us to reconsider and fundamentally reconstruct the mechanisms of learning itself.

Research findings demonstrating the scientific competence of young children have emerged globally, reshaping perspectives on science learning in early childhood (e.g., Trundle & Saçkes, 2015). These studies, supported by compelling empirical evidence, reveal that early scientific concepts—strongly influenced by innate factors—are acquired and transformed within domains such as naïve physics and naïve biology. The long-standing assumption that young children are incapable of conceptual thought and confined to concrete reasoning is increasingly being challenged by contemporary research. Consequently, curricula based on traditional developmental stage theories, which characterise early childhood as incapable of genuine scientific enquiry, are now undergoing critical reconsideration worldwide.

Scientific thinking and expression represent distinctive human intellectual activities that are inherently objective and cumulative, and emerge through interactions with natural objects and phenomena. The value of learning as a human endeavour lies not only in the experiential processes of observation and experimentation but also in the intentional acts of expression, interpretation, and communication that accompany them. This perspective is central to re-envisioning science education for the modern era.

Early childhood science education has gained profound contemporary significance in today's highly technological society. It integrates the innate elements of learning while fostering sustained, expansive, and deep intellectual enquiry, undertaken collaboratively and enriched by emotional engagement. From the earliest stages of life, all children should be afforded access to high-quality science education that both meets global standards and responds to the demands of an increasingly complex and rapidly changing century.

## 22.2 What is STEM? Why STEM for the Gifted?

Globally, STEM education has been a focus of numerous studies and practices across multiple disciplines. A simple Google search for 'STEM', 'STEM education', or 'STEM education research' yields over a billion results. However, serious academic interest only began around 2010, gaining real momentum in the mid-to-late 2010s. In the early stages, Dugger (2010) offered the following summary of STEM:

> Science, which deals with and seeks the understanding of the natural world.
>
> Technology, the modification of the natural world to meet human wants and needs.
>
> Engineering, the profession in which a knowledge of the mathematical and natural sciences gained by study experience, and practice is applied with judgement to develop ways to utilize economically the materials and forces of nature for the benefit of mankind.
>
> Mathematics, the science of patterns and relationships. (Dugger, 2010, pp. 118–119).

The significance of STEM education extends beyond preparing the future workforce to drive economic growth and sustain national leadership. It also lies in its ability to break down traditional disciplinary boundaries, fostering integrated approaches to real-world challenges. STEM emphasises innovation and application of technology to design creative solutions to complex problems. Maker (2020) defined exceptional talent in STEM disciplines and emphasised that performance-based assessment is both effective and essential for fairly evaluating gifted learners' abilities.

Exceptional talent in STEM has two essential components: (a) a highly integrated and interconnected knowledge structure and (b) the ability and willingness to solve a variety of types of problems, from well-structured and known to ill-structured and novel, in science, technology, engineering, and mathematics in the most effective, efficient, elegant, or economical ways. (Maker, 2020, p. 174).

Now, we can recognise three key characteristics of 21st-century science: it is interdisciplinary, collaborative, and international (Sumida & Ohashi, 2015). Fields such as biochemistry are on the rise, reflecting the growing importance of interdisciplinary research. The 2024 Nobel Prize in Physiology or Medicine has been awarded to Professors Victor Ambros and Gary Ruvkun. The Nobel Committee has highlighted their achievement with the statement, 'They have uncovered a completely new principle of gene regulation that is essential for multicellular organisms, including humans. MicroRNAs are proving to be of fundamental importance in understanding how living organisms develop and function.' With the aid of artificial intelligence and machine learning to analyse data, such as gene expression patterns, it is possible to comprehensively investigate the complex interplay between microRNAs and other genes. Thirty years have passed since the discovery of microRNAs in a small nematode, giving rise to a field of biology that now appears to be poised for even greater development. Modern science is increasingly conducted through international and collaborative efforts using technologies.

STEM education is aimed at diversifying and socialising learning, making it a collaborative and practical approach for shaping the future. STEM education empowers gifted and talented children to generate and validate ideas and connects them to the community and society, thereby creating new value and driving change. Integrated, practical, and sustainable STEM education can foster psychological and emotional resilience, thereby, strengthening individuals' perseverance by forcing them to overcome difficult experiences. With its focus on problem-solving and innovation, STEM education offers a strong foundation for recognising and supporting gifted children, helping them grow and thrive.

## 22.3 Key Barriers to Educating Gifted Young Children

### *22.3.1 Unlocking the Meaning of Age*

Are young children simple beings? In reality, they possess a latent intellect that is capable of reasoning with extraordinary logic and harbouring profound questions. They demonstrate independence, curiosity, and a willingness to engage boldly with the unknown. Gifted children are found in all contexts, regardless of age or region. These children include those who immerse themselves passionately in a single pursuit; those with rich vocabularies and exceptional linguistic sensitivity; those with a strong sense of justice and deeply held ideals; and those who perceive the world through original and unconventional perspectives. Their gifts and talents are evident in both everyday interactions and school life.

However, within existing educational frameworks, these exceptional abilities are often overlooked, undersupported, or even misunderstood. Such neglect can result in difficulties including imbalances between intellectual and emotional development, heightened sensitivity to external expectations, and increased vulnerability to frustration or hurt. Some children feel alienated in school or social settings, while others are "twice exceptional" (2e), displaying remarkable strengths and significant challenges. Without environments that actively nurture their abilities, these children risk having their potential stifled or lost.

The need to identify and address these challenges has been recognised for over a century. Hollingworth (1926) conducted pioneering research into the characteristics and educational needs of gifted young children, emphasising their socio-emotional vulnerabilities and experiences of isolation. Bloom (1985) traced the development of individuals who achieved excellence in domains such as music and art, mathematics and science, and athletics, while Gagné (1985) distinguished between natural gifts and cultivated talents, underscoring the role of structured support in talent development. More recently, Smutny et al. (2015) developed a guidance manual for universities supporting children aged 4–9, offering concrete strategies to foster balanced cognitive, creative, and socio-emotional development. Similarly, the National Association for Gifted Children (2006), in its Position Statement, advocates flexible learning environments, enquiry-based approaches, and socio-emotional support. Collectively, this body of research demonstrates both the early emergence of gifted characteristics and critical need to create supportive educational environments.

Parallel developments can be observed in the field of early childhood STEM education. Tunnicliffe and Kennedy (2022), in their synthesis of early STEM education research, highlighted young children's keen interest in natural phenomena and their capacity for early concept formation. Their work underlined the importance of conceptual development, play-based learning, and quality educational support. Bagiati and Evangelou (2015) reported one of the earliest integrated STEM initiatives in gifted education, introducing engineering thinking at the kindergarten level and demonstrating the rationale and effectiveness of enquiry-based approaches in early learning. Similarly, Lange et al. (2019) provided practical strategies for introducing STEM concepts to children aged three to five years, illustrating how mathematics, science, and engineering thinking can be integrated into play-based learning.

At the policy level, a re-evaluation of young children's competencies is gaining momentum, along with calls for qualitative improvements in early education (National Academies of Sciences, Engineering, and Medicine, 2022). Early childhood STEM education builds on the long-standing tradition of creative, play-based learning while simultaneously opening new avenues for educational research and practice. Such approaches not only advance equity and inclusion but also lay the foundations for future social innovation.

### *22.3.2 Unlocking Gender Stereotypes*

In the 2015 United Nations statement, Sustainable Development Goal (SDG) 5 addresses gender equality, with the target to "achieve gender equality and empower all women and girls." It seeks to not only eliminate gender-based discrimination and violence but also build a society in which women and girls can fully realise their potential. This includes tackling discriminatory and harmful practices, recognising the value of unpaid domestic and childcare work, and promoting women's participation in politics and the economy. Yet, despite progress, gender equality remains elusive globally—including in Japan. Challenges persist particularly in relation to women's leadership roles and economic opportunities, as reflected in the underrepresentation of female researchers and developers in STEM fields (UNESCO, 2017).

These structural inequalities are closely linked to the formation of unconscious bias. Numerous studies reveal that gender differences in terms of achievement in science and mathematics are minimal and primarily shaped by social, cultural, and educational factors rather than inherent ability. Societal expectations and career pathways, therefore, play an important role in shaping outcomes for talented female students. Subotnik et al. (2017) highlight the effectiveness of policy-level interventions and educational practices—including mentoring and role models—in bridging the gender gap in STEM talent development. Ong et al. (2011) review research on the 'double bind'—the pressure on talented female students to be highly competent while simultaneously conforming to gender role expectations—underscoring the unique social and psychological barriers they face. Bian et al. (2017) provide evidence that stereotypes about intellectual ability emerge as early as six years of age and can shape girls' learning trajectories and career aspirations, serving as a clear warning to educators. Similarly, Swanson et al. (2021) emphasise the importance of mentoring and early research experiences in counteracting unconscious gender biases.

The implications of this body of research for STEM education in the early years are clear. Young children, when provided with appropriate opportunities, often engage with STEM subjects without considering gender stereotypes. Science, mathematics, and the arts belong to all learners. Curiosity knows no gender. At Kids Academia, for example, many girls actively participate in STEAM programmes. A female Year 2 student demonstrated particular interest in potentially hazardous topics, revealing her detailed knowledge of dangerous sea creatures. She chose earthquakes as the theme for her independent research project, constructing various block structures and investigating how readily each collapsed when shaken, before presenting her systematic findings (see Fig. 22.1). Notably, across the five Kids Academia contests held to date, no significant gender disparity was observed among award winners. In activities that foster free enquiry, the focus shifts to each child's unique ideas, encouraging confidence and pride, while simultaneously allowing children to encounter the unexpected brilliance of their peers. Such environments provide fertile ground for innovation to take root.

The global COVID-19 pandemic posed additional challenges for education, particularly for early childhood settings—traditionally grounded in free play and face-to-face interaction. Many schools were compelled to transition rapidly to online learning. Between 2018 and 2020, significant differences were observed in the motor and cognitive development, learning attitudes, and behavioural self-regulation of children aged four to six (González et al., 2022). Nevertheless, research also suggests that for gifted learners, online modes of education can serve as a viable alternative. Chowkase et al. (2022), for example, found no significant differences in

**Fig. 22.1** A girl exploring earthquake resistance through different ways of stacking blocks

motivation by format, level, or gender, highlighting the potential of out-of-school online programmes as an inclusive pathway for gifted children. Unlocking gender stereotypes not only broadens access to STEAM education but also reinforces the foundations for a more inclusive, innovative, and equitable society.

### 22.3.3 *Unlocking the Perfectionism Trap*

Those engaged in nurturing gifted and talented children will be familiar with moments when expectations are surpassed and challenging problems are solved among peers of similar age and background. Alongside such achievements, there may also be a lingering fear of always needing to succeed, a pressure to excel in every endeavour, and an avoidance of failure. These pressures can suppress the demonstration of true potential and diminish the intrinsic joy of learning.

Perfectionism has long been recognised as a psychological trait, frequently found among gifted learners and with the potential to constrain the free-flowing, exploratory processes that are vital for authentic scientific enquiry. Grugan et al. (2021) emphasise the importance of distinguishing between perfectionistic strivings—the pursuit of high standards—and perfectionistic concerns—the fear of making mistakes or falling short. In fact, the history of science is rich with examples of errors leading to important accidental discoveries. Safe, manageable failures are particularly valuable for intellectually mature, gifted young children, as they enable them to view mistakes as opportunities for exploration and growth.

Gifted children are especially vulnerable to the perfectionism trap precisely because many tasks initially come easily to them. For example, a child who consistently achieves high scores in school or competitions with little effort may gradually attract increasing expectations from parents, teachers, and peers. In turn, this can lead the child to demand perfection of themselves, creating a cycle of anxiety and self-imposed pressure. We must remember that every Nobel Prize-winning scientist has encountered failure—often many times—on the path to discovery.

#### 22.3.3.1 Safe Failure in Practice: The '−1' Pancake Experiment

To explore the role of 'safe failure' in STEAM learning, we piloted an 'okay-to-fail' approach during the Kids Academia Winter School of 2022, conducted online with eight participants (two kindergarten children, three first graders, and three second graders). The experiment centred on making pancakes. Initially, all participants followed the correct recipe: combining flour, baking powder, sugar, eggs, and milk to produce a standard pancake. In the second stage, children were invited to conduct a '−1 science experiment' by deliberately omitting one ingredient of their choice.

Before cooking, the children made predictions. Some hypothesised that leaving out eggs would result in whiter pancakes, while one child chose to omit flour, curious about its unpredictable consequences. The outcomes varied widely, prompting lively discussion about texture, flavour, and appearance. This structured but playful exercise encouraged the children to move beyond rigid notions of 'right' and 'wrong' experiments, instead devising their own hypotheses and recognising the value of divergent outcomes.

The parents' feedback underscored the educational significance: "My daughter, who was accustomed to making pancakes, found that removing an ingredient transformed our daily routine into a scientific experiment." "I believe this experience provided valuable insights for my son, who had become overly confident and reluctant to acknowledge mistakes."

Such reflections highlight how small-scale, low-risk failures can not only broaden children's scientific thinking but also nurture resilience, humility, and openness to alternative outcomes.

#### 22.3.3.2 Safe Failure with Generative AI

Another example of 'safe failure' was introduced through the use of generative artificial intelligence (AI). The use of AI can provide advanced content, personalized learning, creative writing and image manipulation, critical thinking and problem solving, collaboration, research skills, and advanced technology (Siegle, 2023). As part of the same Winter School, children engaged in an activity involving AI-generated multiple-choice questions. The AI tool was instructed to generated possible answers for the question 'Why do rice cakes puff up when they are baked?' It produced the following options:

Option 1. Because the air trapped in the rice cake expands.
Option 2. Because the water in the rice cake expands when steam is formed.
Option 3. Because the heat on the surface of the rice cake causes it to expand.
Option 4. Because the ingredients in the rice cake react chemically to produce gas.
Option 5. Because the rice cake is happy.

When instructed to select the correct response, the AI tool chose Option 1. However, the scientifically correct answer was Option 2, as the increase in volume is greater when liquid water turns into gas than when a gas merely expands. Once this was explained, the AI tool 'acknowledged' its mistake and corrected itself. The children were fascinated to see that AI could be wrong. This realisation prompted surprise and curiosity, sparking further questions and leading to an animated discussion. As Tang (2024) stated, writing feedback from AI and dynamic ways of using AI to generate questions and answers are effective. Most studies have shown that AI has significantly improved skills such as creativity, emotion control, collaborative inquiry, literacy skills, and computational thinking (Su & Yang, 2022). The encounter with an AI error served as a safe, shared moment of failure that modelled critical thinking and reinforced the importance of verification in scientific reasoning.

These examples illustrate how engaging gifted children in structured opportunities for safe failure can be a powerful strategy for mitigating the risks of perfectionism. By reframing mistakes as integral to learning, children are encouraged to balance their perfectionistic strivings with a tolerance for uncertainty and error, thereby fostering creativity, resilience, and innovation. Activities such as the '$-1$' pancake experiment and AI-assisted enquiry demonstrate how everyday contexts and emerging technologies alike can be harnessed to cultivate healthy dispositions toward failure in gifted education.

## 22.4 STEAM in Action—STEAM Education Practices for Gifted Young Children: 'Sound'

This section introduces case examples from the Kids Academia Winter School held in 2023. Ten children from across Japan participated online. Each session lasted for approximately one hour, with parents providing support where necessary, including assisting with device operations. The Summer School comprised four sessions. The first session, Science is Fun!, featured introductions by the children, an overview of Kids Academia, and a series of icebreaker activities. Session 2 offered experiential learning centred on Water, while Session 3 focused on Sound. Session 4, You're a Scientist Starting Today!, invited each child to present research based on their personal interests.

This section highlights the practice from Session 3: Let's Play with Sound!. Sound is a theme that can be explored through play using the five senses in familiar settings and one that readily evokes rich emotions. Sounds are present both in nature and indoors, constantly shifting in quality and form. Building on these resources, we designed and implemented an original program in Kids Academia. The materials were distributed in advance, and the activities were conducted synchronously online.

As an introductory activity, the children made string telephones using paper cups and kite string. The process of handling the materials, constructing the device independently, and testing it fostered active engagement and promoted enquiry. The children discovered that altering the tension of the string affected how the sound travelled, and some experimented further by pinching the string with their fingers. They then predicted what might happen if the string were replaced with another material, such as vinyl cord, wire, or rubber, and tested their predictions. This cycle of hypothesising and experimentation represents a fundamental aspect of scientific thinking, and the children found it thoroughly enjoyable.

By using materials to explore phenomena and engaging in trial and error, the children refined their thinking and developed broader perspectives. When asked what else they would like to test, the children generated a wide range of creative ideas: altering the size of the cup, substituting yarn for string, experimenting with rubber bands, changing the material of the cup, varying the point of string attachment, and placing small pieces of paper inside the cup to observe the strength of the sound, attaching the string without tape (e.g. by inserting it directly), dipping the string in water while speaking, replacing the string with a hollow straw, and even extending the string to a length of approximately one kilometre. Their imaginative suggestions illustrate how a simple activity can spark divergent thinking and sustained curiosity.

Subsequently, we introduced the concept of sound as a waveform using a digital application. This enabled the children to visualise how they were exchanging sounds in real time despite being physically separated in the online workshop. They listened with great interest as we explained how information received by the ear is transmitted to the brain. This step illustrated the value of connecting children's intuitive play with more formal scientific representations.

For this activity, participants were assigned to record a 30-s 'sound of summer' as preparatory work. At the end of each session, the submitted recordings were collated and arranged into a single musical piece with the guidance of music education specialists. Everyday and seasonal sounds—including the preparation of shaved ice, fireworks, splashing water, cicadas, frogs, festival drums, and electric fans—were integrated to form the final composition. This collaborative activity demonstrated how children's lived experiences and cultural contexts can be transformed into creative outputs that foster scientific curiosity and artistic expression.

In the fourth and final session, one participant selected sound as the focus of her independent research project. She began by comparing the difference in sound when one end of a toilet-roll tube was covered by hand versus when it was left uncovered. Building on this, she employed a digital application to investigate how sound frequency varied when the tube length was doubled or tripled by connecting multiple tubes. She subsequently presented these findings to the group. This case illustrates how a simple, child-initiated question can evolve into a systematic enquiry, highlighting the potential of STEAM approaches in nurturing experimental thinking, creativity, and independent research skills in young learners (Fig. 22.2).

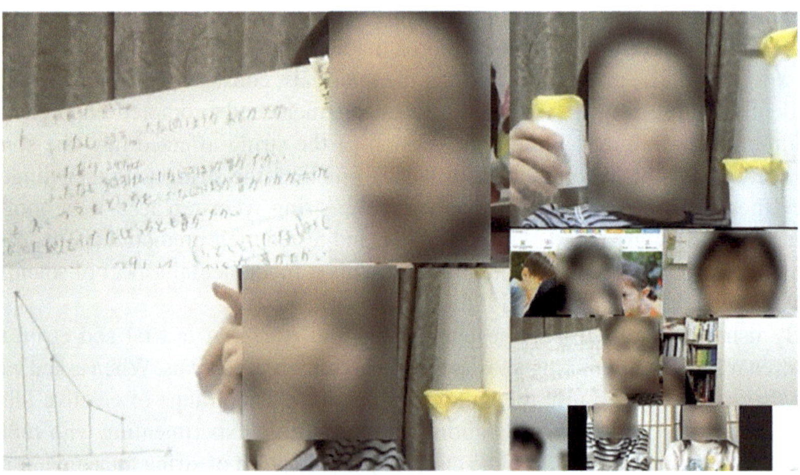

**Fig. 22.2** A girl presenting her research on how paper roll length affects sound vibrations

## 22.5 Conclusion and Way Forward

Providing STEAM education from an early age can contribute to a society in which schools, families, and communities collaborate to nurture each child's strengths, allowing individuality and diversity to flourish. This chapter has explored key dimensions of unlocking children's potential through early STEAM education, focusing on three core aspects: dispelling the assumption that children are primitive and passive beings; eliminating disciplinary and gender stereotypes, such as "science is for boys"; and relinquishing the pursuit of perfectionism. Online platforms have further expanded opportunities to establish learning communities that connect gifted young children across regions, even though they inevitably constrain a range of experiences.

Expert research mentoring provides young children with the experience of "thinking like researchers and developers" from the earliest stages of their education. Such experiences not only nurture critical thinking, creativity, and expressive abilities but also aid in overcoming intellectual isolation. To sustain this development, building a foundation in which homes, schools, and society collaborate to support children's learning is essential. Four key recommendations are proposed as central to enabling children cultivate their strengths:

- Passion: Children should be encouraged to pursue activities from the perspective of their enjoyment and curiosity. Joy is a vital condition for talent to flourish.
- Knowledge and skills: A strong foundation of knowledge is indispensable. Even children with exceptional abilities must continuously engage in practice. This requires not only passion but also educational support to help them understand the long-term value of persistence and mastery.
- Uniqueness: Recognising and supporting each child's individuality and ambition are crucial. Unique goals, when nurtured, can develop into meaningful contributions to society.
- Experiences and opportunities: Through diverse experiences, children make discoveries, learn self-regulation, build resilience, and gain confidence, while forming friendships and learning to recover from setbacks along the way.

STEAM education for gifted young children requires a fundamental reimagining of educational approaches. It can transform learners from passive recipients into active agents invested with the power of shaping the future. Moving beyond the boundaries of the classrooms, it enables "open society learning" and promotes a culture in which talent is valued not merely as an individual trait but as a shared societal asset. Moreover, this approach encourages global collaborative enquiry among young children, forming international learning communities. By actively engaging with the defining challenges of the next generation, such as artificial intelligence, climate change, and disaster resilience, children can experience cutting-edge STEAM education from an early age.

Promoting diversity, equity, and social collaboration in STEAM education not only enriches children's lives but also strengthens society. By educating and supporting gifted young learners, we unlock their potential while simultaneously fostering an inclusive environment that celebrates diversity, nurtures innovation, and facilitates a better future for all.

**Acknowledgements** This work was supported by JSPS KAKENHI Grant Numbers 23K17620, and 23K22338.

## References

Bagiati, A., & Evangelou, D. (2015). Engineering curriculum in the preschool classroom: The teacher's experience. *European Early Childhood Education Research Journal, 23*(1), 112–128. https://doi.org/10.1080/1350293X.2014.991099

Bian, L., Leslie, S.-J., & Cimpian, A. (2017). Gender stereotypes about intellectual ability emerge early and influence children's interests. *Science, 355*, 389–391. https://doi.org/10.1126/science.aah6524

Bloom, B. S. (Ed.). (1985). *Developing talent in young people.* Ballantine Books.

Chowkase, A. A., Datar, K., Deshpande, A., Khasnis, S., Keskar, A., & Godbole, S. (2022). Online learning, classroom quality, and student motivation: Perspectives from students, teachers, parents, and program stuff. *Gifted Education International, 38*(1), 74–94. https://doi.org/10.1177/02614294211060401

Dai, D. Y., & Chen, F. (2013). Three paradigms of gifted education: In search of conceptual clarity in research and practice. *Gifted Child Quarterly, 57*(3), 151–168. https://doi.org/10.1177/00169862134900

Dugger, W. E. (2010). Evolution of STEM in the United States. In *Proceedings of the 6th Biennial international conference on technology education* research (pp. 117–123).

Gagné, F. (1985). Giftedness and talent: Reexamining a reexamination of the definitions. *Gifted Child Quarterly, 29*(3), 103–112. https://doi.org/10.1177/001698628502900302

González, M., Loose, T., Liz, M., Pérez, M., Rodríguez-Vinçon, J. I., Tomás-Llerena, C., & Vásquez-Echeverrí, A. (2022). School readiness losses during the COVID-19 outbreak: A comparison of two cohorts of young children. *Child Development, 93*(4), 910–924. https://doi.org/10.1111/cdev.13738

Grugan, M. C., Hill, A. P., Madigan, D. J., Donachie, T. C., Olsson, L. F., & Etherson, M. E. (2021). Perfectionism in academically gifted students: A systematic review. *Educational Psychology Review, 33*(4), 1631–1673. https://doi.org/10.1007/s10648-021-09597-7

Hollingworth, L. S. (1926). *Gifted children: Their nature and nurture.* The Macmillan Company.

Lange, A. A., Brenneman, K., & Mano, H. (2019). *Teaching STEM in the preschool classroom: Exploring big ideas with 3 to 5 years old.* Teachers College Press.

Maker, C. J. (2020). Identifying exceptional talent in science, technology, engineering, and mathematics: Increasing diversity and assessing creative problem-solving. *Journal of Advanced Academics, 31*(3), 161–210. https://doi.org/10.1177/1932202X20918203

National Academies of Sciences, Engineering, and Medicine. (2022). *Science and engineering in preschool through elementary grades: The brilliance of children and the strengths of educators.* The National Academies Press.

National Association for Gifted Children (2006). *Early childhood: Creating contexts for individualized learning in early childhood education.* Position Statement. Retrieved from https://cdn.ymaws.com/portal.nagc.org/resource/resmgr/knowledge-center/position-statements/early_childhood_position_sta.pdf

OECD. (2006). *Starting strong II: Early childhood education and care.* OECD Publishing.

Ong, M., Wright, C., Espinosa, L. L., & Orfield, G. (2011). Inside the double bind: A synthesis of empirical research on undergraduate and graduate women of color in science, technology, engineering, and mathematics. *Harvard Educational Review, 81*(2), 172–209. https://doi.org/10.17763/HAER.81.2.T022245N7X4752V2

Siegle, D. (2023). A role for ChatGPT and AI in gifted education. *Gifted Child Today, 46*(3), 211–219. https://doi.org/10.1177/10762175231168443

Smutny, J. F., Walker, S. Y., & Honeck, El. I. (2015). *Teaching gifted children in today's preschool and primary classrooms: Identifying, nurturing, and challenging children ages 4–9*. Free Spirit Publishing.

Su, J., & Yang, W. (2022). Artificial intelligence in early childhood education: A scoping review. *Computers and Education: Artificial Intelligence, 3*, Article 100049. https://doi.org/10.1016/j.caeai.2022.100049

Subotnik, R. F., Stoeger, H., & Olszewski-Kubilius, P. (2017). Talent development research, policy, and practice in Europe and the United States: Outcomes from a summit of international researchers. *Gifted Child Quarterly, 61*(3), 262–269. https://doi.org/10.1177/00169862177018

Sumida, M. (2015). Kids Science Academy: Talent development in STEM from the early childhood years. In M. S. Khine (Ed.), *Science education in East Asia: Pedagogical innovations and research-informed practices* (pp. 269–295). Springer.

Sumida, M., & Ohashi, A. (2015). Chemistry education for gifted learners. In J. Garcia-Martine & E. Serrano-Torregrosa (Eds.), *Chemistry education: Best practices, opportunities and trends* (pp. 469–487). Wiley-VCH.

Swanson, J. D., & Sickle, M. V. (Eds). (2021). *Talent development in school: An educator's guide to implementing a culturally responsive talent identification and development program*. Routledge.

Tang, K.-S. (2024). Informing research on generative artificial intelligence from a language and literacy perspective: A meta-synthesis of studies in science education. *Science Education, 108*(5), 1329–1355. https://doi.org/10.1002/sce.21875

Tunnicliffe, S. D., & Kennedy, T. J. (2022). *Play and STEM education in the early years: International policies and practices*. Springer.

Trundle, K. C., & Saçkes, M. (2015). *Research in early childhood science education*. Springer.

United Nations. (2015). *Transforming our world: The 2030 agenda for sustainable development*. Retrieved from https://sdgs.un.org/sites/default/files/publications/21252030%20Agenda%20for%20Sustainable%20Development%20web.pdf

United Nations Educational, Scientific and Cultural Organization. (2017). *Cracking the code: Girls' and women's education in science technology, engineering and mathematics (STEM)*. UNESCO. https//doi.org/https://doi.org/10.54675/QYHK2407

**Manabu Sumida** is a professor in the Faculty of Education at Ehime University, Japan, director of the Centre for Gifted Education and Talent Development, and a researcher on STEAM education for gifted learners. He received the Ryoji Noyori Education Award in 2013. He is president of the Japan Society for Science Education, the Asia Regional Representative of the International Council of Associations for Science Education, and an executive committee member of the Asia-Pacific Federation on Giftedness.

# Chapter 23
# Potential and Limitations of Virtual Field Trips in promoting Nature-Based Learning and Inclusion

**Clara Vasconcelos, Tiago Ribeiro, and André Ramos**

## 23.1 Introduction

Geology, a historical and hermeneutic science, was early on recognized as a field science. Until the nineteenth century, the primary task of the geologist was to travel to the natural environment to collect rock samples and map the studied area, attempting to find evidence that would support the theories being analysed. Geology would be impossible without travel, and the geologist of that period is identified as a traveling naturalist (Brandstetter, 2011).

According to Frodeman (2014), there is no true philosophy of Geology or the Earth Sciences, as exists with fields such as Physics and Biology, because it is believed that examining the Earth Sciences will not reveal relevant characteristics to understand the nature of sciences. However, Geology is a science that deals with processes and forces that cannot be directly observed. The very notion of time, on the geological scale, places this science in a disciplinary field with a peculiar nature that goes beyond the use of mathematical applications and physical or chemical laws. According to Schiappa and Smith (2018), the discipline of geology is a product of early exploration, so

C. Vasconcelos (✉) · T. Ribeiro
Unit of Science Teaching, Faculty of Sciences, Porto University & Interdisciplinary Centre of Marine and Environmental Research, Matosinhos, Portugal
e-mail: csvascon@fc.up.pt

T. Ribeiro
e-mail: tribeiro@fc.up.pt

C. Vasconcelos
Department of Geoscience, Environment and Spatial Plannings, Faculty of Sciences, Porto University, Porto, Portugal

A. Ramos
Unit of Science Teaching, Faculty of Sciences, Porto University, Porto, Portugal
e-mail: up202300995@edu.fc.up.pt



L. Leite et al. (eds.), *Practical Work in Science Education in the Digital Era*, Advances in Technology-Rich Science Education 2, https://doi.org/10.1007/978-3-032-13788-3_23

fieldwork has been a long tradition among undergraduate programs, and field investigations courses offer exciting alternatives to traditional learning. Schiappa and Smith (2018) even mention that field courses broaden student experiences, enhance peer-based and collaborative learning, build team-working skills, and foster a curiosity about the natural environment. Field trips help bridge formal and informal learning, and prepare students for lifelong learning (Tuthill & Klemm, 2002). However, as stated, unlike other sciences, the objects of study in geology are highly complex systems that operate on a large temporal scale with very particular characteristics. Thus, when faced with the difficulties of working with the geological past, whether due to temporal and spatial scale issues or the uniqueness and complexity of geological processes, fieldwork remains the main strategy of geologists. After careful literature review Ståle Holgersen (2021) states that: (i) while field trips have a rich and long-standing history, they have not remained static over time; (ii) according to Anđelković and collaborators (2017), due to new international relations, such as the Bologna Process, and the influence of the constructivist educational paradigm, which places students at the centre of the learning process, there is a growing need to improve the quality of fieldwork and increase its integration into curricula; (iii) roles of teachers have significantly impacted field studies and teachers are shifting away from traditional, instrumental approaches focused on knowledge transfer and efficiency (Golubchikov, 2015); (iv) teachers are now viewed as knowledge transmitters and facilitators of learning, encouraging students to become more self-managed and reflective in their learning processes. However, fieldwork, when used in the teaching process, is quite complex and presents cognitive challenges that are difficult to solve, because what is not directly observable requires abstract reasoning and analogy. For this reason, organizing field trips became crucial for their success as a learning strategy. Based on the above, the purpose of this chapter is: (i) to discuss fieldwork organization models used both in onsite field trips and online field trips; (ii) to highlight the importance of nature-based learning and the relevance of field trips in promoting it; and (iii) to explain the value of virtual field trips as complementary to onsite outings, while also serving as a means of promoting inclusive education.

## 23.2 Fieldwork Organization Models for Onsite and Online Field Trips

Outdoor education involves learning experiences that extend beyond the traditional classroom, engaging students directly with the natural world. This approach enhances comprehension and facilitates the learning of scientific concepts (Orion, 1993). Outdoor learning is understood as consistent, organized learning experiences that take place in both on-campus and off-campus outdoor settings for school-aged children (Mann et al., 2021). In contrast, out-of-school learning refers to educational activities that occur outside the school environment, such as field trips to various locations, including museums and industrial sites, which can be either indoors or outdoors and industrial plants (Mann et al., 2021). Activities such as field trips, nature walks, and hands-on investigations in natural environments are central to both approaches, highlighting the importance of experiential learning. These activities provide students with the opportunity to observe and analyse real-world examples of geological and ecological processes (O'Brien et al., 2011; Vasconcelos & Ribeiro, 2026). For this reason, field trips are highly relevant in helping both teachers and students learn about natural heritage. They are strongly recommended in geoparks to promote both formal and informal education, as well as to enhance public outreach and scientific communication to the public. The literature has underscored the important contribution of fieldwork at geoheritage sites (Morais et al., 2021; Reynard & Brilha, 2017; Vasconcelos & Ribeiro, 2026), focusing on their scientific, educational, cultural, and aesthetic significance. These sites present valuable educational opportunities, enabling hands-on learning that reinforces classroom concepts and fosters a deeper interest in Earth sciences. Several fieldtrips have been undertaken to enhance the teaching and learning of geological heritage and provide educators with specialised training.

There are several organizational models for field trips known in the literature as those presented by Brusi (1992), Compiani and Carneiro (1993), and Pedrinaci and collaborators (1994). Nevertheless, the one most likely to be used is Orion's (1993) model, both for its simplicity and its effectiveness. This organizational model divides field trips into three stages: preparatory unit, visit, and summary unit (Fig. 23.1). The preparatory unit is built around lab inquiry-based activities that cover subjects that will be covered in the field (e.g., modelling exercises or using identifying skills to manage the materials that students encounter on the fieldtrip, including rocks and minerals). As students are introduced to the subsequent stages, the operationalization of these activities enables the development of concepts in tangible subjects, addressing cognitive, psychological, and geographical elements, diminishing space's uniqueness (Orion, 1993; Orion & Hofstein, 1994; Orion et al., 2020). Orion (1993) argues that fieldwork should occur early in the learning process, but not as the first activity. It should be preceded by a preparatory phase aimed at reducing the "novelty space". This preparatory phase, according to the author, consists of three components: cognitive, psychological, and geographical factors. When novelty space is kept to a minimum during field excursions, learning becomes more important. Regarding cognitive factors, these are related to the type and level of prior knowledge and skills of the students, which may make it difficult for them to perform the fieldwork activity. Worksheets, a review of theoretical concepts before the fieldtrip, or the manipulation of samples and materials that students may encounter during the fieldtrip can all be used to address cognitive aspects. By outlining the goals of the fieldtrip, the activities that will be created, or the stops that will be made, psychological issues can be minimized. By providing maps of the fieldtrip's location and even utilizing tools like GIS, which allows one to track how natural systems operate and adapt, geographic factors can be reduced.

By applying system thinking to actual situations and raising the degree of abstraction, the fieldtrip integrates the skills learned in the preceding unit with the natural world. To achieve this, the field exercise ought to give students a chance to connect

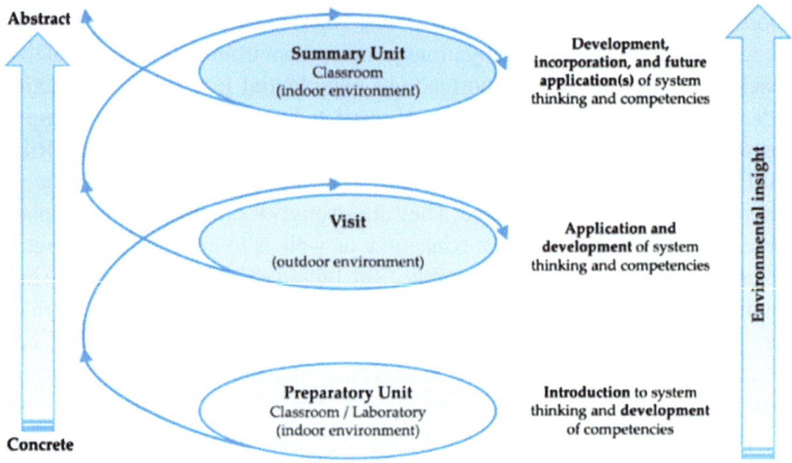


**Fig. 23.1** The three stages of a fieldtrip (Ribeiro & Orion, 2021, p. 8)

with nature by observing lithological and geomorphological features as well as interactions among the various subsystems. The efforts to integrate the multiple intelligences developed in the field are finally finished in the summary unit, increasing the degree of abstraction through the application of information to novel contexts (Orion, 1993; Ribeiro & Orion, 2021). Ribeiro and Orion (2021) argue that field trips are an excellent strategy for promoting environmental insight (Fig. 23.1), a competence that refers to the understanding of Earth as a complex system composed of interrelated subsystems that exchange mass and energy, maintaining equilibrium among themselves.

A key element in achieving this environmental insight, particularly within the Earth Systems Education (ESE) approach, is the ability to think systemically. System thinking involves a high level of abstraction, enabling learners to connect and relate concepts across different scientific domains. The development of system thinking is crucial for fostering a deeper understanding of environmental dynamics and ecosystems. According to the literature, environmental insight is shaped by three main factors: (i) recognizing that we live on a planet where subsystems interact cyclically through the exchange of matter and energy; (ii) understanding that humanity is an integral part of the Earth's system; and (iii) aligning our actions with the dynamics of the Earth system (Ribeiro & Orion, 2021; Vasconcelos & Orion, 2021). These factors work together to help learners not only grasp the interconnectedness of natural systems but also to live in harmony with their complex, dynamic processes. Field trips, in this context, are an invaluable strategy, because they offer students hands-on experience and direct engagement with the environment, enhancing their ability to recognize and understand the interconnections between the natural subsystems. These experiences bridge abstract theoretical knowledge with real-world observations, making the learning process more tangible and meaningful. The model presented has shown evidence of successful onsite field trips in terms of student skill development (Orion et al., 2020), but it has also proven to be enriching for virtual field trips (Ramos, 2025). The chapter outlines successful organizational strategies for both in-person and virtual field trips, reaffirming nature-based learning as a core educational approach to be fostered for future generations.

## 23.3 Nature-Based Learning and Field Trips

As mentioned by Jordan and Chawla (2019), Nature-Based Learning (NBL), or learning through exposure to nature and nature-based activities, occurs in natural settings and where elements of nature have been brought into built environments, such as plants, animals, and water. The same authors say that it encompasses the acquisition of knowledge, skills, values, attitudes, and behaviours in realms including, but not limited to, academic achievement, personal development, and environmental stewardship. It includes learning about the natural world, but extends to engagement in any subject, skill, or interest while in natural surroundings. Geology enriches nature-based learning by grounding ecological exploration in an understanding of Earth's physical foundations. The Earth system encompasses its subsystems—biosphere, atmosphere, hydrosphere, geosphere, and cryosphere—demonstrating how nature-based learning can support students in developing system thinking (Ribeiro & Orion, 2021). Although evidence regarding the impact of NBL on children's outcomes is steadily increasing, there is still much that remains unknown (Jucker & von Au, 2022). Additionally, while many studies point to its immediate benefits, there is still much to learn about its long-term impact on children's cognitive, emotional, and social development. This type of learning can take place in a variety of settings, ranging from formal educational environments to informal or non-formal educational activities. As such, NBL can occur with varying levels of guidance and structure, and can involve individuals of different age groups, whether learning independently or collaboratively. It can also take place in a wide range of settings, including urban, suburban, rural, or wilderness environments. Pelfini and collaborators (2016) refer that fieldwork is considered important for allowing students to engage directly with the discipline of geology, particularly in reconstructing landform and landscape evolution and change. According to the authors, this contact with nature is a positive experience that leaves a lasting impression and contributes to the development of knowledge and skills. According to Brilha (2005) the organization of field trips to nature sites, such as visits to geoheritage locations, helps promote the conservation of nature, particularly its geological aspects.

However, despite the growing body of research highlighting the positive outcomes associated with NBL, significant gaps remain in our understanding of how to best implement this approach across varied educational systems and more evidences are needed to conclude the significant health and wellbeing benefits of nature connection (Mann et al., 2021, 2022). When considering children's NBL, it encompasses a broad spectrum of experiences, each contributing uniquely to their development. Informally, it can take place during children's free play or spontaneous exploration of natural environments, whether in their own backyards, nearby green spaces, schoolyards, or the grounds of childcare centres. These informal interactions with nature allow children to engage in self-directed learning, fostering curiosity and a deep connection to the natural world (MacQuarrie et al., 2015). NBL is present across informal, non-formal, and formal educational contexts (Jordan & Chawla, 2019; La Belle, 1982). This underscores the remarkable flexibility and adaptability of NBL, revealing that it transcends traditional educational boundaries and cannot be confined to a single model or framework. NBL offers the potential to be customized to suit the diverse needs and developmental stages of children, allowing it to take place in a wide array of environments, each presenting distinct opportunities for engagement and personal growth. Whether in natural outdoor spaces, urban parks, or even classroom settings, NBL can provide rich, hands-on experiences that foster creativity, problem-solving, and a deeper connection to the world around them. Additionally, NBL extends into nonformal learning environments, such as after-school programs, summer camps, or family visits to parks and nature centres. In these settings, children have the opportunity to explore nature in a less structured manner, often guided by an adult but still allowing for greater freedom of discovery. These experiences can spark interest in environmental issues, provide opportunities for teamwork and problem-solving, and promote physical activity, all of which contribute to well-rounded development. Furthermore, NBL also plays a role in formal educational settings, where children interact with nature as part of structured activities. This may occur in schools, preschools, and childcare centres, or during planned outdoor field trips. In these contexts, educators often integrate nature into the curriculum, creating opportunities for hands-on learning in subjects such as science, art, and environmental education. Through these formal interactions with nature, children gain a more structured understanding of the world around them, while benefiting from the emotional, cognitive, and social advantages that nature provides. Ultimately, NBL spans a wide range of experiences, from informal and nonformal to formal settings, each offering its own unique benefits.

Regardless of the setting, NBL serves as a powerful tool in promoting holistic development, fostering a deeper sense of responsibility for the environment, and enhancing children's physical and emotional well-being. A study conducted by Supriyoko and colleagues (2022) found that by integrating a nature-based curriculum, students become more engaged and excited to attend school. The authors concluded that the nature-based curriculum can complement the regular curriculum in terms of moral education, particularly in areas such as morality, leadership, entrepreneurship, and the Student Profile. Students are given the freedom to express opinions, solve problems, and make decisions, which significantly impacts their growth and development, benefiting their future lives. This effect can be further enhanced by integrating the four pillars: logic, leadership, morality, and entrepreneurship. This freedom allows both students and teachers to feel more comfortable in the learning process. Additionally, the knowledge gained in class becomes more meaningful as it can be applied to everyday life (Supriyoko et al., 2022). Virtual field trips can be a powerful tool for students to learn about nature, offering several advantages even without being physically present in the natural environment, such as: (i) access to inaccessible locations; (ii) detailed observation; (iii) interactive learning; (iv) promotion of geoheritage awareness; (v) inclusion and accessibility for students with reduced mobility or access difficulties (Ramos, 2025; Vasconcelos & Ribeiro, 2026). During a virtual field trip, students can explore different perspectives in nature and the interactions between Earth's subsystems. Although they do not replace in-person experiences, they can be an excellent way to complement and expand learning, making environmental education more accessible, engaging, and diverse.

## 23.4 Potential and Limitations of Virtual Field Trips

Questions persist about how to effectively integrate NBL into existing curricula, especially in formal schooling contexts that may be heavily structured or standardized. These unanswered questions underscore the need for continued exploration and innovation in the field, as educators, researchers, and policymakers work to optimize the potential of NBL for future generations. Virtual field trips (also known as electronic field trips) arise in this context as a potential resource to learn about nature in the classroom (indoor environment). This means that students stay in the classroom and have digital contact with nature through a virtual experience. But why are Virtual field trips relevant?

Traditional field trips do not resolve some issues related to the organization and implementation of a fieldtrip. The most difficult challenges to overcome include factors such as travel costs, safety, time, distance, mobility difficulties, among others (Hou & Fang, 2012). Virtual field trips (VFTs) thus emerge as an alternative to traditional field trips (Ribeiro et al., 2024). They are available anytime and anywhere and are accessible to students with specific needs that might hinder participation in traditional field trips, making them more inclusive (Meezan & Cuffey, 2012; Ribeiro et al., 2024).

Inclusive education, which seeks to provide equitable opportunities for learning and development to all students, has become a foundational principle in contemporary educational discourse (Ainscow, 2020). In recent years, it has been increasingly understood not only as a guiding policy but also as a pedagogical practice that must be embedded within the everyday functioning of educational institutions (Slee, 2018). As a movement, inclusive education promotes the creation of learning environments where diversity is not only recognized but actively valued and celebrated (Ainscow, 2020). This perspective requires schools to critically reassess and transform their teaching approaches to accommodate the diverse needs of all learners. The goal is to ensure equitable access to education, full participation in school life, and the opportunity for academic success for every student. Furthermore, the same author argues that inclusive education enhances the learning experience for all students by fostering empathy, solidarity, and social justice. It also contributes to the development of learners who are more open-minded, respectful, and responsive to diversity. Nonetheless, despite significant advances in inclusive education policies and frameworks, the effective implementation of inclusive practices continues to face substantial challenges and systemic barriers. According to Lindsay (2003), mainstream schools that adopt an inclusive approach are among the most effective means of addressing discriminatory attitudes, fostering inclusive communities, and advancing the goal of education for all. Furthermore, such schools not only deliver quality education to most students but also enhance the overall efficiency and cost-effectiveness of the education system.

VFTs contribute to the democratization of education by reducing logistical and economic barriers. Furthermore, they promote digital literacy and the development of essential skills for 21st-century education and the job market (Ribeiro et al., 2024). VFTs are more inclusive and offer greater flexibility, allowing for the overcoming of logistical, financial, and accessibility limitations associated with outdoor education (Pugsley et al., 2022; Zhao et al., 2020). The integration of digital tools such as Google Earth enables the exploration of remote environments, offering educational opportunities that were previously limited by geographical and economic constraints (Bond & Cawood, 2021). The need to reformulate traditional teaching and include digital strategies becomes evident in a context where technological resources are still underused. VFTs thus represent an opportunity to develop new teaching strategies, promoting greater access to knowledge. Many students struggled to visualize abstract concepts and understand natural phenomena that are not part of their daily lives. The use of VFTs helps to overcome these challenges by providing an immersive and interactive experience that facilitates the understanding of geological content and promotes more inclusive and engaging teaching (Bond & Cawood, 2021; Zhao et al., 2020). Advancements in technology enable the integration of isolated communities, support the study of region-specific topics, and foster a deeper understanding of global perspective (Tuthill & Klem, 2002). VFTs can take various forms, ranging from a simple presentation or visiting locations through Google Earth, to a fully immersive experience in virtual reality. They can be categorized into two types: location-based or thematic trips (Pugsley et al., 2022). The first type most closely resembles traditional field trips and is linked to the geology of a specific geographical area. The second involves visiting several locations with a common theme (Pugsley et al., 2022). Some of the potential and limitations of VFTs in comparison to traditional field trips are presented in Fig. 23.2. Although they do not replace conventional field trips, VFTs offer valuable indirect field experience. They allow for the visualization of hard-to-reach locations and geological formations, understanding the evolution of a study area, and constructing geological and geomorphological maps. Supported by other materials such as documents, maps, and field guides, they offer students the opportunity to explore a specific geological area (Evelpidou et al., 2022). As such VFTs can be used independently as a strategy to develop competences or used in the preparatory unit of a traditional fieldtrip, where they can serve to reduce the novelty space.

| Virtual Field trips | | Traditional Field trips | |
|---|---|---|---|
| Potential | Limitations | Potential | Limitations |
| **Accessibility** | | | |
| • Allows the exploration of physically inaccessible locations, such as the ocean floor, the Earth's interior, or other planets (Han, 2020).<br>• Enables the participation of students with physical or geographical limitations, making learning experiences more inclusive (Bond & Cawood, 2021; Evelpidou et al., 2021; Ribeiro et al., 2024). | • Limitations in conveying the full sensory experience, such as touch and smell (Zhao et al., 2020).<br>• Requires reliable technological infrastructure and digital skills from both teachers and students (Pugsley et al., 2022). | • Enables direct interaction with the environment, which is especially important for subjects like geology (Bond & Cawood, 2021; Tibaldi et al., 2020). | • May be unfeasible due to high costs, long distances, or safety concerns (Barth et al., 2022; Tibaldi et al., 2020).<br>• May be challenging for students with mobility impairments or those living in remote areas (Mead et al., 2019; Zhao et al., 2020). |
| **Costs** | | | |
| • Significantly reduces costs related to transportation, admission, and logistics (Pugsley et al., 2022; Zhao et al., 2020).<br>• Contributes to the democratization of education by lowering logistical and economic barriers (Ribeiro et al., 2024). | • Dependence on technological equipment, which can be costly for schools (Lege & Bonner, 2020; Schott et al., 2024). | • Offers a fully immersive experience in the environment without the need for additional technology (Whitmeyer & Dordevic, 2021). | • High costs associated with transportation and organization (Bond & Cawood; Pugsley et al., 2022). |
| **Interactivity and Engagement** | | | |
| • Provides dynamic, interactive, and 360-degree visualizations, fostering exploration, inquiry, and increasing student engagement (Mead et al., 2019; Ribeiro et al., 2024; Tibaldi et al., 2020). | • Limited interaction with the physical environment, resulting in reduced sensory and emotional engagement (Han, 2020; Tibaldi et al., 2020). | • Encourages interactivity and engagement collaboration and interaction between students and their surrounding environment (Barth et al., 2022; Bond & Cawood). | • The environment may be less controlled, making it harder for students to stay focused on the activities (Pugsley et al., 2022; Whitmeyer & Dordevic, 2021). |
| **Learning Outcomes** | | | |

**Fig. 23.2** Potential and limitations of VFTs compared to traditional systems (Ramos, 2025, p. 18)

## 23.5 Conclusion and Ways Forward

Nature-based learning plays a key role in supporting the teaching and learning of geology, enabling students to better understand geological content, develop environmental insight, and foster system thinking—competencies that are essential for grasping geological processes, particularly given the broad spatial and temporal scales at which these phenomena occur. Moreover, direct contact with nature has been shown to significantly enhance children's well-being, reinforcing the value of outdoor field trips as a central pedagogical strategy in geoscience education. In this context, virtual field trips have emerged as promising tools for promoting inclusion, by providing access to immersive geological learning experiences for students who may face physical, geographical, or socioeconomic barriers to participating in traditional fieldwork. Nevertheless, further research is required to assess and validate the effectiveness of virtual field trips in promoting inclusive education within the field of geology.

**Funding** This research was supported by the Strategic Funding (refs. UIDB/04423/2020 and UIDP/04423/2020) through national funds provided by the Portuguese National Funding Agency for Science, Research, and Technology (*Fundação para a Ciência e a Tecnologia* [FCT]).

## References

Ainscow, M. (2020). Promoting inclusion and equity in education: Lessons from international experiences. *Nordic Journal of Studies in Educational Policy, 6*(1), 7–16.

Anđelković, S., Dedjanski, V., & Pejic, B. (2017). Pedagogical benefits of fieldwork of the students at the Faculty of Geography in the light of the Bologna process. *Journal of Geography in Higher Education, 42*(1), 110–125. https://doi.org/10.1080/03098265.2017.1379058

Bond, C. E., & Cawood, A. J. (2021). A role for virtual outcrop models in blended learning—Improved 3D thinking and positive perceptions of learning. *Geoscience Communication, 4*(2), 233–244. https://doi.org/10.5194/gc-4-233-2021

Brandstetter, T. (2011). Time machines: Model experiments in geology. *Centaurus, 53*(2), 135–145. https://doi.org/10.1111/j.1600-0498.2011.00216.x

Brilha, J. B. (2005). *Património geológico e geoconservação: A conservação da natureza na sua vertente geológica*. Palimage.

Brusi, D. (1992). Reflexiones en torno a la didáctica de las salidas de campo en geología (II): Aspectos metodológicos. In *VII Simposio Nacional sobre Enseñanza de la Geología* (pp. 391–407).

Compiani, M., & Carneiro, C. D. R. (1993). Os papéis didáticos das excursões geológicas. *Enseñanza De Las Ciencias De La Tierra, 1*(2), 90–97.

Evelpidou, N., Karkani, A., Komi, A., Giannikopoulou, A., Tzouxanioti, M., Saitis, G., Spyrou, E., & Gatou, M.-A. (2022). GIS-based virtual field trip as a tool for remote education. *Geosciences, 12*(9), 327–339. https://doi.org/10.3390/geosciences12090327

Frodeman, R. (2014). Hermeneutics in the field: The philosophy of geology. In B. Babich, & D. Ginev (Eds.), *The multidimensionality of hermeneutics phenomenology*. Springer.

Golubchikov, O. (2015). Negotiating critical geographies through a "feel-trip": Experiential, affective and critical learning in engaged fieldwork. *Journal of Geography in Higher Education, 39*(1), 143–157. https://doi.org/10.1080/03098265.2014.1003800

Holgersen, S. (2021). How to incorporate theory in (urban) field trips: The built environment as concrete abstraction. *Journal of Geography in Higher Education, 45*(3), 361–379. https://doi.org/10.1080/03098265.2020.1833317

Hou, Y. & Fang, A. (2012). Analysis and design on virtual field trips system. In *2012 7th International conference on computer science & education (ICCSE)* (pp. 1679–1681). https://doi.org/10.1109/ICCSE.2012.6295387

Jordan, C., & Chawla, L. (2019). A coordinated research agenda for nature-based learning. *Frontiers in Psychology, 10*, 766. https://doi.org/10.3389/fpsyg.2019.00766

Jucker, R., & von Au, J. (2022). *High-quality outdoor learning: Evidence-based education outside the classroom for children, teachers and society* (p. 386). Springer Nature.

La Belle, T. J. (1982). Formal, nonformal and informal learning: A holistic perspective on lifelong learning. *International Review of Education, 28*, 159–175. https://doi.org/10.1007/BF00598444

Lindsay, G. (2003). Inclusive education: A critical perspective. *British Journal of Special Education, 30*(1), 3–12. https://doi.org/10.1111/1467-8527.00275

MacQuarrie, S., Nugent, C., & Warden, C. (2015). Learning with nature and learning from others: Nature as setting and resource for early childhood education. *Journal of Adventure Education & Outdoor Learning, 15*(1), 1–23. https://doi.org/10.1080/14729679.2013.841095

Mann, J., Gray, T., Truong, S., Brymer, E., Passy, R., Ho, S., Sahlberg, P., Ward, K., Bentsen, P., Curry, C., & Cowper, R. (2022). Getting out of the classroom and into nature: A systematic review of nature-specific outdoor learning on school children's learning and development. *Frontiers in Public Health, 10*, Article 877058. https://doi.org/10.3389/fpubh.2022.877058

Mann, J., Gray, T., Truong, S., Sahlberg, P., Bentsen, P., Passy, R., Ho, S., Ward, K., & Cowper, R. (2021). A systematic review protocol to identify the key benefits and efficacy of nature-based learning in outdoor educational settings. *International Journal of Environmental Research and Public Health, 18*(3), Article 1199. https://doi.org/10.3390/ijerph18031199

Meezan, K. A. L., & Cuffey, K. (2012). Virtual field trips for introductory geoscience classes. *The California Geographer, 52*(1), 1–18.

Morais, M. J. P., Mateus, N. B., Oliveira, S. S., & Carneiro, V. A. (2021). Trabalho de campo e ensinagens sobre a geodiversidade no Parque Estadual da Serra Dourada (Estado de Goiás). *Revista Territorial, 10*(1), 224–236.

O'Brien, L., Burls, A., Bentsen, P., Hilmo, I., Holter, K., Haberling, D., Pirnat, J., Sarv, M., Vilbaste, K., & McLoughlin, J. (2011). Outdoor education, life long learning and skills development in woodlands and green spaces: the potential links to health and well-being. In Nilsson, K., Sangster, M., Gallis, C., Hartig, T., de Vries, S., Seeland, K., & Schipperijn, J. (Eds.), *Forests, trees and human health* (pp. 343–372). Springer Netherlands. https://doi.org/10.1007/978-90-481-9806-1_12

Orion, N. (1993). A model for the development and implementation of field trips as an integral part of the science curriculum. *School Science and Mathematics, 93*, 325–331. https://doi.org/10.1111/j.1949-8594.1993.tb12254.x

Orion, N., & Hofstein, A. (1994). Factors that influence learning during a scientific field trip in a natural environment. *Journal of Research in Science Teaching, 31*(10), 1097–1119. https://doi.org/10.1002/tea.3660311005

Orion, N., Ben-Shalom, R., Ribeiro, T., & Vasconcelos, C. (2020). Geoethics in field-trips: A global geoethics perspective. In C. Vasconcelos, S. Schneider-Voß, & S. Peppoloni (Eds.), *Teaching geoethics: Resources for higher education* (pp. 111–132). U. Porto Edições.

Pedrinaci, E., Sequeiros, L., & García de la Torre, E. (1994). El trabajo de campo y el aprendizaje de la Geología [Fieldwork and the learning of geology]. *Alambique. Didáctica De Las Ciencias Experimentales, 2*, 37–45.

Pelfini, M., Bollati, I., Pellegrini, L., & Zucali, M. (2016). Earth sciences on the field: Educational applications for the comprehension of landscape evolution. *Rendiconti Online della Società Geologica Italiana, 40*, 56–66. https://doi.org/10.3301/ROL.2016.72

Pugsley, J. H., Howell, J. A., Hartley, A., Buckley, S. J., Brackenridge, R., Schofield, N., Maxwell, G., Chmielewska, M., Ringdal, K., Naumann, N., & Vanbiervliet, J. (2022). Virtual field trips utilizing virtual outcrop: Construction, delivery and implications for the future. *Geoscience Communication, 5*(3), 227–249. https://doi.org/10.5194/gc-5-227-2022

Ramos. A. (2025). *Virtual field trip to São Miguel Island in the Azores: Action research with 10th grade students [Internship Report]*. Faculty of Sciences, Porto University. https://hdl.handle.net/10216/167485

Reynard, E., & Brilha, J. (Eds.). (2017). *Geoheritage: Assessment, protection, and management*. Elsevier. https://doi.org/10.1016/C2015-0-04543-9

Ribeiro, T., & Orion, N. (2021). Educating for a holistic view of the earth system: A review. *Geosciences, 11*, 485. https://doi.org/10.3390/geosciences11120485

Ribeiro, T., Paz, M., & Vasconcelos, C. (2024). Virtual field trips in natural sciences education: Advantages, challenges, and a practical application in Portugal. In O. Sparitis (Ed.), *Proceedings of 11th SWS international scientific conference on social sciences—ISCSS 2024* (pp. 349–356). https://doi.org/10.35603/sws.iscss.2024/s08/69

Schiappa, T. A., & Smith, L. (2018). Field experiences in geosciences: A case study from a multi-disciplinary geology and geography course. *Journal of Geoscience Education, 67*(2), 100–113. https://doi.org/10.1080/10899995.2018.1527618

Slee, R. (2018). *Defining the scope of inclusive education. Paper Commissioned for the 2020 Global Education monitoring report. Inclusion and education.* https://unesdoc.unesco.org/ark:/48223/pf0000265773

Supriyoko, S., Nisa, A. F., & Uktolseja, N. F. (2022). The nature-based school curriculum: A solution to learning-teaching that promotes students' freedom. *Journal Cakrawala Pendidikan, 41*(3), 643–652. https://doi.org/10.21831/cp.v41i3.47903

Tuthill, G., & Klemm, E. B. (2002). Virtual field trips: Alternatives to actual field trips. *International Journal of Instructional Media, 29*(4), 453.

Vasconcelos, C., & Orion, N. (2021). Earth science education as a key component of education for sustainability. *Sustainability, 13*(3), 1316. https://doi.org/10.3390/su13031316

Vasconcelos, C., & Ribeiro, T. (2026). The interplay of geoheritage and outdoor experiences through education and communication. In E. Reynard & J. Brilha (Eds.), *Geoheritage: Assessment, protection, and management* (2nd ed.) (pp. 411–430). Elsevier. https://doi.org/10.1016/B978-0-443-28997-2.00016-9

Zhao, J., LaFemina, P., Carr, J., Sajjadi, P., Wallgrün, J. O., & Klippel, A. (2020, March 22–26). Learning in the field: Comparison of desktop, immersive virtual reality, and actual field trips for place-based STEM education [Conference session]. In *2020 IEEE conference on virtual reality and 3D user interfaces (VR)* (pp. 893–902). IEEE. https://doi.org/10.1109/VR46266.2020.158 1091793502

**Clara Vasconcelos** is an associate professor at the Faculty of Science, University of Oporto, Portugal. Her research focuses on science education, sustainability, environmental education, and science communication. She serves as commissioner of IUGS-COGEducation, national co-coordinator of IAPG-Portugal, and member of FEG's Education Panel. She is also a Portuguese council member of IGEO, editor of Geosciences' Education in Geosciences collection, and PI of the SEI Group at CIIMAR.

**Tiago Ribeiro** , PhD, is an invited assistant professor at the Faculty of Science, University of Oporto, Portugal. He is secretary of the International GeoScience Education Organisation, Portuguese Junior Ambassador of USERN, and an associated member of the IUGS Commission on Geoscience Education. As Co-PI of the SEI Group at CIIMAR, his research explores education in formal, non-formal, and lifelong contexts, fostering innovation, social impact, and inclusivity.

**André Ramos** is a middle and high school biology and geology teacher in Portugal, with master's degrees in Environmental Engineering (University of Coimbra) and Biology and Geology Teaching (University of Porto). Formerly an environmental consultant in Portugal, Spain, and Australia, he now focuses on innovative teaching strategies. His research analysed the impact of a virtual field trip to São Miguel Island (Azores), highlighting its potential for inclusive, interactive science learning.

# Part V
# Teaching Practice Issues in Practical Work Contexts

# Chapter 24
# Towards more Research-Informed Practice in the use of Practical Work in Science Education

**James de Winter and Robin Millar**

## 24.1 Introduction

In our view, the fundamental aim of science education is to expand students' knowledge of the natural world and help them develop an understanding of the ideas and models that scientists use to explain and predict its behaviour. And we also want to do this in a way that can give learners some insight into the ways scientific knowledge has been achieved and into the grounds for accepting it. Practical work (sometimes called laboratory/lab work) is widely seen as an essential tool for achieving this and is a prominent feature of science education in many countries (Gericke et al., 2023; Lunetta et al., 2007). By practical work, we mean a teaching activity in which students handle or observe the objects, materials or phenomena they are studying.

The widely held view that practical work is a necessary part of science education can lead to teachers taking it for granted and including it in lessons as a matter of routine. As a result, they may not always fully question its value and purpose, including it in lessons based on expectation rather than with a clear pedagogical rationale, leading to less effective outcomes for students.

Our view is that practical work can become a more effective teaching and learning strategy if its design and use draws more extensively on findings and insights from research. To enable research-informed practice in the classroom use of practical work, however, knowledge from research studies has to be transformed into forms that connect to teacher knowledge and facilitate specific and feasible changes in practice. We will present a model for this process of knowledge transformation in relation to the educational use of practical work.

J. de Winter (✉)
Faculty of Education, University of Cambridge, Cambridge, UK
e-mail: Jad26@cam.ac.uk

R. Millar
Department of Education, University of York, York, UK
e-mail: robin.millar@york.ac.uk



L. Leite et al. (eds.), *Practical Work in Science Education in the Digital Era*, Advances in Technology-Rich Science Education 2, https://doi.org/10.1007/978-3-032-13788-3_24

## 24.2 What Research Has to Say About Effective Practical Work

If we wish to support teachers to make more effective use of practical work in science, then it is important to consider what is meant by the term 'effective'. The *Labwork in Science Education* project considered what this may mean, acknowledging that it may encompass conceptual, procedural and epistemological objectives (Psillos & Niedderer, 2002). Emerging from that work was a model to evaluate effectiveness that could be of value to researchers and teachers. The foundation of this model is that the fundamental purpose of labwork is to support students in making links between two domains: the domain of real objects and observable things, and the domain of ideas. And so, for it to be effective, practical work needs to be 'minds on' as well as 'hands on'. From this, two levels of effectiveness were identified. Effectiveness 1 attended to the things that students actually did in the laboratory and Effectiveness 2 related to the learning outcomes of the activity (Millar et al., 2002). The framework allows teachers to plan and evaluate whether or not the design features of the task enable students to carry out the task successfully (Effectiveness 1), and whether they learned what was hoped for (Effectiveness 2). For example, it may be that students were able to correctly build a DC circuit with a component such as a resistor or lamp from a circuit diagram and collect readings from ammeters in the circuit loop and voltmeters across the component, but not develop any understanding of the nature of the relationship between potential difference, current and resistance. In this case, the activity might be considered effective in terms of Effectiveness 1 but not so with respect to Effectiveness 2.

Subsequent research suggests that practical work often focusses on the domain of observable things rather than the domain of ideas, and is often rather ineffective in developing learners' understanding of scientific ideas (Abrahams & Millar, 2008). A recent study of the views of pre-service teachers also noted that while many were aware of the distinction between these two domains, and its importance, they were not always able to apply this distinction correctly (de Winter & Millar, 2023). For example, it was common for respondents to suggest that ideas such as force and acceleration were in the domain of observables rather than being mental constructs and thus in the domain of ideas. That study also identified that it is common for policy documents, research papers and advice for teachers to refer to practical work in school 'science' rather than in the individual science disciplines and so subject and topic level considerations may not gain the attention that they deserve. As a result, this may lead teachers away from considering how generalised findings from the research on practical work in science may need further adaptation or refinement in the context of a particular subject in science or topic within that subject.

In considering broader issues relating to teacher planning, others have noted that it is important for teachers to be clear about the intended learning outcomes of any practical task (Hart et al., 2000) and that the intended learning outcome should inform the task design (Welzel et al., 1988). In making these design decisions and matching tasks to learning outcomes, there is a range of types of practical work that are commonly carried out in class from which teachers can choose. These include the exploration of relationships between variables ('fair testing'), guided discovery, parameter measurement, hypothesis testing, and predict/observe/explain tasks (Roberts, 2004). A thoughtful and deliberate choice of which type to use requires teachers to be familiar with the range of types of practical work available to them, something that is not always the case (de Winter & Millar, 2023). Erduran and Wooding (2021) suggest that a view held by some teachers, that there is a single 'scientific method' exemplified through hypothesis testing, may dominate classroom practice, restricting teachers from drawing from a wider range of types of practical work open to them.

Another consideration at the design stage is the level of guidance and support given to students. This might relate to the extent to which a task is more open or closed, or the nature and type of discussion that happens before and/or after the task (Millar, 2010).

Many of these insights and recommendations from research, however, have rather little clearly observable impact on everyday classroom practice. One reason is that research findings are published in locations that teachers may be unable to access, or in language that is rather different from that of classroom practice. We will argue in the next section that *knowledge transformation* is necessary to render research knowledge into forms that can more directly inform the decisions of practitioners.

## 24.3 Knowledge Transformation and Bridging Between Researchers and Practitioners

It is widely accepted that disciplinary knowledge (e.g., of topics in physics) is inaccessible to learners unless it is transformed into more simplified structures that can build on the ideas and understandings that learners bring to the learning situation. In the school context, examples might include the different models of the atom used at different stages of a student's education, as opposed to a quantum mechanics view, or Newtonian rather than Einsteinian models of gravity for students under 16. This process is variously referred to as recontextualisation (Bernstein, 1990), didactic transposition (Chevallard & Bosch, 2020) or didactical reconstruction (Mäntylä & Nousiainen, 2014). Across different conceptualisations of this process, there is a broadly consistent view that *taught knowledge* is different from, whilst originating in, the *scholarly knowledge* produced in research environments.

A similar process, relating to knowledge from educational research and how this needs to be transformed in order to support teachers to use it to inform their classroom practice, is equally important but less well recognised and explored, or conceptualised in this way. A summary of research findings stated in more accessible language for teachers may help them understand the research findings. However, it does not necessarily provide knowledge in a form that teachers can easily operationalise in their classroom practice or use to inform their decision-making. In this section, we propose a model for the process of transforming knowledge from research into a form that is more able to directly inform the classroom practice of teachers. Later in the chapter, this process will be exemplified in relation to research on practical work in science.

McIntyre (2005) suggests that one reason for the lack of practical impact of educational research findings is the differences in the nature of the knowledge produced by academic researchers and the knowledge that teachers want. He suggests that researcher knowledge is propositional and abstracted and, whilst it can contribute to pedagogical knowledge, it cannot simply be translated into it. In contrast, the primary concern for teacher knowledge is pedagogical and how it can inform classroom action. Practicality is prioritised with a focus on how it can inform learning in a classroom context (McIntyre, 2005). As a top-level distinction, researchers deal in *knowledge that* and teachers focus on *knowledge how,* and these fundamental differences are what creates a challenge for teachers in using classroom research to inform their practice. These differences are potentially more significant for science graduates entering teaching who are used to a highly structured, hierarchical form of knowledge in their undergraduate discipline and as they move into teaching and the field of education, they find the knowledge there more disjointed and, to their eyes, unscientific (Airey & Larsson, 2018). These differences can contribute to a scepticism amongst education stakeholders and STEM teachers in particular of the limited contribution that educational research can make to the practice of science teachers (Booher et al., 2020). In some cases, these epistemological differences can lead to an even stronger, negative view, seeing educational research as "wishy-washy nonsense" (Guilfoyle et al., 2020, p. 8).

Bridging the gap between researcher knowledge and teacher knowledge, whilst attending to the value of both, requires *knowledge transformation.* Knowledge transformation is the process by which knowledge from the field of research is transformed into a form that is useful, applicable, and accessible for teachers. We do not make value judgements or privilege any one type of knowledge as 'better' or 'more valuable'. Our model acknowledges the contributions that researcher knowledge and teacher knowledge can make to high-quality teaching, and that an active process of knowledge transformation is needed to help to bridge the gap between research and practice, which was McIntyre's aim in exploring the differences between researcher knowledge and teacher knowledge (2005).

One example of the process of knowledge transformation relating to practical work in science starts from the chapter *'Mapping' the domain—Varieties of Practical Work* (Millar et al., 1999) in an academic book that is unlikely to be read by most teachers. This was subsequently transformed into a teacher's guide which included a checklist for analysing practical tasks that teachers use or might consider using, drawing attention to precision about intended learning outcomes, choices of design and presentation in the classroom (Millar, 2010). Here, and with other examples of this kind of work (e.g. Angell et al., 2019; de Winter & Hardman, 2021; Education Endowment Foundation, 2017), the language, and the level of detail and academic citation, vary considerably, acknowledging the background and context of the reader. This is of particular relevance to the current English context, where almost half of all pre-service teachers are educated on school-based post-graduate courses, rather than university-based ones, and so their engagement with, and access to, academic research literature and findings may be very limited. The process of knowledge transformation, its theoretical foundations and further examples are explored in much greater depth by de Winter (2025).

The process of knowledge transformation will inevitably require professional judgments as to what is included and the level of academic detail and citation, but we propose that it can support teachers to make their practice more research-informed. The next section will exemplify this process by illustrating how some of the findings from research on practical work outlined earlier in the chapter can be presented to teachers.

## 24.4 Supporting Research-Informed Use of Practical Work in Science

Drawing on the ideas presented in the earlier sections, we have developed an online course for pre- and in-service physics teachers in England to support them in using research knowledge about practical work to inform their classroom practice. The course expands upon work that the first author was commissioned to produce for the Institute of Physics in the United Kingdom as part of their national program to support early career teachers of physics (Institute of Physics, 2022). One of the primary intentions of the course is to support teachers in reflecting on the design and implementation of specific practical tasks. We wish them to appreciate that there is not a common, generalised set of principles suitable for all practical work in science and that some decisions relating to learning outcomes at a subject and topic-specific level are needed.

This course represents the outcome of a process of knowledge transformation of the body of knowledge from research on practical work, with a particular focus on five themes from the literature, each of which forms the focus of one of the taught sessions of the course. Some of the key messages from the literature that have informed the sessions are outlined below.

*Session 1—The aims and purposes of practical work:* The fact that practical work is a prominent feature of science education in many countries (e.g. Gericke et al., 2023) may lead teachers to see this as a taken for granted aspect of their teaching rather than a deliberate design decision. This session is designed to introduce teachers to some of the reasons that have been identified as to how practical work can support learning (e.g. Bennett, 2003; Hofstein & Lunetta, 2004) from which they can make choices about specific practical tasks that they may use of consider using. The list of some of the possible cognitive outcomes of practical work used by de Winter and Millar (2023) may provide a helpful starting point for discussion and reflection.

- To develop students' competence in using laboratory equipment and carrying out laboratory procedures;
- To encourage accurate observation and description of natural objects, materials, phenomena and events;
- To develop students' ability to design and implement a scientific approach to investigating an issue or solving a problem;
- To enhance understanding of scientific ideas (theories, models, explanations).
- To develop students' ability to present, analyse and interpret data.

*Session 2—How to evaluate effective practical work*: Having supported teachers to consider the reasons why they might include practical work in their teaching, we would want to support teachers to evaluate how successful their implementation of this might be. The effectiveness framework from Millar et al. (2002) provides a way for teachers to consider and reflect upon the differences between the domain of objects/observables and the domain of ideas—the 'hands on' and 'minds on' aspects of practical work and to apply this to their decisions about the design and implementation of specific practical tasks that they may use. This framework can be presented to teachers as practitioner-focused (transformed) knowledge (Millar & Abrahams, 2009), making it accessible to teachers who may not be familiar with or have access to academic journals. This session also supports them to reflect on the distinction between exactly what is 'observable' and what is 'an idea' in specific practical tasks, identified as challenging for some teachers (de Winter & Millar, 2023). These sessions draw from a range of examples across Biology, Chemistry and Physics as well as different topics and learning outcomes within each subject, to highlight that design and implementation decisions need to be made on a task specific level.

*Session 3—The range of types of practical work*: The dominance of hypothesis testing in classroom practice (Erduran & Wooding, 2021) may mean that other types of practical work may not always be considered by teachers, thus limiting the range of experience for students of what constitutes the 'scientific method'. This session introduces teachers to a wide range of possible types of practical work to broaden their repertoire and become more confident in making deliberate design decisions to support learning when planning the design and implementation of specific practical tasks. The range of types of practical work could include, but not be limited to relationship exploration, guided discovery, parameter measurement, hypothesis testing, predict/observe/explain (Roberts, 2004).

*Session 4—Designing effective practical work*: The introduction of the effectiveness framework (Millar et al., 2002) in Session 2 aims to support teachers to consider what 'effective' may mean. However, the design of effective practical work that is well matched to the learning outcomes (Welzel et al., 1988) can be challenging for teachers. They will need time and support to become confident in matching task design to the intended learning outcomes of specific practical tasks (Hart et al., 2000). Initially, this will draw from existing work written with this aim (Abrahams & Reiss, 2017) to scaffold teachers' thinking as they develop their own confidence in this area.

*Session 5—The importance of subject and topic-specific considerations:* Having established some key ideas around learning intentions, effectiveness and task design, this final session supports teachers to consider how the principles that may inform their choices may vary across topics and subjects (de Winter & Millar, 2023). They compare and contrast examples that they had used or produced in the preceding sessions to identify common broad principles as well as identifying where subject- and topic-specific differences may exist.

This course has been designed for teachers, many of whom will not have had a university-based pre-service teacher education. As many may have had limited or no engagement with academic literature, references are limited to one book written for teachers and two articles from practitioner journals that they are much more likely to have encountered or have access to. The outline of the course is shown in Table 24.1. Some of the details from the research base have been refined and adapted in the knowledge transformation process but, with the aim of providing a body of knowledge that teachers wish for and feel able to use, this is inevitable.

Longstanding work on professional development has noted the challenges that all teachers face in implementing and embedding any changes to their practice (Loucks-Horsley et al., 2010). To support teachers in their work after they have completed the course, two aide-mémoires have been developed as part of the course materials. The first is a flowchart for decision making when planning practical work, shown in Fig. 24.1.

The second aide-mémoire draws in some additional concerns that teachers may have concerning the pragmatic and organisational aspects of practical work, such as helping students focus on the right things and the organisation of the room. This is shown in Fig. 24.2. The design of the second aide-mémoire deliberately foregrounds the importance of the 'hands-on, minds-on' aspects of effective practical work.

Concerns around equipment and behaviour management have been identified as barriers that prevent some teachers from including as much practical work in their lessons as they might like (SCORE, 2008). By building these into the planning summary in the second aide-mémoire (Fig. 24.2), the aim is to acknowledge and attend to day-to-day pragmatic concerns whilst maintaining a focus on the aspects of practical work that will make it most effective for student learning. This second aide-mémoire also highlights the importance of teacher questioning and discussion as a vital way to support students in engaging with practical work to support their learning (Abrahams & Reiss, 2017). The emphasis on teachers considering expected student answers and their responses to these in advance highlights that the development of good questions should be a proactive rather than solely reactive aspect of lesson planning and teaching.

**Table 24.1** Outline of course to support teachers to use research knowledge about practical work to inform their classroom practice

| Session | Theme | Notes |
|---|---|---|
| 1 | The aims and purposes of practical work (What is the point of practical work?) | Helping teachers be clear about the aims and purposes of practical work, and to apply this to specific practical tasks that they may use. Core resource is video "Thinking about physics… Practical Work: Purpose and attention" https://www.youtube.com/watch?v=P5zkttkd_GY (adapted with examples from across science) |
| 2 | How to evaluate effective practical work (What does the research say about practical work?) | The session introduces the two levels of effectiveness (Millar & Abrahams, 2009) and supports teachers to consider what observable features in the lesson might provide evidence to make judgments about the effectiveness of specific practical tasks on both levels. Core resource is video "Thinking about physics… Practical Work: Research" https://www.youtube.com/watch?v=1KKSN0zPiBQ (adapted with examples from across science) |
| *Interim task:* Teachers write a short reflective piece on their use or observation of a practical task in a recent lesson using ideas from sessions 1 and 2 | | |
| 3 | The range of types of practical work (Is practical work just about following instructions?) | A session focusing on different types of practical work including guided discovery, parameter measurement, 'fair testing', predict/observe/explain, and hypothesis testing (Erduran & Wooding, 2021) |
| *Interim task:* Teachers write a reflective piece on a practical activity that they have taught or observed recently. They consider how 'effective' they think it was in achieving the specific learning outcomes and what adaptations could be made to improve it | | |
| 4 | Designing effective practical work (What should I think about when planning and teaching?) | The sessions will focus on planning for effective practical work. Core resource is video "Thinking about physics… Practical Work: Pragmatic and planning" https://www.youtube.com/watch?v=TWxi1cDRhqc (adapted with examples from across science) |
| 5 | The importance of subject and topic-specific considerations (From general to the specific—practical work and *science teaching*) | The session will focus on the way in which different topics and learning outcomes in each of the sciences may require different types of practical work, guidance for students and teacher questioning. Using multiple examples, including those in Abrahams and Reiss (2017) |
| *Consolidation/follow-up task:* Arrange a discussion between teachers about aspects of their use of practical work that might be common and/or different across multiple topics and learning outcomes in science | | |

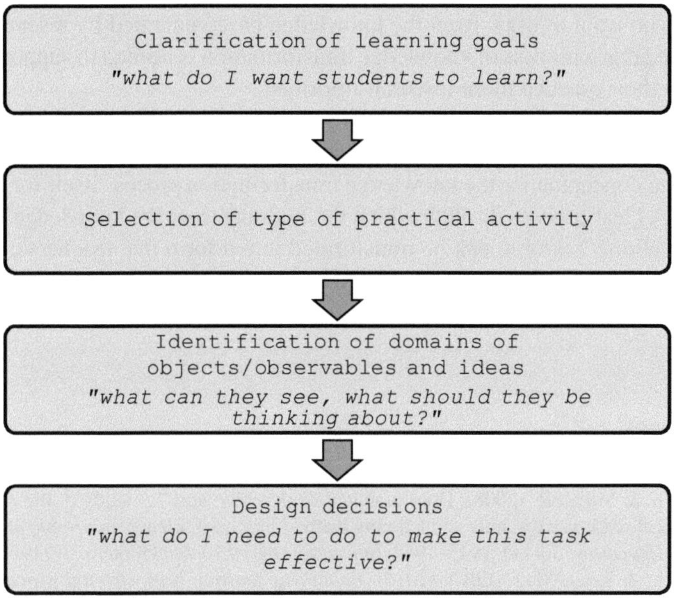


**Fig. 24.1** Simplified flowchart to support planning for effective practical work

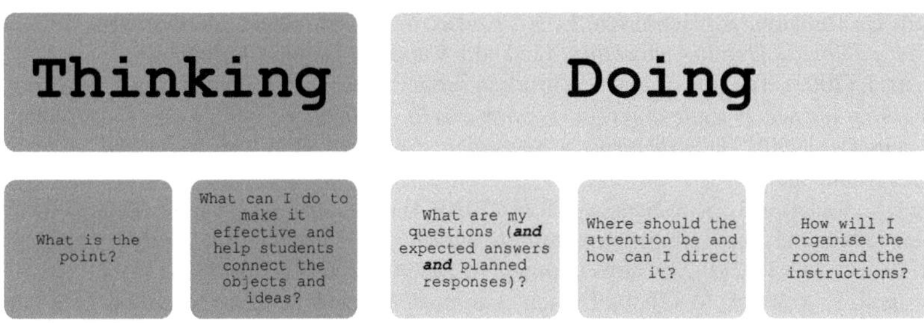


**Fig. 24.2** Aide-mémoire to support teachers in the planning of effective practical work

## 24.5 Conclusion and Ways Forward

There is a large and wide range of academic research knowledge and scholarship relating to practical work in school science. Although there may be differing views and perspectives in some areas, there is general agreement on the value and importance of practical work as part of a student's science education—alongside widespread agreement that, as currently implemented, it is frequently less effective than we would wish in terms of students' conceptual development. Findings and insights from research may offer ways of improving this. However, the nature of knowledge from research is different from what teachers desire, and this creates a challenge both for researchers who want teachers to read and use their work and for teachers who want to draw from the knowledge base generated by researchers. We have argued that a process of knowledge transformation is needed to support teachers in making their practice more research-informed.

Examples of such knowledge transformation already exist, some of which we have referred to above. However, we suggest that more work and effort is needed to explore and conceptualise the knowledge transformation process itself for education research. A clear process through which the findings from the knowledge generated from educational research can be transformed into a form that teachers can access, understand and use in their classroom decision making can enable more of the large body of research knowledge to make its way into the life of classrooms.

## References


Abrahams, I., & Millar, R. (2008). Does practical work really work? A study of the effectiveness of practical work as a teaching and learning method in school science. *International Journal of Science Education, 30*(14), 1945–1969. https://doi.org/10.1080/09500690701749305

Abrahams, I., & Reiss, M. J. (Eds.). (2017). *Enhancing learning with effective practical science 11–16*. Bloomsbury.

Airey, J., & Larsson, J. (2018). Developing students' disciplinary literacy? The case of university physics. In K.-S. Tang & K. Danielsson (Eds.), *Global developments in literacy research for science education* (pp. 357–376). Springer. https://doi.org/10.1007/978-3-319-69197-8_21

Angell, C., Bungum, B., Henriksen, E. K., Kolstø, S. D., Persson, J., & Renstrøm, R. (2019). *Fysikkdidaktikk [Physics education]* (2nd ed.). Cappelen Damm Akademisk.

Bennett, J. (2003). The role of practical work in school science. In J. Bennett (Ed.), *Teaching and learning science: A guide to recent research and its applications* (pp. 73–98). Continuum.

Bernstein, B. (1990). *The structuring of pedagogic discourse: IV (class, codes and control)* (1st ed.). Routledge.

Booher, L., Nadelson, L. S., & Nadelson, S. G. (2020). What about research and evidence? Teachers' perceptions and uses of education research to inform STEM teaching. *The Journal of Educational Research, 113*(3), 213–225. https://doi.org/10.1080/00220671.2020.1782811

Chevallard, Y., & Bosch, M. (2020). Didactic transposition in mathematics education. In S. Lerman (Ed.), *Encyclopaedia of mathematics education* (2nd ed., pp. 214–218). Springer International Publishing. https://doi.org/10.1007/978-94-007-4978-8_48

de Winter, J. (2025). *Educating pre-service physics teachers in England: The need for knowledge transformation* [Acta Universitatis Upsaliensis]. https://urn.kb.se/resolve?urn=urn:nbn:se:uu:diva-552916

de Winter, J., & Hardman, M. (Eds.). (2021). *Teaching secondary physics* (3rd ed.). Hodder Education.

de Winter, J., & Millar, R. (2023). From broad principles to content-specific decisions: Pre-service physics teachers' views on the usefulness of practical work. *International Journal of Science Education, 45*(13), 1097–1117. https://doi.org/10.1080/09500693.2023.2187673

Education Endowment Foundation. (2017). *Improving secondary science*. Education Endowment Foundation. https://educationendowmentfoundation.org.uk/tools/guidance-reports/improving-secondary-science

Erduran, S., & Wooding, S. J. (2021). A project calibrate approach to summative assessment of practical science. *School Science Review, 102*(381), 71–77.

Gericke, N., Högström, P., & Wallin, J. (2023). A systematic review of research on laboratory work in secondary school. *Studies in Science Education, 59*(2), 245–285. https://doi.org/10.1080/03057267.2022.2090125

Guilfoyle, L., McCormack, O., & Erduran, S. (2020). The 'tipping point' for educational research: The role of pre-service science teachers' epistemic beliefs in evaluating the professional utility of educational research. *Teaching and Teacher Education, 90*, Article 103033. https://doi.org/10.1016/j.tate.2020.103033

Hart, C., Mulhall, P., Berry, A., Loughran, J., & Gunstone, R. (2000). What is the purpose of this experiment? Or can students learn something from doing experiments? *Journal of Research in Science Teaching, 37*(7), 655–675. https://doi.org/10.1002/1098-2736(200009)37:7%3c655::AID-TEA3%3e3.0.CO;2-E

Hofstein, A., & Lunetta, V. N. (2004). The laboratory in science education: Foundations for the twenty-first century. *Science Education, 88*(1), 28–54. https://doi.org/10.1002/sce.10106

Institute of Physics. (2022). *Early career teaching CPD videos*. https://spark.iop.org/early-career-teaching-cpd-videos

Loucks-Horsley, S., Stiles, K. E., Mundry, S. E., Love, N. B., & Hewson, P. W. (2010). *Designing professional development for teachers of science and mathematics*. Corwin.

Lunetta, V. N., Hofstein, A., & Clough, M. P. (2007). Learning and teaching in the school science laboratory: An analysis of research, theory, and practice. In S. Abell, N. Lederman, K. Appleton, & D. Hanuscin (Eds.), *Handbook of research on science education* (pp. 393–441). Routledge.

Mäntylä, T., & Nousiainen, M. (2014). Consolidating pre-service physics teachers' subject matter knowledge using didactical reconstructions. *Science & Education, 23*(8), 1583–1604. https://doi.org/10.1007/s11191-013-9657-7

McIntyre, D. (2005). Bridging the gap between research and practice. *Cambridge Journal of Education, 35*(3), 357–382. https://doi.org/10.1080/03057640500319065

Millar, R. (2010). *Analysing practical science activities to assess and improve their effectiveness*. Association for Science Education.

Millar, R., & Abrahams, I. (2009). Practical work: Making it more effective. *School Science Review, 91*(334), 59–64.

Millar, R., Le Maréchal, J., & Tiberghien, A. (1999). 'Mapping' the domain—Varieties of practical work. In J. Leach & A. Paulsen (Eds.), *Practical work in science education: Recent research studies* (pp. 33–59). Roskilde University Press.

Millar, R., Tiberghien, A., & Le Maréchal, J. (2002). Varieties of labwork: A way of profiling labwork tasks. In D. Psillos & H. Niedderer (Eds.), *Teaching and learning in the science laboratory* (2002nd ed., pp. 9–20). Springer.

Psillos, D., & Niedderer, H. (Eds.). (2002). *Teaching and learning in the science laboratory* (2002nd Ed.). Springer.

Roberts, R. (2004). Using different types of practical within a problem-solving model of science. *School Science Review, 84*(312), 113–119. https://www.ase.org.uk/journals/school-science-review/2004/03/312/

SCORE. (2008). *Practical work in science: A report and proposal for a strategic framework*. Science Community Representing Education (SCORE).

Welzel, M., Haller, K., Bandiera, M., Hammelev, D., Koumaras, P., Niedderer, H., Paulsen, A., Robinault, K., & von Aufschnaiter, S. (1988). *Teachers' objectives for labwork. Research tool and cross-country results* (p. 109). European Commission. https://physikdidaktik.uni-bremen.de/pubs/Niedderer/1998-LSE-WP6.pdf

**James de Winter** formerly a secondary school physics teacher, is Associate Teaching Professor of Science Education at the University of Cambridge, UK, where he has led the pre-service physics teacher education course for many years. James also works with national physics and science education organisations in the UK, including The Ogden Trust, in various roles which he characterises as "being nice to physics teachers".

**Robin Millar** is Emeritus Professor of Science Education at the University of York, UK. He taught physics in secondary schools prior to university teaching on initial and in-service teacher education programmes. His main research interests are in teaching and learning in science (especially physics), science curriculum design, and assessment. He was president of ESERA from 1999–2003, and a member of the PISA Science Expert Group for 2006 and 2015.

# Chapter 25
# Teachers' Challenges and Proposed Strategies in Educational Robotics and STEM: A Review and Content Analysis

**Eleni Vlachoudi, Sophia Rapti, and Theodosios Sapounidis**

## 25.1 Introduction

In recent years, Educational Robotics (ER) has attracted researchers' interest. "As an essential branch of educational technology implemented by constructionist activities" (Sophokleous et al., 2021, p. 2), it may enhance students' cognitive and social skills (Leoste & Heidmets, 2020). Based on existing research, during robotics activities, students may collaborate within a group to search for information, brainstorm, and create (Lotriet & Gouws, 2025). Accordingly, it is claimed that robotics may enrich traditional educational methods while developing students' challenging skills and fostering learning during hands-on activities (Bakala et al., 2023).

Over the past few decades, the integration of ER into STEM—Science, Technology, Engineering, and Mathematics—education has expanded its potential. It combines these disciplines in a learning environment that prepares students for the rapidly evolving, technology-driven world (Mallik et al., 2023). Integrating robotics within the STEM framework can support learning across various domains (Tzagaraki et al., 2022) and help develop skills such as collaboration, problem-solving, and creativity (MacAnn & Yadav, 2023).

However, incorporating ER into educational settings may also bring challenges. For instance, the administrative burden associated with using robotics, as well as

E. Vlachoudi (✉) · S. Rapti · T. Sapounidis
Department of Education, School of Philosophy and Education, Aristotle University of Thessaloniki, Thessaloniki, Greece
e-mail: evlachok@edlit.auth.gr

S. Rapti
e-mail: raptsofi@edlit.auth.gr

T. Sapounidis
e-mail: teo@edlit.auth.gr



L. Leite et al. (eds.), *Practical Work in Science Education in the Digital Era*, Advances in Technology-Rich Science Education 2, https://doi.org/10.1007/978-3-032-13788-3_25

the time and effort required to design robotics activities, may cause discomfort to educators. Furthermore, concerns have been expressed regarding the robots' inability to fully respond during social interactions (Reich-Stiebert & Eyssel, 2016). Additionally, inadequate technological training (Papadakis et al., 2021), the shortage of pedagogical knowledge and skills (Friend et al., 2022), and insufficient access to equipment and resources can hinder teachers from integrating ER into everyday practice (Nur Aksu et al., 2019).

As a consequence, researchers have examined the potential obstacles faced in various domains, such as Mathematics (Leoste & Heidmets, 2020), Physics (Saad et al., 2023), Technology (Nur Aksu et al., 2019) and Computer Science (Carlborg et al., 2019; Dorotea et al., 2021). Notwithstanding, a few studies have examined the potential challenges related to ER within a comprehensive STEM framework (Possaghi & Papavlasopoulou, 2024).

While these challenges have barely been identified for in-service teachers, there are even fewer studies investigating the obstacles faced by pre-service teachers, who represent the future of education. For instance, la Hoz et al. (2024) focused on pre-service teachers' beliefs about integrating ER into Mathematics and Physics, and Zha et al. (2020) about their challenges related to programming.

Additionally, although a limited number of studies concentrate on elementary education, there is even a greater gap in understanding the challenges arising at other educational levels, such as kindergarten, secondary, and higher education. Last but not least, there is also a lack of practical strategies that educators can adopt to overcome barriers in ER usage. This deficiency underlines the need for a mapping of the related field. This study employed a literature review to explore and group the challenges pre-service and in-service teachers face when integrating educational robotics into STEM activities. It also proposes effective strategies to overcome these obstacles. Furthermore, a content analysis was conducted to provide useful insights into educators' targeted support for ER's successful usage in STEM education.

## 25.2 Background

### *25.2.1 ER Benefits*

The integration of ER into daily practice can provide children with hands-on experiences that may bridge theoretical concepts with practical applications through tangible programming (Morrison & Hughes, 2024; Sapounidis & Demetriadis, 2017). By engaging with ER, children can familiarize themselves with technology and get used to coding in an interactive and enjoyable way (MacAnn & Yadav, 2023), which facilitates their participation in the learning process.

Through ER activities, students are encouraged to construct new, helpful knowledge based on prior experiences (Anwar et al., 2019). This process can allow children to easily divert complex, abstract concepts into concrete ones (Ackermann, 2001).

Based on existing research, children might have the opportunity to build their robot and can either use a ready-written program or create their own by connecting the robot to a computer, thereby controlling its movements (Kucuk & Sisman, 2017). Furthermore, engaging with ER tools may lead to the enhancement of students' technical and cognitive skills while offering opportunities for creative thinking.

Researchers have found that the utilization of ER might support students in gaining knowledge in various domains (Sapounidis et al., 2022, 2024). For instance, in the maths field (Boz & Allexsaht-Snider, 2022) and language (Hu et al., 2023). Within these, a range of capacities is promoted, such as engagement, teamwork, problem-solving, and communication skills (Rapti & Sapounidis, 2023). Overall, the use of ER in education has the potential to foster student-centered learning and the 4Cs (Communication, Collaboration, Critical thinking, Creativity), and enrich traditional teaching methods (Palioura et al., 2024).

### *25.2.2 ER and STEM*

Educational robotics kits can serve as useful tools for teaching within the STEM framework (Benitti, 2012; Sapounidis & Alimisis, 2020). Alimisis (2013) argues that these kits can promote a deeper understanding of complex ideas by enabling students to apply technology, mathematics, and engineering principles in real-world contexts. By integrating robotics into education, researchers claim that teachers might enhance students' skills while focusing on STEM domains (Atmatzidou & Demetriadis, 2016).

Earlier research has indicated that robotics activities in a STEM framework might enhance learners' motivation (Zaharin et al., 2020) and can increase students' interest and shape positive attitudes toward science (Boz & Allexsaht-Snider, 2022; Ley et al., 2022). ER might also support the development of cognitive, metacognitive, and social skills at various educational levels within STEM domains (Ley et al., 2022).

In this context, robotics has been integrated into STEM education to encourage hands-on activities (Khanlari, 2016), that may contribute to the development of creativity and innovation, competencies often linked to success in science and technology careers (Weng & Li, 2020).

### *25.2.3 Teachers' Challenges*

ER may be an innovative, effective teaching tool able to enrich traditional practices and empower the learning process. However, teachers may face various challenges hindering them from utilizing robotics successfully in STEM education. These may refer to self-esteem, knowledge, skills, training, costs, equipment, and the time needed to design and implement ER activities (Liao et al., 2024; Luo et al., 2023). Notwithstanding, a mapping and a presentation in groups of the potential challenges haven't been done yet. Furthermore, step-by-step, practical strategies for successful coping with challenges while using ER haven't been provided so far (Saad et al., 2023).

Consequently, neither pre-service teachers nor in-service ones feel comfortable enough to integrate ER into their teaching at all educational levels. Thus, this study conducts a literature review and a content analysis to map the related field and investigate the potential challenges faced by educators regarding ER in STEM. In addition, these barriers are grouped and demonstrated to facilitate relevant stakeholders' better understanding and their successful coping. Finally, this review explores and presents the strategies recommended in existing research related to the effective usage of ER in the STEM framework. As a result, educational policymakers may be supported in creating suitable training programs for educators and school curricula tailored to the related field. Finally, researchers can be inspired by this study to design useful future works focused on ER's holistic and successful integration into STEM and school reality.

## 25.3 Methodology

Literature reviews can provide the latest developments in a specific research field, effectively map it, and identify research gaps (Snyder, 2019). Therefore, this was chosen as the most suitable type of study for this paper. Initially, three researchers independently coded the data and then compared their results to validate the consistency of their work. Additionally, to develop guidelines for best practices and inspire future research, a content analysis of the data was conducted using Atlas.ti (version 25). This software enhances the Inter-Rater Reliability process by providing coding for text segments. As a result, a clear picture emerges through an inductive thematic analysis approach, highlighting key findings related to teachers' potential challenges during robotics activities in STEM education. Moreover, practical strategies for effective coping are recommended based on empirical evidence. Five databases and one register (Scopus, Science Direct, ACM, IEEE, Springer Link, and Taylor & Francis) served as sources for relevant publications included in this review. The key search terms were identified through an initial review of the literature and previous systematic reviews on educational robotics. The search string used was: ("problems of practice" OR "practical problems") AND ("educational robotics" OR "robotics in education" OR "robots") AND "Teachers". Additionally, specific inclusion and exclusion criteria were applied to select the most relevant papers: they needed to examine potential challenges faced by teachers when implementing ER at various educational levels, but should not focus on special populations. The papers also had to be written in English and published between 2014 and 2025. Based on these criteria, an initial set of 782 articles was extracted, with 25 papers ultimately selected for thorough analysis (see Diagram 25.1).

Accordingly, the following Research Questions (RQ) have been addressed:

**Diagram 25.1** PRISMA—flow diagram

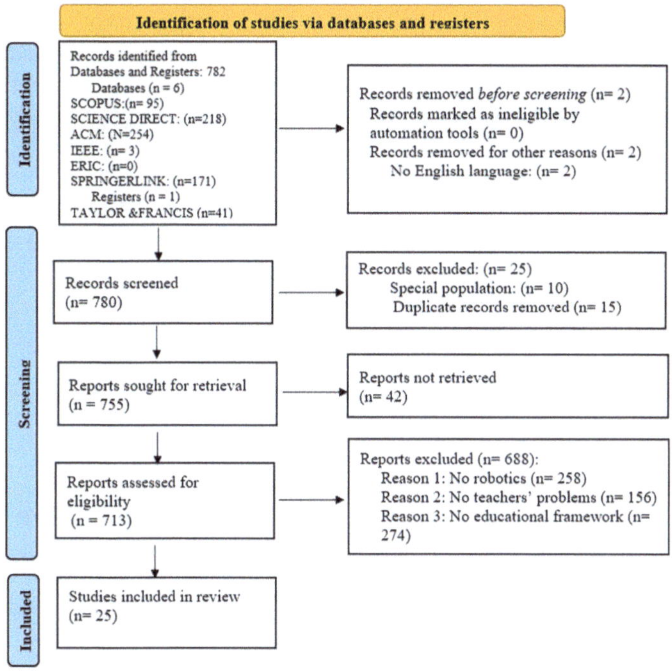


- RQ1: What are the special traits of the studies exploring the teachers' (pre-service and in-service) obstacles while integrating ER into STEM education at various levels (method, duration, sample size)?
- RQ2: What are the potential challenges pre-service and in-service teachers may face related to ER usage in STEM activities?
- RQ3: What strategies are proposed to support educators in coping with ER's challenges and integrating them successfully into the learning process?

## 25.4 Results

### *25.4.1 Special Traits of the Studies in the Finally Selected Papers*

As for pre- and in-service teachers, the studies primarily focus on in-service, with twenty-three (92%) addressing their challenges, while only two (8%) explore the difficulties faced by pre-service teachers (see Fig. 25.1). Related to the distribution of papers by educational level, the bulk of studies examined elementary school teachers (40%). Related to the research method chosen, most of the researchers used the qualitative method. As for the sample size used, Fig. 25.2 indicates that in most studies (56%), the number of participating educators ranged from 1 to 50.

Related to the duration of interventions reported in the studies reviewed, Fig. 25.3 shows that most researchers did not specify the duration of the intervention, while 24% of them conducted one lasting from 4 to 8 weeks.

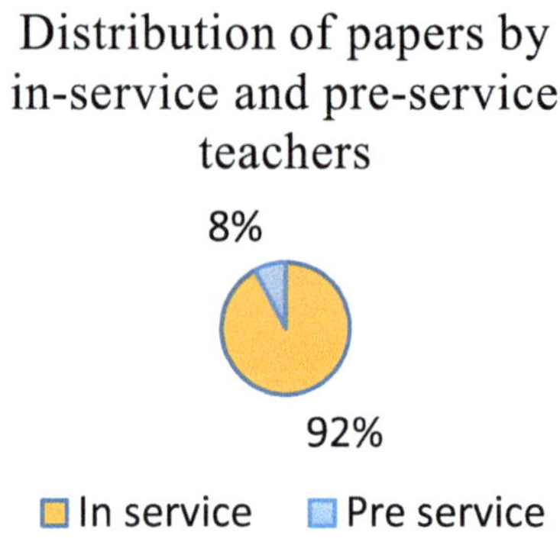


**Fig. 25.1** Types of teachers facing challenges

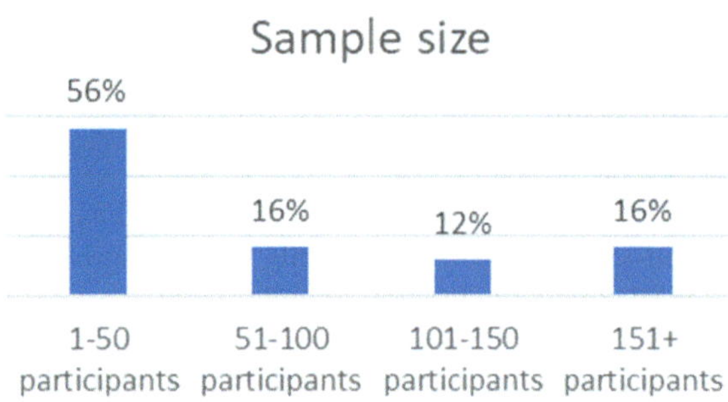


**Fig. 25.2** Sample size

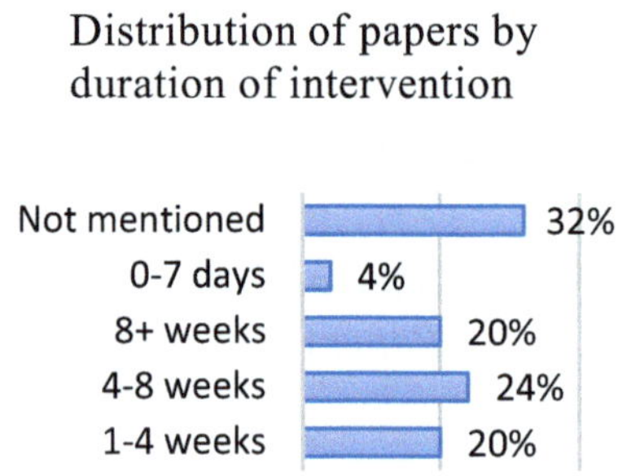


**Fig. 25.3** Duration of intervention in studies

### *25.4.2 The Challenges Associated with the Learning Domain and Robots Used*

Based on the researchers of the studies examined, the challenges encountered can be grouped in three categories defined by their origin: internal, external, and mixed, as illustrated in Fig. 25.4. Specifically, internal challenges refer to personal or psychological factors related to the teacher, such as low self-confidence, limited self-efficacy, attitudes, and (mis)beliefs toward ER learning. External challenges, on the other hand, involve factors beyond an individual's control, typically related to the educational environment or institutional context. These include time constraints, limited access to ER resources and materials, lack of school curriculum integration, and high equipment costs. Finally, mixed challenges combine elements from both internal and external dimensions. These challenges often arise when individual limitations, such as insufficient theoretical knowledge or lack of teacher experience, intersect with systemic constraints, including inadequate training opportunities or lack of access to practical knowledge (Fig. 25.5).

Figure 25.5 shows the links between teachers' reported challenges and the educational areas of STEM, coding, and computational thinking. The commonest obstacle was insufficient theoretical knowledge (20/25 studies), mostly related to STEM, followed by coding and computational thinking. Insufficient training and practical skills (16/25) and teacher inexperience (12/25) were also significant, mostly tied to coding, STEM, and computational thinking. Internal challenges were less frequent: low self-confidence (11/25) mainly appeared in STEM-related situations, while attitudes toward ER learning and beliefs (4/25) were mainly connected to coding. External challenges included the cost of equipment (10/25), especially in STEM projects, while time restrictions (8/25) and limited access to resources (5/25) were reported across all three areas. Finally, integration into school curricula was mentioned in 4/25 studies, again with a stronger focus on STEM.

Related to the association between the potential challenges and the robots used, Fig. 25.6 demonstrates that the wheeled robots are mainly linked to mixed challenges (inadequate theoretical knowledge), while modular robots may lead to a wider range of challenges, including mixed, internal, and external. Besides, humanoid robots are

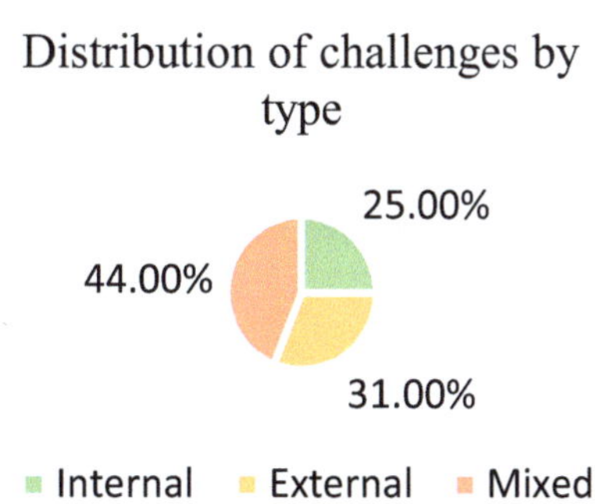


**Fig. 25.4** Types of challenges

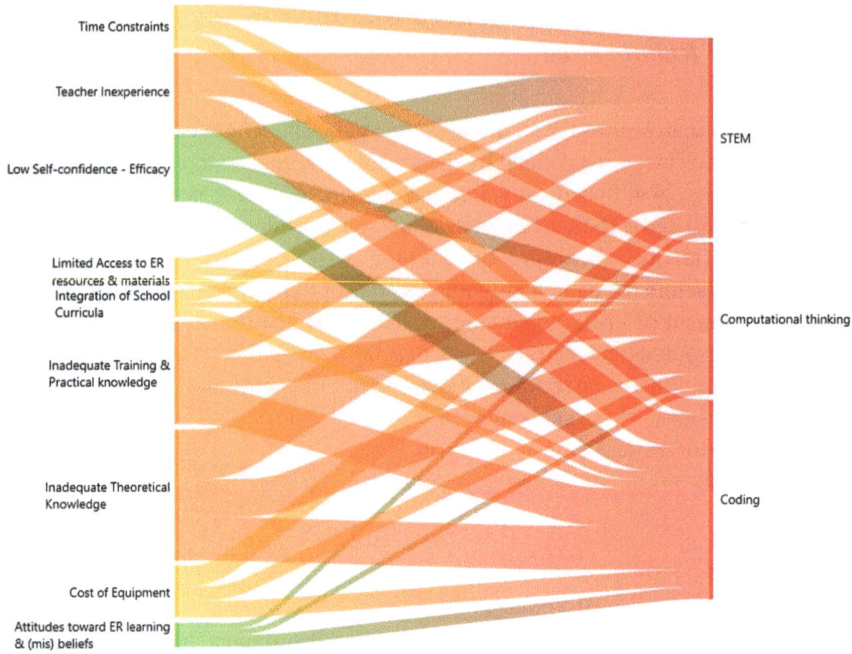


**Fig. 25.5** Flow of challenges per domain

associated with the external challenge of a high cost of purchase. Finally, in numerous studies, robots of an unspecified type can connect to all kinds of challenges.

### *25.4.3 Strategies for Handling Potential Challenges*

The strategies identified in the reviewed studies to overcome barriers in ER integration are classified as extrinsic or intrinsic. Among extrinsic strategies, guided workshops and seminars were the most frequently reported (12/25), followed by instructional school support (7/25) and institutional policies (2/25) (see Fig. 25.7). These were predominantly associated with mixed and external barriers, such as limited access to resources and materials, lack of ER curricula, cost of equipment, and time constraints. Intrinsic strategies included professional development (11/25) and peer mentoring—collaboration (8/25), which were linked to mixed and external challenges, while also addressing internal difficulties such as low self-confidence and teacher inexperience (see Fig. 25.8). Finally, eight studies did not propose any form of strategy, indicating persistent gaps.

**Fig. 25.6** Flow of challenges per type of robot

**Fig. 25.7** Flow of challenges per extrinsic strategies

## 25.5 Discussion

This book chapter outlines the challenges teachers may face while integrating ER into STEM education across all educational levels and explores the proposed strategies to overcome these. Our research indicates that most studies examine the difficulties encountered by in-service teachers (see Fig. 25.1). This may be due to barriers common among all teachers while incorporating ER into their teaching (Kim et al.,

**Fig. 25.8** Flow of challenges per intrinsic strategies

2015). Therefore, future research focused on pre-service teachers is also recommended. Furthermore, most challenges are spotted at the elementary educational level. This can be attributed to several factors. It is claimed that useful skills such as critical thinking and collaboration should be developed from an early age, and ER could enhance these skills through problem-based learning activities (Zhong et al., 2022). Hence, many countries have incorporated STEM education and ER into their official elementary school curriculum (Benitti & Spolaôr, 2017; Mannila, 2023). Conversely, in secondary education, ER is often integrated into specific domains like computer science and technology, or may be offered as an extracurricular activity (Arís & Orcos, 2019; Torres & Inga, 2025). These subjects are typically taught by computer science teachers, who may not encounter the same challenges as elementary school educators because of a lack of technical support or low confidence in using robotics (Yadav et al., 2016).

Related to RQ1 and the special traits of the studies included in this review, it is revealed that most researchers preferred qualitative research methods, thus non-numerical data analysis. This may be explained by the need to capture educators' diverse perspectives and identify concepts that quantitative scales might overlook (Hammarberg et al., 2016; Maxwell, 2012). Regarding the duration of interventions, 32% of the studies did not specify the intervention duration (see Fig. 25.3). Besides, only five studies described interventions lasting more than 8 weeks, perhaps because these interventions usually took place either after school hours or during summer vacations. As for the sample size, the data analysis depicted that most studies utilized small sample sizes ranging from 2 to 50 educators, while only 16% included more than 151 teachers (see Fig. 25.2). These findings highlight the need for long-term research examining more participants to generalize the emerging results more easily.

Related to the RQ2 and the challenges potentially faced by educators while integrating ER into STEM activities at various educational levels, we found that these could be divided into three groups: internal, external, and mixed (see Fig. 25.4).

In detail, internal challenges may include teachers' attitudes, (mis)beliefs, low self-efficacy, and limited confidence. Based on existing literature, teachers' self-efficacy in integrating robotics is a common challenge and varies across educational levels (Ortega-Ruipérez & Lázaro Alcalde, 2023). For instance, elementary school teachers tend to have a more negative attitude toward incorporating educational robots and are less willing to integrate them into their teaching compared to secondary school educators (Reich-Stiebert & Eyssel, 2016). This may be attributed to their limited familiarity with such technologies (Mangina et al., 2024). As for the external challenges, these include the high cost of equipment, limited access to ER resources and materials, integration of school ER curricula tailored to the students' needs and developmental traits, and time constraints. This aligns with existing research depicting that teachers' challenges are further compounded by practical difficulties in designing robotics activities and assessing students' performance (Vinnervik, 2022). Consequently, extra workload and time may be added to educators' school routines (Boz & Allexsaht-Snider, 2022). Besides, our content analysis reveals that the high cost of humanoid robots may hinder teachers from using them (see Fig. 25.6). Finally, regarding mixed challenges, the teachers' shortage of training and personal theoretical and practical knowledge, together with prior experience and familiarity with ER, refer to this group. Based on our findings, teachers may face mixed challenges (44%) most often, then external ones (31%), and internal challenges (25%) following as illustrated in Fig. 25.4. This may be attributed to the fact that mixed barriers are easier to identify whereas internal ones are difficult to "externalize" as personal conceptions (Vongkulluksn et al., 2018). Furthermore, teachers' perceptions of external barriers could lead them to better ER integration into their teaching practices (Xie et al., 2023). On the other side, the intrinsic deficiencies may be harder to identify and alter. In addition, it should be highlighted that the challenges explored in the papers included in this review are the same for both pre-service and in-service teachers. All teachers raise issues about low self-confidence related to ER usage stemming from inadequate training and support. Furthermore, most of the teachers regard the lack of knowledge and the shortage of robotics curricula and suitable equipment as the most important barriers they have to handle. Finally, the time required to design and implement ER into various teaching domains arises as a serious barrier to integration. Based on our content analysis, all types of challenges are associated with all kinds of robots used (see Fig. 25.6). These findings align with previous research indicating that teachers often feel unprepared to integrate robotics meaningfully into their practice without prior training, regardless of the type of robot used (Kopcha et al., 2022). Thus, teachers need professional development and support.

Regarding RQ 3, the proposed strategies for effective coping with potential challenges related to ER usage can also be divided into two groups: intrinsic and extrinsic (Figs. 25.7 and 25.8). Based on our review, these strategies should be applied by educational institutions, schools, and individual teachers. They should bridge theory and practice and pertain to both pre-service and in-service teachers. Regarding extrinsic strategies, long-term, step-by-step guided training, including instructional methods, should be offered to teachers. Existing research indicates that teachers often struggle to utilize ER due to a lack of professional development programs (Aksu &

Durak, 2019). Consequently, educators may not feel sufficiently comfortable using innovative practices like robotics. Thus, we recommend designing free, unstructured exploration activities, collaborative workshops providing access to various ER resources and materials, and thorough guidance regarding theoretical and practical knowledge for implementing robotics in plugged/unplugged sessions. Additionally, researchers recommend offering tips for creating hands-on, maker-learning environments for both teachers and students. This approach would bridge theory and relevant pedagogies with practice in enjoyable settings, promoting maximum engagement in ER school curricula. This aligns with existing literature that emphasizes the importance of a comprehensive curriculum format that clarifies its goals, ER resources, activity phases, and the student evaluation process (Boz & Allexsaht-Snider, 2022; Luo et al., 2023). Furthermore, it is proposed to train teachers through personalized programs enriched with scaffolding related to ER. Teachers should also be supported in creating challenge-based learning scenarios utilizing robotics, encouraging students' autonomy within an educational context (Hu et al., 2023). This aligns with the identified need for learning activities that develop students' decision-making and self-awareness, which can be fostered through challenge-based learning (Gallagher & Savage, 2023). Activities should be tailored to students' interests and personalized as much as possible through projects connected to the local community (Liao et al., 2024). This agrees with findings indicating that the more student-centered an activity is, the more effective it can be. Concerning intrinsic strategies, teachers need to develop twenty-first-century skills for themselves before integrating innovative technologies such as ER into classrooms. Therefore, our content analysis reveals that suitable institutional policies, instructional support from schools, peer mentoring from experienced educators, and collaboration during guided workshops and seminars could significantly contribute to teachers' acquisition of essential skills to handle internal challenges (Fig. 25.7) (Thomas et al., 2021). Besides, learning communities among teachers may help them address barriers like low self-confidence and insufficient theoretical knowledge (McDonald & Mercieca, 2021). In contrast, external challenges are more frequently linked to professional development programs and institutional support, indicating that systemic changes are necessary to address these issues. Finally, the fact that several challenges remain unconnected to any strategy suggests that many teachers still lack adequate support or clear guidance (Orlando et al., 2024). Targeted interventions designed to fill these gaps could facilitate a more inclusive and effective integration of educational robotics across diverse learning environments.

## 25.6 Limitations

Regarding the limitations of this book chapter, we searched for suitable evidence in 5 databases and 1 register, thus, papers in other databases may have been missed. Moreover, the fact that small sample sizes were examined in the studies investigated cannot lead to easily generalized conclusions.

## 25.7 Conclusion and Ways Forward

In the past, several attempts have been made to illustrate the ER's potential impact on children's development. However, they didn't systematically examine the potential challenges faced by teachers or provide clear strategies for handling these successfully. Thus, this book chapter focuses on teachers' barriers and offers useful information and guidelines about them to facilitate their successful coping. Regarding recommendations for future research, it would be useful a. to connect ER with curriculum content that connects schools to the community, b. to shape systemic changes regarding educational policies to support teachers and c. to provide teachers with long-term, step-by-step scaffolding training to foster their 21st-century skills to cope with ER without fear and discomfort.

**Funding** This chapter was prepared under the scope of Project 101194869 — TA Maker Education, funded by the European Union. Views and opinions expressed are, however, those of the authors only and do not necessarily reflect those of the European Union or European Education and Culture Executive Agency. Neither the European Union nor the granting authority can be held responsible for them."This research was funded by ERASMUS Lump Sum Grant number: 101194869, Acronym: TA Maker Education.

## References

Ackermann, E. (2001). Piaget's constructivism, Papert's constructionism: What's the difference? *Future of Learning Group Publication, 5*(3), 438.

Aksu, F. N., & Durak, G. (2019). Robotics in education: Examining information technology teachers' views. *Journal of Education and E-Learning Research, 6*(4), 162–168. https://doi.org/10.20448/journal.509.2019.64.162.168

Alimisis, D. (2013). Educational robotics: Open questions and new challenges. *Themes in Science and Technology Education, 6*(1), 63–71.

Anwar, S., Bascou, N. A., Menekse, M., & Kardgar, A. (2019). A systematic review of studies on educational robotics. *Journal of Pre-College Engineering Education Research (J-PEER), 9*(2), 2. https://doi.org/10.7771/2157-9288.1223

Arís, N., & Orcos, L. (2019). Educational robotics in the stage of secondary education: Empirical study on motivation and STEM skills. *Education Sciences, 9*(2), 73. https://doi.org/10.3390/EDUCSCI9020073

Atmatzidou, S., & Demetriadis, S. (2016). Advancing students' computational thinking skills through educational robotics: A study on age and gender relevant differences. *Robotics and Autonomous Systems, 75*, 661–670. https://doi.org/10.1016/J.ROBOT.2015.10.008

Bakala, E., Pires, A. C., Tejera, G., & Hourcade, J. P. (2023). *"It will surely fall": Exploring Teachers' perspectives on commercial robots for preschoolers*. In ACM International Conference Proceeding Series (Vol. 23, pp. 477–486). https://doi.org/10.1145/3582515.3609570

Benitti, F. B. V. (2012). Exploring the educational potential of robotics in schools: A systematic review. *Computers and Education, 58*(3), 978–988. https://doi.org/10.1016/J.COMPEDU.2011.10.006

Benitti, F. B. V., & Spolaôr, N. (2017). How have robots supported STEM teaching? In *Robotics in STEM education: Redesigning the learning experience* (pp. 103–129). https://doi.org/10.1007/978-3-319-57786-9_5/FIGURES/6

Boz, T., & Allexsaht-Snider, M. (2022). How do elementary school teachers learn coding and robotics? A case study of mediations and conflicts. *Education and Information Technologies, 27*(3). https://doi.org/10.1007/s10639-021-10736-4

Carlborg, N., Tyrén, M., Heath, C., & Eriksson, E. (2019). The scope of autonomy when teaching computational thinking in primary school. *International Journal of Child-Computer Interaction, 21*, 130–139. https://doi.org/10.1016/J.IJCCI.2019.06.005

Dorotea, N., Piedade, J., & Pedro, A. (2021). Mapping k-12 computer science teacher's interest, self-confidence, and knowledge about the use of educational robotics to teach. *Education Sciences, 11*(8). https://doi.org/10.3390/EDUCSCI11080443

Friend, M., McGill, M., & Reinking, A. (2022). *Solve this! K-12 CS education teachers' problems of practice.* In ACM International Conference Proceeding Series. https://doi.org/10.1145/3564721.3564738

Gallagher, S. E., & Savage, T. (2023). Challenge-based learning in higher education: An exploratory literature review. *Teaching in Higher Education, 28*(6). https://doi.org/10.1080/13562517.2020.1863354

Hammarberg, K., Kirkman, M., & De Lacey, S. (2016). Qualitative research methods: When to use them and how to judge them. *Human Reproduction, 31*(3), 498–501. https://doi.org/10.1093/HUMREP/DEV334

Hu, C. C., Yeh, H. C., & Chen, N. S. (2023). Teacher development in robot and IoT knowledge, skills, and attitudes with the use of the TPACK-based support-stimulate-seek approach. *Interactive Learning Environments, 31*(9), 5811–5830. https://doi.org/10.1080/10494820.2021.2019058

Khanlari, A. (2016). Teachers' perceptions of the benefits and the challenges of integrating educational robots into primary/elementary curricula. *European Journal of Engineering Education, 41*(3), 320–330. https://doi.org/10.1080/03043797.2015.1056106

Kim, C., Kim, D., Yuan, J., Hill, R. B., Doshi, P., & Thai, C. N. (2015). Robotics to promote elementary education pre-service teachers' STEM engagement, learning, and teaching. *Computers & Education, 91*, 14–31. https://doi.org/10.1016/J.COMPEDU.2015.08.005

Kopcha, T. J., Wilson, C. Y., & Yang, D. (2022). Improving teacher use of educational robotics to teach computer science in K-5 mathematics. In *Computational thinking in PreK-5* (pp. 47–54). Association for Computing Machinery. https://doi.org/10.1145/3507951.3519287

Kucuk, S., & Sisman, B. (2017). Behavioral patterns of elementary students and teachers in one-to-one robotics instruction. *Computers & Education, 111*, 31–43. https://doi.org/10.1016/J.COMPEDU.2017.04.002

la Hoz, A. D., Melo, L., Cañada, F., & Cubero, J. (2024). Educational robotics for science and mathematics teaching: Analysis of pre-service teachers' perceptions and self-confidence. *Heliyon, 10*(21), e40032. https://doi.org/10.1016/J.HELIYON.2024.E40032

Leoste, J., & Heidmets, M. (2020). Bringing an educational robot into a basic education math lesson. *Advances in Intelligent Systems and Computing, 1023*, 237–247. https://doi.org/10.1007/978-3-030-26945-6_21

Ley, T., Tammets, K., Sarmiento-Márquez, E. M., Leoste, J., Hallik, M., & Poom-Valickis, K. (2022). Adopting technology in schools: Modelling, measuring and supporting knowledge appropriation. *European Journal of Teacher Education, 45*(4), 548–571. https://doi.org/10.1080/02619768.2021.1937113

Liao, Y.-C., Kim, J., Ottenbreit-Leftwich, A. T., Karlin, M., & Guo, M. (2024). Voices of Elementary computer science teachers: Computer science integration rationales and practices. *ACM Transactions on Computing Education.* https://doi.org/10.1145/3688854/ASSET/49FC0462-2844-46ED-BE71-70E11EB42ADC/ASSETS/GRAPHIC/TOCE-2023-0133-F14.JPG

Lotriet, H. H., & Gouws, P. M. (2025). Educational robotics in physics education: A systematic review. *Studies in Science Education.* https://doi.org/10.1080/03057267.2025.2455334

Luo, F., Ijeluola, S. A., Westerlund, J., Walker, A., Denham, A., Walker, J., & Young, C. (2023). Supporting elementary teachers' technological, pedagogical, and content knowledge in computational thinking integration. *Journal of Science Education and Technology, 32*(4). https://doi.org/10.1007/s10956-023-10045-0

MacAnn, V., & Yadav, A. (2023). *Debugging beyond the code: Teachers' perceptions of debugging as a CT practice impacting interdisciplinary teaching and learning.* In CompEd 2023 - Proceedings of the ACM Conference on Global Computing Education (Vol. 1, pp. 119–125). https://doi.org/10.1145/3576882.3617919

Mallik, A., Liu, D., & Kapila, V. (2023). Analyzing the outcomes of a robotics workshop on the self-efficacy, familiarity, and content knowledge of participants and examining their designs for end-of-year robotics contests. *Education and Information Technologies, 28*(6), 7225–7264. https://doi.org/10.1007/S10639-022-11400-1/FIGURES/12

Mangina, E., Psyrra, G., Screpanti, L., & Scaradozzi, D. (2024). Robotics in the context of primary and preschool education: A scoping review. *IEEE Transactions on Learning Technologies, 17*, 342–363. https://doi.org/10.1109/TLT.2023.3266631

Mannila, L. (2023). Integrating programming in other subjects at primary level: Tool, glue or ideation. In *Teaching coding in K-12 schools: Research and application* (pp. 127–144). Springer. https://doi.org/10.1007/978-3-031-21970-2_9/FIGURES/10

Maxwell, J. A. (2012). The importance of qualitative research for causal explanation in education. *Qualitative Inquiry, 18*(8), 655–661. https://doi.org/10.1177/1077800412452856

McDonald, J., & Mercieca, B. M. (2021). The value of communities of practice for early career teachers. In *Sustaining communities of practice with early career teachers: Supporting early career teachers in Australian and international primary and secondary schools, and educational social learning spaces* (pp. 21–43). Springer. https://doi.org/10.1007/978-981-33-6354-0_2/TABLES/2

Morrison, L., & Hughes, J. (2024). Promising practices for online professional learning. *Computers and Education Open, 7*, 100209. https://doi.org/10.1016/J.CAEO.2024.100209

Nur Aksu, F., Durak, G., & Author, C. (2019). Robotics in education: Examining information technology teachers' views. *Journal of Education and E-Learning Research, 6*(4), 162–168. https://doi.org/10.20448/journal.509.2019.64.162.168

Orlando, S., Gaudioso, E., & De La Paz, F. (2024). Toward embedding robotics in learning environments with support to teachers: The IDEE experience. *IEEE Transactions on Learning Technologies, 17*, 874–884. https://doi.org/10.1109/TLT.2023.3339882

Ortega-Ruipérez, B., & Lázaro Alcalde, M. (2023). Teachers' perception about the difficulty and use of programming and robotics in the classroom. *Interactive Learning Environments, 31*(10), 7074–7085. https://doi.org/10.1080/10494820.2022.2061007

Palioura, M., Rapti, S., & Sapounidis, T. (2024). *Educational robotics integration into storytelling activities to promote kindergarteners' communication, collaboration, creativity, and critical thinking. A scoping review: 2014–2024.* In 2024 9th South-East Europe Design Automation, Computer Engineering, Computer Networks and Social Media Conference (SEEDA-CECNSM) (pp. 134–139).

Papadakis, S., Vaiopoulou, J., Sifaki, E., Stamovlasis, D., Kalogiannakis, M., & Vassilakis, K. (2021). *Factors that hinder in-service teachers from incorporating educational robotics into their daily or future teaching practice.* In International Conference on Computer Supported Education, CSEDU - Proceedings (Vol. 2, pp. 55–63). https://doi.org/10.5220/001041390055 0063

Possaghi, I., & Papavlasopoulou, S. (2024). *Challenges in digital STEM education delivery: A case study from the teachers' perspective.* In IEEE Global Engineering Education Conference, EDUCON. https://doi.org/10.1109/EDUCON60312.2024.10578908

Rapti, S., & Sapounidis, T. (2023). "Critical thinking, Communication, collaboration, creativity in kindergarten with educational robotics": A scoping review (2012–2023). *Computers & Education, 210*, Article 104968.

Reich-Stiebert, N., & Eyssel, F. (2016). *Robots in the classroom: What teachers think about teaching and learning with education robots.* Lecture Notes in Computer Science (Including Subseries Lecture Notes in Artificial Intelligence and Lecture Notes in Bioinformatics), 9979 LNAI (pp. 671–680). https://doi.org/10.1007/978-3-319-47437-3_66/TABLES/2

Saad, D., Verner, I., & Rosenberg-Kima, R. B. (2023). *Validation of teachers' competencies for applying robotics in science education*. Lecture Notes in Networks and Systems, 747 LNNS (pp. 27–35). https://doi.org/10.1007/978-3-031-38454-7_3/TABLES/2

Sapounidis, T., & Alimisis, D. (2020). Educational robotics for STEM: A review of technologies and some educational considerations. In *Science and mathematics education for 21st century citizens: Challenges and ways forward* (pp. 167–190). Nova Science Publishers.

Sapounidis, T., & Demetriadis, S. (2017). *Educational robots driven by tangible programming languages: A review on the field*. In Advances in Intelligent Systems and Computing (Vol. 560). https://doi.org/10.1007/978-3-319-55553-9_16

Sapounidis, T., Rapti, S., & Vaiopoulou, J. (2024). Effects of educational robotics on kindergarteners' collaboration, communication, critical thinking, and creativity: A meta-analysis. *Journal of Science Education and Technology, 34*, 73–87.

Sapounidis, T., Tselegkaridis, S., & Stamovlasis, D. (2022). Educational robotics and STEM in primary education: A review and a meta-analysis. *Journal of Research on Technology in Education*. https://doi.org/10.1080/15391523.2022.2160394

Snyder, H. (2019). Literature review as a research methodology: An overview and guidelines. *Journal of Business Research, 104*. https://doi.org/10.1016/j.jbusres.2019.07.039

Sophokleous, A., Christodoulou, P., tsidis, L., & Chatzichristofis, S. A. (2021). Computer vision meets educational robotics. *Electronics, 10*(6), 730. https://doi.org/10.3390/ELECTRONICS10060730

Thomas, L., Tuytens, M., Devos, G., & Vanderlinde, R. (2021). The collegial networks of early career teachers in primary education. In *Sustaining communities of practice with early career teachers: Supporting early career teachers in australian and international primary and secondary schools, and educational social learning spaces* (pp. 99–131). Springer. https://doi.org/10.1007/978-981-33-6354-0_5/FIGURES/3

Torres, I., & Inga, E. (2025). Fostering STEM skills through programming and robotics for motivation and cognitive development in secondary education. *Information, 16*(2), 96. https://doi.org/10.3390/INFO16020096

Tzagaraki, E., Papadakis, S., & Kalogiannakis, M. (2022). Teachers' attitudes on the use of educational robotics in primary school. Lecture Notes in Educational Technology (pp. 257–283). https://doi.org/10.1007/978-981-19-0568-1_13/TABLES/5

Vinnervik, P. (2022). Implementing programming in school mathematics and technology: Teachers' intrinsic and extrinsic challenges. *International Journal of Technology and Design Education, 32*(1), 213–242. https://doi.org/10.1007/S10798-020-09602-0/TABLES/2

Vongkulluksn, V. W., Xie, K., & Bowman, M. A. (2018). The role of value on teachers' internalization of external barriers and externalization of personal beliefs for classroom technology integration. *Computers and Education, 118*. https://doi.org/10.1016/j.compedu.2017.11.009

Weng, J., & Li, H. (2020). Early technology education in China: A case study of Shanghai. *Early Child Development and Care, 190*(10), 1574–1585. https://doi.org/10.1080/03004430.2018.1542383

Xie, K., Nelson, M. J., Cheng, S. L., & Jiang, Z. (2023). Examining changes in teachers' perceptions of external and internal barriers in their integration of educational digital resources in K-12 classrooms. *Journal of Research on Technology in Education, 55*(2). https://doi.org/10.1080/15391523.2021.1951404

Yadav, A., Gretter, S., Hambrusch, S., & Sands, P. (2016). Expanding computer science education in schools: Understanding teacher experiences and challenges. *Computer Science Education, 26*(4), 235–254. https://doi.org/10.1080/08993408.2016.1257418

Zaharin, N. L., Singh, S. S., Talin, R., Mariappan, M., Mohanaraj, N., Jusup, Y., & Suppiah, P. (2020). Promoting students' interest, attitude and intrinsic motivation towards learning stem through minimalist robot education programme. *International Journal of Service Management and Sustainability, 4*(1). https://doi.org/10.24191/IJSMS.V4I1.8054

Zha, S., Jin, Y., Moore, P., & Gaston, J. (2020). A cross-institutional investigation of a flipped module on preservice teachers' interest in teaching computational thinking. *Journal of Digital Learning in Teacher Education, 36*(1), 32–45. https://doi.org/10.1080/21532974.2019.1693941
Zhong, B., Liu, X., Xia, L., & Sun, W. (2022). A proposed taxonomy of teaching models in STEM education: Robotics as an example. *SAGE Open, 12*(2). https://doi.org/10.1177/21582440221099525

**Eleni Vlachoudi** is a PhD candidate at the Department of Philosophy and Education, Aristotle University of Thessaloniki, Greece. Her research focuses on the impact of educational robotics on primary school students' motivation. She holds two master's degrees from the International Hellenic University, in Management and Organization of Educational Units and in Digital and Soft Skills in Behavioural Sciences.

**Sophia Rapti** is a PhD candidate at the Department of Philosophy and Education at the Aristotle University of Thessaloniki, Greece, studying the impact of Robotics on the development of 4Cs to Kindergarteners. She graduated from the Department of Kindergarten of Aristotle University's Pedagogic School, and has two postgraduate degrees ('Management and Organization of Educational Units' and 'Digital and Soft Skills in Behavioral Sciences') from the International Hellenic University.

**Theodosios Sapounidis** is an associate professor at the Aristotle University of Thessaloniki, Greece, specializing in educational technologies and educational robotics. He is co-founder and CEO of the spin-off DuckyCode P.C., a company developing educational robots and tangible programming languages. He has filed multiple patents and won multiple national and international awards.

# Chapter 26
# Assessing Practical Work in Laboratory Settings

**Nikolaos Fotou, Michael J. Reiss, and Rachael Sharpe**

## 26.1 Introduction

This chapter is concerned with the assessment of practical work in laboratories. Our focus is on primary and secondary school settings, though what we write largely applies to older students too, for example to undergraduates studying science subjects in universities. We concentrate on laboratory settings rather than general classrooms (where primary science is often taught) or out-of-school settings (e.g., some ecology teaching). This is not because we are uninterested in other settings but because there is a limit to how much one chapter can cover and because the school laboratory setting is a particularly significant place in which practical work is undertaken. Laboratories are of interest because they present simplifications of reality. If one thinks of a typical school practical—for example, to determine the effect of light intensity on the rate of photosynthesis—it is fairly straightforward for secondary students to undertake within twenty minutes a little investigation to study this. For example, they can move a lamp various distances from a submerged aquatic plant such as *Cabomba* in a darkened room and counting the rate at which bubbles of oxygen-containing gas are produced. However, in the field, it is very difficult to determine the effect of light on the rate of photosynthesis, partly because of the difficulty of controlling the many variables that affect the rate of photosynthesis. Certainly, such a study is not possible at school level.

N. Fotou · R. Sharpe (✉)
School of Education and Communication, University of Lincoln, Lincoln, UK
e-mail: RSharpe@lincoln.ac.uk

N. Fotou
e-mail: NFotou@lincoln.ac.uk

M. J. Reiss
Institute of Education, UCL, London, UK
e-mail: m.reiss@ucl.ac.uk



L. Leite et al. (eds.), *Practical Work in Science Education in the Digital Era*, Advances in Technology-Rich Science Education 2, https://doi.org/10.1007/978-3-032-13788-3_26

Given the subject matter of the book in which this chapter appears, our interest is particularly on the extent to which the assessment of practical work in laboratory settings is any different in an increasingly digital age. To address this issue, we begin with a brief overview of the present state of practical work in school science before turning to the present state of practical work assessment. We then examine how the concepts of posthumanism and new materialism connect with the assessment of practical work before producing a new model for practical work in school science and then a new model for the assessment of practical work in school science.

## 26.2 The Present State of Practical Work

There is a large and growing academic literature on practical work in school science (Abrahams & Millar, 2008; Oliveira & Bonito, 2023). This literature has examined precisely what is meant by 'practical work', what its goals might be, and how it might be undertaken so that its goals are more likely to be attained. In addition, there is a large 'professional' literature that looks at such things as why school practical work is not undertaken more often in science lessons and what the material needs are if it is to be undertaken successfully (e.g., Cramman et al., 2019; Tsakeni, 2022). Overall, these two literatures agree that practical work entails the manipulation by learners, whether individually or in small groups, of tangible objects during or after which observations are made and/or measurements are taken. These literatures also show that while practical work is widely seen as a core component of school science, its place in school science education is not entirely secure. In part this is because of such pragmatic considerations as limitations on the availability of school laboratories, the cost of equipment and renewables, the shortage of school science technicians, and the reluctance of some schools to allow students to undertake much practical work in science. It is also because of the increasing acknowledgement that practical work may have too many goals. In particular, the evidence that it helps students develop conceptual understanding is not as strong as science educators might hope (Abrahams & Reiss, 2012).

## 26.3 The Present State of Assessment of Practical Work

The literature on the assessment of practical work is more limited than the literature on other aspects of practical work. Lunetta et al. (2007) concluded that a significant problem in the assessment of laboratory performance is that such assessment rarely requires actual practical work, relying instead on testing what students write. Lunetta et al. (2007) also pointed out that when assessment does require actual practical work, this results in the use of those laboratory activities that can be undertaken easily in a restricted time.

Abrahams et al. (2013) introduced the distinction between Direct Assessment of Practical Skills (DAPS) and Indirect Assessment of Practical Skills (IAPS):

> Direct Assessment of Practical Skills (DAPS) refers to any form of assessment that requires students, through the manipulation of real objects, to *directly* demonstrate a specific or generic skill in a manner that can be used to determine their level of competence in that skill. An example of this would be if a student was assessed on their skill in using an ammeter and this was determined by requiring them to manipulate a real ammeter and use it within a circuit to take readings and for these readings to need to be within an acceptable range for the student to be credited. In contrast, Indirect Assessment of Practical Skills (IAPS) refers to any form of assessment in which a student's level of competency, again in terms of a specific or generic skill, is *inferred* from their data and/or reports of the practical work that they undertook; for example, when a student writes up an account of the reaction between hydrochloric acid and calcium carbonate chips in a way that the marker could not be certain if the student is faithfully writing what they have just done or simply remembering what they have previously done or been told about this reaction. (p. 244)

Abrahams et al. (2013) concluded that each of DAPS and IAPS had advantages and disadvantages. They also noted that an example of the use of both DAPS and IAPS to assess practical skill and conceptual understanding respectively is provided by the Driving Test in the UK and a number of other countries. In this case, not only does the candidate have to demonstrate a sufficient level of competency in terms of practical driving skills out on the road (DAPS) but they must also pass an on-line test to assess their understanding of how to drive a car safely and competently (IAPS). The question then arises as to whether the increasing richness of digital technologies might afford a way of blurring the distinction between DAPS and IAPS, providing new ways of conceptualising learning through practical work, and of assessing practical work. To address this possibility, we draw on ideas from posthumanism and new materialism.

## 26.4 Integrating Posthumanism and New Materialism in Education

Posthumanism and new materialism are philosophical positions that challenge anthropocentric ways of understanding human relationships with and in the world. Both de-centre the human, and reconceptualise agency as distributed across human and non-human entities (e.g., Higgins, 2016). The application of posthumanism and new materialism to education advocates a countering both of Platonic approaches to didacticism, in which entities act upon other entities external to them, and to traditional constructivist approaches, that often maintain a separation between the teacher and the student of how and what is to be learnt (Roth, 2021). Instead, they emphasise an approach that foregrounds embodied and affective dimensions of learning, thus highlighting the importance of alternative ways of knowing.

A posthuman learning approach redefines traditional roles of teachers as the knowledge-providers and students as the recipients of this knowledge by recognising non-human actors, like digital technologies and AI, as integral to the learning process (Braidotti, 2013). This implies a distributed nature of agency between technology and human actors where learning takes place *with* technology (as in previous traditional approaches to learning) but also *within* technology, through an interactive and immersive engagement that cannot be achieved without it. In a similar way, and rather than seeing human minds as sole disseminators of knowledge, a new materialist approach views learning as the dynamic interplay of human and non-human entities, like digital technologies and AI, that are actively entangled in shaping the learning experience (Mannion, 2023). Both approaches therefore advocate an expanded understanding of education where humans and technology as a more-than-human entity coexist and are inter- and intra-related (Fawcett, 2002).

## 26.5 Beyond Cartesian Dualism: Embracing Relationality in Science Education

More than other subjects, for example the humanities or the arts, science, as an inseparable product of the Cartesian dualist rationality—a paradigm that emphasises objectivity and a separation of the knower from the known and the observer from the observation—faces particular challenges in adopting posthuman and new materialist approaches (Barad, 2007). This also holds true for practical work in science where (all but the smallest) objects are traditionally seen to possess inherent, measurable properties independent of and separate from any observer (Barad, 2007). This influence of Cartesian dualism also manifests in science education which has traditionally been based on standardised assessments and curricula that prioritise single ways of knowing, factual knowledge and an emphasis on tangible outcomes over the exploratory processes central to posthuman and new materialist approaches. Indeed, practical work in school science is traditionally structured to validate or refute pre-determined hypotheses rather than to explore learning processes, whereas, for posthumanism and new materialism, humans, materials and technologies are all entities that are not isolated but rather dynamically and relationally entangled, in that they influence and shape one another (Jeong et al., 2021).

A paradox emerges as the emphasis on exploration and relationality, key characteristics in posthumanism and new materialism, is also inherent in the authentic practice of scientific inquiry itself (Roth, 2021). In true scientific inquiry, the process is dynamic and relational in that it entails cycles of investigation, where scientists revisit, refine and adapt both their ideas and their methods based on interactions with the materials, tools, and environments they work with. This environment is not passive but rather actively shapes the inquiry process that practical work in school science too often fails to recognise (Bennett et al., 2018; Oliveira & Bonito, 2023). There is a need, therefore, for a transformative shift in practical work in science education that not only goes beyond validation but also embraces the assemblage of humans, materials and technologies which is central to knowledge creation (Puzio, 2024).

## 26.6 Intra-Action and Assemblage: Redefining Practical Work in Science Education

Barad's (2007) concept of *intra-action*, as opposed to *interaction*, can be utilised to help decentre humans and, in doing so, re-conceive practical work in science education and the way it is assessed. In discussing knowledge creation through laboratory experimentation, Barad (2007) explains that there is no inherent, objective, pre-determined "cut" that delineates the "object" from the "agencies of observation" or distinguishes the observer from an observation. Instead, both are part of a larger "apparatus" (p. 114), in which divisions are enacted and boundaries created as a form of mattering. The term intra-action therefore refers to a relation between entities (human and non-human) that are created temporarily, as a result of the relation, in a dynamic, emergent fashion. This is in contrast to an interaction, where the entities are seen as existing prior to the interaction, and are distinct, even if there is a prior relationship between them.

The notion of assemblage (Deleuze, 1994) complements Barad's (2007) concept of intra-action in emphasising the heterogeneous and dynamic nature of a science classroom. As in new materialism, assemblage refers to a collection of material, human and non-human entities that affect and are affected by each other. The assemblage can comprise "corporeal, technological, mechanical, virtual, discursive, and imaginary" (Ivinson & Renold, 2013, p. 707) elements that come together for knowledge to be co-constructed. This implies that science learning is shaped by the intra-action of all these various elements, like classroom materials, technologies, laboratory equipment, teacher intentions and strategies, students' prior knowledge, and assessment strategies that all come together to form an entangled whole. The focus thus shifts from single ways of knowing and factual knowledge development (even, accrual) to a deeper understanding of science as a dynamic, interconnected practice that transcends the boundaries of humans, technologies and the environment.

## 26.7 Transformative Practical Work in Science Education: Intra-Action and Technology Assemblage

The adoption of a posthuman and new materialist approach to practical work in science education can transform both the teaching and learning process by fostering and assessing the dynamic intra-actions within assemblages of students, teachers, technological tools and the material world. For example, students can use technology to design solutions to environmental problems, forming hypotheses and testing their utility. This might involve such practices as tracking daily activities and their environmental impact, using devices such as laptops, smartphones and tablets to analyse data and estimate their environmental footprint. This could be done in outdoor practical work (admittedly, somewhat beyond the focus of this chapter), such as monitoring local wildlife or water quality, or in the laboratory, such as analysing transportation choices, energy consumption at home, and dietary habits (e.g., veganism, vegetarianism, etc.) using digital tools in conjunction with more standard laboratory activities (e.g., food tests and determining the energy content of foods). In this example, several layers of intra-action and assemblage occur as students' conceptual understanding is assembled through the use of technological devices, which intra-act in the learning process as they provide real-time data and facilitate their analysis for students to reach conclusions and make informed decisions. Additionally, students' prior knowledge intra-acts with their environment. This enriches their understanding of the real-world implications of their actions, thus helping in the development of a holistic understanding of relevant scientific concepts through the entangled intra-action of all entities involved.

Such an approach shifts the focus from factual knowledge acquisition to the development of a deeper understanding of science as an on-going, interconnected practice that transcends the boundaries of humans, technologies and the environment. The observer and the observed thus become entangled within technologies and broader laboratory apparatuses, making objectivity not absolute but relational (Barad, 2007). Mannion (2020) describes how this can be applied to practical work in a place-responsive heritage education project where students engage with both human and more-than-human elements. In that project, students' prior knowledge and understanding was assembled with GPS devices to map local heritage sites, digital cameras to document their observations, and data-logging equipment to record and analyse environmental data. Through the intra-action of all elements, students formulated hypotheses and tested them to make informed decisions about sustainable practices, thus highlighting the active role of material aspects in co-constructing knowledge and problem-solving skills development.

## 26.8 Assemblage and Intra-Action in Practical Work Assessment

The shift to a posthuman and new materialist approach not only challenges traditional approaches to how practical work is conducted but how it is assessed. Technologies, such as video recording, digital portfolios, simulation software and Augmented Reality (AR) and Virtual Reality (VR) platforms, can be integrated into a cohesive assessment framework. Such a platform can be utilised to capture the complexity of students' interactions with the assemblage of the technological tools themselves, scientific concepts and their understanding and prior knowledge of these, materials and each other. Hetherington (2024), for example, explores the utility of a multimodal interaction analysis framework which can be utilised in assessing practical work in science education through technology. This framework analyses how meaning is constructed using multiple semiotic modes, like verbal language, gestures and material objects, that can be applied in the analysis of classroom interactions. Integrating such frameworks with technology, like video recording, digital portfolios, simulation software and AR/VR platforms, can be utilised in the assessment of students' practical skills and conceptual understanding. This can be done by capturing the multifaceted intra-actions between these tools, the students themselves, their bodies and their teachers that mediate and inform the students' conceptual understanding and skills development during practical work.

To some, this may all sound futuristic, but it is worth emphasising how standard such approaches are increasingly becoming for STEM (Science, Technology, Engineering and Medicine) professionals. For example, simulators have been used for decades in the training of pilots. Indeed, there were sufficient articles, research reports and technical reports for a 1992 statistical meta-analysis to be undertaken of the effectiveness of simulator training for pilots (Hays et al., 1992). This analysis concluded that "the use of simulators combined with aircraft training consistently produced improvements in training for jets compared to aircraft training only" (p. 63). What we find particularly noteworthy, even if it is somewhat unsurprising, is that current training practices continue to make use both of flight simulation and of actual flights (e.g., Paul et al., 2025). Interestingly, in surgery the story seems to be more complicated. In orthopaedic surgery, a recent systematic review concluded that while it would be premature to draw conclusions across the whole field of orthopaedics, in certain specialisms, notably knee arthroscopy, there is good evidence of the benefits that have followed from the introduction of simulation training (Woodward et al., 2025). However, in plastic surgery, a recent systematic review concluded that the recent shift from traditional surgical teaching to the incorporation of simulation training in plastic surgery had not resulted in "reductions in surgical complications or costs" (Lastoria et al., 2025, p. 1).

We also note that were practices for the assessment of school practical work ever routinely to draw on the diversity of tools that we have outlined above, this would have implications both for the distinction between DAPS and IAPS and for the distinction between formative and summative assessment. We could envisage a time when the assessment of practical work routinely drew on both direct and indirect approaches to the assessment of practical skills, and when the need for formal 'big bang' summative assessment at the end of a course or module was reduced as a result of the rich data available on student performance throughout the course or module.

Haus (2020) provides another example of the intra-action of human and non-human entities in the learning process that can be also applied in assessing practical work with technology. The study focuses on a young child and their interactions with a paper aeroplane, highlighting how both human (the young child) and non-human (a sheet of paper that becomes a plane) entities mutually transform each other through their intra-actions. The sheet of paper changes the child as much as the child changes the sheet of paper, with both becoming different (maturing) after their encounter. The plane influences the child's understanding of relevant concepts like lift, drag and thrust, while the child's manipulation of the sheet and then the plane affects its flight patterns. In this example, video data analysis was used to assess how the child and materials intra-act as the child experiences and experiments with the quality of the paper plane and engages with the relevant scientific concepts.

Such video data can be employed for the analysis of embodied actions during practical work and can help reveal how relevant concepts are understood and internalised and skills developed through an intertwined process of play, learning and becoming. AI technology can be used to complement such an analysis in that it can detect patterns in the behaviours of students during practical work that indicate conceptual understanding and skill development (cf. Kotze, 2025). It can also become a tool assembled in the learning process through intra-action as AI-generated feedback from video data analysis can contribute to the mutual transformation of the student, the AI and the learning materials. This occurs as students can receive and act on the AI-generated feedback, while the AI system learns from the students and improves its ability to provide relevant and effective guidance for students' conceptual understanding and skills development.

Compared to traditional direct or indirect methods of assessment, such as lab reports or observations of students when doing practical work, the use of technology within a posthuman and new materialist approach to learning can therefore provide a richer and more comprehensive assessment framework. This supports a multimodal and more nuanced understanding of how students intra-act with material and non-material elements to make sense of scientific concepts, thus enhancing both the teaching process as well as learning outcomes.

## 26.9 Conclusion and Ways Forward

There is no doubt that as we move through the digital age, Solomon's (1980) acceptance of the proposition that science lends itself to the laboratory setting as naturally as cooking does to the kitchen needs re-visiting. Now is the time to question whether the laboratory needs to be solely a physical materialistic environment or whether, in this era of continuous digital advancement, we could more profitably consider the affordances of new technologies and the connectedness between entities—human and non-human.

In this chapter, we have proposed that embracing a shift to posthumanist and new materialist approaches may allow for a transformative direction that not only encourages discussion and inquisitiveness, but also the opportunity to engage in the formative process of learning and, perhaps more fundamentally, a love of learning itself, as opposed to simply focussing on assessment-driven accountability measures.

Comparing the traditional direct or indirect assessment approaches, posthumanism and new materialism open opportunities for learners to explore and connect in a more thinking *and* doing manner. Whilst there is little doubt that science teachers sometimes question the use of practical work, for all that they espouse it, moving the focus from mere *doing* to the actual intra-action between the entities of human, technology and environment may have the potential to allow for more open-ended

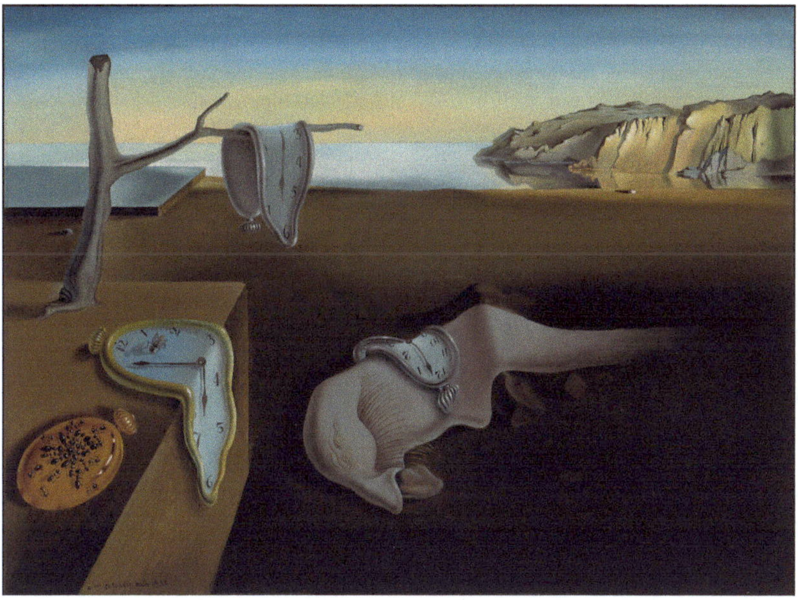

**Fig. 26.1** Salvador Dali's 1931 The Persistence of Memory (https://www.moma.org/collection/works/79018)

learning that moves away from the rhetoric of prescribed or proof-determining practical work.

Current regimes for the assessment of practical work typically presume a separation of the observer from the observed. If we were to consider exploring a more nuanced connectedness, in the sense that the observer and the observed have an intra-action with the environment and the technology around them, we may find a more connected learning environment where the learning becomes one with the learner. This might have benefits for student learning (cf. Fig. 26.1) and for motivation.

Appreciating the potential pragmatic difficulties this may face, such as accountability measures and constraints within curricula, this notion of the oneness of learning may lead to more adaptive teaching and learning that truly puts the learning first. Indeed, science is fundamentally about curiosity and finding clarity and simplicity in very complex ideas. Adapting our thinking around a nuanced posthuman and new materialist approach may enable learners to think at a deeper level than that afforded by much school science:

> I have a friend who's an artist and has sometimes taken a view which I don't agree with very well. He'll hold up a flower and say "look how beautiful it is," and I'll agree. Then he says "I as an artist can see how beautiful this is but you as a scientist take this all apart and it becomes a dull thing," and I think that he's kind of nutty. First of all, the beauty that he sees is available to other people and to me too, I believe. Although I may not be quite as refined aesthetically as he is… I can appreciate the beauty of a flower. At the same time, I see much more about the flower than he sees. I could imagine the cells in there, the complicated actions inside, which also have a beauty. I mean it's not just beauty at this dimension, at one centimeter;

> there's also beauty at smaller dimensions, the inner structure, also the processes. The fact that the colors in the flower evolved in order to attract insects to pollinate it is interesting; it means that insects can see the color. It adds a question: does this aesthetic sense also exist in the lower forms? Why is it aesthetic? All kinds of interesting questions which the science knowledge only adds to the excitement, the mystery and the awe of a flower. It only adds. I don't understand how it subtracts. (Feynman, 1999, p. 2)

Moving forward, it may be worth considering how the assemblage and intra-action of the assessment of practical work may provide learners with increased resilience in their journey of science exploration (and other aspects of their everyday lives). The innovative use of technology, such as VR or AR, to allow learners to explore and question in ways beyond the mundane, can help contribute to knowledge creation as opposed to merely acquisition of knowledge. Seeing the process as a learning journey and setting up the environment that allows for connectedness across multi-agencies may allow for such effective learning, as Haus (2020) suggests. By focussing on the intra-action of technology, environment and learner, the inquisitive exploration of science may be further strengthened. Rather than science being learnt as a single response or didactic process, we propose a dynamic interconnected practice that transcends the conventional boundaries of such entities to provide a deeper understanding where the student embodies the learning and the intra-action itself, a new bespoke learning assemblage.

Such a paradigm shift may not only allow for flexibility in a student's learning, it may also facilitate the development of their inquisitive nature that science should invoke. Instead of the notion of 'dumbing down science', this approach could open up an avenue for greater accessibility and student-led learning accompanied by the agencies that could embellish and guide students to envisage new understandings. Of course, the teacher plays an important part in creating this environment where students have the time, space and accessibility to materials in their surrounding world to explore learning as opposed to the increasing focus on narrow learning goals (Areljung, 2020). We wish to propose that this may well be the start of changing the rhetoric from science being seen as difficult to a science that is all about exploring; and where assessment becomes part of the journey and not *the* journey.

## References

Abrahams, I., & Millar, R. (2008). Does practical work really work? A study of the effectiveness of practical work as a teaching and learning method in school science. *International Journal of Science Education, 30*(14), 1945–1969. https://doi.org/10.1080/09500690701749305

Abrahams, I., & Reiss, M. J. (2012). Practical work: Its effectiveness in primary and secondary schools in England. *Journal of Research in Science Teaching, 49*, 1035–1055. https://doi.org/10.1002/tea.21036

Abrahams, I., Reiss, M. J., & Sharpe, R. M. (2013). The assessment of practical work in school science. *Studies in Science Education, 49*, 209–251. https://doi.org/10.1080/03057267.2013.858496

Areljung, S. (2020). Capturing the world with verbs: Preschool science education beyond nouns and objects. *Contemporary Issues in Early Childhood, 21*(1), 70–82. https://doi.org/10.1177/1463949118805438

Barad, K. (2007). *Meeting the universe halfway: Quantum physics and the entanglement of matter and meaning*. Duke University Press.

Bennett, J., Dunlop, L., Knox, K. J., Reiss, M. J., & Torrance-Jenkins, R. (2018). Practical independent research projects in science: A synthesis and evaluation of the evidence of impact on high school students. *International Journal of Science Education, 40*, 1755–1773. https://doi.org/10.1080/09500693.2018.1511936

Braidotti, R. (2013). *Metamorphoses: Towards a materialist theory of becoming*. John Wiley & Sons.

Cramman, H., Kind, V., Lyth, L., Gray, H., Younger, K., Gemar, A., Eerola, P., Coe, R., & Kind, P. (2019). *Monitoring practical science in schools and colleges*. https://durham-repository.worktribe.com/output/1605305

Deleuze, G. (1994). *Difference and repetition*. Columbia University Press.

Fawcett, L. (2002). Children's wild animal stories: Questioning interspecies bonds. *Canadian Journal of Environmental Education, 7*(2), 125–139.

Feynman, R. P. (1999). *The pleasure of finding things out: The best short works of Richard P. Feynman*. Perseus Books.

Haus, J. M. (2020). Performative intra-action of a paper plane and a child: Exploring scientific concepts as agentic playmates. *Research in Science Education, 50*, 1305–1320. https://doi.org/10.1007/s11165-018-9733-8

Hays, R., Jacobs, J., Prince, C., & Salas, E. (1992). Flight simulator training effectiveness: A meta-analysis. *Military Psychology, 4*(2), 63–74. https://doi.org/10.1207/s15327876mp0402_1

Hetherington, L. (2024). Material-dialogic space as a framework for understanding material and embodied interaction science education. *Research in Science & Technological Education, 42*(1), 202–218. https://doi.org/10.1080/02635143.2024.2328766

Higgins, M. (2016). Decolonizing school science: Pedagogically enacting agential literacy and ecologies of relationships. In C. Taylor & C. Hughes (Eds.), *Posthuman research practices in education* (pp. 186–205). Springer.

Ivinson, G., & Renold, E. (2013). Valleys' girls: Re-theorising bodies and agency in a semi-rural post-industrial locale. *Gender and Education, 25*(6), 704–721. https://doi.org/10.1080/09540253.2013.827372

Jeong, S., Sherman, B., & Tippins, D. J. (2021). The Anthropocene as we know it: Posthumanism, science education and scientific literacy as a path to sustainability. *Cultural Studies of Science Education, 16*(3), 805–820.

Kotze, B. (2025). Artificial intelligence applied in innovative assessments in the world of work scenarios. *Global Journal of Engineering Education, 27*(1), 58–63.

Lastoria, D., Rehman, S., Ahmed, F., Jasionowska, S., Salibi, A., Cavale, N., Dasgupta, P., & Aydin, A. (2025). A systematic review of simulation-based training tools in plastic surgery. *Journal of Surgical Education, 82*(1), Article 103320. https://doi.org/10.1016/j.jsurg.2024.103320

Lunetta, V. N., Hofstein, A., & Clough, M. P. (2007). Learning and teaching in the school science laboratory: An analysis, of research, theory, and practice. In S. K. Abell & N. G. Lederman (Eds.), *Handbook of research on science education* (pp. 393–441). Lawrence Erlbaum.

Mannion, G. (2020). Re-assembling environmental and sustainability education: Orientations from new materialism. *Environmental Education Research, 26*(9–10), 1353–1372. https://doi.org/10.1080/13504622.2018.1536926

Mannion, G. (2023). Fieldnotes and situational analysis in environmental education research: Experiments in new materialism. *New Materialisms and Environmental Education*. Routledge.

Oliveira, H., & Bonito, J. (2023). Practical work in science education: A systematic literature review. *Frontiers in Education, 8*, 1151641. https://doi.org/10.3389/feduc.2023.1151641

Paul, N., Moncion, B., & Cao, S. (2025). An experimental comparison on the effectiveness of various levels of simulator fidelity on ab initio pilot training. *Ergonomics*, 1–17. https://doi.org/10.1080/00140139.2024.2449110
Puzio, A. (2024). The entangled human being—A new materialist approach to anthropology of technology. *AI and Ethics, 5*, 2339–2356. https://doi.org/10.1007/s43681-024-00537-z
Roth, W. M. (2021). Gardener-becoming-tree, tree-becoming-gardener: Growing-together as a metaphor for thinking about learning and development. *Cultural Studies of Science Education, 16*, 915–930. https://doi.org/10.1007/s11422-021-10032-0
Solomon, J. (1980). *Teaching children in the laboratory*. Croom Helm.
Tsakeni, M. (2022). STEM education practical work in remote classrooms: Prospects and future directions in the post-pandemic era. *Journal of Culture and Values in Education, 5*(1), 144–167. https://doi.org/10.46303/jcve.2022.11
Woodward, C. J., Khan, O., Aydın, A., Dasgupta, P., & Sinha, J. (2025). Simulation-based training in orthopedic surgery: A systematic review. *Current Problems in Surgery, 63*, Article 101676. https://doi.org/10.1016/j.cpsurg.2024.101676

**Nikolaos Fotou** is a senior lecturer and the MA of Arts in Education program, lead at the University of Lincoln, UK. He studied physics at the University of Crete and earned his MA and PhD in science education from the University of Leeds. He has taught from primary school to postgraduate programs in various institutions. He advocates for evidence-based education, with research interests including analogical reasoning and interactive pedagogies.

**Michael J. Reiss,** PhD and post-doc in evolutionary biology, is professor of science education at UCL Institute of Education (UK), honorary visiting professor at the Royal Veterinary College, honorary fellow of the British Science Association, fellow of the Academy of Social Sciences, and a member of the Nuffield Council on Bioethics. He was the Director of Education at the Royal Society, president of the International Society for Science and Religion, and president of the Association for Science Education.

**Rachael Sharpe,** PhD and PGCE (secondary physics), is a senior lecturer and programme leader for the PGCE in secondary education, alongside teaching on undergraduate and postgraduate courses at the University of Lincoln, UK. Her expertise is in science education, teacher training, educational attainment, and students' attitudes. She worked on numerous funded research projects, was actively involved in local authority, academy, and private school governance, and acted as an external examiner in teacher education.

# Part VI
# Ethical and Safety Issues in Practical Work

# Chapter 27
# Ethical Issues in the Science Laboratory

**Kelly Laas and Elisabeth Hildt**

## 27.1 Introduction

Ethics education is an integral part of science education in general, and education related to scientific research practice in particular (Zandvoort et al., 2013). STEM research is a complex process where issues of data management, authorship, supervisor-trainee relationships, power imbalances, and factors beyond the scientific research process can significantly impact the research process and the success of researchers and students. Students new to the research process confront several challenging situations, which can be a valuable opportunity to learn the skills needed to navigate and practice new ethical challenges in real life. This includes practical and ethical issues in collecting and analyzing data, what to do when something goes wrong in the research process, negotiating complex collaborations, and potentially authorship on a research paper. This chapter will discuss why ethics education matters, what critical issues researchers are facing in the lab, including issues around the use of artificial intelligence in the research process, and discuss some of the benefits and challenges of engaging senior and junior researchers in exploring and developing ethics guidelines of this kind in a digital setting.

Various approaches towards ethics in STEM research and education exist, with different countries focusing on other concepts. The ethical aspects of STEM research and questions surrounding research ethics have been explored through various concepts, including research integrity, the Responsible Conduct of Research (RCR), and Responsible Research and Innovation (RRI). What counts as "research integrity" topics and how it is taught in higher education varies worldwide.

---

K. Laas (✉) · E. Hildt
CSEP, Illinois Institute of Technology, Chicago, IL, USA
e-mail: laas@illinoistech.edu

E. Hildt
e-mail: ehildt@illinoistech.edu



L. Leite et al. (eds.), *Practical Work in Science Education in the Digital Era*, Advances in Technology-Rich Science Education 2, https://doi.org/10.1007/978-3-032-13788-3_27

In the United States, the responsible conduct of research (RCR) is of central importance for ethics in STEM research. The United States National Institutes of Health (NIH) defines responsible conduct of research as "the practice of scientific investigation with integrity. It involves the awareness and application of established professional norms and ethical principles in performing all activities related to scientific research" (National Institutes of Health, 2009). In many RCR courses in the United States, this topic is broken down into the following themes: research misconduct; protection of human subjects; research involving animals; authorship and publication practices; mentor/training responsibilities; data issues, e.g., ownership, management, acquisition; and conflicts of interest (Office of Research Integrity, 2000).

In Europe, the concept of responsible research and innovation (RRI) emerged as a key focus in research integrity studies, stemming from a policy-driven initiative by the European Union's Framework Program for Research and Innovation, known as "Horizon 2020." Its purpose was to foster the design of inclusive and sustainable research and innovation with "an emphasis on co-creation and co-productivity in society" (Owen & Pansera, 2019). As defined by René von Schomberg, RRI "is a transparent, interactive process by which societal actors and innovators become mutually responsive to each other with a view to the (ethical) acceptability, sustainability and societal desirability of the innovation process and its marketable products (to allow a proper embedding of scientific and technological advances in our society)" (2013). This includes scientific and technological development processes, taking into account the effects and potential impacts on the environment and society. Others have defined RRI as "Responsible innovation means taking care of the future through collective stewardship of science and innovation in the present (Stilgoe et al., 2013, 1570).

Responsible Conduct of Research (RCR) and Responsible Research and Innovation (RRI) share a focus on research and ethics, but differ in a few key ways. RCR is narrower in scope than RRI, focusing on the conduct of research. RCR, following its more traditional definition, is not about innovation, but rather about avoiding certain kinds of misconduct in research (Davis & Laas, 2014). Responsible Research and Innovation, however, has a much broader purview, focusing on both ethical issues in research and including concepts like public engagement in science, open access to science/open science (including access to publicly funded research), and issues of gender equality in science, to name only a few.

These initiatives, along with a growing realization of the vital role of ethical standards in research, have helped produce a series of internationally recognized guidelines that set the standard for research integrity. These include high-level guidelines, such as the European Code of Conduct for Research Integrity, published by the All European Academies (ALLEA, 2023), and the Singapore Statement on Research Integrity, developed in 2010 by the World Conference on Research Integrity (WCRI, 2010). Both guidelines seek to provide members of the research community (European and international) with a framework for self-regulation across all scientific disciplines and research institutions. International standards, such as these, paired with discipline-specific codes of ethics (e.g., the U.S. National Society of Professional Engineers or the American Physical Society), serve as an excellent starting point for understanding the ethical principles common across research fields and how they are applied in various ways.

One final helpful method for considering research integrity issues is the distinction between micro and macro ethics. Microethics examines the individuals and internal relationships involved in research, as well as the standards and norms that guide individual behavior. In this way, the U.S. NIH definition of Responsible Conduct of Research deals primarily with issues of microethics. Macroethics, on the other hand, applies to the collective social responsibility of researchers, societal decisions about technology, and the social implications of technology (Herkert, 2005). Over the past twenty years, the U.S. has joined its international colleagues in advocating for the inclusion of research integrity education in curricula for young scientists and engineers, including discussions about the ethical implications of their work in society (Bird, 2014).

Recent developments in the area of research integrity education are pushing this curriculum to include more diverse subjects, such as issues of diversity and inclusion, collaboration—both internally in the research group and externally with colleagues from around the world—, and sexual harassment, as well as new, emerging issues such as the use of artificial intelligence and big data in research (Watts et al., 2017). There is also an immense variety in the ethical topics and questions relevant to different disciplines. Research ethics in psychology, biomedical engineering, and medicine differ significantly from the ethical issues that arise in chemical and mechanical engineering research labs.

Along with this variety of ethics topics, approaches to integrating ethics into STEM curricula vary, with techniques ranging from a full-semester course in research ethics to micro-insertions of ethics modules or discussions in methods courses, as well as stand-alone workshops and online education modules. All these techniques offer benefits in terms of ease of delivery, focus on topics and ethical issues pertinent to specific disciplines, and increased engagement levels. A one-size-fits-all approach has proven to be impossible. Universities, departments, and research groups must develop a program tailored to the unique needs of their researchers and students. The delivery of RCR education in the U.S. has been thoroughly analyzed, and studies have found that traditional RCR education approaches, such as classroom-based courses, workshops, and online courses, have been ineffective (Holsapple et al., 2012; Kalichman, 2014; Plemmons & Kalichman, 2013).

In general, three factors should be considered when developing an effective program to engage researchers in learning about research ethics. (1) The content should extend beyond merely learning about laws and rules governing research, (2) it should include discussion of ethical issues in their relevant context and involve all stakeholders, including students' peers, mentors, and supervisors (Watts et al., 2017). Third, ethics education is most effective when it occurs within the respective *institutional culture,* which comprises both the *organizational context* and the *peer environment* (Holsapple et al., 2012).

In meta-analyses of ethics education, some of the most successful methods include engaging a multidisciplinary expert teaching team, involving students in active case-study discussions of scenarios that reflect real life, and including activities that ask students to practice and actively engage with the material at hand individually (Martin et al., 2021; Mulhearn et al., 2017; Mumford et al., 2015; Watts et al., 2017).

## 27.2 Ethical Issues Related to the Use of Digital Technologies in Labs

The use of digital technologies within physical labs raises a host of ethical issues that are largely undiscussed, except in the broadest terms, by prevailing national and international codes of ethics and value statements. Ethical issues surrounding the integrity of data, including its production, storage, confidentiality, and reproducibility, are significant concerns in all research labs. As labs increasingly rely on third-party cloud-based systems to store, analyze, and share data, common issues of data security, confidentiality, and privacy are raised (Jahnke & Asher, 2013). Lab-specific or university-specific storage systems may mitigate large-scale data breaches. Any cloud-based storage is vulnerable to cyberhacking, however, and depends on the effective maintenance of staff assigned to safeguard these systems. Equitable access to necessary technologies, storage, and the ability to run complex programs on shared servers is often determined by seniority in the lab, decisions made by lab supervisors, and deadlines, and can significantly impact the success of new researchers.

Combined with lab-based classes, new researchers learn basic skills and techniques in the lab through collaboration with their supervisors and superiors, as well as the norms and values integral to sound scientific research. The COVID-19 pandemic shifted much hands-on lab teaching and research to an online environment, resulting in the expansion of hybrid models that aim to blend the benefits of e-learning with those of in-person instruction and experience. In a physical lab, in-person training has traditionally enabled students to develop the necessary skills, including proficiency with lab equipment, experience in data acquisition, analysis, and interpretation. Online labs and virtual reality labs have the potential to encourage teamwork, facilitate active learning, and foster creativity (Qin et al., 2020; Serrano-Perez et al., 2023). Virtual labs have the potential sometimes to meet or exceed the effectiveness of traditional labs for certain learning goals or under specific conditions. This includes reducing the need for animal use in teaching labs, as well as the fact that in many virtual labs, experiments can be repeated multiple times with few or no time constraints, allowing students to develop the necessary skills (Gaganis et al., 2021; Yap et al., 2021). Downsides of virtual labs include a lack of real-life practical experience, however.

Along with the potential benefits of this shift, ethical issues inevitably emerge. This includes issues related to student inequalities in digital access and student data security (Gaganis et al., 2021; Maglio et al., 2025). The overall cost of third-party platforms that make virtual laboratories and online instruction possible can lead to higher course fees for students, and software that requires higher video quality and fast internet connectivity can be out of reach for some students, depending on their income level, location, and technology access (Moran et al., 2023). This is particularly important for universities working with and in developing countries (Kalolo, 2019). The security of student data (including name, email, passwords, etc.) is a significant concern, as it ensures that third-party vendors do not have access to or store this data on external cloud-based servers, thereby rendering them vulnerable to cyber hacking and data theft. Universities and instructors using these platforms must ensure the safety and security of students' data (Ulven & Wangen, 2021).

## 27.3 Our Approach—Situating Ethics in a Research Lab

The Illinois Institute of Technology has a long history of integrating ethics into practical STEM courses. Founded in 1976, the Center for the Study of Ethics in the Professions (CSEP) helped pioneer the field of engineering ethics in the late 70s and early 80s and taught a "train the trainer" course from 1997 to 2004 which invited science and engineering faculty from around the country to come and learn how to integrate ethics discussions into their existing curriculum. From 2004 to the present, the Ethics Center faculty and staff have worked with Illinois Tech instructors to develop ethics content for their existing classes, sometimes serving as guest lecturers and sometimes co-teaching one to three course sessions. In each session, all the designed activities ideally engage the students in an active learning activity that requires them to discuss, analyze, and make and defend ethical decisions about a fictional case study or a real-life scenario.

Drawing from the history of the Ethics Center's teaching on this topic and several formative sessions held with colleagues from the biomedical engineering department at Illinois Tech that centered around informal discussions with colleagues about questions and ethical issues that arise as part of the research process, the research team began discussing the possibility of convening groups of graduate students and researchers who work in the same lab, and running instructional sessions that would help them workshop possible guidelines the address existing ethical issues.

This approach, which was generously funded within the National Science Foundation's program "Cultivating Cultures for Ethical STEM (CCE STEM)", project number 1635661, has two benefits. First, it allows stakeholders in the lab the time and structure to identify factors they consider relevant to their specific research environment. Second, it promotes the cultivation of an ethical culture in the lab by integrating these factors into a bottom-up approach to develop context-specific ethical guidelines. Central to the strategy is moving away from traditional classroom-based ethics education to integrating ethics discussions with the laboratory experience. In other words, education that directly addresses specific issues found in the laboratory or departmental environment. Furthermore, active involvement of graduate students is accomplished by tasking them with leading the ethics discussion and development. For the students, their goal is to develop guidelines that address the issues they face, guidelines they consider valuable for their laboratory situation, and to discuss and further develop them with other students and faculty (Hildt et al., 2019; Laas et al., 2020, 2021).

In 2016, work on this project began with two parallel activities. First, the research team conducted a survey that asked all members of the research community (undergraduates, graduate students, faculty, and staff) to identify the ethical issues they considered most important in their field. Second, the team conducted semi-structured interviews with 32 graduate students, exploring their experiences of joining and working in research labs, as well as confronting and resolving ethical issues during their research. Findings from this background study highlighted that while faculty and students hold similar views about key issues, such as avoiding research misconduct, ensuring data integrity, and maintaining the overall integrity of research, they also identified relevant differences. Students, especially graduate students, identified communication, fairness, respect, and tolerance issues as critical ethical concerns in the research environment (Laas et al., 2022). In the interview study, students expressed the vital role lab supervisors play in building ethical, equitable lab cultures, and how a research environment that promotes open communication, collaboration, and the ability to ask questions assists students in learning about, recognizing, and ultimately attempting to solve ethical issues that come up in the lab (Haven et al., 2023; Hildt et al., 2022, 2024; Laas et al., 2024a, 2024b).

Second, the research team developed a six-session interactive workshop that engaged sixteen students at a time (teams of eight from two different disciplines) in actively reflecting on and articulating what they believe are the crucial ethical issues they encounter in their research environments. After discussing and analyzing existing ethical codes within their discipline, the student participants were asked to develop guidelines that would be useful in addressing ethical issues in research laboratories. This workshop was conducted during the fall and spring 2016 semesters at Illinois Tech and during the summer of 2019 with a group of biomedical engineering students at the University of Texas at San Antonio. An analysis of the guidelines developed found that students were most concerned about the fairness of the research environment (fair distribution of projects, work, funds, ability to attend and present at conferences), having clear guidance around expectations and responsibilities in the lab, fair access to shared lab resources such as lab equipment, the need for safety training, and the ability to meet with the lab supervisor regularly to discuss research progress (Laas et al., 2022). The guidelines, while also addressing many traditional research ethics issues such as data management and publication ethics, focused mainly on the mentor/mentee relationship and how defining these roles would improve the research environment.

This project aimed to develop a flexible workshop series that could be delivered in various methods and time frames. Variations of the workshop were conducted, as mentioned previously, at the University of Texas, San Antonio, and with research labs at Illinois Institute of Technology. When the pandemic hit in 2020, the research team had to pivot on how the developed ethics workshop could be delivered to Illinois Tech students, faculty, and staff. This change aligned with the need for Illinois Tech's Research Compliance Office to revise the university's Responsible Conduct of Research education policy to incorporate both online training through CITI and new options for in-person training.

## 27.4 Virtual Research Integrity Workshop

In collaboration with Illinois Tech's Research Compliance Staff, the Ethics Center set to work developing a workshop module that can be delivered in person or via video chat. The research team decided that a synchronous experience would best facilitate the open discussion that had been the hallmark of the graduate student workshops conducted from 2016 to 2019. Inspired by a presentation given by Kory Trott during his tenure as the Director for Research Integrity at Virginia Tech, the Ethics Center decided to offer both a research integrity discussion series, where researchers at Illinois Tech had the chance to present and discuss ethical issues encountered in their work, and alongside this discussion series, short, online versions of the ethics workshops.

The first virtual ethics workshop was held in February 2024 and was attended by 23 university faculty members, all of whom were engaged in federally funded research. The participants came from a diverse range of departments, including biomedical engineering, computer science, chemical engineering, and the Food Safety Laboratory. The workshop began with a 20- to 30-min introduction to the former in-person workshops held with graduate students, findings from the research conducted as part of that project, and an overview of our NSF project's findings, and an introduction to how professional codes of ethics and guidelines can influence decision-making in the research lab.

Participants in the virtual workshop were then invited to share experiences that they had encountered either in their educational or professional careers at Illinois Tech or elsewhere. Despite some hesitation at the beginning, participants soon became eager to discuss issues such as authorship and publication, collaborating with labs outside the university, troubleshooting problems that arise when supervising graduate students, and difficulties navigating federal and institutional policies when reimbursing human subject participants. Although the original plan was to document the topics that occurred during the initial discussion and then break into discussion rooms focused on different topics, the participants opted to stay in the same room and individually brainstorm possible solutions and guidelines for each story or question shared. The facilitator from the Ethics Center helped guide the discussion, took notes on resources and the names of contact people to share with the whole group after the meeting, and otherwise allowed the discussants to use the bulk of the available time.

Similar virtual workshops have been held since February 2024, averaging 2–3 faculty workshops per semester, with some labs opting to hold a workshop that includes all their lab members, including undergraduate, graduate, staff, and faculty participants. The dynamics of the workshop discussions tend to follow the same sequence in either case, with workshop participants initially sharing relatively conventional research issues. Then other participants chimed in more eagerly as another person brought up an experience that resonated with their background, interests, or a problem they were currently grappling with.

One of the benefits of this open, virtual format is the feedback fellow participants provide about an issue shared with the group. Initially, workshop discussions primarily focused on issues related to mentoring, student-supervisor relationships, and establishing an ethical research environment. Recent workshops have focused on the application of artificial intelligence in research, concerns regarding intellectual property and the use of generative AI, as well as the utilization of proprietary AI systems in research. Having experts in artificial intelligence from the university's computer science department has been incalculably helpful in helping participants understand the complexities of these emerging issues.

Topics identified as most important in our interview studies and supporting literature were reflected in all workshop discussions, both in-person and online. Both faculty and students who have participated in the virtual talks have shared many stories about the impact positive and negative mentoring relationships have had on their research careers. In lab-specific workshops that involve both the faculty supervisor and their graduate and undergraduate students, the story-sharing section of the workshop often leads to detailed discussions about how these kinds of relationships can be more readily managed, and expectations are shared among new lab members on both sides.

Beyond the benefits of group discussions that include a wide array of participants—faculty, staff, and students—and the ability to learn from colleagues beyond a single lab group or department, the virtual environment also made it easier to schedule multiple sessions per semester, resulting in a higher attendance rate. The virtual format also introduced the drawbacks commonly experienced in most video chat interactions. Participants who were less engaged were able to turn off their videos and multitask during the session. The use of breakout rooms, where individuals are placed in small discussion groups based on the topic, helped address this challenge. Issues of privacy are a factor during in-person sessions, and this challenge is only exacerbated in an online setting. In online settings, participants can never be entirely sure of who is listening to what they are saying. In contrast, with in-person sessions, the audience is clearly defined by those in the room, and participants may feel a stronger urgency to keep confidentiality. None of the sessions were recorded, and no AI note-taking software was utilized during the meetings. However, attendance was taken for use by the Illinois Tech Office of Research Integrity. While participants were reminded that they could change their screen name if they wished, Illinois Tech is a relatively small school. Participants were easily recognized either by their screen name or appearance. Faculty working in their office or lab, or students using their computers in a lab, dorm room, or public space, may have colleagues, roommates, or others overhearing their discussions. The virtual environment also diminishes the overall sense of connection that often occurs more readily in face-to-face communication and might have led some participants to be more guarded in their remarks. As with all virtual education, several factors need to be considered in choosing whether to engage in an entirely virtual, hybrid, or in-person format. Ideally, students, researchers, faculty, and staff can choose the format they prefer.

## 27.5 Conclusion and Ways Forward

Awareness and meaningful discussion of ethical issues among members of laboratory research groups and peer populations (graduate students engaged in research, faculty) is imperative for the responsible pursuit of research. While there are many different ways of integrating education around research integrity into a curriculum, key points to consider in developing any program include (1) tailoring the program to cover ethical issues and questions relevant to the discipline and the research the program participants are currently engaged in, (2) involving participants in active deliberation and discussion about these ethical topics, and (3) helping them develop knowledge means of resolving ethical questions, either through knowledge of guidelines or policies, or through building connections with colleagues, research administrators, and others who can help answer questions, or even working as a team to develop guidelines and practices that can help resolve issues in their research environment.

The workshops, initially developed for graduate students at Illinois Tech, and the shorter faculty and lab group workshops, designed and run from 2024 to the present, provide an engaging ethics education experience that complements the more traditional ethics training concurrently offered at Illinois Tech. Running these discussions alongside online training courses and the Discussions in Research Integrity series allows Illinois Tech researchers to share their experiences and challenges faced in addressing emerging ethical issues, such as the use of artificial intelligence in research, and to share promising solutions that others have developed for meeting these challenges. Faculty have found these workshops helpful in building connections with colleagues across disciplines, discussing systemic challenges faced by research groups, often related to supervising graduate students, and identifying policies and systems that may need reassessment. Students find the discussions reaffirming, as faculty advisors share their experiences with the challenges they faced as graduate students, and the lab group collectively grapples with the ethical issues integral to their research environment.

Ultimately, these workshops and discussions, in their various forms, help reaffirm efforts to establish an ethical and inclusive research culture at the university. Open discussion of issues can help identify areas where more attention is needed and increase awareness of ethical challenges researchers are facing in an ever-changing environment. Our workshops will serve as a model for other universities seeking to expand their research integrity programs.

**Acknowledgements** The U.S. National Science Foundation, Award Number 1635661 generously funded this project.

## References


All European Academies. (2023). *European code of conduct for research integrity.* Retrieved June 9, 2025 from https://allea.org/code-of-conduct/

Bird, S. J. (2014). Social responsibility and research ethics: Not either/or but both. *Professional Ethics Report, 27*(2), 1–4.

Davis, M., & Laas, K. (2014). "Broader impacts" or "responsible research and innovation"? A comparison of two criteria for funding research in science and engineering. *Science and Engineering Ethics, 20*, 963–983. https://doi.org/10.1007/s11948-013-9480-1

Gaganis, V., Beckett, E., Choate, J., Aguilar-Roca, N., Etherington, S., Haigh, C., Scott, D., Sweeney, T., Zubek, J., & French, M. (2021). Virtual delivery: A panacea for the financial and ethical challenges associated with physiology laboratory classes? *Advances in Physiology Education, 45*(4), 744–748. https://doi.org/10.1152/advan.00242.2020

Haven, T., Bouter, L., Mennen, L., & Tijdink, J. (2023). Superb supervision: A pilot study on training supervisors to convey responsible research practices onto their PhD candidates. *Accountability in Research, 30*(8), 574–591. https://doi.org/10.1080/08989621.2022.2071153

Herkert, J. R. (2005). Ways of thinking about and teaching ethical problem solving: Microethics and macroethics in engineering. *Science and Engineering Ethics, 11*, 373–385. https://doi.org/10.1007/s11948-005-0006-3

Hildt, E., Laas, K., Miller, C., Taylor, S., & Brey, E. (2019). Empowering Graduate students to address ethics in research environment. *Cambridge Quarterly of Healthcare Ethics, 28*(3), 542–550. https://doi.org/10.1017/S096318011900046X

Hildt, E., Laas, K., Miller, C. Z., & Brey, E. (2024). *Building inclusive ethical cultures in STEM.* Springer.

Hildt, E., Laas, K., Miller, C. Z., & Taylor, S. (2022). Student views on the culture of stem research laboratories: Results from an interview study. *Accountability in Research, 31*(2), 1–38. https://doi.org/10.1080/08989621.2022.2109018

Holsapple, M. A., Carpenter, D. D., Sutkus, J. A., Finelli, C. J., & Harding, T. S. (2012). Framing faculty and student discrepancies in engineering ethics education delivery. *Journal of Engineering Education, 101*(2), 169–186. https://doi.org/10.1002/j.2168-9830.2012.tb00047.x

Jahnke, L. M., & Asher, A. (2013). Dilemmas of digital stewardship: Research ethics and the problems of data sharing. *Research Data Management, 80*, 21–28. https://doi.org/10.1016/j.jclinepi.2025.111680

Kalichman, M. (2014). Rescuing responsible conduct of research (RCR) education. *Accountability in Research, 21*(1), 68–83. https://doi.org/10.1080/08989621.2013.822271

Kalolo, J. F. (2019). Digital revolution and its impact on education systems in developing countries. *Education and Information Technologies, 24*(1), 345–358. https://doi.org/10.1007/s10639-018-9778-3

Laas, K., Miller, C. Z., Brey, E. M., & Hildt, E. (2024a). Student interactions with ethical issues in the lab: Results from a qualitative study. *Research Ethics, 21*(1). https://doi.org/10.1177/17470161241252431

Laas, K., Miller, C. Z., Brey, E. M., & Hildt, E. (2024b). In situ ethics education within research laboratories: Insights into the ethical issues important to research groups and educational approaches. In E. Hildt, K. Laas, E.M. Brey, & C. Z. Miller (Eds.), *Building inclusive ethical cultures in STEM* (Vol. 42). Springer. https://doi.org/10.1007/978-3-031-51560-6_13

Laas, K., Miller, C. Z., Brey, E. M., Taylor, S., & Hildt, E. (2020). Infusing ethics into research groups: A bottom-up approach. *Advances in Engineering Education, 8*(3), 1–14.

Laas, K., Taylor, S., Miller, C. Z., Brey, E. M., & Hildt, E. (2021). Views on ethical issues in research labs: A university-wide survey. *Accountability in Research, 29*(3). https://doi.org/10.1080/08989621.2021.1910503

Laas, K., Taylor, S., Miller, C. Z., Brey, E. M., & Hildt, E. (2022). Views on ethical issues in research labs: A university-wide survey. *Accountability in Research, 29*(3), 178–201. https://doi.org/10.1080/08989621.2021.1910503

Maglio, C., Williams, M., & Camponeschi, A. (2025). Biology wet lab e-learning during and after the COVID-19 pandemic: A review of student learning and experiences. *Biochemistry and Molecular Biology Education, 53*(4), 445–454. https://doi.org/10.1002/bmb.21897

Martin, D. A., Conlon, E., & Bowe, B. (2021). Using case studies in engineering ethics education: The case for immersive scenarios through stakeholder engagement and real life data. *Australasian Journal of Engineering Education, 26*(1), 47–63. https://doi.org/10.1080/22054952.2021.1914297

Moran, M. J. I., Paul A., May, D., & Hussein, R. (2023). *RHLab: Digital inequalities and equitable access in remote laboratories*. In 2023 ASEE Annual Conference & Exposition.

Mulhearn, T. J., Steele, L. M., Watts, L. L., Mederios, K. E., Mumford, M. D., & Connelly, S. (2017). Review of instructional approaches in ethics education. *Science and Engineering Ethics, 23*, 883. https://doi.org/10.1007/s11948-016-9803-0

Mumford, M. D., Steele, L., & Watts, L. L. (2015). Evaluating ethics education programs: A multilevel approach. *Ethics and Behavior, 25*, 37–60. https://doi.org/10.1080/10508422.2014.917417

National Institutes of Health. (2009). *Update on the requirement for instruction in the responsible conduct of research*. Retrieved April 10, 2025 from https://grants.nih.gov/grants/guide/notice-files/NOT-OD-10-019.html

Office of Research Integrity. (2000). *Final PHS policy for instruction in the responsible conduct of research*. Federal Register, 65, 76647. Retrieved April 11, 2025 from http://www.ori.dhhsgov/policies/RCR_Policy.shtml

Owen, R., & Pansera, M. (2019). Responsible innovation and responsible research and innovation. In D. Simaon, S. Kuhlmann, J. Stamm, & W. Canzler (Eds.), *Handbook on science and public policy* (pp. 26–48). Elgar.

Plemmons, D. K., & Kalichman, M. W. (2013). Reported goals of instructors of responsible conduct of research for teaching skills. *Journal of Empirical Research on Human Research Ethics, 8*(2), 95–103. https://doi.org/10.1525/jer.2013.8.2.95

Qin, T., Cook, M., & Courtney, M. (2020). Exploring chemistry with wireless, PC-less portable virtual reality laboratories. *Journal of Chemical Education, 98*(2), 521–529. https://doi.org/10.1021/acs.jchemed.0c00954

Serrano-Pérez, J. J., González-García, L., Flacco, N., Taberner-Cortés, A., García-Arnandis, I., Pérez-López, G., Pellin-Carcelén, A. Y., & Romá-Mateo, C. (2023). Traditional vs. virtual laboratories in health sciences education. *Journal of Biological Education, 57*(1), 36–50. https://doi.org/10.1080/00219266.2021.1877776

Stilgoe, J., Owen, R., & Macnaghten, P. (2013). Developing a framework for responsible innovation. *Research Policy, 42*(9), 1568–1580. https://doi.org/10.1016/j.respol.2013.05.008

Ulven, J. B., & Wangen, G. (2021). A systematic review of cybersecurity risks in higher education. *Future Internet, 13*(2), 39. https://doi.org/10.3390/fi13020039

Von Schomberg, R. (2013). A vision of responsible innovation. In R. Owen, M. Heintz, & J Bessant (Eds.), *Responsible innovation: Managing the responsible emergence of society and innovation in society* (pp. 51–74). Wiley.

Watts, L. L., Medeiros, K. E., Mulhearn, T. J., Steele, L. M., Connelly, S., & Mumford, M. D. (2017). Are ethics training programs improving? A meta-analytic comparison spanning 35 years of ethics instruction in the sciences. *Ethics and Behavior, 5*(27), 351–384. https://doi.org/10.1080/10508422.2016.1182025

World Conferences on Research Integrity. (2010). *Singapore statement on research integrity*. Retrieved May 11, 2025 from https://www.wcrif.org/guidance/singapore-statement

Yap, W. H., Teoh, M. L., Tang, Y. Q., & Goh, B. H. (2021). Exploring the use of virtual laboratory simulations before, during, and post COVID-19 recovery phase: An Animal Biotechnology case study. *Biochemistry and Molecular Biology Education, 49*(5), 685–691. https://doi.org/10.1002/bmb.21562

Zandvoort, H., Børsen, T., Deneke, M., & Bird, S. J. (2013). Editors' overview perspectives on teaching social responsibility to students in science and engineering. *Science and Engineering Ethics, 19*(4), 1413–1438. https://doi.org/10.1007/s11948-013-9495-7

**Kelly Laas** is a librarian and ethics instructor at the Center for the Study of Ethics in the Professions, Illinois Institute of Technology, USA. Her research interests include the history and application of codes of ethics in professional fields, ethics education in STEM disciplines, research ethics, and integrating ethics into technical curricula.

**Elisabeth Hildt,** PhD, is professor of philosophy and director of the Center for the Study of Ethics in the Professions, Illinois Institute of Technology, USA. Her research focuses on bioethics, ethics of technology, and research ethics. Her research interests include research ethics, philosophical and ethical aspects of neuroscience, and artificial intelligence.

# Chapter 28
# Ethics Issues in Outdoors Science Education in an increasingly Digital Age

**Michael J. Reiss**

## 28.1 Introduction

When I look back on my school science education, I recall sitting in school science laboratories with Bunsen burners, racks of test tubes, oscilloscopes and electric circuits. I suspect I am not alone in my memories of the centrality of the laboratory to the science I learnt. Indeed, both students (Sharpe & Abrahams, 2019) and teachers (Millar, 2011) see laboratory practical work as playing a key role in school science. But we learn about science outside the lab too and outdoors science education has traditionally been of particular importance in biology when teaching ecology. In this chapter, I focus on ethical issues that can arise when undertaking outdoors science education and ask whether the increasing affordances of digital learning can help address these.

## 28.2 An Anecdote

From 1983 to 1988, I taught at a Sixth Form College in Cambridge, England. In England, the compulsory end of schooling was then, as it still is, 16 years of age. Sixth Form Colleges take students for one or two years from the ages of 16 to 18. The large majority of our students were with us for two years taking what were, and still are, called 'Advanced levels'. Most students took three of these and my main job at the College was to teach Advanced-level biology. The second half of the summer terms was less busy than the rest of the year as the Advanced-level students in their second year finished their courses at half-term (the end of May), only coming in thereafter for their end-of-course examinations.

M. J. Reiss (✉)
Institute of Education, UCL, London, UK
e-mail: m.reiss@ucl.ac.uk



L. Leite et al. (eds.), *Practical Work in Science Education in the Digital Era*, Advances in Technology-Rich Science Education 2, https://doi.org/10.1007/978-3-032-13788-3_28

One of the things we did with the extra time was to take the students coming to the end of the first year of their Advanced-level biology on a field trip to Wicken Fen (we took half the students one Saturday and half the students the next Saturday). Wicken Fen is a 250-hectate Site of Special Interest, National Nature Reserve and Ramsar Wetland that lies some 25 km northeast of Cambridge. It is one of the UK's oldest nature reserves and one of the few surviving wild areas of fenland as 99.9% of the fenland in East Anglia is now cultivated for crops. I know Wicken Fen well and from 1990 to 1995 I was a member of its Local Management Committee.

Wicken Fen is a wonderful place. Figure 28.1 gives an impressive of its flat, wind-swept nature and shows common reed (*Phragmites australis*) in the foreground and carr (waterlogged shrubs and small trees) on the right. The students, and we as staff, almost always enjoyed our long day—and here is the reason I am relating this tale. At the end of each Saturday, we would bring back in plastic bags samples of the organisms we had collected from the water, put them into aquaria at the College and then spend the Sunday identifying the animals and trying to work out their feeding relationships. I can't remember for certain what we did with the organisms at the end of our studies, though I am certain we did not return them to Wicken Fen. I think they simply remained in their aquaria at the College for the rest of their lives.

**Fig. 28.1** The Lode (a man-made watercourse) at Wicken Fen National Nature Reserve, Cambridgeshire, England (https://commons.wikimedia.org/wiki/File:Wicken_Lode1.JPG)

## 28.3 Arguments for Outdoors Science Education

Many people associate school science with laboratories. Indeed, one of my favourite articles in the literature on science education is titled 'In the beginning was the Bunsen: The foundations of secondary school science' (Delamont et al., 1988). In this article, the authors (who are not science educators) take an ethnographic approach to the science lessons that the observed students having at the start of their secondary school education. The researchers noted that these early lessons tended to emphasise the importance of danger in the laboratory and that therefore it is necessary for students to behave safely, precisely in accordance with laboratory rules (no movement without permission, long hair tied back, and so forth). The other common feature of these early secondary science lessons is an introduction to the Bunsen burner—hence the title of their article. They concluded:

> The initial encounters with "science" in our schools seems to embody the following thought-styles: Science is an esoteric domain of experience which deals with dangerous objects and substances. The laboratory is an alien space in which special care and precautions must be taken. Practical work in the laboratory consists of the exploration and manipulation of special objects, some of which - like the bunsen burner - are accorded special, even reverential, significance. The extraordinary work of the laboratory demands special expertise. It leads to measurements - the technicalities of which have to be mastered - and the "successful" generation of such measurements is a warrant for the methods. (p. 325)

What school science students are not always helped to reflect on is why it is that the laboratory is a special place. Of course, it is a part of the real world so that the laws of science hold as much inside the laboratory as outside it, but in Foucauldian terms (Foucault, 1967/1986), a laboratory is an instance of a heterotopia. A heterotopia can be thought of as a world within a world. Foucault gives the example of a cemetery, and school laboratories, similarly, are treated with a certain respect and operate under different rules from the places that enclose them. In terms of science, they function as stripped-down versions of reality, and that is where outdoors science education plays an important role. For a rich understanding of how science operates, a student wants (needs) both to see the process of oxidation in a school laboratory as, for example, iron wool is heated, *and* examples of rust in 'the real world' on, for example, a chemistry trail. In everyday language, students sometimes complain that they can't see the relevance of what they are doing in science. Such students may need to be helped to make links between the science they learn in the laboratory and from their textbooks and teachers and the world in which they live outside of school. As one of the adult participants in a workshop I once ran with David Lambert put it: "not all science happens in test tubes and young people need to realise this" (Lambert & Reiss, 2014, p. 8).

Somewhat more prosaically, there is much science learning that can more easily (or only) take place outside of school than within it—think, observation of the night sky, launching a rocket, visiting a science museum, recording vegetation using quadrats, determining a soil profile. As Martin Braund and I argued twenty years ago:

> laboratory-based school science teaching needs to be complemented by out-of-school science learning that draws on the actual world (e.g. through field trips), the presented world (e.g. in science centres, botanic gardens, zoos and science museums) and via the virtual worlds that are increasingly available through information and communications technologies (ICT). (Braund & Reiss, 2006, p. 213)

In addition, learning in the field can lead to deeper learning and has many social and health benefits for students (Amos & Reiss, 2012; Berman, 2025; Sheldrake et al., 2019).

## 28.4 Problems with Outdoors Science Education

Despite the strong arguments in favour of outdoors education being a key part of school science, there are problems with its implementation in science as in other subjects, including related ones such as geography (Scott et al., 2014; Tilling, 2018). Both surveys and interviews in various studies show that teachers cite health and safety concerns, the bureaucracy involved in getting parental and other permissions to take students on a fieldtrip, greater pressures for students to do well in high-stakes examinations (which rarely require fieldwork) and the additional workload entailed (see my anecdote above which involved teaching for two consecutive weekends). It also seems likely that many teachers of school biology now have less experience of biological fieldtrips at school or undergraduate level than they did in the past, and so lack self-efficacy (Ordon et al., 2021).

## 28.5 Ethical Issues with Fieldwork

A different 'problem' to do with fieldwork, hinted at in the anecdote that begins this chapter, comes from a greater awareness in recent decades of our responsibilities and duties to organisms and the natural environment. Many people have no doubt always been sensitive to and respectful of animals. However, it is only in recent decades that academic philosophy has given serious consideration to how humans should regard and treat non-humans (Reiss, 2017). As is well known, René Descartes, that seventeenth century intellectual giant and one of the founders of modern philosophy, held that animals (taken from now on to mean non-human animals) were automata, incapable of thought or self-consciousness. While Descartes owned a pet dog, on whom he apparently lavished much affection, the standard presumption is that he held that animals lacked the ability to feel (e.g., pleasures or pains). In fact, there is a view that this is to misunderstand him (Harrison, 1992). Nevertheless, the standard interpretation has largely held sway, certainly among non-specialists. Similarly, a century later, another giant of modern philosophy, Immanuel Kant, held that we have no duties to animals and that the only reason we should avoid being cruel to them is because if we were to get into the habit of being cruel to animals, we might be more likely to be cruel to humans, as indeed is widely thought to be the case (Merz-Perez et al., 2001), though the evidence is less clear-cut than is commonly supposed (Tomlinson et al., 2022).

Perhaps the first philosopher, at least in the West, to argue that animals should be regarded, in terms of how we treat them, in much the same way that we regard humans was Jeremy Bentham, one of the founders of utilitarianism. Utilitarianism is the ethical framework that holds that the right way to decide on the ethical appropriateness of a possible action is to determine (easier in theory than in practice) all the possible consequences of that action and then, using a form of moral calculus (again, easier in theory than in practice), calculate whether the world, overall, would be a better place if the action took place or not. Animals enter into such calculations since, as Bentham famously put it, "the question is not, Can they *reason?* nor, Can they *talk?* but, Can they *suffer?*" (Bentham, 1789/1907, p. 311, note 1).

Another utilitarian, Richard Ryder, coined the word 'speciesism', subsequently popularised by Peter Singer (Singer, 1975). The term intentionally echoes 'racism' and 'sexism', similarly protesting against the presumption that an entity's position in a classificatory system is relevant. Instead, what is relevant are the entity's capacities. Nor should it be assumed that humans necessarily have a greater capacity to experience pleasure or pain that do animals. Take something as simple as receiving an injection from a sharp needle. If I know that I am receiving an injection for my benefit—for example, in vaccination—I am presumably likely to experience less pain and therefore suffer less than my cat may at the vet, particularly if she hates being caught and put in a carrier.

Some contemporary Kantians also maintain that animals are worthy of moral consideration. Korsgaard (2018) argues that, like humans, animals are 'ends-in-themselves'; in other words, we should not treat them merely as means to our ends.

These arguments all rely on animals being sentient, that is, capable of experiencing pleasures and pains. While it is true that biologists are now more alert to the possibility that many more animal species may be capable of this than previously assumed—there is, for example, active debate about whether insects can experience pleasures and pains (Elwood, 2023; Klein & Barron, 2016)—an argument that relies on sentience is not likely to apply to all animals (think sponges, for example), let alone to plants, fungi and many other organisms. A number of philosophers have argued on very different grounds that all species deserve our respect. Taylor (1986) took what is sometimes called a biocentric perspective and argued that all living beings possess inherent value and deserve our moral concern. A related position is that of 'deep ecology', which dates back to Rachel Carson's seminal *Silent Spring* (Carson, 1962) and the arguments of Arne Næss (1973), and views all of life on Earth holistically and existing in relationship. It can also be argued that there are ethical arguments against despoiling the environment even when there is no possibility of life being harmed (cf. Elliot, 2008). This point therefore applies as much to chemistry, earth science or physics taught outside the classroom as it does to biology.

School students, science teachers and others will differ in the extent to which they consider non-humans to be worthy of moral consideration. Nevertheless, few would now defend the position that only humans are worth such consideration. Given this, and the various difficulties to do with fieldwork discussed above, what is the place of virtual reality in outdoors science education?

## 28.6 Affordances of the Virtual World

It is obvious that no animals are harmed in the use of teaching materials in the virtual world but it is worth emphasising that the benefits of the virtual world are not only to do with the avoidance of animal suffering. Take over-collecting, for instance. The Society for the Protection of Birds was founded by Emily Williamson in 1889 because the fashion for feathers was driving a number of bird species, including little egrets, great crested grebes and birds of paradise, to extinction. Indeed, a number of bird species, including the huia and the Auckland Islands merganser are thought to have gone extinct as a result of over-collecting (Gill & Martinson, 1991). It is not only birds that can be driven to the verge of extinction through over-collecting. Norton et al. (1994) identify a number of instances where over-collecting by botanists may have driven New Zealand plants either to extinction or to the verge of it.

The actions of schools seem unlikely to drive species to extinction but there are a lot of schools, and their actions might reduce local populations and/or damage vegetation or organisms in other habitats, such as rocky shores. So how might the virtual world help? It is worth bearing in mind learning as well as ethics. Take the topic of succession, for instance. Figure 28.2 shows a typical example of the colonisation of sand dunes by vegetation. This is an example of primary succession, and I used to love teaching it to school students in the field. In the case of the sand dunes I know best—on the East Coast of England—the pioneer plant species include sand couch, sea sandwort, sea rocket and various *Atriplex* species (Field Studies Council, 2025). The crucial point is that as these plants grow and then die, organic matter begins to accumulate in the otherwise barren sand. This allows other plant species to colonise, and so the succession gets underway. Fast forward fifty or a hundred years and there can be shrubs and even trees.

However, some students out in the field don't fully grasp what is going on. Primary succession (i.e., succession on bare ground, such as on volcanic larva, a glacial moraine or a new sand dune) takes place over relatively long periods of time—typically, many years. For this reason, it cannot be observed directly in nature in the lifetime of a course taken by a school student. Accordingly, in the case of sand dunes, distances from the sea is used as a measure of the length of time that the sand has been available for colonisation by organisms, since younger 'embryo' dunes are typically formed at the base (seaward side) of a dune system, as a result of which the dune system moves towards the sea. Unsurprisingly, not all students understand the way in which distance serves as a measure of time.

There are now a number of computer packages that simulate succession. For example, back in 2002, Berardi described the development of a software framework for predicting vegetation change over time within the Astroni crater, a state nature

**Fig. 28.2** Plant colonisation of sand dunes at Neuhaus Beach, Dierhagen, Germany (Matti Blume). (https://commons.wikimedia.org/wiki/File:Dune,_Dierhagen_(P1100758-Pano).jpg)

reserve near Naples, Italy (Berardi, 2002). The computer programme was designed to allow the analysis of environmental and management scenarios in the crater, thus enabling the user to help the reserve's manager in deciding how to manage the reserve effectively.

I admit, though, that I was surprised in doing the research for writing this chapter that what is available on the market for teaching succession through the virtual world seems no better than was available twenty years ago. Broadening the search to the teaching of environmental issues, I found more recent examples. For example, Bodzin et al. (2022) developed a virtual-reality field trip to help pre-service and in-service secondary science teachers learn about the environmental changes that occurred during the past two centuries because of a zinc smelting plant operation in the Lehigh River watershed, Pennsylvania, USA. Their reason for developing the virtual-reality field trip was "that there are always a few students who are unable to attend a Saturday field trip" (p. 2). The designed the trip so that it would "serve as a learning experience prior to the actual field trip to the Lehigh Gap and would also provide an alternative experience for students who were unable to attend the actual site field trip" (p. 5). In the event, they concluded that the virtual-reality field trip was a success. In particular, with regards to the students who 'attended' the virtual-reality field trip and then came on the actual trip field trip:

> all students commented that viewing the dVFT [desktop virtual-reality field trip] enhanced their actual physical field-trip experience at the Lehigh Gap. They commented that it helped prime them for the field trip and assisted them with connecting with the actual physical location. They discussed how the dVFT made them feel immersed in the virtual-learning experience and that their sense of place for the Lehigh Gap was initially developed through that experience. Their sense of place was then augmented by being at the actual physical location. The students noted important experiential activities during the in-person field trip that enhanced their learning experience. These included walking on a trail and picking up a piece of anthracite coal, viewing test plot signs of different warm-season grasses, and

> visually seeing the land footprint of the former zinc smelting plant. This was a surprise for us to hear since we were not anticipating that the dVFT would serve as a learning experience that would then serve to further augment the students' sense of place when they visited the actual field site. (p. 15)

A second example is provided by Mead et al. (2019) who devised a lesson on life and environments during the Ediacaran Period (roughly 550 million years ago), centring on a series of fossil beds from a site in the Flinders Ranges of South Australia:

> Each "screen" of the activity includes a short text introduction that may ask the student to make an observation or navigate to a particular viewpoint. Additionally, some screens include short video clips, multiple-choice questions, or open text response boxes. The complete lesson - including instructions, activities, and assessment - is provided in the web environment, which allows the iVFT [immersive, interactive virtual field trip] to be completed by students independently. Students work through a series of activities and progressively gain an understanding of the conditions at the Nilpena site during the Ediacaran Period. Students are tutored in this journey with automatic, adaptive feedback provided by the Smart Sparrow learning environment. This feedback anticipates common errors, such as looking at the wrong fossil bed while answering a question. Feedback text may also reiterate information from earlier in the lesson to help students on assessments. (p. 133)

Both high school students and undergraduates showed large and statistically significant gains in content knowledge as a result of taking the lesson and were reported as enjoying it.

My own view, I admit as yet a prediction rather than a conclusion backed up by solid empirical evidence, is that the best of these simulation packages will be found to allow students to vary various parameters and see the consequence of so doing. For example, retuning to sand dune succession, what happens if there is a 'blow out' (when the wind takes away part of a sand dune and the accompanying vegetation), or if there is more rain than usual, or if there is less grazing than usual. Given the positive effects of Independent Research Projects' (Bennett et al., 2018). I would expect such features to motivate students and enhance their conceptual understanding.

A final point is that virtual experiences may be of particular value for students with certain disabilities—for instance for students with mobility issues (whether temporary, e.g., a broken leg, or longer-term). For such students, virtual experiences may help science education to be more inclusive (e.g., Chiarella & Vurro, 2020).

## 28.7 Conclusion and Ways Forward

I have written for some years about the cognitive, affective, social and health benefits of undertaking outdoors science education (Reiss, 1993, 2020). The last thing I want to do in this chapter is to give the impression that virtual fieldwork can replace actual, authentic fieldwork. It can't and I don't want it to. What it can do is to complement the experience of being in the field. It can also enable those who would otherwise miss out on the benefits of fieldwork—for reasons of difficulties with access—to derive more benefit from their studies that would otherwise be the case. As I discussed above, it's actually quite difficult for many students to understand sand dune succession when they are in the middle of an actual dune system—and that is even more the case if it is windy (which it often is, so that it's not easy to make notes or do a drawing), if it's raining (which it sometimes is at the coast), if you are tired (which students often are, unused as they may be to standing and walking for several hours) and so on. Accordingly, some prior learning about sand dune succession can probably help.

Indeed, there is a growing literature on the educational value of pre-visit preparation and post-visit follow-up bookending field trips (e.g., Lee et al., 2020). We are, though, still in the early days of virtual field trips and it would be premature to provide guidance as to the best way to combine virtual and face-to-face experiences to maximise the benefits of fieldwork. There is room for further research here.

A final point about pedagogy, given the emphasis in this chapter on ethical arguments. Ethics is not an intellectual free-for-all but it is more subjective as a discipline than science is. There is therefore particular merit in encouraging students to discuss ethical issues (cf. Wishart et al., 2011) and to think about the ethical validity of alternative courses of action—e.g. taking organisms from the wild into the school lab and then returning them as opposed to not taking them in the first place. Such activities are likely to develop their understanding of ethics in science and to enhance their ethical sensitivity (Reiss, 2017).

## References


Amos, R., & Reiss, M. J. (2012). The benefits of residential fieldwork for school science: Insights from a five-year initiative for inner-city students in the UK. *International Journal of Science Education, 34*, 485–511. https://doi.org/10.1080/09500693.2011.585476

Bennett, J., Dunlop, L., Knox, K. J., Reiss, M. J., & Torrance-Jenkins, R. (2018). Practical independent research projects in science: A synthesis and evaluation of the evidence of impact on high school students. *International Journal of Science Education, 40*, 1755–1773. https://doi.org/10.1080/09500693.2018.1511936

Bentham, J. (1789/1907). *Introduction to morals and legislation.* Clarendon Press. Retrieved September 14, 2025 from https://oll-resources.s3.us-east-2.amazonaws.com/oll3/store/titles/278/0175_Bk.pdf

Berardi, A. (2002). ASTROMOD: A computer program integrating vegetation dynamics modelling, environmental modelling and spatial data visualisation in Microsoft Excel. *Environmental Modelling & Software, 17*(4), 403–412. https://doi.org/10.1016/S1364-8152(02)00011-7

Berman, M. G. (2025). *Nature and the mind: The science of how nature improves brain, body and community.* Vermilion.

Bodzin, A., Araujo Junior, R., Schwartz, C., Anastasio, D., Hammond, T., Birchak, B., & Center, L. G. N. (2022). Learning about environmental issues with a desktop virtual reality field trip. *Innovations in Science Teacher Education, 7*(1), 1–20.

Braund, M., & Reiss, M. (2006). Validity and worth in the science curriculum: Learning school science outside the laboratory. *The Curriculum Journal, 17*(3), 213–228. https://doi.org/10.1080/09585170600909662

Carson. R. (1962). *Silent spring.* Houghton Mifflin.

Chiarella, D., & Vurro, G. (2020). Fieldwork and disability: An overview for an inclusive experience. *Geological Magazine, 157*(11), 1933–1938. https://doi.org/10.1017/S0016756820000928

Delamont, S., Beynon, J., & Atkinson, P. (1988). In the beginning was the Bunsen: The foundations of secondary school science. *Internation Journal of Qualitative Studies in Education, 1*(4), 315–328. https://doi.org/10.1080/0951839880010302
Elliot, R. (2008). *Faking nature: The ethics of environmental restoration.* Routledge.
Elwood, R. W. (2023). Behavioural indicators of pain and suffering in arthropods and might pain bite back? *Animals, 13*(16), 2602. https://doi.org/10.3390/ani13162602
Field Studies Council. (2025). *Sand dune succession.* Retrieved September 14, 2025 from https://www.field-studies-council.org/resources/16-18-biology/succession/sand-dunes/.
Foucault, M. (1967/1986). Of other spaces. *Diacritics, 16*(1), 22–27. https://doi.org/10.2307/464648
Gill, B., & Martinson, P. (1991). *New Zealand's extinct birds.* Vintage New Zealand.
Harrison, P. (1992). Descartes on animals. *The Philosophical Quarterly (1950-), 42*(167), 219–227. https://doi.org/10.2307/2220217
Klein, C., & Barron, A. B. (2016). Insects have the capacity for subjective experience. *Animal Sentience, 1*(9), 1. https://doi.org/10.51291/2377-7478.1113
Korsgaard, C. M. (2018). *Fellow creatures: Our obligations to the other animals.* Oxford University Press.
Lambert, D., & Reiss, M. J. (2014). *The place of fieldwork in geography and science qualifications.* Institute of Education, University of London. Retrieved September 14, 2025 from https://discovery.ucl.ac.uk/id/eprint/10022102/1/Lambert_Reiss_2014_Fieldwork_Report.pdf
Lee, H., Stern, M. J., & Powell, R. B. (2020). Do pre-visit preparation and post-visit activities improve student outcomes on field trips? *Environmental Education Research, 26*(7), 989–1007. https://doi.org/10.1080/13504622.2020.1765991
Mead, C., Buxner, S., Bruce, G., Taylor, W., Semken, S., & Anbar, A. D. (2019). Immersive, interactive virtual field trips promote science learning. *Journal of Geoscience Education, 67*(2), 131–142. https://doi.org/10.1080/10899995.2019.1565285
Merz-Perez, L., Heide, K. M., & Silverman, I. J. (2001). Childhood cruelty to animals and subsequent violence against humans. *International Journal of Offender Therapy and Comparative Criminology, 45*(5), 556–573. https://doi.org/10.1177/0306624X01455003
Millar, R. (2011). Practical work. In J. Osborne & J. Dillon (Eds.), *Good practice in science teaching: What research has to say* (pp. 108–134). Open University Press.
Næss, A. (1973). The shallow and the deep, long-range ecology movement. A summary. *Inquiry, 16*(1–4), 95–100. https://doi.org/10.1080/00201747308601682
Norton, D. A., Lord, J. M., Given, D. R., & De Lange, P. J. (1994). Over-collecting: An overlooked factor in the decline of plant taxa. *Taxon, 43*(2), 181–185. https://doi.org/10.2307/1222876
Ordon, K. J., Bartelheimer, M., & Asshoff, R. (2021). Biology student teachers' interest and self-efficacy in planning and conducting field trips after participation in a university course. *Environmental Education Research, 27*(1), 88–109. https://doi.org/10.1080/13504622.2020.1849565
Reiss, M. J. (1993). Organizing and running a residential fieldtrip. *School Science Review, 74*(268), 132–135.
Reiss, M. J. (2017). A framework within which to determine how we should use animals in science education. In M. P. Mueller, D. J. Tippins, & A. J. Stewart (Eds.), *Animals and science education—Ethics, curriculum and pedagogy* (pp. 243–259). Springer.
Reiss, M. J. (2020). The educational benefits of out-of-classroom learning. In T. Burt & D. Thompson (Eds.), *Curious about nature: A passion for fieldwork* (pp. 116–130). Cambridge University Press.
Scott, G. W., Boyd, M., Scott, L., & Colquhoun, D. (2014). Barriers to biological fieldwork: What really prevents teaching out of doors? *Journal of Biological Education, 49*(2), 165–178. https://doi.org/10.1080/00219266.2014.914556
Sharpe, R., & Abrahams, I. (2019). Secondary school students' attitudes to practical work in biology, chemistry and physics in England. *Research in Science & Technological Education, 38*(1), 84–104. https://doi.org/10.1080/02635143.2019.1597696

Sheldrake, R., Amos, R., & Reiss, M. J. (2019). *Children and nature: A research evaluation for the wildlife trusts*. The Wildlife Trusts. Retrieved September 14, 2025 from https://www.wildlifetrusts.org/sites/default/files/2019-11/Children%20and%20Nature%20-20UCL%20and%20The%20Wildlife%20Trusts%20Full%20Report.pdf

Singer, P. (1975). *Animal liberation: A new ethics for our treatment of animals*. HarperCollins.

Taylor, P. (1986). *Respect for nature: A theory of environmental ethics*. Princeton University Press.

Tilling, S. (2018). Ecological science fieldwork and secondary school biology in England: Does a more secure future lie in geography? *The Curriculum Journal, 29*(4), 538–556. https://doi.org/10.1080/09585176.2018.1504315

Tomlinson, C., Murphy, J., Matijczak, A., Califano, A., Santos, J., & McDonald, S. (2022). The link between family violence and animal cruelty: A scoping review. *Social Sciences, 11*(11), 514. https://doi.org/10.3390/socsci11110514

Wishart, J., Green, D., Joubert, M., & Triggs, P. (2011). Discussing ethical issues in school science: An investigation into the opportunities to practise and develop arguments offered by online and face-to-face discussions. *International Journal of Science Education, 1*(1), 47–69. https://doi.org/10.1080/21548455.2010.543863

**Michael J. Reiss,** PhD and post-doc in evolutionary biology, is professor of science education at UCL Institute of Education (UK), honorary visiting professor at the Royal Veterinary College, honorary fellow of the British Science Association, fellow of the Academy of Social Sciences, and member of the Nuffield Council on Bioethics. He was the Director of Education at the Royal Society, president of the International Society for Science and Religion, and president of the Association for Science Education.

# Chapter 29
# Ethical Issues with Extended Reality Technology in Schools and how Teachers can respond to these in STEM Education

**Erica Southgate, Toni Maddock, Ella Camporeale, and Robertson Burgess**

## 29.1 Introduction

Extended reality (XR) technologies immerse users in digital environments that can blend the physical and virtual worlds (Kosko et al., 2021). XR technologies provide opportunities for simulated learning in safe environments, virtual travel to different times and places that would not normally be possible, the manipulation of the size and scale of objects to develop understanding, and the ability to create high fidelity or very imaginative objects and worlds to demonstrate learning (Southgate, 2020).

The term XR encompasses augmented reality (AR), mixed reality (MR), and virtual reality (VR). AR allows virtual objects to be overlayed on the real world via a wearable device such as smart glasses or by using a smart phone (We Protect Global Alliance, 2024). There is generally limited interaction between virtual and real-world elements. MR has a more complex history. Historically, it referred to the introduction of a person or actual object in a virtual world. Nowadays, it refers to a virtual environment where digital elements are superimposed into the real-world with these interacting with actual objects (Tremosa, 2025). VR involves the blocking out of the real world by using a headset which replaces reality with a virtual environment that can allow different degrees of user interaction with virtual objects and avatars that can be controlled by real people or computing software (Dede et al., 2017).

E. Southgate (✉) · T. Maddock · E. Camporeale · R. Burgess
University of Newcastle, Newcastle, NSW, Australia
e-mail: Erica.Southgate@newcastle.edu.au

T. Maddock
e-mail: Toni.maddock@uon.edu.au

E. Camporeale
e-mail: Ella.camporeale@uon.edu.au

R. Burgess
e-mail: Robertson.burgess@newcastle.edu.au



L. Leite et al. (eds.), *Practical Work in Science Education in the Digital Era*, Advances in Technology-Rich Science Education 2, https://doi.org/10.1007/978-3-032-13788-3_29

Many XR technologies also have artificial intelligence (AI) integrated into them. AI is "designed with varying levels of autonomy… that may exhibit adaptiveness after deployment, and… from the input it receives… generate[s] outputs such as predictions, content, recommendations, or decisions" (European Union, 2024, n.p.). AI further complicates ethical implications of using XR with students (We Protect Global Alliance, 2024).

Despite substantial advancements in educational technology, ethical considerations remain inadequately explored (Marín & Tur, 2024; Pulijala, 2024). In response, this chapter explores the ethical complexity of using XR technologies in schools and how teachers can respond to this in pedagogical and practical ways so that the human rights of students are protected. The chapter is organised into three sections. The first section outlines human rights, educational technology and their ethical implications for children (those under 18 years of age). The next section details XR risks for schools and ways to mitigate these. This is followed by a section that provides two examples of ethical teacher practice related to the use of VR in Australian middle schools for STEM education. The chapter concludes with an overview of key ideas for ethical integration of XR.

## 29.2 Human Rights, Ethics and Technology for Children

The *Universal Declaration of Human Rights* (UDHR) (United Nations, 1948) establishes the right to education (article 26) and the right to access and to participate in science (article 27). Some have argued that science education is itself a human right (for a review, see Schuck & Feser, 2022). Contemporary human rights documents such as the UDHR were, in part, a response to the horrors of World War II (Annas & Grodin, 1999) with the Nuremburg trial of Nazi physicians prompting the creation of the *Nuremberg Code* (1947) (British Medical Journal, 1996). This Code stipulates prerequisites for the moral and legal use of human beings for scientific and medical research. The Nuremburg Code influenced the development of important guidelines on ethical scientific conduct that highlighted principles such as respect for the person, promoting wellbeing in others, and fairness. Such principles align with human rights ethics which are premised on the inherent dignity of a person, empowered participation, non-discrimination, legality and accountability (Australian Human Rights Commission, n.d.).

The subfield of child rights reflects these ethical principles while acknowledging the special vulnerability of children. The United Nations (1989) *Convention on the Rights of the Child*, provides children with the right to privacy, anti-discrimination, and anti-exploitation. In 2021, the United Nations Committee on the Rights of the Child (2021) released *General comment No. 25 (2021) on children's rights in relation to the digital environment* (hereafter referred to as *General Comment No. 25*). This document identifies that technologies used for leisure and learning can collect vast amounts of different types of information on "children's identities, activities, location, communication, emotions, health and relationships. Certain combinations of personal data, including biometric data, can uniquely identify a child" (United Nations Committee on the Rights of the Child, 2021, p. 11). The document recognises that automated data processing and profiling may affect children at latter stages of their lives and that there should be no arbitrary collection of children's data. Fundamentally, the design and use of digital products, including those in schools, should reflect the best interests of the child within a context of rigorous accountability with older children given the right to object to data practices that may impact their human rights (United Nations Committee on the Rights of the Child, 2021).

The risks outlined in General Comment No. 25 not only apply to, but can be amplified by, XR technologies. The accelerated integration of AI into XR technologies has heightened ethical concerns. AI can be used to collect, analyse and connect vast amounts of data from individuals, often without their awareness. This can be used to profile individuals and personalise their online interactions in terms of content exposure and communication (for example, AI companions). There are varying levels, and in some cases no regulation, of AI in different national contexts and this has led to mounting ethical concerns about consent, privacy, bias, surveillance and manipulation (Brennan et al., 2025). The next section of this chapter highlights some key areas of risk that span XR technologies as well as concerns specific to VR.

## 29.3 Understanding and Mitigating XR Risk in Schools

To ensure the ethical integration of XR in schools—that is, using the technology in ways that support learning and the best interests of the child—three areas need attention. These are: (1) choosing the technology and content most appropriate to the developmental stage/s and special needs of the child; (2) understanding the implications of the data practices of companies that develop the technology (hardware and software); and (3) developing effective safety protocols and learning material on technology including technology ethics, for the classroom. For informed decision-making, it is recommended that prior to purchasing and integrating XR technologies into classrooms, teachers locate online XR hardware manufacturer health and safety advice and software (games, XR experiences) content ratings.

### *29.3.1 Developmental Considerations*

Child development includes physical (motor and perceptual), cognitive, linguistic, affective, social and moral domains and is usually conceptualised in stages, although there are often individual differences between children within the same developmental window (Berk, 2022). XR technologies are designed to immerse the user into a different version of reality or entirely new (often fantastical) realities as is the case with VR. There are different types of immersion. These include actional immersion that empowers users to control aspects of virtual environments, semantic immersion that involves content that can trigger strong physical and emotional reactions, and social immersion where users share virtual spaces as avatars to communicate and act in the virtual world together (Dede et al., 2017).

There is limited research on how children and adolescents experience immersion in XR technologies especially from a developmental perspective. A recent systematic review of 26 studies found that that VR use in the under 14 cohort appears to have few safety concerns or adverse events but that most of this research does not assess safety as a primary objective with the quality of adverse event reporting being generally poor (Bexson et al., 2024). VR hardware companies provide online health and safety material that advise on age limits for the technology which are generally between 10 and 12 years of age with adult supervision. VR headsets are made for adult-sized heads, and even with adjustable sizing, they are not a comfortable fit for children and adolescents (and even for some adults).

Furthermore, XR environments can create an intense feeling of immersion that may be difficult for children and adolescents to cognitively and emotionally regulate. Research has found that when preschool and early primary school children were given an immersive experience through a VR headset, 50% believed the experience to be real after one week (OECD, 2024). In a study of the use of an educational AR game, Alien Contact, conducted in the outdoor areas of several schools with children aged between 11 and 16 years, researchers found that some students became so engrossed that they lost track of their real environment, sometimes in an unsafe way, with some students asking the researchers if they were FBI agents (as in the game) and if aliens had actually landed (Dunleavy et al., 2008).

## *29.3.2 Safety*

Even before the inclusion of XR technologies in schools, science teachers were very familiar with their obligation to ensure the physical safety of students in classrooms and laboratories (Love et al., 2023). Lack of knowledge (Gudyanga, 2020), or proper implementation (Hussein & Shifera, 2022), of safety measures by teachers is a contributing factor in the ongoing rates of accidents. Research indicates that school students who receive safety training are less likely to experience an injury than those who have not received training at school (Love et al., 2023). Such research points to the need for a risk assessment approach to XR integration in schools with students and teachers undertaking training on XR risks.

Teachers and students should be trained to recognise and respond to cybersickness in VR. This includes symptoms such as nausea, dizziness, and disorientation (Bexson et al., 2024). Eyestrain can also occur with strict time limits and regular breaks recommended (OECD, 2025). There is also a need to set up the virtual play areas in classrooms away from obstacles, stairs and other students to accommodate for movement in highly interactive types of VR and MR experiences. Even though some VR headsets have ‘pass through’ cameras built into the headsets that change the view to the real world once a user steps outside of the virtual play area, it is still important that students be supervised by a teacher or trained peer who can act as a spotter for potential hazards (Southgate et al., 2019). Students can also become distracted by AR applications experienced via mobile device screens, becoming less aware of hazards in the real world, making supervision necessary even when using less immersive forms of the technology.

Checking content ratings for XR experiences is vital for child protection. When using XR social applications, that is multi-user environments, teachers should ensure there is a means to "lock down" environments so that user membership is controlled. Social XR represent serious child protection risks. The We Protect Global Alliance (2024) suggest that social XR platforms can expose children to harassment, inappropriate content, be used to simulate child abuse, and initiate it and stream it in real life. As they explain:

> This anonymity, combined with the open access of these platforms, makes it easier for offenders to contact and exploit children. Abusers might use XR to initiate child grooming or entice victims into progressively more harmful forms of exploitation. ... Participants reflected that XR environments can be difficult to monitor due to the multitude of ways in which users can interact with each other. (We Protect Global Alliance, 2024, p. 3).

### *29.3.3 Data Practices*

In the current machine age, individual's data can be automatically collected through a range of mechanisms, some of which are not obvious to, or under the control of, humans, including teachers and students (Livingstone et al., 2024). This data can be rapidly shared, combined and analysed within digital ecosystems that are predominately commercial in nature (Laird & Dwyer, 2023). There is high variability amongst countries in the regulation of data collection (UN Trade and Development, n.d.). This presents several types of risks for students and teachers as XR technologies can harvest vast amounts of different types of identifying data.

Like many digital applications, XR can collect data inputted by users, metadata, location data, and data gleaned by tracking user interactions in and across applications and platforms. The technology also collects contextual data through sensors and when devices within environments interact with each other. Headsets and smart phones can automatically record the real-world and virtual environments of the user. Biometric data can also be collected, especially in headsets that rely on the tracking of human movement to render real time, interactive virtual experiences. Biometrics are data of and about bodies and can be behavioural or physiological (North-Samardzic, 2020). Some of the types of biometrics VR systems can capture are verbal communication, hand and limb movements and associated kinesiological patterns, eye tracking and pupil dilation to understand patterns of attention, facial recognition to infer emotion, and other biometrics involved in authentication processes (Heruatmadja et al., 2023; OECD, 2024).

Biometric data is granular and uniquely identifying. For example, research on one popular interactive VR game indicated that significant amounts of personal attributes could be accurately and consistently inferred from motion data alone such as a player demographics, environment, health and disability status (Nair et al., 2024, p. 482). Even with anonymised data sets, students' personal data, including biometric information, can be re-identified and shared to create profiles and predictions about individuals that might potentially affect opportunities for future employment and insurance. As the OECD (2024) suggests of VR, it is the "unprecedented collection of data directly related both to users' environment and their bodies (that) creates even higher risks for the fundamental rights and freedoms of individuals" (p. 126). Teachers should read the health and safety and privacy information provided by XR hardware manufacturers and software developers so that they are cognisant of the types of data collected through XR technologies, how this can be managed or not, and where and who it is being shared with, and for what purposes. With XR technologies, the issue of user data consent is often complicated and this should be clearly communicated to students and their families.

## 29.4 Examples of Teacher Practice in Addressing XR Risk in Middle School STEM Education

Given the myriad concerns outlined in the previous section, teachers should critically evaluate technology appropriateness, engage in professional learning, undertake risk assessment, and develop pedagogical approaches and resources to ensure ethical integration of technology (Marín & Tur, 2024; Pulijala, 2024). To illustrate ethical practice, this section provides two examples from Australian teachers who have ethically integrated VR into STEM middle school classrooms. These examples are drawn from ongoing research on integrating virtual reality into Australian schools (University of Newcastle Human Research Ethics Committee Approval Number 2017–0229). The first example involves a curriculum design approach to teaching about the ethics of biometric data practices in XR before introducing VR into the STEM classroom. The second example provides a case study reflecting practical approaches for VR classroom integration.

### *29.4.1 Example 1—Curriculum Design for Ethical Learning About Biometrics and XR in School STEM Classrooms*

It is imperative that school curricula instil ethical understanding and critical reasoning related to technologies in students (Landesman et al., 2024). This includes developing an understanding and skills about data practices and privacy, particularly related to emerging technologies that automate the harvesting of different types of data, especially biometrics.

The Australian Curriculum (Australian Curriculum, Assessment and Reporting Authority, 2025) is structured to provide opportunities to learn about technology ethics in ways that can be included in the units of work (lesson sequences) that teachers develop for their students. It does this in three ways. Firstly, through its General Capabilities framework which define knowledge, skills and behaviours designed to develop lifelong learners who can successfully navigate their future. There are seven General Capabilities, each scoped to the developmental stage of students, with the most relevant being Critical and Creative Thinking, Digital Literacy and Ethical Understanding. Secondly, the Learning Areas (disciplines), for example Science or Technology, often have ethical components reflected in their content descriptors. Thirdly there are cross-curriculum priorities that provide for Indigenous, regional and sustainability perspectives. The inclusion of ethical components in this tripartite curriculum arrangement allows Australian teachers to design units of work that support students in developing the skills and knowledge required to approach complex ethical issues related to technology design and use.

One example of this, is a unit of work on biometrics developed for middle school students that aligns with the knowledge and understanding content descriptions in the Australian Curriculum Design and Technologies Learning Area and the General Capabilities of Critical and Creative Thinking, Digital Literacy and Ethical Understanding. This unit of work is implemented prior to students first using VR in class for creative design tasks. The learning objective of the unit was: "Investigate how biometric data in technologies, particularly in XR, can impact people and society, and explain your own viewpoint (using references) on the benefits and ethical issues involved". The unit of work aligned with the following Australian Curriculum Version 9 content descriptors: (1) Analyse how people in design and technologies occupations consider ethical, security and sustainability factors to innovate and improve products, services and environments (ref:AC9TDE10K01); and (2) Analyse the impact of innovation, enterprise and emerging technologies on designed solutions for global preferred futures (ref:AC9TDE10K02). It also reflected the following General Capabilities component of the Australian Curriculum: (1) Critical and Creative Thinking by identifying the objective and subjective aspects of a complex concept or problem, with sensitivity to context; (2) Digital Literacy by developing knowledge to manage digital privacy, safety and wellbeing (3) Ethical Understanding by evaluating the meaning of ethical concepts, such as trust, freedom and rights and responsibilities in different contexts.

The unit of work was undertaken over four fifty-minute lessons. It begans by introducing students to the concept of biometrics and biometric data capture through a lesson incorporating instruction and short videos on the history and core concepts. Students were then presented with an assignment to individually investigate issues about biometrics in XR technologies, including VR. Students must report back on their research in written (600 words) or multimodal format (3–4 min). They are provided with three web articles that they could cite whilst also being encouraged to do their own research. A scaffold (Table 29.1) for the assignment was provided with question prompts to guide inquiry.

**Table 29.1** Scaffold for biometric data assignment

| Assignment sections | Question prompts for inquiry |
|---|---|
| What is biometric data? | Explain in your own words what biometric data is |
| | List some types of biometric data that can be collected |
| What/who collects biometric data? | Provide examples of technologies and/or companies who collect biometric data |
| | What is it used for? |
| | How are biometrics used in XR, particularly VR? |
| Benefits of Biometrics | How can this benefit companies, governments, or users of different technologies? |
| | What are the benefits to society? |
| Issues around Biometrics | What are the social/ethical/legal implications of technology companies collecting biometric data? |
| | Specifically, what are the issues surrounding XR, particularly VR, and biometrics? |
| Bibliography | Provide a Harvard style bibliography |

Implementation of this unit of work has indicated that middle school students can successfully identify what biometric data is and methods of capture. In their assignments it was evident that students showed understanding of both positive and negative impacts of biometrics. They explored issues surrounding privacy, consent, data storage and personal implications. Importantly, the unit of work laid a foundation for students to build their knowledge about biometrics and critically engage with its ethical implications before using VR in class.

### *29.4.2 Example 2—Ethical Approaches to Using VR to Learn About Biomes*

Example two illustrates how 360-degree VR was introduced into a middle school classroom with a focus on practical ethical considerations. VR was incorporated in an integrated unit of work called 'The Future of Food' which spanned the Australian Curriculum learning areas of Science (biology), Humanities and Social Science (geography) and the Technologies and the cross-curriculum priority of Sustainability. The core question guiding the unit of work was: How do we secure food for our future? The focus was on studying global and local biomes and the effect humans have on them.

The unit of work involved a pedagogical mix of explicit instruction, independent student research, fieldwork, and 360-degree content creation using a platform called VRTY which did not require students to have coding knowledge to create their own interactive virtual world to demonstrate learning. The part of the unit of work using 360-degree VR was sequenced so that students were initially asked to add information to an existing 360-degree image of a biome before being tasked with creating their own interactive virtual biome. Working individually or in small groups, they created their own virtual biomes using a 360-degree camera during a fieldtrip to a local biome where they took their own images of different aspects of the environment and created other content to embed in their creation. These 360-degree scenes were then storyboarded in the platform and linked to create linear or non-linear navigation paths through different scenes of the virtual biome. Students sourced and created a range of text, image and video content to insert as pop-up information markers in each 360-degree scene that were triggered through gaze-based interaction. Gaze-based portals allowed students to navigate between scenes. The students' virtual worlds could be experienced in immersive mode using a VR headset or on a computer screen. Having both options created an inclusive environment for students who did not like feelings of immersion or experienced cybersickness.

After an initial equipment on-boarding session, students were asked help develop a safety protocol. This activity was followed up by a class discussion about safety in VR with students brainstorming ideas. This led to the development of a safety protocols on what do if students experienced cyber sickness, setting time limits (15 min) in a headset, and the need to respect the personal space of other students while they are in VR. This was very relevant for younger and neurodiverse students where sudden physical contact or loud interruptions might provoke distress or confusion (Ramirez, 2021; Wehrmann & Zender, 2024). The safety ideas from the student and teacher discussion were incorporated into a classroom safety poster. To further ensure physical safety, students were usually seated while in VR with educator supervision.

Moving outside the classroom, the fieldtrip to a local freshwater biome required education about ethically capturing images and video of the real-world scenes, especially since the students' virtual worlds were to be shared and experienced beyond their classroom at parent and caregiver and community forums. Developing an understanding in students of the need to not record identifiable individuals or private property was paramount. Students were guided to focus on landscape and environmental features, ensuring their work aligned with data minimisation principles and maintained the integrity of public space ethics in educational media use (Bicalho et al., 2025; Ramirez, 2021).

Collectively, these practical and pedagogical strategies reflect the importance of ethically grounded planning when integrating VR into schools, where physical and psychological wellbeing and privacy remain a central consideration.

## 29.5 Conclusion and Ways Forward

Ethical principles emphasise respect for the person, acting in ways that promote well-being in others, and fairness (World Medical Association, 1966). Considering the best interests of the child is central to human rights and core to child rights frameworks. Schools should uphold human rights by ethically integrating technology into the classrooms. This includes understanding the range of risks related to the unique properties of XR technologies that can psychologically transport users to new realities, especially from a developmental perspective. It also involves developing knowledge of, and effective responses to, the safety implications of XR and the range of data practices of the XR industry that may negatively impact child privacy and potentially exploit school students. Close examination of the health and safety information published online by XR hardware and software companies is vital.

With increased understanding, teachers can mitigate potential harm. Research suggests that students who receive training can also increase classroom safety (Love et al., 2023). As illustrated by the two examples of STEM teacher practice presented in this chapter, teachers should be empowering students to understand XR risk and respond to it in ethical and practical ways before and during integration of the technologies in classrooms. It is vital that teachers create and share curriculum resources on XR ethics and provide authentic case studies of how these ethics apply in classrooms. Upholding human rights in schools in a digital era of fast-moving innovation is only possible when teachers and students are actively involved in the ethical integration of technology.

**Competing Interests** Erica Southgate has received a research grant (G2200229) from the Association of Independent Schools of South Australia.

## References

Annas, G. J., & Grodin, M. A. (1999). Medical ethics and human rights: Legacies of Nuremberg. In *L. Hofstra & Pol'y Symposium*. https://scholarlycommons.law.hofstra.edu/cgi/viewcontent.cgi?article=1029&context=hlps

Australian Curriculum, Assessment and Reporting Authority. (2025). *The Australian Curriculum*. https://v9.australiancurriculum.edu.au/

Australian Human Rights Commission. (n.d.). *Human rights based approaches*. https://humanrights.gov.au/our-work/rights-and-freedoms/human-rights-based-approaches

Berk, L. E. (2022). *Development through the lifespan*. Sage.

Bexson, C., Oldham, G., & Wray, J. (2024). Safety of virtual reality use in children: A systematic review. *European Journal of Pediatrics, 183*(5), 2071–2090. https://doi.org/10.1007/s00431-024-05488-5

Bicalho, D. R., Piedade, J., & Matos, J. F. (2025). IVRPM: Conceptual proposal of an immersive virtual reality pedagogical model. *Applied Sciences, 15*(4), 2162. https://doi.org/10.3390/app15042162

Brennan, K., Kak, A., & Myers West, S. (2025, June 3). *Artificial power: AI Now 2025 landscape*. AI Now Institute. https://ainowinstitute.org/2025-landscape

British Medical Journal. (1996). Nuremberg Code (1947). *British Medical Journal*, N7070 V313, 1448. https://media.tghn.org/medialibrary/2011/04/BMJ_No_7070_Volume_313_The_Nuremberg_Code.pdf

Dede, C. J., Jacobson, J., & Richards, J. (2017). Introduction: Virtual, augmented, and mixed realities in education. In D. Liu, C. Dede, R. Huang, & J. Richards (Eds.), *Virtual, augmented, and mixed realities in education* (pp. 1–16). Springer.

Dunleavy, M., Dede, C., & Mitchell, R. (2008). Affordances and limitations of immersive participatory augmented reality simulations for teaching and learning. *Journal of Science Education and Technology, 18*(1), 7–22. https://doi.org/10.1007/s10956-008-9119-1

European Union. (2024). *EU Artificial Intelligence Act*. https://artificialintelligenceact.eu/article/3/

Gudyanga, R. (2020). Probing physical sciences teachers' chemical laboratory safety awareness in some South African high schools. *African Journal of Research in Mathematics, Science and Technology Education, 24*(3), 423–434. https://doi.org/10.1080/18117295.2020.1841960

Heruatmadja, C. H., Meyliana, Hidayanto, A. N., & Prabowo, H. (2023). Biometric as secure authentication for virtual reality environment: A systematic literature review. In *2023 International Conference for Advancement in Technology (ICONAT)* (pp. 1–7). https://doi.org/10.1109/ICONAT57137.2023.10080713

Hussein, B. A., & Shifera, G. (2022). Knowledge, attitude, and practice of teachers and laboratory technicians toward chemistry laboratory safety in secondary schools. *Journal of Chemical Education, 99*(9), 3096–3103. https://doi.org/10.1021/acs.jchemed.2c00043

Kosko, K. W., Ferdig, R. E., & Roche, L. (2021). Conceptualizing a shared definition and future directions for extended reality (XR) in teacher education. *Journal of Technology and Teacher Education, 29*(3), 257–277. https://doi.org/10.70725/002527odhhrp

Laird, E., & Dwyer, M. (2023). *Report–off task: EdTech threats to student privacy and equity in the age of AI. Center for Democracy and Technology*. https://cdt.org/wp-content/uploads/2023/09/091923-CDT-Off-Task-web.pdf

Landesman, R., Salac, J., Lim, J. O., Ko, A. J., Hamilton, M., Porter, L., Morrison, B., & Denny, P. (2024). Integrating philosophy teaching perspectives to foster adolescents' ethical sensemaking of computing technologies. In *Proceedings of the 2024 ACM Conference on International Computing Education Research - Volume 1* (pp. 502–516).

Livingstone, S., Pothong, K., Atabey, A., Hooper, L., & Day, E. (2024). The Googlization of the classroom: Is the UK in protecting children's data and rights? *Computers and Education Open*, 100195. https://doi.org/10.1016/j.caeo.2024.100195

Love, T. S., Roy, K. R., & Sirinides, P. (2023). A national study examining safety factors and training associated with STEM education and CTE laboratory accidents in the United States. *Safety Science, 160*, Article 106058. https://doi.org/10.1016/j.ssci.2022.106058

Nair, V., Rack, C., Guo, W., Wang, R., Li, S., Huang, B., Cull, A., O'Brien, J. F., Latoschik, M., Rosenberg, L., & Song, D. (2024, March). Inferring private personal attributes of virtual reality users from ecologically valid head and hand motion data. In *2024 IEEE Conference on Virtual Reality and 3D User Interfaces Abstracts and Workshops (VRW)* (pp. 477–484). IEEE.

North-Samardzic, A. (2020). Biometric technology and ethics: Beyond security applications. *Journal of Business Ethics, 167*(3), 433–450. https://doi.org/10.1007/s10551-019-04143-6

Marín, V. I., & Tur, G. (2024). Ethical issues in the use of technologies in education settings. *Education in the Knowledge Society (EKS), 25*, e31301. https://doi.org/10.14201/eks.31301

OECD. (2024). *Digital economy outlook 2024* (Volume 1). https://www.oecd.org/en/publications/oecd-digital-economy-outlook-2024-volume-1_a1689dc5-en.html

OECD. (2025). *How's life for children in the digital age?* https://www.oecd.org/en/publications/2025/05/how-s-life-for-children-in-the-digital-age_c4a22655.html

Pulijala, S. K. (2024). Ethical considerations in educational technology: Balancing innovation and responsibility. *International Journal of Scientific Research in Computer Science, Engineering and Information Technology, 10*(5), 315–325. https://doi.org/10.32628/cseit241051023

Ramirez, E. J. (2021*). The ethics of virtual and augmented reality: Building worlds* (1st ed.). Routledge.

Schuck, P., & Feser, M. S. (2022). Science education as a human right: A systematic review of the literature. *European Journal of Science and Mathematics Education, 10*(3), 338–351. ISSN: 2301-251X (Online).

Southgate, E. (2020). *Virtual reality in curriculum and pedagogy: Evidence from secondary classrooms*. Routledge.

Southgate, E., Smith, S. P., Cividino, C., Saxby, S., Kilham, J., Eather, G., Scevak, J., Summerville, D., Buchanan, R., & Bergin, C. (2019). Embedding immersive virtual reality in classrooms: Ethical, organisational and educational lessons in bridging research and practice. *International Journal of Child-Computer Interaction, 19*, 19–29. https://doi.org/10.1016/j.ijcci.2018.10.002

Tremosa, L. (2025, March 12). Beyond AR vs. VR: What is the difference between AR vs. MR vs. VR vs. XR? *Interaction Design Foundation.* https://www.interaction-design.org/literature/article/beyond-ar-vs-vr-what-is-the-difference-between-ar-vs-mr-vs-vr-vs-xr

UN Trade and Development. (n.d.). *Data protection and privacy legislation worldwide.* https://unctad.org/page/data-protection-and-privacy-legislation-worldwide

United Nations. (1948). *Universal declaration of human rights.* https://www.un.org/en/about-us/universal-declaration-of-human-rights

United Nations. (1989). *Convention on the rights of the child.* General Assembly resolution 44/25. https://www.ohchr.org/en/instruments-mechanisms/instruments/convention-rights-child

United Nations Committee on the Rights of the Child. (2021, May 2). *General comment No. 25 (2021) on children's rights in relation to the digital environment. CRC/C/GC/25.* https://www.ohchr.org/en/documents/general-comments-and-recommendations/general-comment-no-25-2021-childrens-rights-relation

Wehrmann, F., & Zender, R. (2024). Inclusive virtual reality learning: Review and "Best-Fit" Framework for Universal Learning. *Electronic Journal of E-Learning, 22*(3), 74–89. https://doi.org/10.34190/ejel.21.6.3265

We Protect Global Alliance. (2024). *Beyond the headset: Charting a course for safer experiences for children in extended reality environments.* https://www.weprotect.org/resources/library/beyond-the-headset-extended-reality/

World Medical Association. (1966). *Declaration of Helsinki.* https://www.wma.net/policies-post/wma-declaration-of-helsinki/

**Erica Southgate** , Ph.D., is associate professor of Emerging Technologies for Education, University of Newcastle, Australia. She is a teacher educator, maker of computer games for literacy, and a technology ethicist for education. Erica is the lead researcher on the VR School Study, a long-running investigation on the ethical, technological, and pedagogical aspects of embedding immersive virtual reality into classrooms.

**Toni Maddock** is a Ph.D. candidate in education at the University of Newcastle, Australia. She investigates the pedagogical decision-making of teachers using immersive virtual reality in schools and the student experience. She is the Head Teacher of Middle School at Southern Montessori School, Adelaide, Australia.

**Ella Camporeale** Ella Camporeale is a Ph.D. candidate in Education, University of Newcastle, Australia. She conducts research on immersive learning using augmented and virtual reality and artificial intelligence in schools by learning about architecture and the built environment. She is the curriculum leader in Art, Design and Digital Technologies, at Westminster School, Adelaide, Australia.

**Robertson Burgess** , Ph.D., is a lecturer of secondary science and mathematics teaching at the University of Newcastle, Australia. Originally working as a physics researcher before transitioning to teaching science education for preservice and graduate teachers. His research combines emerging technologies, including artificial intelligence and virtual reality, to better understand the impact that these transformative technologies will have on the future of classrooms.

# Chapter 30
# Redefining School Laboratory Safety in the Digital Age

**Remeredzayi Gudyanga**

## 30.1 Introduction

Bayrak and Ağaoğlu (1999, as cited by Chen et al., 2024) define laboratory safety as "…the precautions that must be taken to identify the potential threats posed by the tools, equipment, machines, and devices used in conducting experiments and related preparations and the scientific solutions that must be used to face safety problems (p. 506). Laboratory safety continues to be a critical issue in science education globally (Ménard & Trant (2020). In school science laboratories, the handling of chemicals, apparatus, and various experimental procedures comes with safety risks that require a sustained and initiative-taking commitment to safety practices (Ménard & Trant (2020). Oftentimes, safety concerns have resulted in some science teachers shying away from conducting experiments with learners for fear of litigation in case a learner gets involved in an accident (Sedghpour et al., 2013). Subsequently, concerns revolving around safety may have limited learners' experiences in laboratories. Yet there is consensus that laboratory experiments must be at the epicentre of science learning experiences if learners are to understand scientific principles, develop critical thinking and comprehend the nature of science (García-Carmona, 2020).

Laboratory safety protocols have previously focused on physical hazards, such as chemical spills, fire risks, and exposure to toxic substances (Abedsoltan & Shiflett, 2024), while giving less attention to broader systemic or ethical dimensions of laboratory work. However, the emergence of innovative technologies and the increasing integration of digital tools into educational practice, require a redefinition of what constitutes laboratory safety in the 21st century. Some of these technologies include online simulations, Virtual Reality (VR), augmented reality (AR), immersive reality and Mixed Reality (MR). Extended Reality (XR) is the umbrella term used to refer

R. Gudyanga (✉)
Faculty of Education, University of the Free State, Bloemfontein, South Africa
e-mail: GudyangaR@ufs.ac.za



L. Leite et al. (eds.), *Practical Work in Science Education in the Digital Era*, Advances in Technology-Rich Science Education 2, https://doi.org/10.1007/978-3-032-13788-3_30

to all these technologies (Chuah, 2018). While education systems are introducing these tools to enhance access, improve pedagogical effectiveness, or circumvent infrastructural constraints, these changes simultaneously present new categories of safety risks (Chontas et al., 2024; Gugenheimer, et al., 2022). Some of these risks are physical, while are others ethical or epistemological. As laboratory safety has increasingly become a multi-dimensional construct, it is imperative to interrogate how policy makers, science teachers and safety officers may redefine traditional laboratory safety frameworks and adapt them to fit contemporary, technology- enhanced learning environments.

To respond to these complexities, there is growing interest in developing hybrid safety strategies that integrate both physical and digital environments. Already it has been reported that the use of pre-laboratory simulations and safety training modules embedded in virtual platforms serve as effective tools for reinforcing safety awareness among learners and teachers alike (Chen et al., 2024). The present digital era has witnessed the introduction of chemical Sensors based on the Internet of-Things, automated inventory systems, and AI-supported hazard detection to support safety in laboratory environments. Though the introduction of these new approaches has been mostly limited higher education and research institutions laboratories, in due course they may be a welcome introduction in school laboratories.

This chapter is a treatise on how laboratory safety is shifting in tandem with new technologies in the digital age. It is grounded in a combination of empirical research drawn from international scholarship and the authors' theoretical and reflective engagements with the field. The chapter is situated within a broader discourse on science education, ethical responsibility and technological advancement. It draws on a range of practical experiences and perspectives from traditional safety management models to emerging approaches grounded in digital pedagogy and sustainability. While the focus is global, it occasionally reflects on lessons drawn from under resource contexts such as some schools in South Africa, where the dual challenges of safety and access intersect in revealing ways. The argument is that laboratory safety in the digital age is a deeply pedagogical concern which should be confronted both in physical and virtual environments. Regulatory safety must be embedded into the very fabric of science education as part of the epistemological and ethical foundation of the discipline.

## 30.2 Redefining Safety as Integral to Science Thinking and Practice: A Philosophical Shift

Research suggests that laboratory safety is frequently treated as a secondary concern rather than as an integral part of science education (National Research Council, 2010). Safe practices must not simply be one of the peripheral concerns in science teaching and learning; they must be woven into the very fabric of science. Most activities in a science lesson, such as testing a hypothesis, taking a measurement or every discussion on a chemical reaction, provide opportunities for students to learn about the nature of science alongside the safety risks which scientists encounter. Recognizing, quantifying, and mitigating those hazards, and exploiting these teaching moments must not be an optional extra but must be an integral part of scientific reasoning and responsible safe inquiry.

Redefining laboratory safety as an integral part of the nature of science calls upon science teachers to make references to safety issues in their everyday teaching, and not merely during practical periods. For example, when teaching about properties of organic compounds such as their volatility, science teachers could draw students' attention to the risks of inhaling vapours from substances like ether or acetone. The exothermic reaction of acids and like sulfuric acid could provide opportunities for teachers to discuss why proper dilution techniques are important. This would easily link the content to the importance of using fume hoods. The history of science provides appropriate opportunities to integrate safety issues and content and may deepen students' understanding of the nature of science. Marie Curie's unfortunate death after prolonged exposure to polonium and radium is a powerful example (Abergel et al., 2022). More recently, the 1996 Dartmouth mercury poisoning incident which resulted in the death of a professor after small amounts of dimethylmercury sifted through her gloves (Pawlaczyk et al., 2018). These examples can be powerful teaching moments in that they illustrate how the drive for knowledge without a safety mindset, can have lethal consequences. By foregrounding such stories, teachers can help students see that every scientific breakthrough must be accompanied by foresight about unintended risks. Through such stories, students can appreciate science as a human endeavour. Furthermore, this integration of a snapshot of the history of science enhances students' understanding of the nature of science.

By presenting laboratory safety as an integrated part of the nature of science, there may be other benefits. When students build mental models of chemical reactions, they must simultaneously model any potential hazards. "What could go wrong here?", becomes a central equation to "Why does this reaction proceed?" This dual model cultivates anticipatory reasoning, preparing learners to transfer safe practice habits to new contexts, whether in a high school laboratory or in a future research career.

The United States' Next Generation of Science Standards curriculum is quite progressive in terms of this philosophical shift, as it interweaves safety with the three-dimensional model to 'doing science' (NGSS Lead States, 2013). Science and engineering are coupled with 'Disciplinary core ideals and crosscutting concepts to mirror real science' (Feder & Rhodes, 2014; Fick & Arias, 2022). Under the NGSS students are explicitly required to "plan and carry out investigations" and this explicitly entails hazard and risk assessments before, during and after experiments (Fick & Arias, 2022). This may be a good example of efforts to embed laboratory safety with the teaching of the nature of science and scientific reasoning.

Integrating safety discussions across topics need not overburden science teachers. A simple 'safe pause' at the start of any lesson, posing one key question about possible harm, frames students' thinking in a risk-aware mindset. For example, when dealing with the topic of magnetism, teachers may ask students questions such as "How might a strong neodymium magnet pose safety risk to someone handling it" or "How do engineers deal with safety issues when they work with strong magnets while at the same time they might be working with other electronic devices", which effectively links abstract concepts to real world design and safety protocols.

The digital age requires science teachers to change the way they perceive laboratory safety. Laboratory safety must be viewed as an integral part of the science content rather than just a bureaucratic compliance procedure. Such a shift may assist learners to view science as a human endeavour in which curiosity and caution are interconnected. Ultimately, laboratory safety must be a core scientific pedagogical value.

## 30.3 Rethinking School Laboratory Safety in Digitally Mediated Virtual Learning Environments

Digitally mediated environments mimic experimental procedures that enable learners to conduct practical science in virtual environments (Wang, 2025). Platforms like Labster and PhET allow learners to conduct chemical reactions, manipulate digital apparatus and rehearse processes without physical risks (Reyes et al., 2024). These tools have demonstrated value in context where resources are limited or where traditional labs are inaccessible (Reyes et al., 2024). While these innovations offer unprecedented educational opportunities and reduce certain physical hazards, they also introduce new pedagogical, ethical, and infrastructural challenges, and new safety risks.

Head-set mounted screen-based environments such as VR, AR may result in safety risks that revolve around simulator sickness, spatial awareness, posture fatigue risks, instructional design weaknesses, Data Privacy, and Integrity risks (Chen et al., 2024; Gugenheimer et al., 2022). Appendix A shows a table outlining XR related safety risks and ways to mitigate each risk. One category of safety risks stems from the learner's physical surroundings and ergonomic design of the headsets. The use of head-mounted displays (HMD) can obscure learners' awareness of their immediate environment. This may lead to collisions or students tripping. Extended use of HMD has been reported to cause repetitive strain injuries if posture is not managed properly. Virtual reality environments may also pose cognitive, physical, and sensory risks. Simulator sickness is often triggered by a mismatch between visual and bodily cues, and is evidenced by the occurrence of nausea, dizziness, and disorientation (Allcoat et al., 2021; Conner et al., 2022). Simulator sickness may negatively impact the usability of virtual laboratories as it may reduce learning engagement. Oftentimes, simulator sickness may distract learners from the tasks and learners may abandon tasks altogether. Educators and designers in the digital age may mitigate the development of simulator sickness and its impact on students through a series of strategies: limiting rotational motion, dynamically adjusting field view, and adjusting exposure times (Allcoat et al., 2021; Conner et al., 2022).

Learners engaging solely with virtual environments may develop a conceptual understanding of scientific procedures without acquiring real-world practical habits related to safe laboratory practices. Even with their sophistication, virtual environments are not like "real world" environments. The absence of sensory engagement such as olfactory cues and visual risk indicators may lead to a diminished awareness of real-world hazards. Kelly (2020) argues such "Decontextualised learning" has its short comings. There is a risk of cultivating theoretical proficiency at the expense of real-world safety awareness.

Immersive reality tools attempt to bridge the gap between virtual and physical realities by overlaying digital instructions, alerts, and procedural prompts onto real world laboratory setups (Bhambri, 2025). Augmented reality can enhance learner's situational awareness during direct tasks (Chen et al., 2024). For example, such applications may guide a student step by step through titration while simultaneously warning about incorrect handling of reagents. This blended experience offers the potential to reinforce safe practices, while retaining the authenticity of physical real-world experimentation.

According to Chen and others (2024), mixed reality experiments must have an "as if it was real" approach when conducting experiments virtually. Safety is taken as seriously as if the experiment was being conducted in the physical environment. Safety in digital and emissive laboratories should not be presumed. The assumption that virtual environments are inherently safe overlooks the critical role of instructional design, teacher facilitation and cultivating responsible scientific habits. In both physical or digital settings, learners must be explicitly provided with opportunities to recognize hazards and evaluate the risks, use protective measures, interpret safety data sheets, and respond to emergencies.

The digital era calls for the fusion of virtual safety protocols with traditional physical protocols in science education, rather than replacing traditional practices (Lisachenko et al., 2017). These tools should be embedded within pedagogical models that prioritise both cognitive understanding and embodied practice. Laboratory safety in the digital age must blend the affordances of emissive technology with principles of sustainable safe science.

### *30.3.1 Digital Pre-Lab Tools for Safety Preparation*

Pre-laboratory preparation is a critical phase in minimising safety risks during practical work (Vagle, 2024). In the digital age, this phase has continued to be boosted with interactive simulations, instructional videos, online quizzes and gamified safety models (Vagle, 2024). These tools are designed not only to introduce learners to the theory behind an experiment, but also to sensitise them to possible hazards and safety protocols before entering a physical laboratory. Digital pre-lab tools allow exposure to laboratory procedures in a low-risk environment (Chen et al., 2024). For example, simulations can demonstrate the correct method of heating a substance or the sequence of steps in a titration, while simultaneously alerting learners to common mistakes or unsafe practices. When well-designed, such tools serve to reinforce procedural accuracy and hence conceptual clarity and instil a proactive safety mindset. Gamified platforms, which present safety rules as interactive challenges or competitions, have shown promise in increasing learner engagement and retention of critical information (Zainuddin et al., 2020). These approaches are particularly effective in context where safety training is typically passive, or lecture based. By converting safety protocols into experiential learning opportunities, teachers can ensure that students become more attuned to the importance of risk mitigation, not merely as compliance but as part of scientific responsibility.

To achieve their potential, digital pre-laboratory tools must be situated within structural safety education frameworks. Teachers' instruction on safety must be explicit and linked directly to science content. Pre-laboratory digital tools can enhance learner preparedness, reduce accidents, and enforce safety culture.

## 30.4 Leveraging Emerging Technologies for Safety Monitoring in the Digital Age

Some of the solutions to laboratory safety challenges (both old and new) in the digital age may be found in the utilisation of technology itself. In their article on smart education, Zeeshan and others (2022) postulate that conventional ways must give way to innovative approaches in education in line with the digital era. Artificial intelligence, IoT and smart sensors have already emerged as part of safety solutions in other sectors. These novel technologies permit for real-time tracking of laboratory environmental variables such as temperature, humidity, gas leaks or chemical storage circumstances (Zeeshan et al., 2022). Traditionally, these conditions have been monitored manually, if at all.

Adopting safety monitoring systems in school laboratories would reduce the burden on teachers and improve early detection of safety risks. Timely interventions may be conducted before incidents occur. IoT-enabled sensors can, for example, trigger alerts when flammable substances are stored at unsafe temperatures or when laboratory air quality falls below acceptable thresholds. These alerts can be sent to the science teacher, laboratory manager, or school safety officers equipped with digital detection devices, ensuring that safety oversight does not rely solely on human vigilance. AI-driven systems further enhance laboratory safety by analysing data trends over time. This may assist personnel to identify potential safety risks and monitor parameters that may indicate systemic risks or recurring safety lapses. Bar code readers may be deployed in laboratory storerooms to keep track of chemicals, and alerts may be set when chemicals expire.

There are challenges to the adoption of these digital monitoring tools in improving laboratory safety. Economic disparity among countries or even among different schools in the same country means that some schools may not have access to such technological resources. Furthermore, lack of expertise among laboratory personnel and limited internet connectivity in an area may be another impediment to the implementation of such monitoring systems in poorly resourced parts of the globe. Moreover, digital systems require regular calibration, software updates and maintenance tasks, which schools are often ill equipped to manage without external support.

Despite these challenges, there are approaches that may ameliorate disparities. In contexts where resources are scarce, authorities should begin with basic digital tools such as barcoded chemical inventories or mobile apps for incident reporting. By building capacity for more advanced integration over time, and responsibly implementing the available technologies, schools have the potential to transform laboratory safety from a reactive task into a proactive intelligence system. These actions could enable science education to align with broader global trends in innovation by reinforcing the foundational objective of all laboratory work and protecting those who engage in the pursuit of scientific understanding.

## 30.5 Laboratory Safety Training in the Digital Age

Research suggests that teacher education programs across many countries do not devote sufficient attention to laboratory safety as a structured area of competence for pre-service teachers (Cofré et al., 2022). Instead, safety is assumed to be intuitive, or teachers are supposed to learn about safety issues through experience. As a result, many science teachers step into classrooms without the necessary preparation to work with chemicals or to keep learners safe.

Teacher training institutions must formalise laboratory Safety training and integrate this training across science education curriculum. This includes modules focused on hazard identification, chemical waste disposal, safe equipment handling, and emergency response. It is also important for teacher training institutions to expose future teachers to both virtual and physical safety environments. This would sufficiently prepare them for the increasingly hybrid workplace.

School laboratory safety should not be the responsibility of science teachers alone. Laboratory assistants and technicians must play a fundamental role in maintaining safe and functional laboratory environments. In many well-resourced school systems, these professionals are responsible for preparing reagents, managing inventories, calibrating equipment, and ensuring that laboratories are compliant with safety regulations. In some parts of the of the globe where some schools do not even have proper laboratory facilities, having laboratory assistants could be considered far-fetched. In these circumstances, teachers assume laboratory management responsibilities with little to no formal training. The extra burden on some science teachers increases the occurrence of safety incidences, compromise the quality of practical activities or even limit students' exposure to science practical. Policymakers must increase the urgency for trained laboratory assistants to assist teachers in schools.

Improving school laboratory safety challenges teacher preparation programs to explicitly prepare science teachers and laboratory assistants for their responsibilities in the lab. Laboratory safety needs to be integrated into science teacher training through hands on experiences in both physical and virtual environments Laboratory assistants are vital for the smooth running of safe laboratories while at the same time they allow sciences teachers to focus on their teaching. Where resources are limited in affording each school its laboratory assistant, shared regional support and mobile safety units could fill the gaps. Ultimately, building a strong safety culture entails ensuring that everyone involved in science teaching is sufficiently prepared, supported and value.

## 30.6 Conclusion and Ways Forward

While the emergence of XR technologies presents opportunities in making science learning more engaging, it also presents new safety risks. New risks such as simulator sickness, spatial awareness, posture fatigue risks, instructional design weaknesses, Data Privacy and Integrity risks have emerged alongside digital advancements. Bayrak and Ağaoğlu's (1999) definition (1999, as cited by Chen et al., 2024) requires modification. Laboratory safety in the digital age includes not only precautions related to physical tools and materials but also the identification and mitigation of risks associated with digital platforms, virtual labs and other XR technologies. It encompasses scientific technological and instructional measures to manage safety in both physical and virtual learning environments. At the same time lab safety should be an integral part of scientific thinking and the nature of science.

While guidelines may be provided to mitigate these safety risks, the way these new risks may be confronted must continue evolving as the digital space continues developing. Academics generally agree that these XR experiences, as afforded in VR, AR and MR-enhanced technologies should be integrated with or used in conjunction with physical environments for full benefits to be derived from both environments. This integration of the physical and the virtual environments therefore calls for laboratory safety to be considered holistically from both physical and virtual perspectives. The Internet of Things and other AI driven dashboards, and pre-lab safety simulations, offer fresh ways to prepare students and reduce hazards.

Teacher training programs could also develop modules on laboratory safety to pre-service teachers instead of leaving these novice teachers to have to learn laboratory safety through trial and error. The sourcing of trained laboratory assistants into school laboratories could reduce the burden on teachers and ultimately improve laboratory safety.

Ultimately, in this digital age laboratory safety belongs at the epicentre of science education, not at the periphery. As digital technologies fuse with science classrooms, there must be an integration of laboratory safety into every aspect of how we teach and how students learn to think like scientists. When safety becomes part of the daily logic of science classrooms, it will reflect the kind of thoughtful, responsible scientists and citizens that science education aims to cultivate.

# Appendix A

See Table 30.1.

**Table 30.1** Safety risks issues in Extended Reality (XR) technologies

| Safety issue | Definition of safety risk | Description | Safety precaution |
|---|---|---|---|
| Spatial awareness | Physical risk—Danger of tripping or colliding into real world objects. | Head mounted displayers and controllers can block a user's view of the physical environment, increasing the chances of bumping into furniture or walls. | Designate a 2x2 m play area. Remove obstacles before activities. |
| Posture and fatigue | Ergonomic risk—strain from poor posture or extended exertion. | Prolonged time using MR tools may cause neck pain or repetitive strain injury due to postures or fixed hand positions. | Encourage short breaks every 15-20 minutes, provide adjustable chairs, and consider alternating between sitting and standing during activities. |
| Simulator sickness | Cognitive/ physiological risk—disruption of sensory process | User may feel nausea, dizziness, and disorientation due to mismatches between visual and bodily cues (this is more common among new users). | Start with short sessions, then gradually increase, monitor learners for signs of discomfort. |
| Instructional design | Cognitive risk: Poorly designed environments may trigger disorientation in learners. | XR environments tasks, with rapid movement, spinning visuals, or unclear cues can confuse users and increase motion sickness. | Use smooth transitions or teleportation-style navigation; add stable visual cues like static horizon lines and virtual hands. |
| Software validation | Technical risks: Errors in software may cause faulty simulations or interactions. | If virtual labs are not validated properly, users might follow unsafe procedures based on inaccurate representations. | Test and validate software, before use in educational settings. Ensure simulations follow correct scientific protocols. |
| Data privacy and integrity | Data risk—sensitive private data may be mishandled or misused. | Virtual environments may collect biometric, location, and behavioural data, which may be exploited if not securely managed. | Apply secure data storage protocols, limit unnecessary data collection, and comply with privacy regulations. |

## References


Abedsoltan, H., & Shiflett, M. B. (2024). Mitigation of potential risks in chemical laboratories: A focused review. *ACS Chemical Health & Safety, 31*(2), 104–120. https://doi.org/10.1021/acs.chas.3c00097

Abergel, R., Aris, J., Bolch, W. E., Dewji, S. A., Golden, A., Hooper, D. A., Margot, D., Menker, C. G., Paunesku, T., Schaue, D., & Woloschak, G. E. (2022). The enduring legacy of Marie Curie: impacts of radium in 21st century radiological and medical sciences. *International Journal of Radiation Biology, 98*(3), 267–275. https://doi.org/10.1080/09553002.2022.2027542

Allcoat, D., Hatchard, T., Azmat, F., Stansfield, K., Watson, D., & von Mühlenen, A. (2021). Education in the digital age: Learning experience in virtual and mixed realities. *Journal of Educational Computing Research, 59*(5), 795–816. https://doi.org/10.1177/0735633120985120

Bayrak, C., & Ağaoğlu, E. (1999). Edt: M. Zor, *Laboratory applications and security in science teaching*. Anadolu University Open Education Faculty Publications, 595.

Bhambri, P. (2025). Virtual and augmented reality for immersive learning experiences: Practical applications and challenges. *Architecting the Digital Future: Platforms, Design, and Application*, 17–46. https://doi.org/10.4018/979-8-3693-2973-3.ch002

Chen, C. M., Li, M. C., & Tu, C. C. (2024). A mixed reality-based chemistry experiment learning system to facilitate chemical laboratory safety education. *Journal of Science Education and Technology, 33*(4), 505–525. https://doi.org/10.1007/s10956-024-10101-3

Chontas, P. E., Iftene, A., & Buraga, S. C. (2024, November). Evaluating the effects XR has on users: An exploratory study. In *International Conference on Computer-Human Interaction Research and Applications* (pp. 150–167). Springer Nature Switzerland. https://doi.org/10.1007/978-3-031-83845-3_9

Chuah, S. H. W. (2018). Why and who will adopt extended reality technology? Literature review, synthesis, and future research agenda. *Literature Review, Synthesis, and Future Research Agenda (December 13, 2018)*. https://doi.org/10.2139/ssrn.3300469

Cofré, H., Vergara, C., Santibáñez, D., & Pavez, J. M. (2022). Preservice science teachers education around the globe: Trends, challenges, and future directions. In J. Luft & M. Jones (Eds.), *Handbook of research on science teacher education* (pp. 163–177). Routledge.

Conner, N. O., Freeman, H. R., Jones, J. A., Luczak, T., Carruth, D., Knight, A. C., & Chander, H. (2022, November). Virtual reality induced symptoms and effects: Concerns, causes, assessment & mitigation. *In Virtual Worlds, 1*(2), 130–146. https://doi.org/10.3390/virtualworlds1020008

Feder, M. A., & Rhodes, H. (Eds.). (2014). *Literacy for science: Exploring the intersection of the Next Generation Science Standards and Common Core for ELA Standards: A workshop summary*. National Academies Press.

Fick, S. J., & Arias, A. M. (2022). A review of literature that uses the lens of the next generation science crosscutting concepts: 2012–2019. *Journal of Research in Science Teaching, 59*(6), 883–929. https://doi.org/10.1002/tea.21747

García-Carmona, A. (2020). From inquiry-based science education to the approach based on scientific practices: A critical analysis and suggestions for science teaching. *Science & Education, 29*(2), 443–463. https://doi.org/10.1007/s11191-020-00108-8

Gugenheimer, J., Tseng, W. J., Mhaidli, A. H., Rixen, J. O., McGill, M., Nebeling, M., Khamis, M., Schaub, F., & Das, S. (2022, April). Novel challenges of safety, security and privacy in extended reality. In *CHI Conference on Human Factors in Computing Systems Extended Abstracts* (pp. 1–5). https://doi.org/10.1145/3491101.3503741

Lisachenko, D. A., Barmasov, A. V., Bukina, M. N., Stankova, E. N., Vysotskaya, S. O., & Zarochentseva, E. P. (2017, July). Best practices combining traditional and digital technologies in education. In *International Conference on Computational Science and Its Applications* (pp. 483–494). Springer International Publishing. https://doi.org/10.1007/978-3-319-62404-4_36

Ménard, A. D., & Trant, J. F. (2020). A review and critique of academic lab safety research. *Nature chemistry, 12*(1), 17–25. https://doi.org/10.1038/s41557-019-0375-x

National Research Council (NRC). (2010). *Promoting chemical laboratory safety and security in developing countries.* The National Academies Press. https://doi.org/10.17226/12857

NGSS Lead States. (2013). *Next generation science standards: For states, by states.* National Academies Press.

Pawlaczyk, A., Maćkiewicz, E., & Szynkowska, M. I. (2018). Speciation of trace elements and its importance in environmental and biomedical sciences. *Recent Advances in Trace Elements,* 501–543. https://doi.org/10.1002/9781119133780.ch25

Reyes, R. L., Isleta, K. P., Regala, J. D., & Bialba, D. M. R. (2024). Enhancing experiential science learning with virtual labs: A narrative account of merits, challenges, and implementation strategies. *Journal of Computer Assisted Learning, 40*(6), 3167–3186. https://doi.org/10.1111/jcal.13061

Sedghpour, B. S., Sabbaghan, M., & Sataei, F. M. (2013). A survey on the pre service chemistry teachers' lab safety education. *Procedia-Social and Behavioral Sciences, 90*, 57–62. https://doi.org/10.1016/j.sbspro.2013.07.065

Vagle, A. (2024). *Effects of a pre-laboratory simulation in the chemistry lab.* The Florida State University.

Wang, C. H. (2025). Education in the metaverse: Developing virtual reality teaching materials for K-12 natural science. *Education and Information Technologies, 30*(7), 8637–8658. https://doi.org/10.1007/s10639-024-13156-2

Zainuddin, Z., Shujahat, M., Haruna, H., & Chu, S. K. W. (2020). The role of gamified e-quizzes on student learning and engagement: An interactive gamification solution for a formative assessment system. *Computers & education, 145*, Article 103729. https://doi.org/10.1016/j.compedu.2019.103729

Zeeshan, K., Hämäläinen, T., & Neittaanmäki, P. (2022). Internet of things for sustainable smart education: An overview. *Sustainability, 14*(7), 4293. https://doi.org/10.3390/su14074293

**Remeredzayi Gudyanga** is a lecturer in the Faculty of Education at the University of the Free State, South Africa. He has worked in science education in Zimbabwe, Cuba and Namibia, and previously served as a chemistry laboratory instructor at South Dakota State University (USA), where his interest in laboratory safety emerged. His research focuses on laboratory safety and ethics, and curriculum implementation.

# Chapter 31
# Safety Issues in STEM Virtual Learning Environments

**Cheryl L. Burleigh**

## 31.1 Introduction

Virtual learning environments (VLEs) have transformed science, technology, engineering, and mathematics (STEM) education by expanding access, enabling scientific inquiry, and fostering collaboration. Accelerated by technological advances and COVID-19, VLEs offer flexible, personalized instruction (Hodges et al., 2020). Students can now use simulations, participate in virtual labs, and connect remotely with global STEM communities. For educators, these platforms provide tools to differentiate instruction and increase student engagement (Trust & Whalen, 2021). However, these benefits introduce challenges to student safety, integrity, and instruction. Unlike traditional classrooms, VLEs lack physical oversight, raising concerns about ethical decision-making, simulation use, and navigation of digital tools. Cybersecurity threats heighten these issues (UNESCO, 2023), and unregulated AI tools further complicate efforts to maintain academic integrity and data security (Eke, 2023).

Many educators report insufficient training and limited support for STEM VLE instruction, while students struggle to assess online scientific resources or collaborate responsibly online (Trust & Whalen, 2021). Institutional policies vary widely, resulting in inconsistent practices and fragmented safety protocols (UNESCO, 2023). Some institutions have adopted practices to promote safer STEM VLEs, such as secure platforms and digital ethics professional development (PD). Yet, most institutions lack clear policy guidance and comprehensive frameworks to address the evolving risks of STEM VLE instruction. This chapter contributes to the discourse by reframing safety not only as a technical concern but also as a pedagogical and ethical imperative, emphasizing the integration of cybersecurity, digital citizenship, and teacher preparation into instructional design and institutional policy. This chapter

C. L. Burleigh (✉)
Sanford College of Education, National University, San Diego, USA
e-mail: cburleigh@nu.edu



L. Leite et al. (eds.), *Practical Work in Science Education in the Digital Era*, Advances in Technology-Rich Science Education 2, https://doi.org/10.1007/978-3-032-13788-3_31

investigates safety issues in STEM VLEs using comparative document content analysis and explores the central question: *What are the prevailing safety concerns in STEM VLE education?*

## 31.2 Literature Review

Historically, safety in STEM education has centered on managing risks in physical classrooms, including chemical exposure, equipment handling, and adherence to procedural protocols (Charteris et al., 2021; Duffy & Roy, 2020; Yuliardi et al., 2024). Educators and administrators relied on direct supervision and institutional oversight to ensure safety and ethical practices. As educators integrate simulation-based labs, augmented and virtual reality (AR/VR) experiences, and AI instruction, new risks emerge, including cybersecurity, data privacy concerns, and ethical challenges in immersive content (Cohney et al., 2020; Steele et al., 2020b).

As STEM learning continues to shift online, safety frameworks must evolve to address both technical safeguards and pedagogical strategies. Cybersecurity now shapes instructional design and managing student data, while ethical challenges within VLEs complicate efforts to foster scientific integrity in digital settings (Burg, 2025; Wang et al., 2025; Zhang et al., 2024). As a core competency in navigating virtual environments, digital literacy enables students to assess sources, safeguard information, and collaborate ethically (Onanuga & Saka, 2025). However, generic digital citizenship curricula often fail to address the unique ethical dilemmas of STEM VLEs (Fielding & Belmont, 2025; Kocoglu et al., 2023).

Recent literature also stresses the importance of embedding ethical foresight into the design of educational technologies. Steele et al. (2019, 2020a, 2020b) advocate for participatory design in AR/VR platforms that prioritize inclusivity, psychological safety, and contextual relevance. This review synthesizes the evolving discourse around STEM safety in VLEs, organized into four key themes: cybersecurity and data privacy, ethics in virtual experimentation and design, digital literacy and citizenship, and institutional safety policies.

### *31.2.1 Cybersecurity and Data Privacy*

STEM VLE platforms often require personal accounts, third-party apps, and data submission, exposing students' personally identifiable information (PII) such as names, IP addresses, and activity logs to risks (Onanuga & Saka, 2025; Yildirim & Öztürk, 2023). Without strong safeguards, this data is vulnerable to breaches, tracking, and exploitation. Yet many institutions lack robust cybersecurity, with gaps such as weak encryption, unsecured APIs, limited firewalls, and poor authentication. As cloud-based services expand, schools face greater risks from phishing, ransomware, and denial-of-service attacks (Kostelic & Etinger, 2025). Risks increase when student data is shared across platforms without clear consent. Some apps share anonymized data with third parties, raising concerns about surveillance and commodification (Creel & Dixit, 2022; Office of Elementary and Secondary Education [OESE], 2024). Institutional policies often lack STEM-specific cybersecurity guidelines, overlooking needs like secure experimental data storage or simulation authentication (Burg, 2025; Hasan, 2023; Kostelic & Etinger, 2025). As digital learning grows, targeted cybersecurity strategies are essential to STEM VLE safety.

### *31.2.2 Ethics in Virtual Experimentation and Design*

Virtual laboratories enhance STEM learning by providing accessible, repeatable experiments that reduce physical risks and logistical barriers (Coban et al., 2022), yet they also introduce ethical challenges often overlooked in instructional design. Unlike traditional labs with established safety protocols, virtual labs frequently lack clear ethical frameworks to guide conduct, leaving students unprompted to consider data integrity or the consequences of manipulating results, which can erode methodological rigor and foster academic dishonesty when assessments emphasize outcomes over process (Williamson & Prybutok, 2024). The use of AI-generated feedback without human oversight further raises concerns about transparency and accountability (Eke, 2023), while design decisions risk reinforcing bias by excluding certain groups, underscoring the need for participatory approaches that involve educators, students, and ethicists to align technologies with ethical and pedagogical values (Steele et al., 2019). Moreover, VLEs create a paradox between surveillance and anonymity, as platforms monitor activity for analytics yet lack the direct oversight needed to prevent plagiarism or data fabrication (Attarbashi et al., 2021). These dynamics highlight the importance of embedding explicit ethical instruction rather than relying solely on external monitoring.

### *31.2.3 Digital Literacy and Citizenship*

Digital literacy is critical for ethical and effective participation in STEM VLEs, extending beyond technical skills to include evaluating online content, safeguarding data, and collaborating responsibly in AI-driven environments (Eke, 2023; Onanuga & Saka, 2025). Misuse of shared code, simulations, or lab outputs can compromise data integrity and academic standards, while structured instruction strengthens ethical decision-making in validating sources, managing data, and communicating virtually (Buchan et al., 2024). Most digital citizenship curricula remain too broad, neglecting STEM VLE-specific challenges such as secure coding, ethical use of open-source tools, and data stewardship in simulations (Fielding & Belmont, 2025; Kocoglu et al., 2023). Immersive technologies further complicate these dynamics as AR/VR platforms raise ethical concerns of consent, representation, and simulation realism; however, embedding ethical reflection in these contexts can foster deeper critical thinking, particularly in higher education (Steele et al., 2020a). Preparing students for ethical engagement in digital STEM fields requires institutions to embed discipline-specific digital literacy and citizenship as central curricular priorities rather than peripheral content.

### *31.2.4 Institutional Safety Policies*

As VLEs become integral to STEM education, institutions must adopt safety policies tailored to the unique demands of digital science instruction. Despite guidance from NSTA (Duffy & Roy, 2020) and UNESCO (2023), many institutions use generic IT policies that ignore STEM-specific risks such as virtual labs, cloud collaboration, and AI-supported learning (Charteris et al., 2021; Yuliardi et al., 2024). Virtual labs and simulations require clear protocols for data handling, ethical experimentation, and collaborative integrity. Yet, most policies treat STEM VLE courses the same as general education classes, overlooking the specific demands of virtual scientific inquiry (Burg, 2025; Wang et al., 2025). The result is a lack of accountability and inconsistent implementation, leaving educators without adequate guidance. Many educators report that they feel unprepared to navigate ethical dilemmas in VLEs. Steele et al. (2020b) found that teachers lacked institutional support for evaluating immersive tools and fostering safe, ethical collaboration. Similarly, Eke (2023) highlights the challenges posed by AI-powered platforms, including bias, opacity, and insufficient content moderation.

## 31.3 Methodology

A comparative document content analysis was used to examine how educational and policy documents conceptualize safety in STEM VLEs. Document content analysis is a qualitative method that allows researchers to interpret textual data systematically to uncover patterns, themes, and contextual meanings (Bowen, 2009) as an approach to enable the identification of similarities and differences across a diverse set of documents.

Following Cooper's (2017) principles of representativeness, relevance, and credibility, a purposive sampling strategy was used to curate 19 of 49 publications from 2020 to 2025, that met three criteria: direct relevance to K-16 STEM education, focus on online or hybrid instructional modalities, and explicit attention to safety-related concerns in VLEs such as cybersecurity, ethics, digital citizenship, and educator training. The documents included nine peer-reviewed articles, four national and international policy reports, two curriculum frameworks, and four best institutional practice guidelines. Sources were retrieved from academic databases, government agencies, non-governmental organizations, and educational research repositories.

Inductive coding was used to identify categories and analyze the data (Saldaña, 2021). Initial open coding identified categories such as data security, ethical simulation use, teacher preparedness, and AI content moderation. These codes were refined through constant comparison, yielding four overarching themes: cybersecurity protocols, virtual lab ethics, digital citizenship, and educator training.

Triangulation enhanced validity by cross-referencing policy documents with academic sources and applying systematic coding (Saldaña, 2021). Nonetheless, the study is limited by its reliance on publicly available documents, which may not fully reflect on-the-ground practices. Interpretive variability across contexts and the subjective nature of qualitative analysis may also introduce researcher bias, despite efforts to ensure analytic rigor.

## 31.4 Findings and Analysis

The findings are based on a comparative content analysis of peer-reviewed journal articles, national and institutional policy guidelines, curricular frameworks, and PD resources. These documents were analyzed to identify how safety is conceptualized, prioritized, and implemented within STEM VLEs. The diversity of sources enabled a comprehensive examination of both theoretical perspectives and practical strategies employed across educational contexts. Peer-reviewed literature contributed insights into the academic discourse surrounding digital ethics, cybersecurity, and pedagogical innovation in STEM VLEs. Policy documents, often issued by governmental and professional bodies, emphasized general safety principles but varied widely in their specificity and applicability to STEM-focused instruction. Curriculum and PD resources illustrate how safety is applied in practice. Four themes emerged from the analysis: cybersecurity protocols, virtual lab ethics, digital citizenship, and educator training. These themes collectively highlight a lack of coherence in how educational systems manage safety in STEM VLE instruction. They also point to opportunities for greater alignment between policy development, curriculum design, and educator support systems.

### *31.4.1 Cybersecurity Protocols*

Analysis revealed significant variation in how institutions address cybersecurity in STEM VLEs. While nearly all documents acknowledged the importance of digital safety, most institutional policies were embedded within broader IT frameworks, lacking specificity for STEM contexts. Institutional policies often cite password management but rarely address lab data security, cloud encryption, or authentication for third-party tools (Archambault, 2021; Hasan, 2023; OESE, 2024). Few documents identified proactive measures such as real-time threat detection, zero-trust security models, or designated personnel responsible for digital safety oversight. While UNESCO (2023) calls for systemic integration of privacy and cybersecurity safeguards into education, its guidance has yet to be fully operationalized at the institutional level. This gap is particularly concerning given the sensitive nature of student data and the growing use of AI-enhanced platforms in STEM education.

### *31.4.2 Virtual Lab Ethics*

Many instructional and policy documents lack clear ethical guidelines for student conduct in VLEs. Educators and developers frequently frame simulations in VLEs as safe alternatives to physical experiments, yet they often omit critical guidance on ethical responsibilities such as data accuracy, reproducibility, and scientific integrity (Zhang et al., 2024). As a result, curriculum frameworks seldom provide instructional materials that promote ethical reflection or encourage responsible scientific practices in digital learning environments. Additionally, guidance on intellectual property, collaborative accountability, and the responsible use of generative AI tools was either absent or inadequately developed (Eke, 2023; Steele et al., 2020b). This gap is critical, as VLEs may normalize unethical practices like falsifying results or copying data, particularly when students perceive fewer consequences in digital settings. Isma'il et al. (2021) emphasize the need to embed ethics instruction within STEM VLEs to cultivate integrity and critical thinking. Without such guidance, students may complete simulations successfully without understanding the ethical standards foundational to scientific inquiry.

### *31.4.3 Digital Citizenship*

Although digital citizenship was mentioned across many documents, it was often framed in general terms—such as promoting respectful communication or preventing plagiarism—without specific reference to STEM VLE contexts. Few documents provided targeted instruction or assessment tools addressing ethical issues unique to digital STEM environments, such as secure collaboration on open-source platforms, responsible use of simulation data, or digital footprint awareness in scientific research (Son & Ha, 2024; Yildirim & Öztürk, 2023; Zhong & Zheng, 2023). Moreover, most curricular materials lacked integration of digital citizenship into the design and evaluation of virtual lab activities. Only a small subset of documents included structured instructional plans or rubrics for teaching ethical digital behavior. Steele et al. (2020b) and Buchan et al. (2024) suggest that immersive technologies can be used as platforms for fostering digital ethics, but this potential remains underutilized.

In the absence of STEM-specific digital literacy instruction, students are left to navigate complex technological tools without adequate guidance or accountability.

### *31.4.4 Educator Training*

The analysis revealed wide disparities in educator training related to VLEs' safety. Some PD materials included comprehensive modules on data privacy, ethical moderation of online learning environments, and integrating digital literacy into instruction. However, many resources treated safety and ethics as peripheral topics, often covered only briefly or in optional segments (Hasan, 2023; Love, 2022). Educators often felt unprepared to use AI platforms, manage ethics in virtual labs, or ensure psychological safety. Steele et al. (2020b) and Eke (2023) reported that educators feel underprepared to address ethical dilemmas in VLEs due to limited institutional support and insufficient training. Inconsistent PD widens gaps in instructional quality and compromises student safety, especially in under-resourced schools (Love, 2022). Documents that included scenario-based training or interdisciplinary collaboration between IT and instructional staff were more likely to report positive outcomes (Hasan, 2023; Love, 2022; Steele et al., 2020b). These findings highlight the importance of embedding digital safety competencies into all levels of ongoing PD.

## 31.5 Discussion

While awareness of safety issues in STEM VLEs is growing, institutional responses remain fragmented, insufficiently detailed, and inconsistently implemented. This gap between awareness and action highlights the urgent need for coordinated strategies that integrate cybersecurity, ethical instruction, digital citizenship, and teacher preparation into a unified framework for STEM VLEs. Without systemic reforms, these safety gaps will continue to pose serious risks to student privacy, academic integrity, and equitable access.

Cybersecurity remains one of the most inconsistently addressed areas. Many institutions rely on broad IT policies that fail to reflect the unique technical demands of STEM instruction. These policies often overlook data security in simulations, authentication for third-party tools, and encryption of student work. As Archambault (2021) and Hasan (2023) argue, STEM VLEs require advanced safeguards that align with the complexity of digital platforms and data systems used in instruction. When these safeguards are absent or vague, students and educators are left exposed to threats including identity theft, surveillance, and algorithmic tracking—issues that directly undermine educational equity and digital trust.

Equally concerning is the limited ethical guidance for VLEs. Although VLEs enhance accessibility and allow for experimentation beyond the limits of physical classrooms, they are rarely accompanied by structured ethical frameworks.

Many curriculum documents and training materials overlook principles of scientific integrity, such as data accuracy, reproducibility, and transparency (Zhang et al., 2024). When ethical expectations are unclear, students may manipulate data, plagiarize results, or bypass collaborative norms without understanding the implications (Eke, 2023). This can lead to superficial learning outcomes, where students gain procedural competence without internalizing ethical reasoning. The unchecked use of AI-assisted tools in simulations further complicates these dynamics, often removing human oversight and limiting opportunities for reflective dialogue (Eke, 2023).

Digital citizenship, though frequently referenced in institutional materials, is poorly aligned with the specific demands of STEM instruction. Most frameworks emphasize general behaviors like respect and plagiarism prevention, but neglect STEM-specific issues such as ethical code use, simulation data privacy, and digital IP rights. As Son and Ha (2024) and Yildirim and Öztürk (2023) suggest, digital citizenship must extend beyond compliance to include critical engagement with ethical dilemmas unique to scientific inquiry. Without structured instruction and meaningful assessment, students may unknowingly violate academic norms or misuse technology, particularly in VLEs lacking consistent oversight.

To address these gaps, digital citizenship training should be fully embedded within STEM VLE curricula rather than treated as an ancillary topic. Students must be taught to handle algorithmic bias, use AI ethically, and maintain simulation integrity with care and accountability. Strategies such as reflective journaling, collaborative projects, and peer evaluations can reinforce these competencies and deepen ethical understanding (Onanuga & Saka, 2025; Zhong & Zheng, 2023). Moreover, curriculum developers should integrate media literacy and critical inquiry into STEM content, enabling students to critically assess the credibility, relevance, and implications of digital information.

Teacher preparation is a critical yet inconsistently addressed component of STEM VLE education. While some institutions provide comprehensive training on data security, ethical facilitation, and the responsible use of digital platforms, others treat these areas as peripheral or omit them entirely (Hasan, 2023; Love, 2022). As a result, many educators feel ill-equipped to model ethical conduct or manage the complexities of AI-driven and immersive technologies (Eke, 2023; Steele et al., 2020b), leading to fragmented practices that jeopardize both student learning and institutional integrity. Addressing these challenges requires a coordinated, multi-tiered approach that integrates policy, curriculum, and PD. Institutions should work collaboratively with STEM educators, technologists, and ethics scholars to develop digital codes of conduct that promote academic honesty, data stewardship, and ethical collaboration, supported through scenario-based learning, interdisciplinary case studies, and formative assessments (Duffy & Roy, 2020; Kocoglu et al., 2023; Onanuga & Saka, 2025). Teacher preparation programs must also include required modules on data ethics, digital safety, and emerging technologies, covering legal frameworks such as the Family Education Rights and Privacy Act (FERPA) and the Children's Online Privacy Protection Act (COPPA) alongside topics like algorithmic transparency and AI moderation (Eke, 2023). Ongoing, practice-based PD that includes mentorship and input from both pedagogical and technical experts is essential to equipping educators to lead safe, ethical, and inclusive STEM VLEs.

At the policy level, national and institutional strategies must be aligned with global standards such as UNESCO's (2023) recommendations on digital education and AI ethics. Institutions that fail to adopt and localize these standards risk reinforcing digital inequities, particularly in underserved communities. Disparities in infrastructure, training, and policy implementation can lead to unsafe or unethical learning environments that disproportionately affect students with limited access to resources or support (UNESCO, 2023). Developing context-sensitive, inclusive policies can mitigate these risks while fostering trust and engagement among educators and learners.

Effective STEM VLE instruction demands more than technology adoption—it requires ethical foresight, pedagogical intentionality, and policy coherence. Safety, equity, and academic integrity must be foundational design elements in any digital learning initiative. To ensure that STEM VLE ecosystems prepare students not only to succeed academically but to lead responsibly in scientific and technological fields, stakeholders must work collaboratively to establish robust, forward-looking frameworks for digital safety and ethical practice.

## 31.6 Recommendations and Implications

STEM VLE education presents critical implications for leaders, curriculum developers, educators, and policymakers. Inconsistent safety protocols, the absence of structured ethical frameworks, and fragmented teacher preparation reveal systemic gaps that compromise both instructional quality and student safety. These shortcomings expose learners to cybersecurity risks, ethical ambiguity, and inequitable access to high-quality VLEs.

### *31.6.1 Establishing Ethical and Safe STEM Virtual Learning Environments*

Institutions must adopt standardized ethical frameworks for VLEs. The lack of formalized virtual lab ethics jeopardizes students' understanding of scientific integrity and weakens learning outcomes (Steele et al., 2020a; Zhang et al., 2024). National and regional authorities should implement discipline-specific safety standards that reflect the complexities of simulations, AI-supported platforms, and collaborative coding environments. Digital codes of conduct, developed with input from educators, technologists, and ethicists, should emphasize academic honesty, data stewardship, and equitable collaboration. Implementation can be reinforced through case studies, guided ethical reasoning, and reflective assessments. Emerging tools such as AR/VR and generative AI must also be evaluated for ethical transparency, inclusivity, and educational value.

### 31.6.2 Embedding Digital Literacy and Ethics into STEM Curricula

Digital ethics and safety must be integrated into STEM curricula as core components, not peripheral topics. Instruction should explicitly address coding practices, simulation fidelity, data privacy, and responsible AI use. These concepts should be introduced early and reinforced across grade levels to build cumulative competence. Interdisciplinary approaches—such as modules on algorithmic bias, data justice, and the social impacts of technology—can deepen understanding. Activities like secure peer collaboration, reflective journaling, and structured assessments strengthen students' ability to apply ethical reasoning in virtual contexts (Buchan et al., 2024; Onanuga & Saka, 2025).

### 31.6.3 Strengthening Teacher Preparation and Professional Development

Teacher education programs must prepare educators for the ethical, technical, and instructional demands of VLEs. Current PD is inconsistent, ranging from comprehensive modules to minimal or outdated content (Hasan, 2023; Love, 2022). Standardized training should include legal frameworks (FERPA, COPPA), cybersecurity fundamentals, AI ethics, and simulation governance. PD should be ongoing and interactive, incorporating workshops, mentorship, and interdisciplinary collaboration. When educators are equipped with both technical knowledge and ethical reasoning, they can create safer, more inclusive digital learning environments (Eke, 2023; Steele et al., 2020b).

### 31.6.4 Institutional Accountability and Cross-Functional Policy Development

Institutions must ensure that digital safety policies are enforced through regular audits, feedback loops, and transparent reporting. Zero-trust cybersecurity frameworks—emphasizing continuous verification and strict data controls—should be standard practice (Cohney et al., 2020). Policymaking must be collaborative, engaging faculty, IT professionals, instructional designers, legal experts, and students. Global guidelines, such as UNESCO's (2023) recommendations, provide a foundation but must be adapted locally. Task forces can create agile policies that evolve with technology while promoting equity and accountability. Vendor compliance with privacy standards and institutional policies should also be monitored through ongoing evaluations.

### 31.6.5 Fostering a Culture of Digital Citizenship in STEM

Promoting digital citizenship within STEM education is essential to preparing students for ethical innovation and leadership in scientific fields. Instruction should explicitly cover responsible digital conduct and ethical technology use, with competencies assessed as rigorously as academic content. Approaches such as case studies, project-based learning, and collaborative evaluations help students recognize the consequences of unethical behavior. Institutional alignment across departments, professional development, and student services is critical, and faculty must be equipped to guide these discussions confidently. Building this culture requires sustained investment in curricula, training, and infrastructure so students understand their rights and responsibilities in VLEs and are prepared to contribute ethically to the scientific community.

### 31.6.6 Toward Equitable, Ethical, and Future-Ready STEM VLE Education

The promise of STEM VLEs can only be realized through systemic reforms that embed safety, ethics, and equity into every layer of instruction and governance. Alignment across curricula, teacher preparation, and institutional policy is essential. These reforms must protect students, support educators, and preserve the integrity of STEM learning in digital environments. By embedding ethical reasoning, strong cybersecurity, and inclusive practices into the foundation of VLEs, institutions can ensure that virtual STEM education is equitable, accountable, and transformative for all learners.

## 31.7 Conclusion and Ways Forward

This chapter affirms that while STEM VLE education continues to expand, institutional frameworks have not kept pace with the ethical, technical, and pedagogical demands of digital learning environments. Through comparative document content analysis, the study identified persistent gaps in cybersecurity readiness, virtual lab ethics, digital citizenship instruction, and teacher preparedness. If left unaddressed, these gaps may compromise the safety, equity, and quality of STEM VLEs.

As virtual platforms become embedded in K-16 STEM education, educators and policymakers must treat safety not only as a technical concern but as a core pedagogical and ethical imperative. While digital tools enable flexibility and innovation, they also introduce risks related to privacy, academic dishonesty, and inequitable access. The inconsistency of safety protocols, the absence of discipline-specific ethical frameworks, and the variability in teacher training reveal a fragmented system needing strategic reform. Embedding ethical reasoning into virtual lab activities, developing cybersecurity standards with educator input, and prioritizing structured teacher preparation are essential steps toward strengthening STEM VLE ecosystems (Buchan et al., 2024; Love, 2022; Onanuga & Saka, 2025). Institutions must also create mechanisms for accountability, such as audits of digital platforms, stakeholder feedback cycles, and public transparency measures (Archambault, 2021; Duffy & Roy, 2020; Hasan, 2023). Without such reforms, the rapid growth of STEM VLEs may amplify inequities and compromise the integrity of science education.

The most critical novel finding of this work lies in reframing safety in STEM VLEs not simply as a technical challenge but as a pedagogical and ethical imperative that must be embedded across policy, curriculum, and professional development. By synthesizing policy, curricular frameworks, and institutional practices, this chapter illuminates the systemic nature of current safety gaps and offers a unified framework connecting digital safety with scientific integrity. This perspective underscores the need for coordinated reforms that emphasize ethical reasoning, teacher preparation, and accountability as essential to sustaining safe, equitable, and future-ready STEM education.

Future research should explore how digital safety protocols are implemented across diverse educational settings, particularly in under-resourced and rural districts where digital infrastructure and training may be limited. Comparative international analyses could provide insights into how global policy frameworks, such as UNESCO's (2023) AI ethics guidelines, are localized and applied in STEM education. Such studies would help identify scalable, culturally responsive strategies for ensuring ethical and secure VLEs.

As VLEs continue to evolve, research must investigate the new risks and benefits associated with these tools. In particular, studies should examine how AI-supported platforms influence equity, transparency, and ethical decision-making in STEM education. Exploring issues such as content moderation, algorithmic bias, and AI-generated feedback can help ensure that emerging technologies align with educational values and ethical standards (Eke, 2023; Usher & Barak, 2024). Moving forward, a proactive and interdisciplinary research agenda must address how digital safety, equity, and ethics intersect within increasingly complex VLEs. Such research will be critical in shaping policies and practices that support inclusive, future-ready STEM education for all learners.

## References

Archambault, S. G. (2021). Student privacy in the digital age. *BYU Education & Law Journal, 1*(6). https://scholarsarchive.byu.edu/byu_elj/vol2021/iss1/6

Attarbashi, Z. S., Hashim, A. H. A., Abuzaraida, M. A., Khalifa, O. O., & Mustafa, M. (2021). Teaching lab-based courses remotely: Approaches, technologies, challenges, and ethical issues. *IIUM Journal of Educational Studies, 9*(3), 37–51. https://doi.org/10.31436/ijes.v9i3.406

Bowen, G. A. (2009). Document analysis as a qualitative research method. *Qualitative Research Journal, 9*(2), 27–40. https://doi.org/10.3316/QRJ0902027

Buchan, M. C., Bhawra, J., & Katapally, T. R. (2024). Navigating the digital world: Development of an evidence-based digital literacy programme for youth. *Smart Learning Environments, 11*(8), 1–24. https://doi.org/10.1186/s40561-024-00293-x

Burg, M. (2025, January 9). *Implementing zero trust cybersecurity in K-12: A proactive approach to digital safety.* Lightspeed Systems. https://www.lightspeedsystems.com/blog/implementing-zero-trust-in-k-12-a-proactive-approach-to-digital-safety/

Charteris, J., Anderson, J., & Page, A. (2021). Psychological safety in innovative learning environments: Planning for inclusive spaces. *Inclusive Education, 27*(5), 540–555. https://doi.org/10.1080/13603116.2021.1974108

Coban, M., Bolat, Y. I., & Goksu, I. (2022). The potential of immersive virtual reality to enhance learning: A meta-analysis. *Educational Research Review, 36*, Article 100452. https://doi.org/10.1016/j.edurev.2022.100452

Cohney, S., Teixeira, R., Kohlbrenner, A., Narayanan, A., Kshirsagar, M., Shvartzshnaider, Y., & Sanfilippo, M. (2020). *Virtual classrooms and real harms: Remote learning at U.S. universities.* arXiv. https://doi.org/10.48550/arXiv.2012.05867

Cooper, H. (2017). *Research synthesis and meta-analysis: A step-by-step approach* (5th ed.). SAGE Publications.

Creel, K., & Dixit, T. (2022, August 26). *Privacy and paternalism: The ethics of student data collection.* MIT Schwarzman College of Computing. https://doi.org/10.21428/2c646de5.b725319a

Duffy, B., & Roy, K. (2020, June 14). *NSTA minimum safety practices and regulations for in-person and virtual sessions.* NSTA Safety Advisory Board. https://static.nsta.org/pdfs/MinimumSafetyPracticesAndRegulations.pdf

Eke, D. O. (2023). ChatGPT and the rise of generative AI: Threat to academic integrity? *Journal of Responsible Technology, 13*, Article 100060. https://doi.org/10.1016/j.jrt.2023.100060

Fielding, K., & Belmont, T. (2025, March 18). *Light lift strategies for incorporating digital citizenship into lesson plans.* Edutopia. https://www.edutopia.org/article/strategies-incorporating-digital-citizenship-lessons/

Hasan, R. (2023). Understanding the perceptions and awareness of educational technologies' privacy and security issues. In *Proceedings on Privacy Enhancing Technologies, 2023*(4), 269–286. https://doi.org/10.56553/popets-2023-0110

Hodges, C., Moore, S., Lockee, B., Trust, T., & Bond, A. (2020, March 27). *The difference between emergency remote teaching and online learning.* EDUCAUSE Review. https://er.educause.edu/articles/2020/3/the-difference-between-emergency-remote-teaching-and-online-learning

Isma'il, A., Suleiman, B., & Abdulrahaman, A. (2021). Ethics and strategies for its integration into science, technology, engineering, and mathematics (STEM) classrooms at the secondary school level. In *Proceedings of the 61st Annual Conference of the Science Teachers Association of Nigeria* (pp. 26–34). Science Teachers Association of Nigeria.

Kocoglu, E., Oguz-Hacat, S., & Gocer, V. (2023). The relationship between digital literacy and digital citizenship levels of STEM teacher candidates: The mediating role of digital teaching material development self-efficacy. *Journal of Education in Science, Environment and Health (JESEH), 9*(3), 194–205. https://doi.org/10.55549/jeseh.1331283

Kostelic, K., & Etinger, D. (2025). Securing the metaverse: A bibliometric analysis of cybersecurity challenges and research trajectories. *IEEE Engineering Management Review, 53*(3), 93–109. https://doi.org/10.1109/EMR.2024.3453974

Love, T. S. (2022). Examining the influence that professional development has on educators' perceptions of integrated STEM safety in makerspaces. *Journal of Science Education and Technology, 31*(3), 289–302. https://doi.org/10.1007/s10956-022-09955-2

Office of Elementary and Secondary Education. (2024). *K-12 cybersecurity.* U.S. Department of Education. https://www.ed.gov/teaching-and-administration/safe-learning-environments/school-safety-and-security/k-12-cybersecurity

Onanuga, P. A., & Saka, A. O. (2025). Emerging digital technologies: Building competencies of STEM pre-service teachers. *Journal of Teaching and Learning, 19*(1), 155–169. https://doi.org/10.22329/jtl.v19i1.8810

Saldaña, J. (2021). *The coding manual for qualitative researchers* (4th ed.). SAGE Publications.

Son, M., & Ha, M. (2024). Development of a digital literacy measurement tool for middle and high school students in the context of scientific practice. *Education and Information Technologies, 30*, 4583–4606. https://doi.org/10.1007/s10639-024-12999-z

Steele, P., Burleigh, C., Bailey, L., & Kroposki, M. (2019). Studio thinking framework in higher education: Exploring options for shaping immersive experiences across virtual reality/augmented reality curricula. *Journal of Educational Technology Systems, 48*(3), 416–439. https://doi.org/10.1177/0047239519884897

Steele, P., Burleigh, C., Kroposki, M., Magabo, M., & Bailey, L. (2020a). Ethical considerations in designing virtual and augmented reality products virtual and augmented reality design with students in mind: Designers' perceptions. *Journal of Educational Technology Systems, 49*(2), 219–238. https://doi.org/10.1177/0047239520933858

Steele, P., Burleigh, C., & Young, L. (2020b). Considering educators' perceptions of ethics and safety in virtual learning environments (VLEs). *Journal of Educational Research and Practice, 10*(1), 404–419. https://doi.org/10.5590/JERAP.2020.10.1.26

Trust, T., & Whalen, J. (2021). K–12 teachers' experiences and challenges with using technology for emergency remote teaching during the COVID-19 pandemic. *Italian Journal of Educational Technology, 29*(2), 10–25. https://doi.org/10.17471/2499-4324/1192

UNESCO. (2023). *Technology in education: A tool on whose terms? Global Education Monitoring Report 2023.* UNESCO. https://unesdoc.unesco.org/ark:/48223/pf0000385723

Usher, M., & Barak, M. (2024). Unpacking the role of AI ethics online education for science and engineering students. *International Journal of STEM Education, 11*, Article 35. https://doi.org/10.1186/s40594-024-00493-4

Wang, X., Bhuse, V., & Cheng, Y. (2025). A zero trust module for cybersecurity education. *Journal of the Colloquium for Information Systems Security Education, 12*(1). https://doi.org/10.53735/cisse.v12i1.193

Williamson, S. M., & Prybutok, V. (2024). The era of artificial intelligence deception: Unraveling the complexities of false realities and emerging threats of misinformation. *Information, 15*(6), 299. https://doi.org/10.3390/info15060299

Yildirim, T., & Öztürk, D. (2023). A mixed-method study on middle school students' digital literacy levels and opinions. *International Journal of Education and Literacy Studies, 11*(2), 70–86. https://doi.org/10.7575/aiac.ijels.v.11n.2p.70

Yuliardi, R., Kusumah, Y. S., Nurjanah, N., Juandi, D., & Suparman, S. (2024). Development of a STEM-based digital learning space platform to enhance students' mathematical creativity in future learning classrooms. *Eurasia Journal of Mathematics, Science and Technology Education, 20*(12), Article e2545. https://doi.org/10.29333/ejmste/15665

Zhang, Y., Yang, Y., Chu, Y., Sun, D., Xu, J., & Zheng, Y. (2024). Virtual laboratories in science education: Unveiling trajectories, themes, and emerging paradigms (2013–2023). *Journal of Baltic Science Education, 23*(5), 990–1009. https://doi.org/10.33225/jbse/24.23.990

Zhong, J., & Zheng, Y. (2023). "What it means to be a digital citizen": Using concept mapping and an educational game to explore children's conceptualization of digital citizenship. *Heliyon, 9*(9), Article e19291. https://doi.org/10.1016/j.heliyon.2023.e19291

**Cheryl L. Burleigh**, EdD, is a professor at National University's Sanford College of Education, USA. She has extensive K-12 and higher education experience in the U.S. and internationally, specializing in educational leadership, science education, and coaching. Her current research interests are educational leadership, adult learning, educational policy and governance, and the role of generative artificial intelligence in education.